THIN FILM PHYSICS AND DEVICES

Fundamental Mechanism, Materials and Applications for Thin Films

Recommended Titles in Related Topics

Physics of Electrons in Solids
by Jean-Claude Tolédano
ISBN: 978-1-78634-972-9

Low-Energy Excitations in Disordered Solids:
A Story of the 'Universal' Phenomena of Structural Tunneling
by Richard B Stephens and Xiao Liu
ISBN: 978-981-121-724-1

Modern Aspects of Superconductivity: Theory of Superconductivity
Second Edition
by Sergei Kruchinin
ISBN: 978-981-123-451-4

Elastic Constants in Heavily Doped Low Dimensional Materials
by Kamakhya Prasad Ghatak and Madhuchhanda Mitra
ISBN: 978-981-122-946-6

Defects in Functional Materials
edited by Francis Chi-Chung Ling, Shengqiang Zhou and
Andrej Kuznetsov
ISBN: 978-981-120-316-9

THIN FILM PHYSICS AND DEVICES

Fundamental Mechanism, Materials and Applications for Thin Films

Jianguo Zhu, Xiaohong Zhu,
Hong Liu and Jie Xing

Sichuan University, China

World Scientific

NEW JERSEY • LONDON • SINGAPORE • BEIJING • SHANGHAI • HONG KONG • TAIPEI • CHENNAI • TOKYO

Published by

World Scientific Publishing Co. Pte. Ltd.
5 Toh Tuck Link, Singapore 596224
USA office: 27 Warren Street, Suite 401-402, Hackensack, NJ 07601
UK office: 57 Shelton Street, Covent Garden, London WC2H 9HE

Library of Congress Control Number: 2021940619

British Library Cataloguing-in-Publication Data
A catalogue record for this book is available from the British Library.

This work is funded by B & R Book Program.

薄膜物理与器件
Originally published in Chinese by National Defense Industry Press

THIN FILM PHYSICS AND DEVICES
Fundamental Mechanism, Materials and Applications for Thin Films

ISBN 978-981-122-398-3 (hardcover)
ISBN 978-981-122-399-0 (ebook for institutions)
ISBN 978-981-122-400-3 (ebook for individuals)

For any available supplementary material, please visit
https://www.worldscientific.com/worldscibooks/10.1142/11929#t=suppl

Contents

Preface

Thin films have wide application in science and technology, industries, military purposes, and so on. They play an especially important role in the field of new material technology. In order to obtain films with various properties for application in important technical fields, it is necessary to master relevant film-forming techniques and be well versed in the fundamentals of film physics. The preparation, physical mechanisms and applications of thin films, such as semiconductor thin films, magnetic thin films, ferroelectric and piezoelectric thin films, superconductor thin films, optical thin films, and so on, are all undergoing rapid development. The mechanism of film formation, the kinetics of film growth, and some important physical properties of films remain a rich area for study and exploration. The preparation technology of thin films continues to develop and improve as we write this. This makes it impossible for a book of this size to cover all the aspects of the field.

Readers will find that this book has focused primarily on introducing the basic principles and methods of the physical and chemical preparation of thin films, the basic contents of the formation, structure and defects of thin films, and the main aspects of film physics. This book would serve as a general introduction for people embarking on the field and also for experts seeking information on other subfields.

The authors make every effort to bring out the richness of the field, highlight on innovative materials and frontiers of thin films, and present the contents in a systematic, comprehensive, and practical fashion to meet the needs of teaching and training staff in many college disciplines, such as material science and engineering, electronic science and engineering, electronic materials and components, semiconductor physics and devices, applied physics, etc. Consequently, this book is written in the interest of both theoretical and practical demands.

Among the aims of this book is to try to summarize the fundamentals and established techniques of the synthesis and processing of thin films. This would provide readers a systematic and coherent picture of the mechanism, synthesis, and processing of thin films. At the same time, each chapter of this book could be read independently, so that teachers and students of different college disciplines, researchers of related scientific and technological enterprises, company management, and technical personnel may selectively study and reference the material with ease. The authors were also attentive to a concise, clear, and informative presentation while avoiding lengthy mathematical derivations. Therefore, this book is well suited as a textbook for advanced undergraduate, graduate, and professional short courses.

The original version of this book was published in Chinese by the National Defense Industry Press of China in December 2010. Many Chinese universities and colleges have been using it as a teaching material for undergraduate and graduate students. This book was translated into English by Professor Jianguo Zhu (Chapters 1, 3, 6, 7, 8, and 9), Professor Xiaohong Zhu (Chapter 2), Associate Professor Hong Liu (Chapter 5) and Associate Professor Jie Xing (Chapter 4). Professor Jianguo Zhu also proofread the book. For collected corrections, data, and illustrations, the authors are grateful to Ms. Jingwen Xi, Mr. Lixue Xie, Mr. Fei Wang, Mr. Yining Xie, Mr. Yuheng Guo, Ms. Shujie Tie, Mr. Deyu Xin, Mr. Hao Chen, Mr. Deqi Tang, Mr. Yuan Chen, Mr. Tao Pu, Mr. Siyu Su, and so on. Among the National Defense Industry Press staff, my thanks go to Mr. Tianming Bai for publication supervision.

The authors are grateful to World Scientific Publishing Co., the National Defense Industry Press of China, and Sichuan University of China for their support in the preparation of this book.

Given the time constraints and the rapid advancement of the field, this book inevitably still has room for improvement. Readers' corrections and suggestions will be gratefully received and may be addressed to the authors at College of Materials Science and Engineering, Sichuan University, Chengdu, China, or by email to nic0400@scu.edu.cn.

Chapter 1

Foundation of Vacuum Technology

In general, the preparation of thin film materials by physical methods (such as vacuum evaporation, sputtering coating, and ion coating) and the subsequent processing and characterization of thin films are carried out at relatively low pressure, and the production, transport or reaction of gas phase will also be involved in the preparation of thin films. Therefore, giving a basic knowledge of gas properties, vacuum acquisition methods, and measurement techniques is the foundation of thin film technology. This chapter will introduce a basic knowledge of vacuum, a basic knowledge of rarefied gases, the acquisition of vacuum, and vacuum measurement methods.

1.1 Basic Knowledge of Vacuum

1.1.1 *Definition and unit of measurement of vacuum*

A vacuum is a space of gas below normal atmospheric pressure (one bar), a thin gas state compared to normal air. In the equilibrium state, the gas state equation can be used to describe the properties of the gas:

$$P = nkT \tag{1.1}$$

or

$$PV = \frac{m}{M}RT, \tag{1.2}$$

where P is pressure (Pa); n is the molecular density of the gas (per cubic centimeter); V is volume (m^3); M is the molecular weight of the gas (kg/mol); m is the mass of the gas (kg); T is the absolute temperature (K); k is Boltzmann constant (1.38×10^{-23} J/K);

R is the universal gas constant (8.314 J/mol·K), which can also be expressed by $R = N_A \cdot K$, where N_A is Avogadro's constant (6.023×10^{23} per mol). Therefore, n value can be obtained from Equation (1.1):

$$n = 7.2 \times 10^{22}\frac{P}{T} \text{ (per cubic meter)}. \tag{1.3}$$

From the above equation, it can be seen that in the standard state (1 atmosphere, room temperature), the density of any gas molecule is about $3 \times 10^{19}/\text{cm}^3$. Even at high vacuum with $P = 1.3 \times 10^{-11}$ Pa (1×10^{-13} torr) and $T = 293$ K, $n = 4 \times 10^{13}/\text{cm}^3$. Therefore, a vacuum is only a relative concept, and absolute vacuum does not exist. What is commonly referred to as a vacuum is a "relative vacuum".

In engineering technology, especially vacuum technology, the degree of vacuum can be expressed by a number of physical quantities, the most commonly used ones being "vacuum" and "pressure". In addition, the density of gas molecules, the average free path of gas molecules, and the time required to form a molecular layer can also be expressed. "Vacuum" and "pressure" are two related concepts. Lower pressure means fewer molecules per unit volume and higher vacuum. Conversely, the lower the vacuum, the higher the pressure. Since vacuum is related to pressure, the unit of measure of vacuum is traditionally expressed in units of pressure.

In the international system of units (kilogram, meter, second; SI) the unit of measure of pressure is Pascal (Pa). The commonly used Torr, which is a special unit of vacuum, has been used since the earliest discussion of vacuum. The conversion relationship between Pa and Torr is 1 Torr = 133.322 Pa. In addition, several old units (Torr, mmHg, bar, and atmospheres) are even now occasionally used in practical engineering because it is not easy to completely change the previous data. Therefore, the pressure unit is sometimes called a Torr.

The conversion relationship between several old units and Pa is as follows:

$$1 \text{ Torr} = \frac{1}{760}\text{atm} = 133.322 \text{ Pa}.$$

mmHg and Torr are essentially the same thing: 1 mmHg = 1.00000014 Torr, and 1 atm is the standard atmosphere, 1bar = 10^5 Pa. Table 1.1 lists some of the unit conversions of pressure.

1.1.2 *Classification of vacuum regions*

At present, the vacuum degree which can be achieved with the latest technology is about 10^{-12} Pa, and the atmospheric pressure is about 10^5 Pa.

For the convenience of research and practical engineering application, vacuum is often divided into several general areas artificially: low vacuum (1×10^5 to 1×10^2 Pa), medium vacuum (1×10^2 to 1×10^{-1} Pa), high vacuum (1×10^{-1} to 1×10^{-5} Pa), ultra-high vacuum (1×10^{-5} to 1×10^{-9} Pa) and extreme high vacuum ($< 1 \times 10^{-9}$ Pa). With the increase of vacuum, the properties of vacuum (especially the properties of gas molecules) will gradually change.

1.1.2.1 *Low vacuum*

In the state of low vacuum, the characteristics of gaseous space and atmosphere are not significantly different, the number of gas molecules is large and still dominated by thermal motion, collision between molecules is very frequent, and the average free path of gas molecules is very short. Generally, the main purpose of using vacuum technology in this vacuum region is to obtain the pressure difference without changing the nature of the space. For example, the vacuum impregnation process used in capacitor production is in this region.

1.1.2.2 *Medium vacuum*

In this vacuum region, the number of gas molecules per cubic centimeter is about 10^{16} to 10^{13}. Gas molecular density and atmospheric state are very different; charged ions in the gas under the action of electric field reduce the gas conductivity. In this state, the flow of the gas also gradually decreases from the viscous retention state to the molecular state; when the gas molecules have obvious kinetic characteristics, the convection phenomenon of the gas obviously disappears. Therefore, if the metal is heated in this case, combination

Table 1.1: Unit conversions of pressure.

	Pa	Torr	μbar	atm	at	Inch Hg	psi
1 Pa	1	7.5006×10^{-3}	10	9.869×10^{-4}	1.0197×10^{-5}	2.9530×10^{-4}	1.4503×10^{-4}
1 Torr	1.3332×10^{2}	1	1.3332×10^{3}	1.3158×10^{-3}	1.3595×10^{-3}	3.9370×10^{-2}	1.9337×10^{-2}
1 μbar	10^{-1}	7.5006×10^{-4}	1	9.8692×10^{-7}	1.0197×10^{-6}	29.9530×10^{-4}	1.4503×10^{-5}
1 atm	1.0133×10^{5}	760.0	1.0133×10^{6}	1	1.0332	29.921	14.695
1 at	9.8067×10^{4}	735.56	9.8067×10^{5}	9.6784×10^{-1}	1	28.959	14.223
1 inch Hg	3.3864×10^{3}	25.4	3.3864×10^{4}	3.8421×10^{-2}	3.4532×10^{-2}	1	0.49115
1 psi	6.8948×10^{3}	51.715	6.8948×10^{4}	6.8046×10^{-2}	7.0307×10^{-2}	2.036	1

with the gas can be basically avoided, so the vacuum heat treatment generally takes place in the low vacuum state. In addition, as the pressure in the container drops, the boiling point of the liquid also drops, causing a dramatic evaporation known as vacuum freezing dehydration. In this vacuum state, as the number of gas molecules decreases, the average free path of molecules can be compared with the size of the container, and the number of collisions between molecules decreases, while the number of collisions between molecules and the container wall increases greatly.

1.1.2.3 *High vacuum*

In this vacuum region, the number of gas molecules per cubic centimeter is about 10^{13} to 10^{9}. The molecular density of the gas is even lower, so there are fewer molecules in the container. Therefore, collision between molecules in the process of motion is rare, the average free path of gas molecules is larger than the dimension size of the general vacuum container, and the majority of molecules collide with the container. Thus, the molecules (or particles) of a material vaporized in a high vacuum will basically fly away in a straight line. In addition, due to the high vacuum in the container, the combination of any object in the container space and the residual gas molecules is very weak. In this state, the heat conduction and internal friction of the gas have become independent of pressure. High vacuum is needed for single crystal growth, surface coating, and electron tube production.

1.1.2.4 *Ultra-high vacuum*

In this vacuum region, the number of gas molecules per cubic centimeter is about $10^9 - 10^5$.

There are less than 10^{10} molecules per cubic centimeter of gas. Collisions between molecules are rare, and molecules mainly collide with the walls of the container. The purpose of an ultra-high vacuum is to obtain a pure gas and a pure solid surface. At this time, the gas molecules on the solid surface is mainly adsorption or retention. Film deposition, surface analysis, particle accelerators, cryogenic refrigeration, etc., all require ultra-high vacuum.

1.1.2.5 *Extreme high vacuum*

In this vacuum region, the number of gas molecules per cubic centimeter is less than 10^5, the average degree of freedom of molecules is greater than 10^9 cm, and the frequency of molecules colliding with the container wall is relatively low. The use of extreme high vacuum is mainly for space simulation and nanometer analysis.

1.1.3 *Gases and vapors*

For each gas there is a specific temperature above which the gas will not liquefy no matter how compressed. This temperature is called the critical temperature of the gas. Critical temperatures are often used to distinguish gases from vapors. Gaseous substances whose temperature is above the critical temperature are called gases; gaseous substances whose temperature is below the critical temperature are called vapors. In practice, the distinction between gases and vapors is usually made at room temperature.

Table 1.2 lists the critical temperatures of various substances. As can be seen from Table 1.2, the critical temperature of nitrogen, hydrogen, argon, oxygen, and air is much lower than room temperature, so they are "gases" at room temperature. Carbon dioxide has a critical temperature close to room temperature and is highly liquefied, while water vapor, organic materials and gaseous metals are vapor.

Table 1.2: Critical temperatures of certain substances.

Matter	Critical temperature (°C)	Matter	Critical temperature (°C)
He	−267.8	Ar	−122.4
H_2	−239.96	O_2	−118.0
Ne	−228.0	Kr	−62.5
N_2	−147.0	Xe	14.7
Air	−140.0	CO_2	31.0
Ether	194.0	Fe	3700.0
NH_2	132.4	CH_4	−82.5
Alcohol	243.0	Cl_2	144.0
H_2O	374.2	CO	−140.2
Hg	1450.0		

1.2 Properties of Rarefied Gases

The properties of rarefied gases mainly include velocity distribution, mean free path, collision times of gas molecules, etc., which are closely related to the four parameters of pressure P, volume V, temperature T and mass m. Among them, the parameters P, V and T are mainly described by three ideal gas laws, which can be deduced from Equations (1.1) and (1.2) under certain conditions.

1.2.1 *Ideal gas law*

1.2.1.1 *Boyle's law*

For a certain mass of gas, the pressure times the volume of the gas is constant at constant temperature. That is:

$$PV = C \tag{1.4}$$

or

$$P_1V_1 = P_2V_2. \tag{1.5}$$

1.2.1.2 *Guy Lussac's law*

For a certain mass of gas, the volume of the gas is proportional to the absolute temperature at a certain pressure. That is:

$$V = CT \tag{1.6}$$

or

$$V = \frac{V_O}{T_O} \cdot T. \tag{1.7}$$

1.2.1.3 *Charlie's law*

For a certain mass of gas, if the volume remains the same, the pressure of the gas is proportional to the absolute temperature. That is:

$$P = CT, \tag{1.8}$$

$$P = \frac{P_0}{T_0} \cdot T. \tag{1.9}$$

1.2.2 Velocity distribution of gas molecules

The gas molecules in the container are in perpetual motion due to thermal motion, and there are endless frequent collisions between the gas molecules and the container wall. The velocities (including size and direction) of the molecules are not the same. In the stable state, the molecular velocity satisfies a certain statistical distribution law, usually called the Maxwell–Boltzmann distribution. That is, in the equilibrium state, when the interaction between gas molecules can be ignored, the probability of molecules distributing in any velocity interval is

$$\frac{dN}{N} = 4\pi\left(\frac{m}{2\pi kT}\right)^{3/2}\exp(-mv^2/2kT)v^2dv, \tag{1.10}$$

where N is the total number of gas molecules in the container; m is the molecular mass of gas; T is the gas temperature (K); k is the Boltzmann constant.

When the gas type is determined (m is determined), the temperature is determined,

$$\frac{dN}{N} = 4\pi\left(\frac{m}{2\pi kT}\right)^{3/2}\exp(-mv^2/2kT)v^2dv$$

just depends on the molecular velocity v.

Take equal intervals between different velocities, dN/N are different. When the gas type and temperature are determined, dN/N just depends on the molecular velocity v. Therefore, people define this part as the velocity distribution function and express it as $f(v)$:

$$f(v) = 4\pi\left(\frac{m}{2\pi kT}\right)^{3/2}\exp(-mv^2/2kT)v^2. \tag{1.11}$$

So we can get

$$\frac{dN}{N} = f(v)dv. \tag{1.12}$$

The velocity distribution function $f(v)$, also known as Maxwell's rate distribution law, represents the ratio of the number of molecules distributed within the unit velocity interval near the velocity to the total number of molecules. The distribution curve is shown in

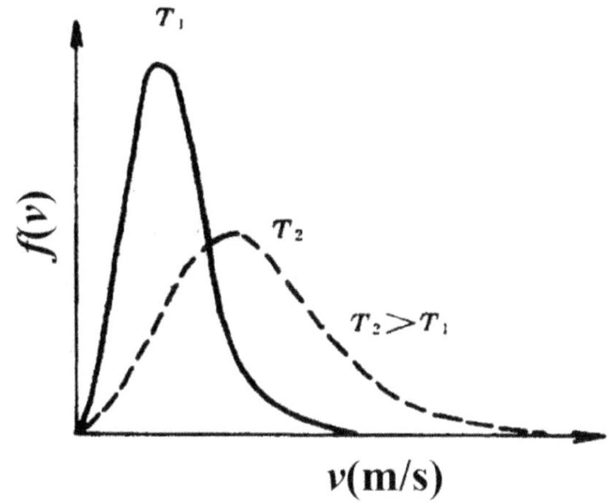

Figure 1.1: Maxwell's rate distribution curve.

Figure 1.1, which can also reflect the change of molecular velocity with temperature.

According to Maxwell's rate distribution law, it can be deduced theoretically that the maximum value of the molecular rate can be obtained at a certain V_m. Therefore, it is called the most probable velocity, whose value is

$$V_m = \sqrt{\frac{2kT}{m}} = \sqrt{\frac{2RT}{M}} = 1.41\sqrt{\frac{RT}{M}}\ \text{(cm/s)} \tag{1.13}$$

The average velocity of a gas molecule V_a is

$$V_a = \sqrt{\frac{8kT}{\pi m}} = \sqrt{\frac{8RT}{\pi M}} = 1.59\sqrt{\frac{RT}{M}}\ \text{(cm/s)} \tag{1.14}$$

The RMS velocity of a gas molecule V_r is

$$V_r = \sqrt{\frac{3kT}{m}} = \sqrt{\frac{3RT}{M}} = 1.73\sqrt{\frac{RT}{M}}\ \text{(cm/s)}. \tag{1.15}$$

As the expression shows, V_r is the largest, V_a is smaller, and V_m is the smallest. When we talk about the velocity distribution, we're going to use V_m. When you calculate the average distance the molecule moves, you use V_a. When calculating the average kinetic energy of a molecule, V_r is used.

1.2.3 *Mean free path*

The distance of each molecule between any two consecutive collisions is defined as the "molecular free path" The statistical average of the "molecular free path" is called the "average free path":

$$\lambda = \frac{1}{\sqrt{2}\pi\sigma^2 n}, \tag{1.16}$$

where σ is the molecular diameter. It can be seen that the mean free path is inversely proportional to the square of the molecular diameter σ and the molecular density n. Substitute Equation (1.1) into Equation (1.16) to obtain

$$\lambda = \frac{kT}{\sqrt{2}\pi\sigma^2 P}. \tag{1.17}$$

Equation (1.17) shows that the average free paths of gas molecules are inversely proportional to pressure and directly proportional to temperature. Therefore, in the case of a fixed gas type and temperature,

$$\lambda \cdot p = \text{constant}. \tag{1.18}$$

For example, $\lambda \cdot p \approx 0.667$ (cm · Pa) in air at 25°C. Obviously, the higher the vacuum, the greater the average free path of the molecule.

1.2.4 *Number of collisions and cosine scattering law*

The number of gas molecules that collide on the container wall per unit area per unit time is called the incident frequency ν. The incident frequency is directly proportional to the gas molecules' density n in front of the wall, and the larger the average velocity of the molecules V_a, the larger the ν, that is:

$$\nu = \frac{1}{4} n V_a. \tag{1.19}$$

Equation (1.19) is called the Hertz–Knudsen formula, which is an important formula for describing the thermal motion of gas molecules. From Equations (1.1) and (1.14), another expression of the incident frequency can be obtained:

$$\nu = \frac{P}{\sqrt{2\pi mkT}}. \tag{1.20}$$

For example, for air at 20°C:

$$\nu = 2.86 \times 10^{18} P(\text{unit/cm}^2 \cdot \text{s}) \tag{1.21}$$

The unit in Equation (1.21) is Pa.

Table 1.3 lists the properties of some important gases related to coatings. Here, m is mass ($\times 10^{-23}$ g), v_a is average speed ($\times 10^4$ cm/s, 0°C), λ is mean free path (cm · Pa, 25°C), σ ($\times 10^{-8}$ cm, 0°C) is the molecular diameter, times is collision times ($\times 10^{14}$ unit/cm^2), t is time required to form monolayer, n is the molecular number of monolayer ($\times 10^{14}$ unit/cm^2), d is thickness (nm/min). As can be seen in Table 1.3, when preparing a thin film at a pressure of 1.33×10^{-4} Pa, if gas molecules are incident at a speed of 50–100 nm/min, it only takes 1–2 s to form a single molecular layer of the gas.

Below we will discuss the reflection of gas molecules incident on the solid surface from the surface. According to Knudsen's research on low-pressure gas flow and molecular reflection studies,

Table 1.3: Properties of several commonly used gases.

	Value					At 1.33×10^{-4} Pa			
Gas	M	m	v_a	σ	λ	Times	$t(s)$	n	d
H_2	2.0	0.3	16.9	2.8	1.2	15.1	1.0	15.3	16.3
O_2	32	5.3	4.3	3.6	0.72	3.8	2.3	8.7	9.5
Ar	40	6.6	3.8	3.7	0.71	3.4	2.5	8.6	8.7
N_2	28	4.7	4.5	3.8	0.67	4.0	2.0	8.1	11.3
Air	29	4.8	4.5	3.7	0.68	4.0	2.1	8.3	10.8
H_2O	18	3.0	5.7	4.9	0.45	5.0	1.1	8.3	26.8
CO	28	4.7	4.5	3.8	0.67	4.0	2.0	8.0	11.5
CO_2	44	7.3	3.6	4.7	0.45	3.2	1.7	5.3	16.8

the following cosine law is proved to be true. The cosine law holds that the gas molecules which collide with the solid surface fly away from the surface in a direction independent of the original incident direction, and are distributed according to the cosine of the angle θ formed with the normal direction of the surface. Therefore, when a molecule leaves the solid surface, the probability that it is in the solid angle $d\omega$ (angle θ with the surface normal) is

$$\mathrm{d}p = \frac{1}{\pi} d\omega \cdot \cos\theta. \tag{1.22}$$

$1/\pi$ in Equation (1.22) is the normalization condition (that is, the probability of being in the 2π solid angle is 1).

It is important that the direction in which the molecule reflects off the surface is independent of the direction of incidence. Because it means that an incident molecule can be viewed as a molecular beam incident from one direction or any direction, the result is the same for both. Cosine law reveals one aspect of the action of solid surfaces on gas molecules; that is, the original directivity of molecules is completely "cleared out" and all are scattered according to the cosine law. In addition, cosine law also indicates that gas molecules must stay on the solid surface for a certain period of time. It is a prerequisite for gas molecules to be able to exchange energy and momentum with solids.

1.2.5 *Role of vacuum in thin film preparation*

The role of vacuum in the preparation of thin films mainly includes two aspects: reducing the collision between evaporating molecules and residual gas molecules, and inhibiting the reaction between evaporating molecules and residual gas molecules. The higher the vacuum, the better the quality of the film.

The evaporating molecules are always hit by the residual gas molecules while they travel. Let the number of N_0 vaporized molecules not impacted by residual gas molecules after travelling a distance d be

$$N_d = N_0 e^{-d/\bar{\lambda}}. \tag{1.23}$$

The ratio of the number of molecules being collided N_1 to the total number of molecules N_0 is

$$f = \frac{N_1}{N_0} = 1 - \frac{N_d}{N_0} = 1 - e^{-d/\overline{\lambda}}. \tag{1.24}$$

According to Equation (1.24), when the average free path is equal to the distance from the evaporation source to the substrate, 63% of the molecules will be scattered. If the average free path increases by 10 times, the number of molecules colliding decreases to 9%. It can be seen that only when the average free path is much larger than the distance from the evaporation source to the substrate can the collision be reduced effectively.

The rate of residual gas molecules reaching the substrate is

$$J = \frac{pN_{\text{A}}}{(2\pi\mu_G RT)^{1/2}}, \tag{1.25}$$

where N_A is Avogadro's constant and μ_G is the molar mass of the residual molecule.

The rate of evaporated molecules reaching the substrate is

$$F = \frac{\rho d N_{\text{A}}}{\mu \cdot t} \tag{1.26}$$

where ρ, d, μ, and t are, respectively, the density, thickness, molar mass of the film layer and evaporation time, respectively. If $J/F < 10^{-1}, p \approx 10^{-4} - 10^{-5}$ Pa for commonly used materials and moderate evaporation rate.

1.3 Acquisition of Vacuum

1.3.1 *Gas flow state*

The flow state of gas varies greatly depending on the geometry of the gas container, the pressure and temperature of the gas, and the type of gas. In the high vacuum environment, except for the collision of gas molecules with the container wall, there is almost no collision between gas molecules. This state of gas flow is called a molecular flow state. The molecular flow state is characterized by the mean free path of the gas molecules exceeding or equivalent to the size of the gas container. High-vacuum thin film preparation systems or various surface analysis instruments work in the state of molecular flow.

When the pressure is high, the average free path of gas molecules is short, and collisions between gas molecules are extremely frequent. This gas flow state is called a viscous flow state of the gas. Chemical vapor deposition with higher working pressure generally works in a viscous flow state. Compared with the molecular flow state, the physical mechanism of the viscous flow state is much more complicated.

In the case of low-velocity flow, the viscous flow is in a so-called laminar flow state, that is, in the direction perpendicular to the gas flow direction, it can be imagined that there are clear laminar flow lines between different gas flow layers, and the flow direction of the stratified gas can always be kept parallel to each other. For example, when the gas flows in the pipeline at a slower speed, near the tube wall, the gas molecules are affected by the resistance of the tube wall and the flow speed is close to zero. As the distance from the pipe wall increases, the gas flow speed increases and is the fastest at the center of the pipe. This state of gas flow is laminar.

In the case of high gas flow rates, the flow directions of various gases will no longer be able to maintain a parallel state with each other, but will present a vortex-like flow form. There are some low-pressure vortices in the flowing gas. At the same time, any small obstruction in the flow path will greatly affect the flow. This flow state is called a stagnant state.

The boundary between gas flow states can be divided by the Knudsen parameter K_n, which is defined as

$$K_n = D/\lambda. \tag{1.27}$$

Among them, D is the size of the gas container, and λ is the average free path of the gas molecules. According to the size of K_n, the gas flow state can be roughly divided into the following three different states:

Molecular flow state	$K_n < 1$,
Intermediate state (transition state)	$K_n = 1 - 110$,
Viscous flow state	$K_n > 110$.

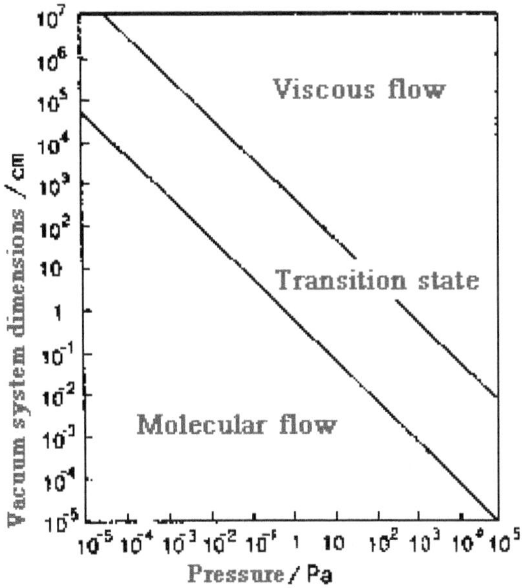

Figure 1.2: Relationship between the gas flow state and the vacuum system size and gas pressure.

In the case of viscous flow, it goes to the turbulent state when the velocity is fast and the laminar state when the velocity is slow. Figure 1.2 schematically illustrates the division of the flow state of gas molecules. In addition, in different parts of the same real gas system, it is possible to have different gas flow states at the same time.

1.3.2 *Acquisition of vacuum*

In order to obtain a vacuum environment, various vacuum systems need to be selected. A typical vacuum system should include the following main parts: the container to be evacuated (vacuum chamber), the equipment to obtain the vacuum (vacuum pump) and the equipment to measure vacuum (vacuum meter), as well as necessary piping, valves, and other ancillary equipment. Among them, the vacuum pump is the main equipment for obtaining a vacuum environment in a vacuum system.

Two important parameters for measuring the performance of a vacuum system are the pumping rate and the ultimate vacuum (limiting pressure). The pumping rate refers to the volume of gas pumped out per unit time at a specified pressure to indicate the time it takes to reach a predetermined vacuum. The pumping rate is defined as

$$S = \frac{Q}{P}, \tag{1.28}$$

where P is the pressure at the inlet of the vacuum pump, and Q is the gas flow rate passing through the unit in unit time.

For any vacuum system, it is impossible to obtain absolute vacuum ($P = 0$), but it has a certain pressure P_u, which is the lowest pressure that the system can achieve, and it is also one of the important indicators of whether the vacuum system can meet the needs of coating.

In theory, the degree of vacuum that can be achieved by any vacuum system can be determined by the following equation:

$$P = P_u + \frac{Q}{S} - \frac{V}{S} \cdot \frac{dP_i}{dt}, \tag{1.29}$$

where P_u is the ultimate pressure (Pa) of the vacuum pump, S is the pumping rate (L/s) of the vacuum pump, P_i is the partial pressure (Pa) of the space gas being pumped, $V$is the volume of the vacuum (L), and t is time (s).

The acquisition of a vacuum environment requires the use of various vacuum pumps, which are the main components of vacuum system. According to the principle of obtaining vacuum, vacuum pumps can be divided into two categories; transport vacuum pumps and capture vacuum pumps. Transport vacuum pumps use gas compression to transport gas molecules outside the vacuum system, while capture vacuum pumps rely on agglutination or absorption of gas molecules in the vacuum system to capture gas molecules and exclude them from the vacuum system. Unlike transport vacuum pumps, some capture vacuum pumps may release captured gas molecules back to the vacuum system after the job is completed.

Transport pumps can be subdivided into mechanical gas transport pumps and air-flow gas transport pumps. Rotary mechanical vacuum pumps, roots pumps, and turbo molecular pumps are typical examples of mechanical gas transport pumps, while oil diffusion pumps are air-flow gas transport pumps. Capture vacuum pumps include cryogenic adsorption pumps, sputtering ion pumps, etc.

Table 1.4 lists the exhaust principle of common vacuum pumps, the working pressure range, and the lowest pressure that can usually be obtained.

Figure 1.3 shows the pumping speed range of several common vacuum pumps, where 1 is a single stage vane pump, 2 is a sputtering ion pump, 3 is a bipolar vane pump, 4 is a roots pump, 5 is a diffusion pump, and 6 is a molecular pump. It can be seen that there is no pump that can work directly from the atmosphere to ultra-high vacuum. Therefore, usually several vacuum pumps are used in combination. Among these, the pump which can make the pressure decrease from 1 atmospheric pressure is often referred to as a "fore-vacuum pump", while the vacuum pump which can only pump air from a low pressure to obtain a lower pressure is called a "secondary pump". Examples include mechanical pump + diffusion pump system and adsorption pump + sputtering ion pump + titanium sublimation pump system; the former is an oil system, and the latter is an oil-free system.

1.3.2.1 *Mechanical pump*

Commonly used mechanical pumps include rotary vane, fixed vane, and slide valve types. Among them, slide valve (also known as reciprocating) vacuum pumps are the most widely used due to their low noise and high operating speed. Figure 1.4 is a schematic diagram of its structure. The working principle of the slide valve pump is to rely on the eccentric connecting rod mechanism to drive the wetting valve. While the slide valve is moving, the gas is drawn into the pump body through the suction port, and then discharged through the exhaust valve under the compression of the slide valve.

Rotary vane mechanical pumps rely on a rotatable rotary vane placed in an eccentric rotor to isolate and compress the gas, and

Table 1.4: Exhaust principle and working range of main vacuum pumps.

	Type	Principle	Working pressure area								
			10^4	10^2	1	10^{-2}	10^{-4}	10^{-6}	10^{-8}	10^{-10}	10^{-12}
Mechanical pump	Oil seal mechanical pump (single stage)	Mechanical force	—	—	—	—	—				
	Oil seal mechanical pump (double stage)		—	—	—						
	Molecular pump					—	—	—	—		
	Roots pump				—	—	—				
Steam jet pump	Mercury diffusion pump	Momentum of steam jet				—	—	—			
	Oil diffusion pump										
	Hole jet pump					—	—	—			
Dry pump	Sputtering ion pump	Sputtering and sublimation				—	—				
	Titanium sublimation pump					—	—				
	Adsorption pump	Low temperature surface		—	—	—					
	Condensate pump					—	—	—			
	Condensation adsorption pump					—	—	—			

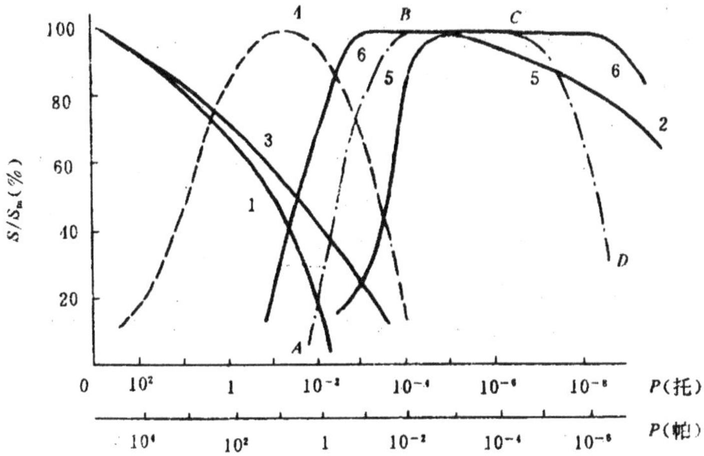

Figure 1.3: Comparison of pumping speeds of several vacuum pumps.

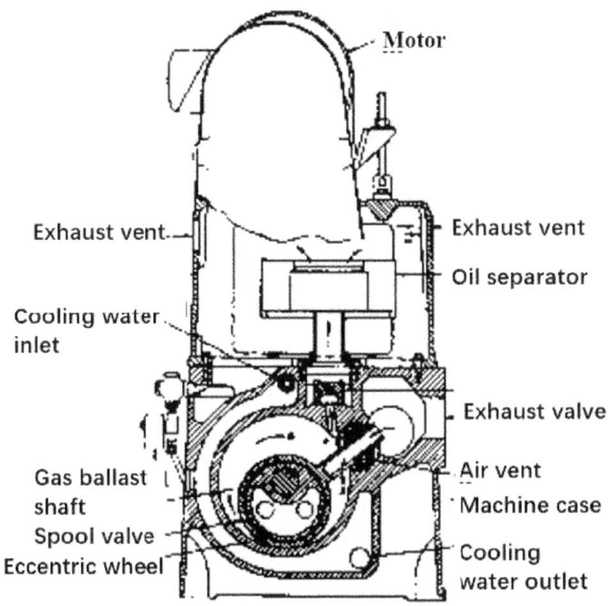

Figure 1.4: Structure of a sliding valve vacuum pump.

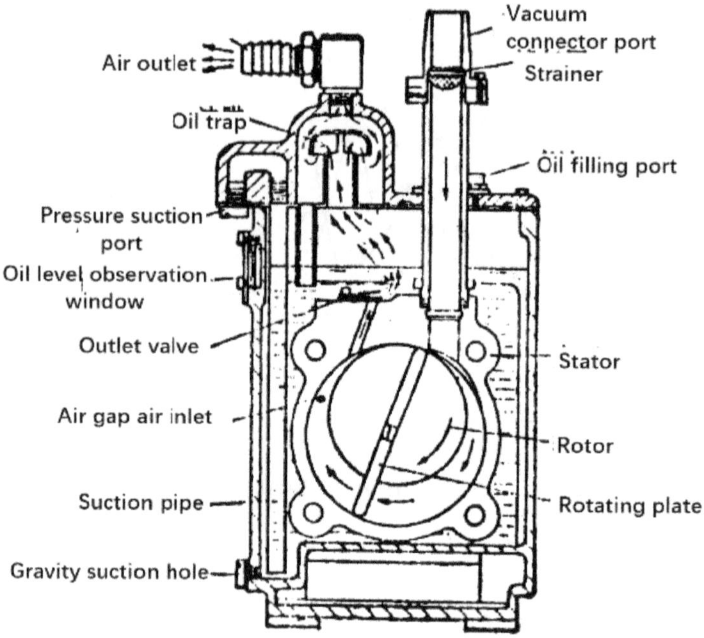

Figure 1.5: Structure of a rotary vane mechanical vacuum pump.

then discharge it from the pump. Figure 1.5 is a schematic diagram of its structure. It can be seen from Figure 1.5 that the rotary vane mechanical pump is mainly composed of a stator, a rotary vane, and a rotor, and all of it is immersed in the mechanical pump oil. Its working principle is based on Boyle's law. Figure 1.6 schematically illustrates its working principle and process. Figure 1.6(a) indicates that the air is being sucked in, and at the same time, the gas sucked in the previous cycle is gradually compressed; Figure 1.6(b) indicates that the suction is cut off. At this time, the pump's suction volume reaches the maximum and will begin to compress; Figure 1.6(c) shows another inhalation of the suction space, while the exhaust space continues to compress; Figure 1.6(d) indicates that the gas in the exhaust space has been compressed so that when the pressure exceeds 1 atmosphere, the gas is pushed out of the exhaust valve and discharged through the exhaust pipe. In this way, the rotor rotates continuously in the direction of the arrow, and suction, compression,

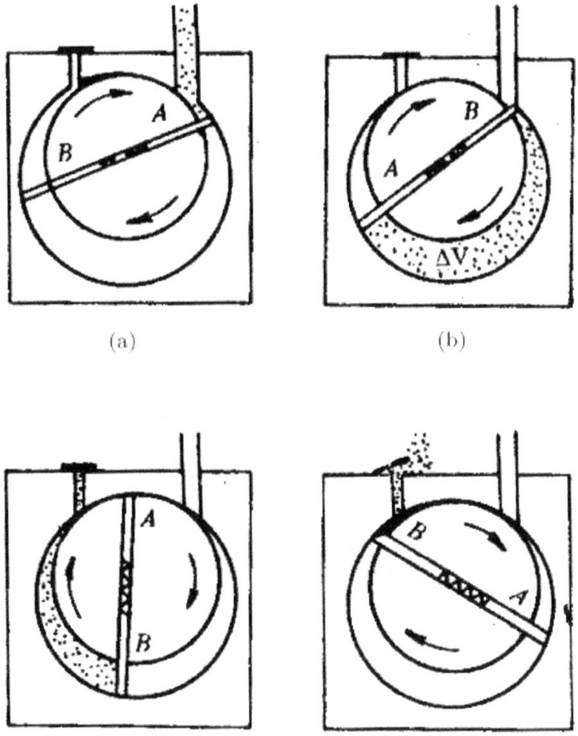

(a) (b)

(c) (d)

Figure 1.6: Working process of a rotary vane mechanical pump.

and exhaust are continuously performed, so the vacuum container connected to the mechanical pump obtains a vacuum.

1.3.2.2 *Diffusion pump*

Unlike mechanical vacuum pumps, there are no rotating or compression parts in the diffusion pump. The diffusion pump realizes the exhaust by utilizing the phenomenon that the pumped gas diffuses into the vapor flow. The structure and working principle of the diffusion pump are shown in Figure 1.7. When the diffusion pump oil is heated, a large amount of oil vapor is generated. The oil vapor is transmitted to the upper part along the vapor duct, and is ejected outward through the umbrella nozzle. Because the pressure outside

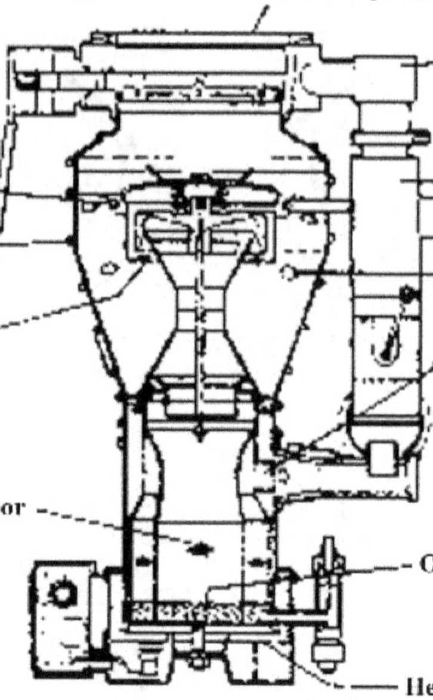

Figure 1.7: Structure and working principle of a diffusion pump.

the nozzle is low, the vapor will be ejected downward for a long distance, forming a high-speed directional vapor flow. Its convection speed can reach about 200 m/s, and its partial pressure is lower than the pressure of the pumped gas above the inlet of the diffusion pump. In this way, the gas molecules in the vacuum chamber will inevitably diffuse toward the nozzle of the diffusion pump with a lower pressure, and collide with the ultra-high-speed vapor molecules with higher energy to cause energy exchange. The pumped gas molecules are driven to move at high speed along the direction of the vapor flow and are taken to the outlet, which is pumped away by the mechanical pump.

The stream of oil vapor emitted from the nozzle is sprayed onto the water-cooled pump wall to be cooled into a liquid, flows back to the bottom of the pump, and is heated again to form steam. In this way, the circulation of oil vapor is ensured in the pump, and the diffusion pump works continuously, so that the pumped container obtains a higher vacuum degree. Diffusion pump oil is the main working substance of diffusion pump. Pump oil should have good chemical stability (non-toxic, non-corrosive), thermal stability, oxidation resistance and low saturated vapor pressure ($\leq 10^{-4}$ Pa). And it should have the highest vapor pressure when working.

A major disadvantage of oil diffusion pumps is that the backflow of oil vapor in the pump will directly cause contamination of the vacuum system. Therefore, oil diffusion pumps are generally not used in precision analysis instruments and other ultra-high vacuum systems.

The diffusion pump must be used in conjunction with a mechanical pump to form a high vacuum system. The use of a diffusion pump alone has no suction effect.

1.3.2.3 *Turbo molecular pump*

The turbomolecular pump is a high vacuum pump produced in order to meet the requirements of modern vacuum technology for oil-free high vacuum environment. Similar to oil diffusion pumps, turbomolecular pumps work by the principle of applying force to gas molecules and moving the gas molecules in a specific direction. Its structure is shown in Figure 1.8. When gas molecules collide with a solid surface moving at a high speed, they always stay on the surface for a short time, and when they leave the surface, they get a relative tangential velocity close to the solid surface velocity. This is the momentum transfer effect. The turbomolecular pump works by using this phenomenon: the high-speed rotor collides with the gas molecules and drives it to the exhaust port, which is pumped away by the foreline pump, so that the pumped container obtains an ultra-high vacuum. Its main features are these: fast startup, low noise, stable operation, high pumping speed, and no working fluid is required.

It can be seen from Figure 1.8 that the rotor blades of the turbomolecular pump have a specific shape, and while the blades rotate at a high speed of 20,000–30,000 r/min, they transmit momentum to gas molecules. At the same time, the turbomolecular pump has many stages of blades, and the gas molecules delivered by the upper stage blades will be compressed by the next stage blades to

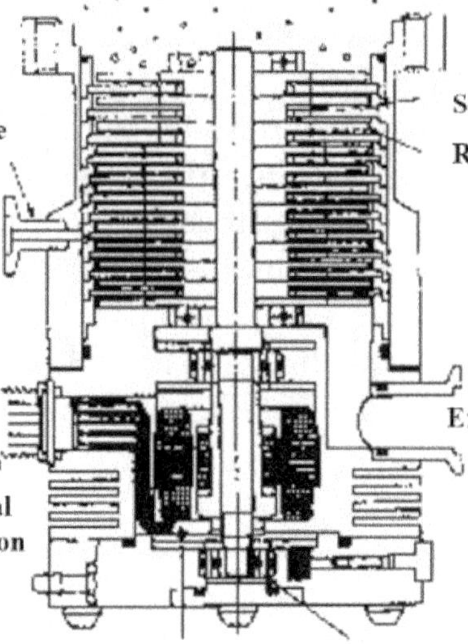

Figure 1.8: Structure of a turbo molecular pump.

the next stage. Therefore, another feature of turbomolecular pumps is that they are extremely effective at pumping gas molecules in general. For N_2, for example, the compression ratio can reach 10^9. However, the turbomolecular pump has a poor ability to extract gas with a relatively low atomic mass. For example, for H_2, its compression ratio is only about 10^3.

Due to the high compression ratio of the turbomolecular pump to the gas, the problem of oil vapor backflow during operation can be ignored. The ultimate vacuum of a turbomolecular pump can reach the order of 10^{-8} Pa, and the pumping speed can reach 1,000 L/s, and the pressure range to reach the maximum pumping speed is between 1 and 10^{-8} Pa. Therefore, a rotary vane mechanical pump is required as the backing pump.

1.3.2.4 *Cryogenic adsorption pump*

A vacuum pump that relies on spontaneous condensation of gas molecules or adsorption on the surface of other substances to achieve

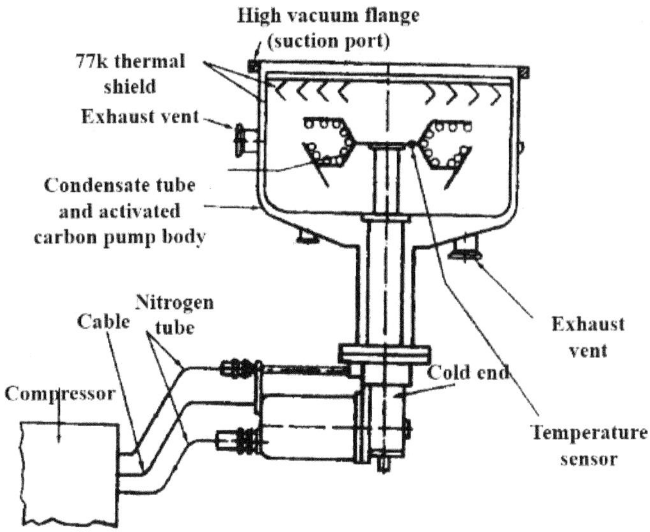

Figure 1.9: Schematic diagram of a cryogenic adsorption pump.

high vacuum is called a cryogenic adsorption pump. The degree of vacuum that can be obtained by this method depends on various factors such as the low temperature reached, the surface area of the adsorbed substance, and the type of adsorbed gas. The ultimate vacuum degree is generally between 10^{-1} and 10^{-8} Pa. Figure 1.9 is a schematic diagram of a low-temperature adsorption pump driven by a circulating refrigerator. To reduce the heat exchange between the low-temperature greenhouse and the outside world, liquid nitrogen is also used as a thermal insulation layer. Some materials often used as gas adsorption surfaces include the following: metal surfaces, low-temperature surfaces covered by condensation of high-boiling gas molecules, and adsorption materials with large specific surfaces, such as activated carbon.

The pre-vacuum required for cryogenic adsorption pump operation should reach below 10^{-1} Pa to reduce the heat load of the pump and avoid the accumulation of excessive gas condensation products in the pump body. The ultimate vacuum P_u of the cryogenic adsorption pump is related to the type of gas to be pumped. For nitrogen, it is generally about 10^{-9} Pa.

H_2, He, and Ne gases with high vapor pressure at low temperature can be easily removed by a low temperature adsorption pump. In addition to the above gases, the pumping speed of the cryogenic adsorption pump for various gases is very large, because it depends on the speed of the gas molecules moving toward the condensation surface and the surface area involved in the condensation process. Low-temperature adsorption pumps have higher operating costs, but as a means to obtain an oil-free high-vacuum environment, they can be equipped with low-vacuum pumps such as rotary vanes as the only high-vacuum pumps, and it can also be used in combination with other high-vacuum pumps, such as turbo molecular pumps.

1.3.2.5 *Sputtering ion pump*

The working principle of the sputtering ion pump is that high-speed electrons emitted from the high-voltage cathode collide with residual gas molecules to cause gas ionization discharge, and the ionized gas molecules will sputter a large number of Ti atoms when they hit the cathode at high speed. Due to the high activity of the Ti atom, it will capture a large number of gas molecules in the form of an adsorption or chemical reaction and deposit them in the pump body, thereby achieving an oil-free high vacuum environment in the vacuum chamber, as shown in Figure 1.10 The structure principle of the sputtering ion pump is shown.

The pumping speed of a sputtering ion pump is different for different gases. For more active gases, the sputtering ion pump has a larger pumping speed. For example, the pumping speed of the sputtering ion pump for H_2 is several times that for O_2, H_2O vapor, or N_2, and its pumping speed for the latter gases is much higher than that for Ar gas.

From the point of view of adsorption of gas molecules, the sputtering ion pump is similar to the cryopump, but the difference is that the gas molecules extracted by the sputtering ion pump will not be released at high temperature. At the same time, the continuous sputtering of the Ti electrode puts a certain limit on the life of the ion pump. In order to extend the life of the electrode, the sputtering

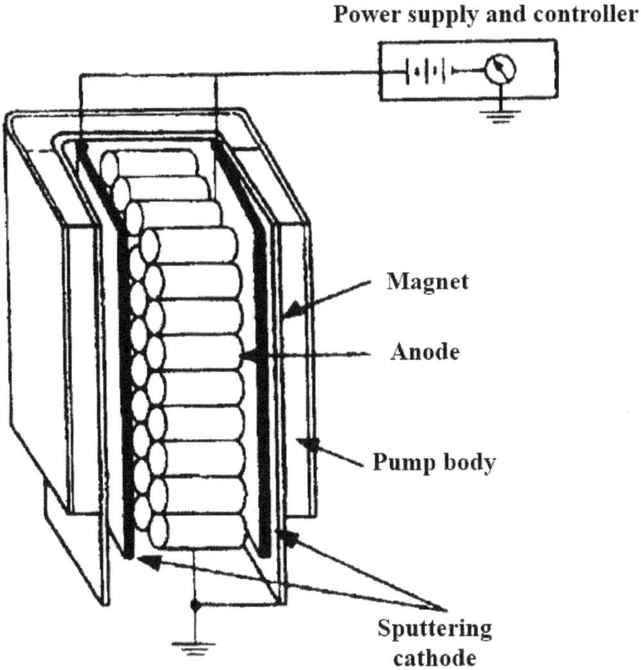

Figure 1.10: Structure of a sputtering ion pump.

ion pump is used in series with other pumps, and its ultimate vacuum can reach about 10^{-8} Pa.

1.4 Vacuum Measurement

The gases pressure encountered in vacuum technology is so low that it is extremely difficult to measure directly. Therefore, it is used to measure some physical quantities related to pressure under low pressure, and then determine the pressure in the container after transformation so as to obtain the vacuum degree in the container. These physical phenomena, in which the properties associated with pressure change as the pressure changes, are the basis of vacuum measurement. In addition, any specific physical property changes significantly over a pressure range. Therefore, any method has its measurement range, which is called the range of the vacuum gauge. At the same time, there may be errors between the vacuum degree measured by the instrument and the actual vacuum degree of the vacuum chamber due to different temperatures. At present, there is no vacuum gauge that can measure the vacuum in the whole range from atmospheric pressure to 10^{-10} Pa. Vacuum gauges can be divided into many types according to different principles and structures. Table 1.5 lists the main characteristics of several commonly used vacuum gauges.

1.4.1 *Thermocouple and thermal resistance vacuum gauge*

The thermocouple vacuum gauge is based on the principle that the heat conduction of gas under low pressure is related to gas pressure. If the pressure is high, the heat that the gas conducts is independent of pressure, but is proportional to pressure if the pressure falls in the low vacuum range.

The heat generated by the heating filament is emitted by three routes: by radiating heat (θ_1), by conducting heat (θ_2) between the filament and the thermocouple, and by the loss of heat when the gas molecules collide with the filament (θ_3):

$$\theta = \theta_1 + \theta_2 + \theta_3. \tag{1.30}$$

When the thermal equilibrium is reached, the filament temperature T is a certain value. In this case, θ_1 and θ_2 are constant, and only θ_3 varies with the number of collisions between gas molecules on the filament, that is, with the number of molecules in the gas, or with the pressure of the gas. The higher the pressure, the more the number of molecules in the gas, the more the number of collisions, the more is the heat taken away from the filament, and the greater the change in filament temperature. The vacuum gauge that measures the resistance value of the hot wire changing with temperature is called the thermal resistance vacuum gauge (as shown in Figure 1.11), also known as the Pirani vacuum gauge. A vacuum gauge that directly measures the temperature of the hot wire with a thermocouple is called a thermocouple

Table 1.5: Working principle and measuring range of several kinds of vacuum gauges.

Name	Working principle	Measuring range/(Pa)
U-tube pressure gauge (oil)	The atmospheric pressure is measured according to the height difference of the liquid column	10^4–10^{-2}
U-tube pressure gauge (Hg)		10^5–10^2
Pirani vacuum gauge	Molecular heat Conduction of a gas	10^2–10^{-2}
Resistance vacuum gauge		10^4–10^{-2} (10^{-3})
Thermocouple vacuum gauge		
Thermal cathode ionization gauge	Using the ionization of hot electrons with residual gas molecules	10^{-1}–10^{-5}
Schultz vacuum gauge		10^2–10^{-2}
B-A-type vacuum gauge		10^{-1}–10^{-8}
Extract vacuum gauge		10^{-1}–10^{-10}
Penning discharge vacuum gauge	Gas ionization in a magnetic field is related to pressure	1–10^{-5}
Magnetron vacuum gauge		10^{-4}–10^{-10}
α-ray ionized vacuum gauge	Using ionization of residual gas molecules with gamma rays	10^5–10^{-2}
Gas discharge tube	Properties of gas discharge in relation to pressure	10^3–1
Knudsen vacuum gauge	Using the momentum difference of the molecules produced by heat	10^{-1}–10^{-5}
Viscous vacuum gauges	Using the viscosity of the gas	10^1–10^{-3}
McCraw gauge	Pressure is measured by the height difference of the compressed liquid column	10^2–10^{-2}
Bourdon gauge	Pressure is measured by electrical or mechanical means of measuring the elastic deformation caused by the pressure difference	10^5–10^3
Diaphragm gauge (mechanical)		10^5–10^{-2}
Diaphragm gauge (electrical)		10^5–10^{-2}

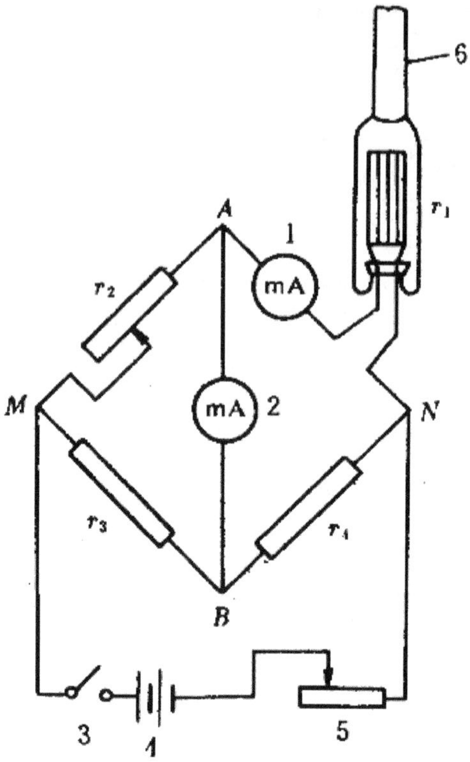

Figure 1.11: Thermal resistance gauge.

vacuum meter (as shown in Figure 1.12). The thermocouple includes nickel–chromium–constantan, iron–constantan, copper–constantan, etc. The thermocouple vacuum gauge is widely used. The surface temperature of the hot wire is closely related to the vacuum state of the hot wire. The higher the vacuum degree, the higher the surface temperature of the hot wire (fewer gas molecules collide with the hot wire), and the higher the thermoelectric potential of the thermocouple output. The lower the vacuum, the lower the surface temperature of the hot wire (more gas molecules collide with the hot wire, taking away more heat), and the lower the output electromotive force of the thermocouple. Figure 1.13 shows the curve of thermoelectromotive force of a thermocouple vacuum gauge versus gas pressure.

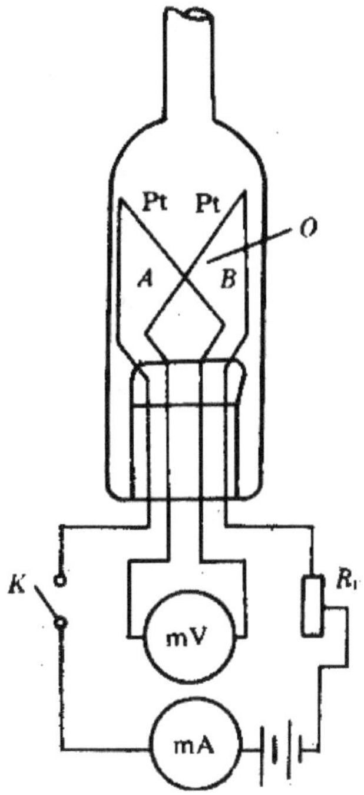

Figure 1.12: Thermocouple vacuum gauge.

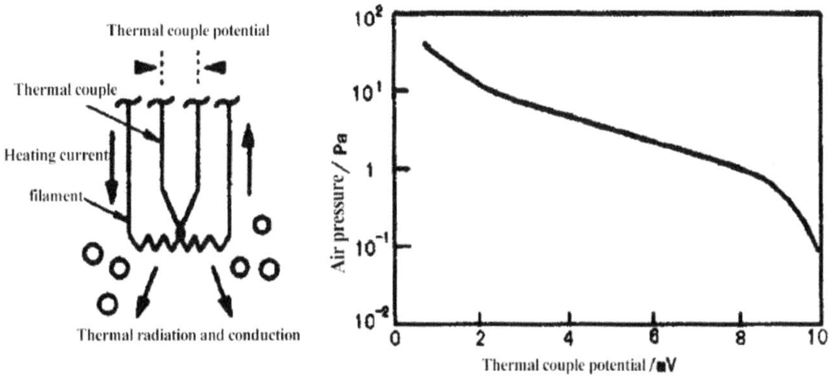

(a) (b)

Figure 1.13: Thermocouple vacuum gauge (a); curve of thermoelectromotive force versus gas pressure (b).

1.4.2 *Ionization vacuum gauge*

The high vacuum gauge that is often used in combination with a thermocouple vacuum gauge is the ionization vacuum gauge. Figure 1.14 shows the structure of a hot-cathode ionization vacuum gauge and the curve of ionic current versus pressure. The ionization vacuum gauge is the main instrument to measure a high vacuum.

The ionization vacuum gauge measures the vacuum by gas molecules ionization. According to the source of gas ionization, it can be divided into the hot cathode ionization gauge and the cold cathode ionization gauge. In a rarefied gas, the electrons emitted by the filament are accelerated by an accelerated electric field and have enough energy to ionize the molecules when they collide with them, producing positive ions and negative electrons. The ionization probability depends on the energy of the electrons. The times the electrons collide with molecules (or the number of positive ions produced) in a given flight path is proportional to the molecular density of the gas. Because $P = nkT$, the number of positive ions produced at a given temperature is also proportional to the gas pressure P. Therefore, the pressure in the measured space can be determined by the number of ions collected by the ion collector of the ionization vacuum gauge, which is the working principle of the ionization vacuum gauge.

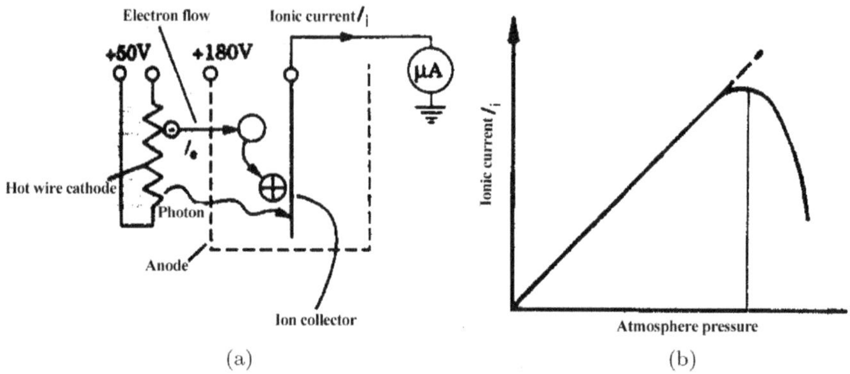

Figure 1.14: (a) Schematic diagram of a hot-cathode ionization vacuum gauge structure and (b) the ion current versus pressure curve.

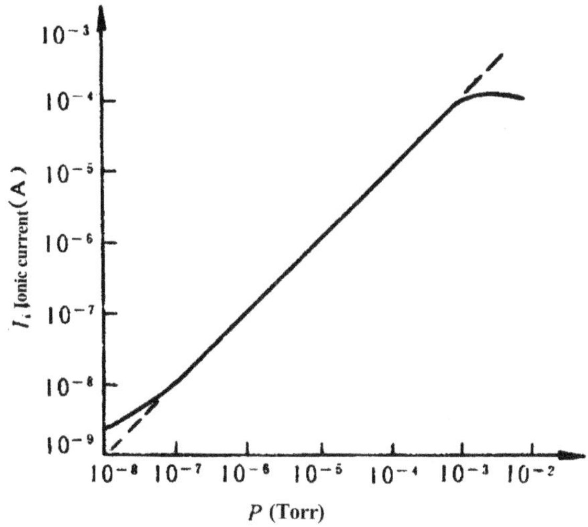

Figure 1.15: Relationship between ionic current and pressure.

The measuring range of a thermal-cathode current vacuum gauge is about 10^{-1}–10^{-6} Pa (see Figure 1.15). When the pressure is greater than 10^{-1} Pa, although the number of gas molecules increases and the number of collisions between electrons and molecules increases, the energy decreases and the probability of ionization decreases. Therefore, when the pressure increases to a certain extent, the ionization reaches saturation, the curve deviating from linearity. Therefore, the upper limit of the measurement is 10^{-1} Pa.

Under low pressure (less than 10^{-6} Pa), high-speed electrons with a certain energy collide with accelerometers to produce soft X-rays. Their energy will transfer to the free electrons in the metal through the radiation to the ion collector, which will make the free electrons escaping the metal form a photocurrent and this will lead to an increased ion flow; that is to say, the ion flow measured by the ion collector is the sum of the ionic current and the photocurrent. When the ionic current and photocurrent are numerically comparable, the curve will deviate from linearity. Thus, 10^{-6} Pa is the lower limit. The B–A ionization vacuum gauge changes the collector pole to a needle-like shape and places the filament outside the accelerator pole to reduce the area of the collector pole exposed to soft X-ray emission; it can be used to measure higher vacuum (about 10^{-10} Pa).

1.4.3 *Film vacuum gauge*

Different from the above two kinds of vacuum gauges, the film vacuum gauge relies on the film to produce mechanical displacement under the condition of gas pressure difference and can be used for the measurement of absolute gas pressure, while the result is independent of the type of gas. Its typical structure is shown in Figure 1.16. The vacuum gauge has two separate vacuum chambers. When the pressure in one chamber is known and the pressure in the other chamber is unknown, the displacement of the film is proportional to the pressure difference between the two. To improve the measurement accuracy, the film displacement is obtained by measuring the change of capacitance C_1 between the film and another metal electrode.

In order to further reduce the mechanical error caused by temperature drift, on the one hand, differential measurement can be adopted. That is, the change of capacitance C_2 between the film and another reference electrode can be measured at the same time, and the difference between C_1 and C_2 can be used as the gas pressure measurement method. On the other hand, we can also take the measure by heating the film vacuum gauge at constant

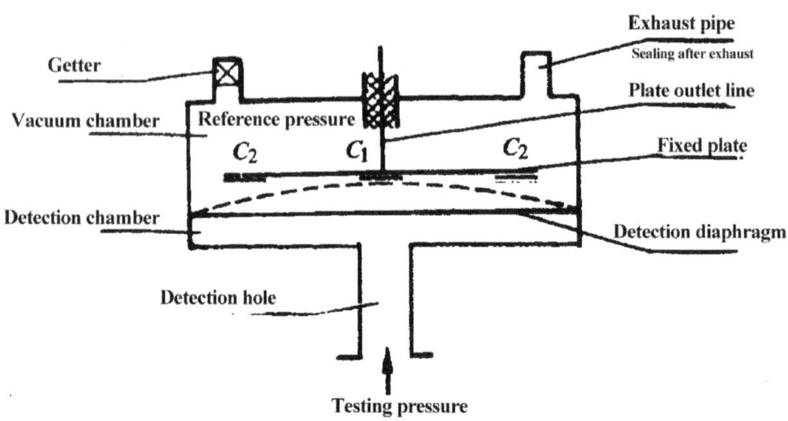

Figure 1.16: Schematic diagram of a film vacuum gauge.

temperature to reduce the influence of temperature drift on the pressure measurement.

The thin film vacuum gauge has good linearity over a wide range, and its lower detection limit is about 10^{-3} Pa, which means that the detected displacement of the thin film is only one atomic size, while the upper measurement limit depends on the failure strength of the thin film material itself or the displacement limit of the thin film.

1.4.4 *Other kinds of vacuum gauges*

To measure more area of low vacuum, the gauge is required to work above 10^{-1} Pa, and to measure higher vacuum, the gauge is required to work below 10^{-10} Pa. For the former, the Schutz ionization vacuum gauge and the impending ionization vacuum gauge are invented, and for the latter, the B–A ionization vacuum gauge and the shield ionization vacuum gauge are invented. In addition, to measure the ingredients of the chamber, the partial pressure gauge is also needed, like the magnetic deflector mass spectrometer and the quadrupole mass spectrometer. Readers may consult the resources.

1.5 Exercises and Questions

1. Briefly describe the definition of a vacuum, the classification of vacuum regions, and the unit of vacuum measurement?
2. What are the main types of vacuum pumps, and what is the working pressure range?
3. Please give a schematic diagram of the structure of a thermocouple vacuum gauge and the relationship between thermoelectromotive force and gas pressure.
4. Please give a schematic diagram of the structure of a hot-cathode ionization vacuum gauge and the relationship between ion flow and gas pressure.
5. For a vacuum pump, describe its basic structure and working mode.
6. There is 1 mol of gas in a container. What is the way to create a vacuum in one half of the container by accumulate the gas in the other half by the self-pumping effect?

7. There is 10^{-4} atm O_2 in a 1 m^3 vacuum chamber at 300 K.
   (1) How many O_2 molecules are in the chamber?
   (2) What is the ratio of the maximum potential energy to the average kinetic energy of the molecules?
   (3) What is the percentage of molecules whose kinetic energy is larger than RT in the x direction? What about 2 RT?
8. Metal flap valves with packing washers are used to divide the chamber and the pump in many systems.
   (1) A sample is poured into the vacuum chamber with 760 Torr. The pump remains at 10^{-4} Torr. How much force the flap needs with a diameter of 15 cm to seal the valve?
   (2) The chamber is vacuumized to 10^{-4} Torr beforehand. In this condition, how much is the force?
9. One of the velocity components of an ultrasonic molecule beam is given as

$$f(v) = Av^3 \exp -\frac{M(v - v_0)^2}{2RT},$$

   where v_0 is the flow speed, related to the Mach number.
   (1) What is the graph of $f(v) - v$?
   (2) Give the average speed of the gas with v_0, M, and T.

   Note: $\int_0^\infty x^{2n} e^{-ax^2} dx = \frac{\Gamma(n+1/2)}{2a^{n+1/2}}$, $\int_0^\infty x^{2n+1} e^{-ax^2} dx = \frac{n!}{2a^{n+1}}$, if $v_0 = 0$.
10. Two pumps are vacuumizing a container with the same speed. The first connect method: parallel connection, or use two pumps at the same time. The second connect method: serial connection, or use one pump once alternately. Please compare the two methods (in speed, maximum vacuum, etc.).
11. Quartz tubes inserted in a pot are usually used to anneal thin films in vacuum. Vacuumize a cylindrical tube with length L, diameter D, and conductivity C, with the speed of q_0 (Torr$\cdot$ L/cm$^2\cdot$ s). Please give the expression of the steady state pressure. (Clue: in the length of dx, the loaded gas is equal to the vented gas).

12. Vacuumize a container of 30 L to 1×10^{-6} Torr, and then take off the pump. The pressure of the container increases to 1×10^{-5} Torr in 3 min.

    (1) What is the leak rate?
    (2) Connect a diffusion pump with an effective pumping speed of 40 L/s. What is the maximum vacuum?
    (3) Select an equipment that uses high vacuum (an electron microscope, an evaporator, a helical spectrometer, etc.). Study the exhausting devices in the system and explain how to measure the working pressure.

Chapter 2

Physical Preparation Technology of Thin Films

2.1 Overview of Film Preparation Methods

According to the type of film-forming mechanism, film preparation methods can be roughly divided into two categories: physical methods and chemical methods. Physical vapor deposition (PVD) is a controllable atomic transfer process from the original material to a thin film through some physical processes, such as thermal evaporation of the material or sputtering of atoms on the surface of the material when bombarded by a particle beam. The main features of PVD are as follows:

(1) The use of solid or molten materials as the original material in the deposition process;
(2) The original material transforms into the gas phase through physical processes;
(3) It requires a relatively low gas pressure environment;
(4) No chemical reaction occurs in the gas phase and on the surface of the substrate.

Another type of thin film deposition technique, but structurally distinct from PVD, is chemical vapor deposition (CVD). CVD technology is a technology that uses a gaseous precursor reactant to form a solid film through atomic and intermolecular chemical reaction processes. A variety of films can be prepared by this method, including the various films required for solid electronic devices, wear resistant coatings for bearings and tools, and high temperature protective layers for engine or nuclear reactor components. CVD technology can control the chemical composition of the film effectively, and it has good compatibility with the other related processes, such as semiconductor processes.

There are some cross-overs between these two film technologies. For example, reaction co-evaporation, reactive co-sputtering, and molecular beam epitaxy are all related to these two methods. In addition, the newly developed pulsed laser deposition (PLD) has become a widely studied, modern film preparation technology. Table 2.1 lists some commonly used film preparation methods.

This chapter focuses on the basic principles and processes of the physical preparation of thin films, related devices, and their characteristics.

2.2 Vacuum Evaporation Coating

Vacuum evaporation coating (referred to as vacuum evaporation) is a method for PVD film formation, during which the raw materials to be formed in the evaporation container are heated in a vacuum chamber to vaporize its atoms or molecules and escape from the surface, forming a vapor stream, and shoot on the surface of a solid (usually used as a substrate), thereby condensing to form a solid film. The main physical process of the vacuum evaporation method is to turn the evaporation material into a gaseous state by heating, so this method is also called thermal evaporation method. This represents the earliest and simplest membrane-making method that was first adopted by Farady in 1857 and is widely used even now. In recent years, the improvement of the method has focused mainly on the evaporation sources. In order to restrain or avoid chemical reactions between the raw materials of the film and the evaporation heating plate, we use heat-resistant ceramic crucibles instead, such as BN crucibles; in order to evaporate low vapor pressure substances, an electron beam heating source or a laser heating source can be used; in order to prepare complex components or multilayer composite thin films, the multi-source co-evaporation or sequential

Table 2.1: Summary of various preparation methods of thin films.

Category	Classification	Detailed division
Physical preparation method	Vacuum evaporation	Reaction evaporation
		Instantaneous evaporation
		Instantaneously evaporation
	Molecular beam epitaxy	Molecular beam epitaxy
		Solid phase epitaxy
		Laser MBE
	Sputtering	DC sputtering
		AC sputtering
		Reactive sputtering
		RF sputtering
		Magnetron sputtering
	Ion beam method	Total ion beam
		Partial ion beam
		Ion beam epitaxy
	Pulsed laser deposition	PLD
Chemical production method	Chemical vapor phase	Hydrogen reduction
		Halogen transport
	Chemical solution	Hydrogen thermal decomposition
		Metal–organic compound thermal decomposition
		Sol–gel method
	Chemical plating	Electrolytic plating
		Electroless plating
Others	Liquid phase epitaxy	Solution delivery
		Solution temperature drop method
	Oxidation method	High temperature oxidation
		Low temperature oxidation
		Anodizing
	Diffusion method	Gas phase diffusion
		Solid phase diffusion
	Ion implantation	Ion implantation

evaporation methods have been developed; in order to prepare compound films or to restrain the deviation of film components from raw materials, methods such as the reaction evaporation method have appeared.

2.2.1 *Principle of vacuum evaporation*

2.2.1.1 *Characteristics of vacuum evaporation*

Compared with the film-forming methods such as chemical vapor deposition and sputtering coating to be described later, vacuum evaporation has the following characteristics: the equipment is simple and easy to operate, the film has high purity and good quality, and the thickness can be controlled accurately; the film-forming rate is fast and the efficiency is high, and clear graphics can be obtained with a mask; the growth mechanism of the thin film is relatively simple. The main disadvantages of this method are as follows: it is not easy to obtain a thin film with a good crystalline structure, the formed film has a small adhesion capacity on the substrate, and the process repeatability is not good enough.

Figure 2.1 shows the principle of vacuum evaporation coating. As can be seen from Figure 2.1, the device is mainly composed of a vacuum chamber, an evaporation source, a substrate, a substrate heater, and a measuring device. The vacuum chamber provides the necessary vacuum environment for the evaporation process; the evaporation source places the evaporation material and heats it; the substrate is used to receive the evaporation substance and form a solid evaporation film on its surface.

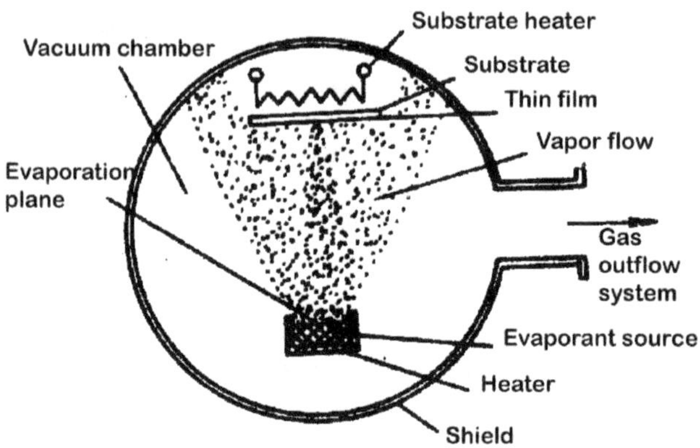

Figure 2.1: Schematic diagram of vacuum evaporation coating.

As can be seen from Figure 2.1, vacuum evaporation coating includes the following three basic processes:

(1) **Heating and evaporation process:** It includes a phase transition from a condensed phase to a gas phase. Each evaporation substance has a different saturated vapor pressure at different temperatures; when a compound is evaporated, reactions occur between its components, some of which enter the evaporation space in a gaseous or vaporous state.
(2) **Transportation of vaporized atoms or molecules between the evaporation source and the substrate:** This is the flight of particles in the ambient atmosphere. The number of collisions with residual gas molecules in the vacuum chamber during the flight depends on the mean free path of the evaporated atoms and the distance from the evaporation source to the substrate. This distance is often called the source–base distance.
(3) **Deposition of evaporated atoms or molecules on the basic surface:** This is the process where the vapor first gathers into a core on the substrate, then the core grows, and then it forms a continuous film. Since the temperature of the substrate is much lower than the temperature of the evaporation source, a phase transition process from the gas phase to the solid phase will occur directly on the surface of the substrate.

All of the three processes must be performed in a high-vacuum environment. If the vacuum is too low, the vapor atoms or molecules will collide with a large number of gas molecules. This will cause the film to be contaminated and even form oxides, or the evaporation source is heated to oxidize or even burned; it may also be difficult to form a uniform continuous film due to the collision barrier of air molecules.

2.2.1.2 *Saturated vapor pressure*

All kinds of solids or liquids will evaporate at any temperature when placed in closed containers, and the vapor that evaporates forms a vapor pressure. At a certain temperature, the vapor pressure measured at the time when the number of molecules evaporated per unit time is equal to the number of molecules condensed on the wall of the vessel and returning to the evaporated substance is called the saturated vapor pressure of the substance at this temperature. At this time, the solid phase or liquid phase and the gas phase of the evaporating substance are in dynamic equilibrium; that is, the number of molecules reaching the surface of the solid phase or liquid phase is equal to the number of molecules leaving the solid phase or liquid phase and returning to the gas phase. In general, the saturated vapor pressure of a substance increases with temperature. At a certain temperature, the saturated vapor pressures of various substances are different, but they all have a constant value. Similarly, a certain saturated vapor pressure must correspond to a certain temperature. Generally, the temperature corresponding to the substance when the saturated vapor pressure is 10^{-2} Torr is the evaporation temperature of the substance. Table 2.2 lists the saturated vapor pressure of various substances.

The mathematical expression between the saturated vapor pressure P_v and the temperature T can be derived from the Clapeylon–Clausius relationship

$$\frac{dp_v}{dT} = \frac{H_v}{T(V_g - V_s)}. \tag{2.1}$$

In Equation (2.1), H_v is the molar heat of vaporization or evaporation (J/mol); V_g and V_s are the molar volume (cm^3) of the gas phase and

Table 2.2: Saturated vapor pressure of various substances.

Substance	Saturated vapor pressure/Torr (at 20°C)	Substance	Saturated vapor pressure/Torr (at 20°C)
Water	17.5	Seal grease	10^{-3}–10^{-7}
Mechanical pump oil	10^{-2}–10^{-5}	General diffuser pump oil	10^{-5}–10^{-8}
Mercury	1.8×10^{-3}	275 Ultra-high vacuum diffusion pump oil	5×10^{-10} (25°C)

the solid phase (or liquid phase), respectively; and T is the absolute temperature (K).

Because $V_g \gg V_s$ and we assume that the vapor molecules meet the ideal gas state equation at low pressure, then

$$V_g - V_s \approx V_g, \quad V_g = RT/P_v. \tag{2.2}$$

In Equation (2.2), R is the gas constant, the value of which is 8.31×10^7erg/K · mol. So Equation (2.1) can be rewritten as

$$\frac{dP_\nu}{P_\nu} = \frac{H_\nu \cdot dT}{RT^2} \tag{2.3}$$

and it can also be written as

$$\frac{d(\ln P_\nu)}{d(1/T)} = \frac{-H_\nu}{R}.$$

Therefore, if the relationship between the natural logarithmic value of P_ν and $1/T$ is plotted, it should be a straight line.

Since the heat of vaporization H_ν usually changes only slightly with temperature, H_ν can be regarded as approximately a constant, so Equation (2.3) is integrated to obtain

$$\ln P_\nu = C - \frac{H_v}{RT}. \tag{2.4}$$

In Equation (2.4), C is the integral constant, and Equation (2.4) is expressed by common logarithms to obtain

$$\lg P_\nu = A - \frac{B}{T}. \tag{2.5}$$

Values of A and B in Equation (2.5) can be confirmed by experiment, in which $A = C/2.3$ and $B = H\text{v}/2.3R$. Table 2.3 lists some constant values used for material vapor pressure calculations. In fact, the relationship between P_v and T is mostly determined by experiment, and there is a relationship of $H_v = 19.123B$ (J/mol). Equation (2.5) is the approximate relationship between the saturated vapor pressure and temperature of the evaporated material. For most materials, Equation (2.5) is a more accurate expression in a relatively

Table 2.3: Calculation constants of vapor pressure equations for some metals.

Metal	A	B	Metal	A	B	Metal	A	B
Li	10.99	8.07×10^3	Sr	10.71	7.83×10^3	Si	12.72	2.13×10^4
Na	10.72	5.49×10^3	Ba	10.70	8.76×10^3	Ti	12.50	2.32×10^4
K	10.28	4.48×10^3	Zn	11.63	6.54×10^3	Zr	12.33	3.03×10^4
Cs	9.91	3.80×10^3	Cd	11.56	5.72×10^3	Tb	12.52	2.84×10^4
Cu	11.96	1.698×10^4	B	13.07	2.962×10^4	Ge	11.71	1.803×10^4
Ag	11.85	1.427×10^4	Al	11.79	1.594×10^4	Sn	10.88	1.487×10^4
Au	11.89	1.758×10^4	La	11.60	2.085×10^4	Pb	10.77	9.71×10^3
Be	12.01	1.647×10^4	Ga	11.41	1.384×10^4	Sb	11.15	8.63×10^3
Mg	11.64	7.65×10^3	In	11.23	1.248×10^4	Bi	11.18	9.53×10^3
Ca	11.22	8.94×10^3	C	15.73	4×10^4	Cr	12.94	2.0×10^4
Mo	11.64	3.085×10^4	Co	12.70	2.111×10^4	Os	13.59	3.7×10^4
W	12.40	4.068×10^4	Ni	12.75	2.096×10^4	Ir	13.07	3.123×10^4
U	11.59	2.331×10^4	Ru	13.50	3.38×10^4	Pt	12.53	2.728×10^4
Mn	12.14	1.374×10^4	Rh	12.94	2.772×10^4	V	13.07	2.572×10^4
Fe	12.44	1.997×10^4	Pd	11.78	1.971×10^4	Ta	13.04	4.021×10^4

narrow temperature range when the vapor pressure is less than 1 Torr (133 Pa).

Table 2.4 and Figure 2.2 show the relationship between the saturated vapor pressure and temperature of some commonly used materials (mainly metal materials). From the approximate curve of $P_v - 1/T$ in Figure 2.2, we can note that the saturated vapor pressure increases rapidly with increasing temperature. Therefore, the evaporation of substances under vacuum is much easier than that at normal pressure. The required evaporation temperature is greatly reduced, the evaporation time is greatly shortened, and the evaporation rate is significantly increased.

2.2.1.3 *Evaporation rate*

According to the theory of gas molecular motion, when the gas with pressure P is in a thermal equilibrium state, the number of molecules that collide with the wall of unit area per unit time is

$$J = \frac{1}{4} n v_a = \frac{P}{\sqrt{2\pi kT}}. \tag{2.6}$$

Table 2.4: The relationship between the vapor pressure and temperature of some commonly used materials.

Metal	Molecular weight	Temperature, T(K) at different vapor pressures, P_ν/Pa						Melting point/(K)	Evaporation rate*
		10^{-8}	10^{-6}	10^{-4}	10^{-2}	10^{0}	10^{2}		
Au	197	964	1080	1220	1405	1670	2040	1336	6.1
Ag	107.9	759	847	958	1105	1300	1605	1234	9.4
In	114.8	677	761	870	1015	1220	1520	429	9.4
Al	27	860	958	1085	1245	1490	1830	932	18
Ga	69.7	796	892	1015	1180	1405	1745	303	11
Si	28.1	1145	1265	1420	1610	1905	2330	1685	15
Zn	65.4	354	396	450	520	617	760	693	17
Cd	112.4	310	347	392	450	538	665	594	14
Te	127.6	385	428	482	553	647	791	723	12
Se	79	301	336	380	437	516	636	490	17
As	74.9	340	377	423	477	550	645	1090	17
C	12	1765	1930	2140	2410	2730	3170	4130	19
Ta	181	2020	2230	2510	2860	3330	3980	3270	4.5
W	183.8	2150	2390	2680	3030	3500	4180	3650	4.5

Note: *Unit: $J \times 10^{17}(\text{cm}^{-2} \cdot \text{s}^{-1})(P \approx 1$ Pa, adhesion coefficient $\alpha \approx 1)$.

In Equation (2.6), n is the molecular density, v_a is the arithmetic average speed, m is the molecular mass, and k is the Boltzmann constant. If it is considered that not all evaporation molecules are condensed during the actual evaporation process, Equation (2.6) can be rewritten as:

$$J_e = \frac{\alpha P_v}{\sqrt{2\pi mkT}}. \tag{2.7}$$

In Equation (2.7), α is the condensation coefficient, generally $\alpha \leq 1$, and P_v is the saturated vapor pressure.

Assuming that the liquid and gas phases on the surface of the evaporation material are in dynamic equilibrium, the evaporation rate can be expressed as

$$J_e = \frac{dN}{A \cdot dt} = \frac{\alpha_e(P_v - P_n)}{\sqrt{2\pi mkT}}, \tag{2.8}$$

where dN is the number of evaporated molecules (atoms), α_e is the evaporation coefficient, A is the evaporation surface area, t is the

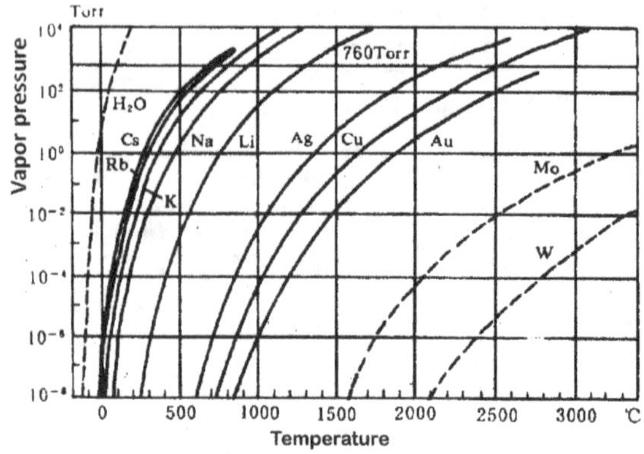

(a) Vapor pressure for group IA and IB elements

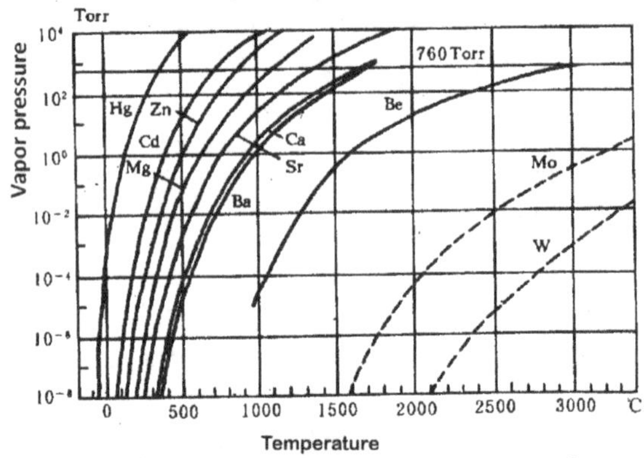

(b) Vapor pressure for group IIA and IIB elements

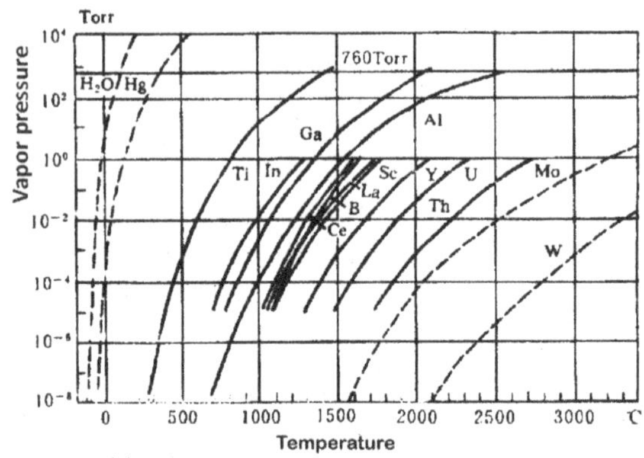

(c) Vapor pressure for group IIIA and IIIB elements

Figure 2.2: Relationship between the vapor pressure and temperature of III-group elements.

time (s), and P_v and P_n are the saturated vapor pressure and liquid pressure (Pa), respectively.

When $\alpha_e = 1$ and $P_n = 0$, the evaporation rate reaches the maximum.

$$J_m = \frac{dN}{Adt} = \frac{P_v}{\sqrt{2\pi mkT}} \approx 2.64 \times 10^{24} P_v \left(\frac{1}{\sqrt{TM}}\right), \tag{2.9}$$

where M is the molar mass of the evaporated substance. Langmuir pointed out that Equation (2.9) is also applicable for evaporation from solid free surfaces. If Equation (2.9) is multiplied by the atomic or molecular mass, the mass evaporation rate G per unit area is obtained.

$$G = mJ_m = \sqrt{\frac{m}{2\pi kT}} \cdot P_v \approx 4.37 \times 10^{-3} \sqrt{\frac{M}{T}} \cdot P_v \tag{2.10}$$

Equation (2.10) determines the relationship between evaporation rate, vapor pressure, and temperature. It's an important expression for describing evaporation rate.

It should be noted that the evaporation rate is related to the molecular weight of the evaporated substance, the absolute temperature, and the saturated vapor pressure of the evaporated substance at the temperature T, as well as the cleanliness of the material itself. In particular, the temperature change of the evaporation source has a great influence on the evaporation rate. If the relationship (2.5) between P_v and T is substituted into (2.10) and differentiated, the relationship between the evaporation rate and temperature can be obtained, that is

$$\frac{dG}{G} = \left(2.3\frac{B}{T} - \frac{1}{2}\right)\frac{dT}{T}. \tag{2.11}$$

For metals, 2.3 B/T is usually between 20 and 30, that is,

$$\frac{dG}{G} = (20 \sim 30)\frac{dT}{T}. \tag{2.12}$$

It can be seen that when evaporation is performed above the evaporation temperature, a small change in the temperature of the evaporation source can cause a large change in the evaporation rate.

Therefore, in the process of film formation, to control the evaporation rate, the temperature of the evaporation source must be accurately controlled, and excessive temperature changes should be avoided as much as possible during heating.

2.2.1.4 *Mean free path and collision probability of evaporated molecules*

There are atoms or molecules of evaporation substances and residual gas molecules in the vacuum chamber, so vacuum evaporation is actually performed in the residual gas with a certain pressure. Obviously, these residual gas molecules will affect the formation of the film and even the properties of the film.

Under thermal equilibrium conditions, the number of gas molecules N_g that pass through per unit area per unit time is

$$N_g = 3.513 \times 10^{22} \frac{P}{\sqrt{TM}} \quad (\mathrm{cm}^{-2} \cdot \mathrm{s}^{-1}), \tag{2.13}$$

where, P is the gas pressure (Torr); M is the molar mass of the gas (g), T is the gas temperature (K), and N_g is the collision rate of the gas molecules on the substrate. In fact, Equation (2.13) is equal to Equation (2.9).

Table 2.5 shows the N_g of several typical gas molecules. It can be seen from Table 2.5 that about 10^{15} gas molecules reach the surface of the unit substrate per second, and the general film deposition rate is several angstroms/second (about 1 atomic layer thickness). Obviously, when the residual gas pressure is 10^{-5} Torr, gas molecules and evaporated substances reach the surface of the substrate almost

Table 2.5: Collision number of gas molecules.

		$N_g(\mathrm{cm}^{-2} \cdot \mathrm{s}^{-1})^*$	
Substance	Molecular weight	10^{-5} (Torr)	10^{-2} (Torr)
H_2	2	1.4×10^{15}	1.4×10^{18}
Ar	40	3.2×10^{15}	3.2×10^{18}
O_2	32	3.6×10^{15}	3.6×10^{18}
N_2	28	3.8×10^{15}	3.8×10^{18}

in a ratio of 1:1. The adhesion coefficient of gas molecules to the surface of the substrate depends on factors such as the nature of the residual gas molecules and the surface of the substrate, and the temperature of the substrate. For substrates with high chemical activity, the adhesion coefficient is approximately one. Therefore, to obtain a high-purity thin film, a high degree of vacuum is required; that is, a very low pressure of the residual gas is required.

Evaporation material molecules fly in the residual gas. These particles collide with each other and with the vacuum wall under the irregular thermal motion, which will change the original direction of motion and reduce its speed. As mentioned earlier, the average distance that a particle flies between two collisions is called the mean free path λ of the evaporated molecules

$$\lambda = \frac{1}{\sqrt{2}\pi n d^2} = \frac{kT}{\sqrt{2}\pi P d^2}, \tag{2.14}$$

where n is the residual gas molecule density, and d is the collision cross section.

Both the mean free path and the collision of evaporation molecules with residual gas molecules have statistical laws. Suppose that after N_0 evaporation molecules fly a distance of x meters, the number of molecules that have not been collided by residual gas molecules is N_x and $N_x = N_0\exp(-x/\lambda)$, so the percentage of molecules that are collided is:

$$f = 1 - (N_x/N_0) = 1 - \mathrm{e}^{-x/\lambda}. \tag{2.15}$$

Figure 2.3 shows the relationship between the collision percentage and the ratio of actual path to mean free path during the flight between the evaporation source and the substrate (source–base) according to Equation (2.15). When the mean free path is equal to the source–base distance, about 63% of the evaporated molecules are collided; if the mean free path is increased by 10 times, the probability of collision will be reduced to about 9%. It can be seen that only when $\lambda \gg L$ (L is the source–base distance), that is, only when the mean free path is much larger than the source–substrate distance, can the

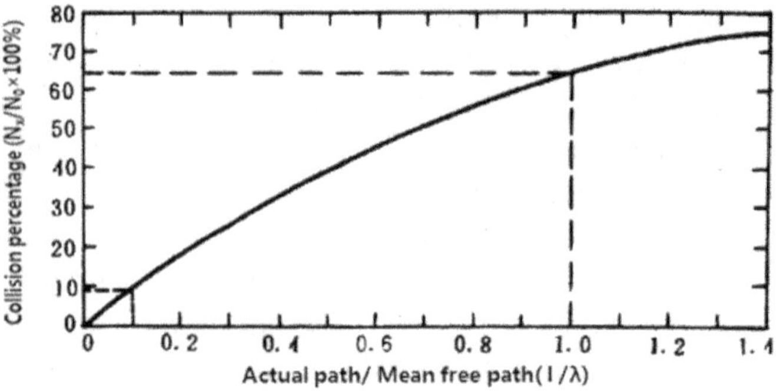

Figure 2.3: Relationship between the collision percentage and the ratio of the actual path to the mean free path.

collision phenomenon of evaporation molecules be effectively reduced during the coating process.

If the degree of vacuum is high enough, the mean free path is large enough, and the condition $\lambda \gg L$ is satisfied, then there is $f \approx L/\lambda$. Substituting Equation (2.14), we obtain that

$$f \approx 1.5\,L \cdot P \tag{2.16}$$

It can be concluded that in order to ensure the quality of the coating, P must satisfy $P \leq 3 \times 10^{-3}$ Pa if the source–substrate distance $L = 25\,\text{cm}$ when $f \leq 0.1$ is required.

The residual gas molecules in the vacuum chamber not only have an effect on the mean free path of the molecules, but also cause some pollution on the membrane layer. The general residual gas is mainly composed of oxygen, nitrogen, water vapor, oil vapor of the diffusion pump, and other polluting gases. When the film is actually deposited, the situation becomes complicated due to the interaction between the residual gas and the evaporated film and the evaporation source.

2.2.1.5 *Required heat for evaporation and energy of evaporating particles*

In addition to heating and evaporating the evaporating material, the heat needed for the evaporation source should also consider the heat radiation generated by the evaporation source during the heating process and the heat lost by the heat conduction; that is, the total heat Q required for the evaporation source is the sum of the above three parts of energy

$$Q = Q_1 + Q_2 + Q_3, \tag{2.17}$$

where, Q_1 is the heat required for the evaporation material to evaporate; Q_2 is the heat lost by the evaporation source due to heat radiation; Q_3 is the heat lost by the evaporation source due to heat conduction.

(1) **Heat required for the evaporation material to evaporate:** If a substance with a molecular weight of M and a weight of Wg is heated from room temperature T_0 to evaporation temperature T requires a heat of Q_1, then there is

$$Q_1 = \frac{W}{M}\left(\int_{T_o}^{T_m} C_s dT + \int_{T_m}^{T} C_L dT + L_m + L_\nu\right), \tag{2.18}$$

where C_s is the specific heat of the solid (J/°C · mol); C_l is the specific heat of the liquid (J/°C · mol); L_m is the heat of fusion of the solid (kJ/mol); L_v is the heat of molecular gasification (kJ/mol); T_m is the melting point (K) of the solid. In addition, L_m and L_v can be ignored for directly sublimated substances. The heat of evaporation required for common metal materials is listed in Table 2.6.

Table 2.6: Evaporation heat of common metals ($P = 1$ Pa).

Name	Q_1 (kJ/g)	Name	Q_1 (kJ/g)
Al	12.98	Cr	8.37
Ag	2.85	Zr	7.53
Au	2.01	Ta	4.60
Ba	1.34	Ti	10.47
Zn	2.09	Pb	1.00
Cd	10.47	Ni	7.95
Fe	79.53	Pt	3.14
Cu	5.86	Pd	4.02

It can be seen from Table 2.6 that the evaporation heat values required at the same pressure are different for different substances. It should be noted that more than 80% of the evaporation heat Q value is consumed as the evaporation heat Q_1. In addition, there is also the heat lost due to radiation and conduction.

(2) **Heat lost due to heat radiation:** The heat loss in this part is related to the shape, structure, and material of the evaporation source; it can be estimated by the following formula:

$$Q_2 = \sigma \cdot a_s \cdot T^4, \tag{2.19}$$

where σ is the Stefan–Boltzmann constant, $\sigma = 5.668 \times 10^{-12}\text{W/cm}^2$ or $\sigma = 1.35 \times 10^{-12}(\text{Cal/cm}^2 \cdot \text{s} \cdot \square^4)$; a_s is the radiation coefficient that can be found in the Handbook of Physics; T is the evaporation temperature.

(3) **Loss of heat due to heat conduction:** If calculated by a single-layer shielded heat condition, then

$$Q_3 = \frac{\xi F(T_1 - T_2)}{S}, \tag{2.20}$$

where ξ is the coefficient of heat conduction of the electrode material; F is the heat conduction area; S is the original height of the heat conduction wall; T_1 is the high-temperature surface's temperature; T_2 is the low-temperature surface's temperature.

2.2.2 *Evaporation characteristics of evaporating sources*

In the vacuum evaporation process, in order to obtain a film with uniform thickness on the substrate, the evaporation source needs to be carefully designed. The film thickness at different evaporation positions on the substrate depends on the evaporation characteristics of the evaporation source, the geometry of the substrate, and the evaporation source relative position, and evaporation amount of the evaporated material. Several of the most commonly used evaporation sources are introduced below.

2.2.2.1 *Point evaporation source*

A micro-spherical evaporation source capable of evaporating the same amount of material in all directions is called a point evaporation source (or point source for short). A small sphere ds evaporates in all directions at the same evaporation rate of m grams per second. In unit time and in any direction, the total amount of evaporation material through the solid angle $d\omega$ as shown in Figure 2.4 is dm; then

$$dm = \frac{m}{4\pi} \cdot d\omega. \tag{2.21}$$

Therefore, when the geometrical dimensions of the evaporation material reaching a small area ds_2 that makes an θ angle with the evaporation direction are known, the thickness and quantity of the film material deposited on this area can be obtained. Figure 2.4 shows that

$$ds_1 = ds_2 \cdot \cos\theta, \qquad ds_1 = r^2 \cdot d\omega.$$

So

$$d\omega = \frac{ds_2 \cdot \cos\theta}{r^2} = \frac{ds_2 \cdot \cos\theta}{h^2 + x^2}, \tag{2.22}$$

where r is the distance between the point source and the observed point on the substrate; then the total amount dm of the evaporation

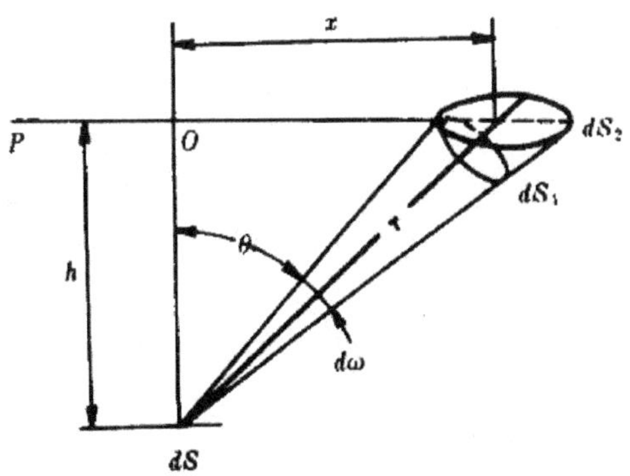

Figure 2.4: Emission characteristics of the point evaporation source.

material reaching ds_2 is

$$dm = \frac{m}{4\pi} \cdot d\omega = \frac{m}{4\pi} \cdot \frac{\cos\theta}{r^2} ds_2. \tag{2.23}$$

Assuming that the density of the thin film produced is ρ, the thickness of the thin film deposited on ds_2 per unit time is t, and the volume of the thin film deposited on ds_2 is $t \cdot ds_2$, then

$$dm = \rho \cdot t.ds_2. \tag{2.24}$$

$$\text{Then} \quad t = \frac{m}{4\pi\rho} \cdot \frac{\cos\theta}{r^2}. \tag{2.25}$$

$$\text{That is,} \quad t = \frac{mh}{4\pi\rho r^3} = \frac{mh}{4\pi\rho(h^2 + x^2)^{3/2}}. \tag{2.26}$$

Equation (2.26) is the film thickness distribution formula of the point evaporation source.

When ds_2 is directly above the point evaporation source, that is, $\theta = 0$, $\cos\theta = 1$, and t_0 represents the film thickness at the origin, then $t_o = \frac{m}{4\pi\rho h^2}$ is the maximum film thickness that can be obtained in the plane of the substrate. The film thickness distribution at any point in the substrate plane can be expressed by the following formula:

$$t = \frac{1}{[1 + (x/h)^2]^{3/2}} t_0. \tag{2.27}$$

2.2.2.2 *Small-plane evaporation source*

As shown in Figure 2.5, a small-plane evaporation source is used instead of a point source. Since the emission characteristics of this evaporation source are directional and follow the cosine angle distribution law; that is, the mass of the material evaporated at the angle of θ is proportional to $\cos\theta$, where θ is the angle between the normal of the plane evaporation source and the connection between the center of the receiving plane ds_2 and the center of the plane source. So when the film material evaporates from the small plane ds at a rate of m grams per second, the evaporation capacity dm of

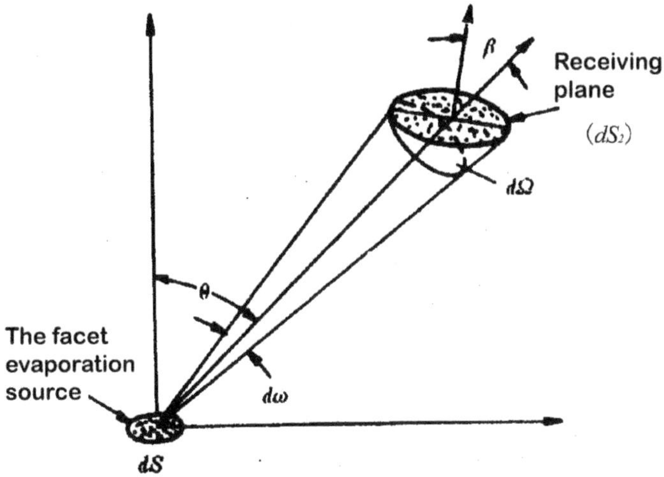

Figure 2.5: Influence of the acceptance angle of the facet evaporation source on the deposited film thickness.

the film material passing through the solid angle $d\omega$ in the direction of θ angle with the normal of the horizontal plane per unit time is

$$dm = (m/\pi) \cdot \cos\theta \cdot d\omega. \tag{2.28}$$

The $1/\pi$ in Equation (2.28) is given because the evaporation range of the facet source is limited to the hemispherical space. Similar to the point source calculation method, the film thickness t of any point on the facet evaporation source substrate is

$$t = \frac{m}{\pi p} \cdot \frac{\cos\theta \cos\beta}{r^2} = \frac{mh^2}{\pi p(n^2 + x^2)^2}. \tag{2.29}$$

When $\theta = 0$ and $\beta = 0$ (the facet source is a square, refer to Figure 2.5), and the film thickness at this point is represented by t_0, then

$$t_0 = m/(\pi\rho h^2), \tag{2.30}$$

where t_0 is the maximum evaporation film thickness obtained in the substrate plane. The film thickness distribution in other parts of the

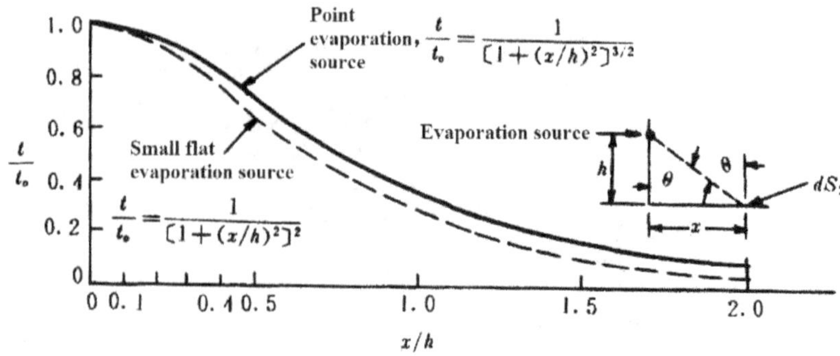

Figure 2.6: Distribution of film thickness on the plane by evaporation.

substrate plane can be calculated by the following formula:

$$t = \frac{1}{[1 + (x/h)^2]^2} t_0. \tag{2.31}$$

Figure 2.6 compares the relative thickness distribution curves of both point evaporation sources and facet evaporation sources. In addition, comparing Equations (2.26) and (2.29), it can be seen that although the thicknesses of the film layers deposited on the substrate by the two evaporation sources are similar, due to the different shapes of the evaporation sources, the maximum thickness of the planar evaporation source can be about four times thicker than that of the point evaporation source when the evaporation material and the distance between the evaporation source and the substrate is certain.

2.2.2.3 *Slim plane evaporation source*

The emission characteristics of an elongated planar evaporation source are shown in Figure 2.7. Assuming that the substrate is parallel to the slender evaporation source of the length of l, the source–base distance is h, and the small evaporation area with a distance S from the center point is dS. The tiny area at any point (x, y) on the x–y plane is $d\sigma$. When the distance between dS and $d\sigma$ is r, according to the geometric relationship in Figure 2.7, we can obtain

$$\cos\theta = h/r, \quad r^2 = (x - S)^2 + a^2, \quad a^2 = h^2 + y^2.$$

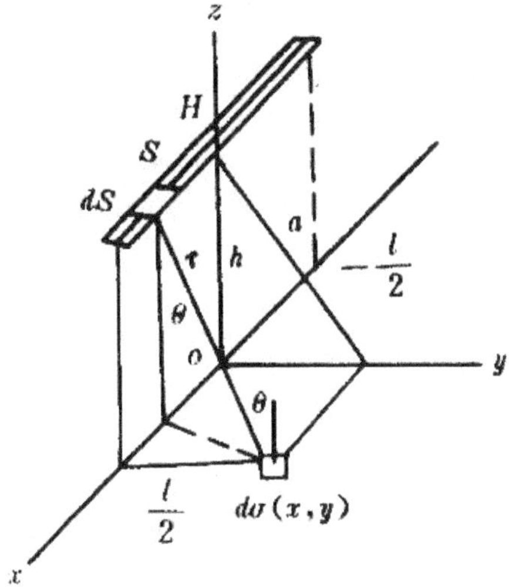

Figure 2.7: Emission characteristics of a slender planar evaporation source.

When the evaporation material is uniformly distributed in the evaporation source, the mass dm on the dS surface of the evaporation source is

$$dm = (m/l)dS. \tag{2.32}$$

At this time, the elongated planar evaporation source is equivalent to the small planar evaporation source. Through the above calculation method, the film thickness distribution formula can be obtained

$$t = \frac{mh^2}{2\pi l P a^2}\left[\frac{l\left(a^2 - x^2 + \frac{l^2}{4}\right)}{(a^2+x^2)^2 + (a^2-x^2)\frac{l^2}{4} + \frac{l^4}{16}} + \frac{1}{a} + \mathrm{tg}^{-1}\frac{la}{a^2+x^2-\frac{l^2}{4}}\right] \tag{2.33}$$

At the origin, since $x = 0$ and $a = h$, the film thickness t_0 is

$$t_0 = \frac{M}{2\pi l \rho a^2}\left[\frac{l}{n^2(l^2/4)} + \frac{1}{h}\mathrm{tg}^{-1}\frac{lh}{h^2+(l^2/4)}\right].$$

2.2.2.4 *Ring evaporation source*

To obtain better film thickness uniformity in a larger area, a ring-shaped evaporation source (referred to as a ring source) can be used. In actual evaporation, when the substrate is in a rotating state, it is similar to a ring source. Figure 2.8 shows the emission characteristics of a ring planar evaporation source.

If a unit area dS_1 is taken on the ring, the mass of the film material evaporated to the receiving surface per unit time is

$$dm = \frac{m}{2\pi} d\varphi. \tag{2.34}$$

According to the geometric characteristics in Figure 2.8, the above method can be used to obtain the film thickness distribution formula

$$t = \frac{mh^2}{2\pi\rho} \cdot \frac{2h^2 + (A+R)^2 + (A-R)^2}{[h^2 + (A+R)^2]^{3/2}\,[h^2 + (A-R)^2]^{3/2}}. \tag{2.35}$$

The film thickness at the origin directly below ds_1 is

$$t_o = \frac{mn^2}{\pi p} \cdot \frac{1}{(n^2 + R^2)^2}. \tag{2.36}$$

The film thickness distribution using a ring planar evaporation source is shown in Figure 2.9. On selecting an appropriate ratio of R to h, the

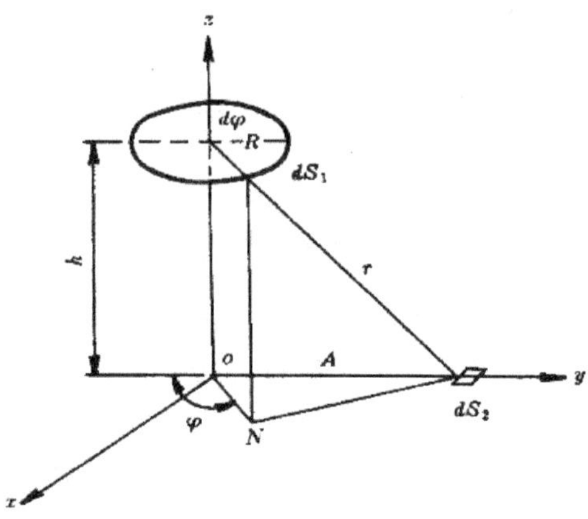

Figure 2.8: Emission characteristics of an annular planar evaporation source.

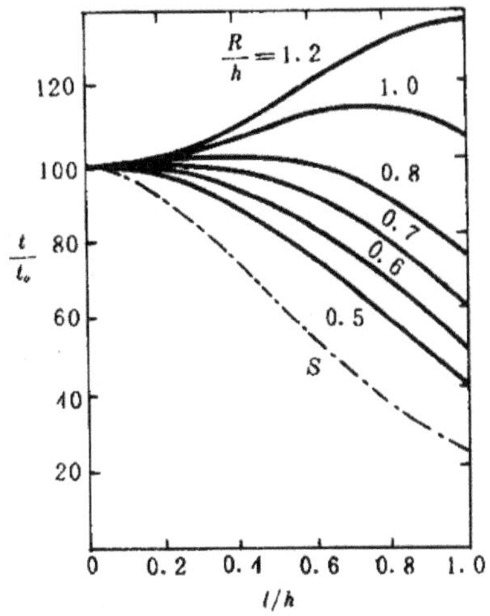

Figure 2.9: Film thickness distribution of the ring planar evaporation source.

film thickness distribution is relatively uniform over a considerable range on the plane of the evaporation source. For a certain R, the film thickness distribution on the plane can be calculated by using Equation (2.35).

In practical applications, the evaporation source is often formed into the shape of a needle, a boat, a cone, a crucible, etc. According to their respective emission characteristics, the film thickness can be approximated by using the above-mentioned distribution equations of the film thickness of the evaporation sources and combining with the specific evaporation source used.

The emission characteristics of the evaporation sources are relatively complicated issues. In order to obtain a more uniform film thickness, attention must be paid also to the configuration of the source and the substrate, or to make the substrate revolve and rotate.

2.2.3 *Several heating methods of the evaporation source*

The most important component in the evaporation deposition film device is the heating device of the evaporation source, which can be divided into the following types according to different heating principles.

2.2.3.1 *Resistance heating device*

The resistance heating method has certain requirements for resistance materials and evaporation materials. For resistance materials, the melting point must be high and must be higher than the evaporation temperature of the evaporation material. The saturated vapor pressure should be low to prevent or hinder the resistance materials from entering the film as the evaporation materials evaporate at high temperatures. The chemical properties of resistance materials should be stable to prevent chemical reactions with evaporated materials at high temperatures. In addition, the resistance material is required to have good heat resistance. When the heat source is changed, the change of power density is small. And the raw materials are rich, economical, and durable. Table 2.7 lists the melting points of commonly used evaporation source materials in the resistance heating method and the temperatures at which the specified equilibrium vapor pressure is reached. Table 2.8 shows the evaporation sources used for the evaporation of some substances. The actual resistance heating materials are generally refractory metals such as W, Mo, Ta, etc. Table 2.9 lists their main physical parameters.

Table 2.7: The melting point of the resistance evaporation source material and the corresponding equilibrium vapor pressure temperature.

Evaporation source temperature	Melting point (K)	Equilibrium temperature (K)		
		10^{-8} Torr	10^{-5} Torr	10^{-2} Torr
W	2683	2390	2840	3500
Ta	3269	2230	2680	3330
Mo	2890	1865	2230	2800
Nb	2741	2035	2400	2930
Pt	2045	1565	1885	2180
Fe	1808	1165	1400	1750
Ni	1726	1200	1430	1800

Table 2.8: Resistance heating evaporation source materials for various elements.

	Temperature (°C)		Evaporation source material		
Elements	Melting point (K)	10^{-2}Torr	filaments, flake	Crucible	Remark*
Ag	961	1030	Ta, Mo, W	Mo, C	Arranged according to the degree of suitability. No infiltration with W
Al	659	1220	W	BN, TiC/C, TiB_2–BN	Can be alloyed with all RM and is difficult to evaporate. It can react with Ti, Zr, Ta, etc. at high temperature
Au	1063	1400	M, Mo	Mo, C	Soaks W, Mo. Forms alloy with Ta. Ta should not be used as an evaporation source
Ba	710	610	W, Mo, Ta, Ni, Fe	C	Can't form alloys. Soaks RM. Reacts with most oxides at high temperatures
Bi	271	670	W, Mo, Ta, Ni	Al_2O_3, C, etc.	Vapor is toxic
Ca	850	600	W	Al_2O_3	Pre-melt degassing in a He atmosphere
Co	1495	1520	W	Al_2O_3, BeO	Forms alloy with W, Ta, Wo, Pr, etc.
Cr	~1900	1400	W	C	
Cu	1084	1260	Mo, Ta, Nb, W	Mo, C, Al_2O_3	Can't infiltrate Mo, W, Ta directly
Fe	1536	1480	W	BeO, Al_2O_3, ZrO_2	Forms alloy with all RM. EBV preferably
Ge	940	1400	W, Mo, Ta	C, Al_2O_3	Low solubility to W. Could soak RM. Couldn't soak C

(*Continued*)

Table 2.8: (*Continued*)

Elements	Temperature (°C)		Evaporation source material		Remark*
	Melting point (K)	10^{-2}Torr	filaments, flake	Crucible	
In	156	950	W, Mo	Mo, C	
La	920	1730	—	—	EBV preferably
Mg	650	440	W, Ta, Mo, Ni, Fe	Fe, C, Al_2O_3	
Mn	1244	940	W, Mo, Ta	Al_2O_3, C	Soaks RM
Ni	1450	1530	W	Al_2O_3, BeO	Forms alloy with W, Mo, Ta, etc. EBV preferably
Pb	327	715	Fe, Ni, Mo	Fe, Al_2O_3	Couldn't soak RM
Pd	1550	1460	W(Coat Al_2O_3)	Al_2O_3	Forms alloy with all RM
Pt	1773	2090	W	ThO_2, ZrO_2	Forms alloy with Ts, Mo, Nb. Forms partial alloy with W. EBV or sputtering preferably
Sn	232	1250	Ni-Cr alloy, Mo, Ta	Al_2O_3, C	Soaks and erodes Mo
Ti	1727	1740	W, Ta	C, ThO_2	Reacts with W. Couldn't react with Ta. Sometimes Ta breaks during melting
Tl	304	610	Ni, Fe, Nb, Ta, W	Al_2O_3	Soaks the left metals but doesn't form alloy. Soaks W, Ta slightly. Doesn't soak Mo
V	1890	1850	W, Mo	Mo	Soaks Mo but doesn't form alloy. The solubility in W is small. Can form alloy with Ta
Y	1477	1632	W		
Zn	420	345	W, Ta, Mo	Al_2O_3, Fe, C, Mo	Soaks RM, but doesn't form alloy
Zr	1852	2400	W		Soaks W. Little solubility

Note: RM — metal with high melting point: EBV — electron beam evaporation.

Table 2.9: Properties of metal materials for resistance evaporation sources.

	Temperature (°C)	27	1027	1527	1727	2027	2327	2527
W (melting point: 3380°C, relative density: 19.3)	Resistivity ($\mu\Omega\cdot$cm)	5.66	33.66	50	56.7	66.9	77.4	84.7
	Vapor pressure (Pa)	—	—	—	1.3×10^{-9}	6.3×10^{-7}	7.6×10^{-7}	1.0×10^{-3}
	Evaporation rate (g/cm$^2\cdot$s)	—	—	—	1.75×10^{-11}	7.8×10^{-11}	8.8×10^{-9}	1.1×10^{-7}
	Spectral emissivity (0.665 μm)	0.470	0.450	0.439	0.435	0.429	0.423	0.449
Mo (melting point: 2630°C, relative density: 10.2)	Resistivity ($\mu\Omega\cdot$cm)	5.63	35.2 (1127°C)	47.0	53.1	59.2 (1927°C)	72	78
	Vapor Pressure (Pa)	—	2.1×10^{-13}	8×10^{-9}	5×10^{-5}	4×10^{-3}	1.4×10^{-3}	9.6×10^{-3}
	Evaporation rate (g/cm$^2\cdot$s)	—	2.5×10^{-17}	1.1×10^{-10}	5.3×10^{-9}	5.0×10^{-7}	1.6×10^{-5}	1.04×10^{-4}
	Spectral emissivity	0.418	—	0.367 (1330°C)	0.353 (1730°C)	—	—	—
Ta (melting point: 2980°C, relative density: 16.6)	Resistivity ($\mu\Omega\cdot$cm)	15.5 (20°C)	54.8	72.5	78.9	88.3	97.4	102.9
	Vapor Pressure (Pa)	—	—	—	1.3×10^{-8}	8×10^{-8}	5×10^{-4}	7×10^{-3}
	Evaporation rate (g/cm$^2\cdot$s)	—	—	—	1.63×10^{-12}	9.8×10^{-11}	5.5×10^{-8}	6.6×10^{-7}
	Spectral emissivity	0.490	0.462	0.432	0.432	(1927°C) 0.409	0.400	0.394

In the resistance heating method, the "wettability" of the evaporation material and the evaporation source material must also be considered. The wettability is related to the surface energy of the evaporation material. It can be said that when the high-temperature melting evaporation material has a tendency to expand on the evaporation source, it is easy to wet. On the contrary, if there is a tendency to agglomerate on the evaporation source and it is close to forming a spherical shape, it can be considered difficult to wet. Figure 2.10 shows several cases of infiltration between the evaporation source materials and coating materials.

In the case of infiltration, since evaporation occurs from a large surface and is relatively stable, it can be considered as evaporation from a surface evaporation source; when the infiltration is little, it can generally be considered as evaporation from a point evaporation source. In addition, if the infiltration easily occurs, the evaporation material has a strong affinity with the evaporation source, so the evaporation state is stable; if it is difficult to infiltrate, when the filamentary evaporation source is used, the evaporation material will easily fall off the evaporation source. For example, Ag will fall off after melting on the gallium wire. According to the nature of the evaporation material and considering the wettability of the evaporation source material, it can be made with different shapes and different evaporation source materials. The tungsten wire can be

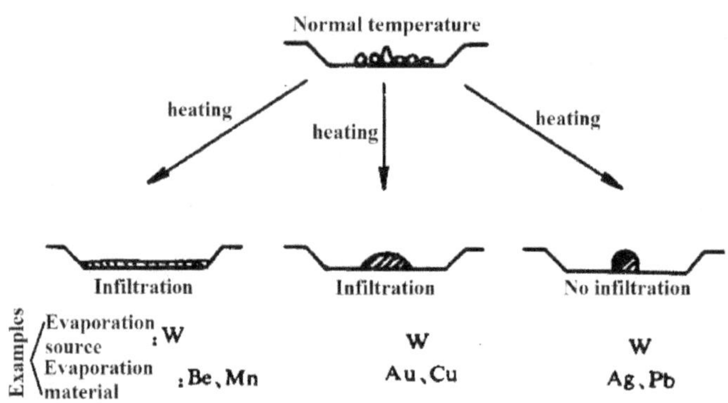

Figure 2.10: Wetting state of the evaporation source material and the coating material.

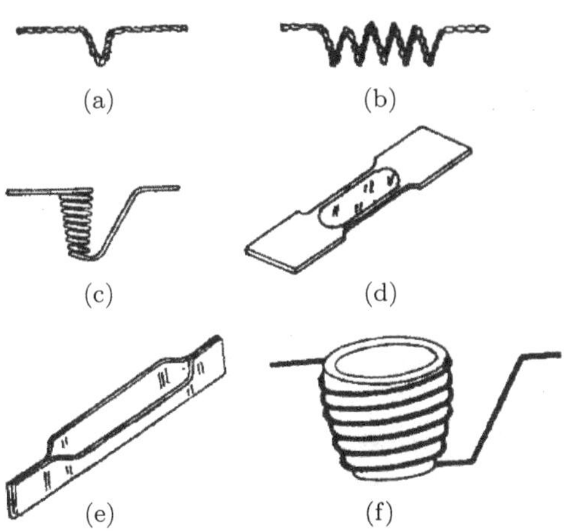

Figure 2.11: Some common resistance heating devices.

formed into a variety of spirals of equal or unequal diameters, which can then be used as the heating source of the substance. A tungsten wire acts as a heater, on the one hand, and as a support for the heated material, on the other hand. Figure 2.11 shows some common resistance heating devices.

For materials that cannot be heated with tungsten wires, such as powders of some materials, a resistance heating device made of refractory metal plates can be considered. At this time, the refractory metal plate can be formed into various shapes to contain the material to be heated. For substances that can be sublimated in a solid state, refractory metals can also be used to make special containers for sublimation. When heating, it is necessary to consider not only the heating and supporting effects, but also the material splash that may be caused by the outgassing process of the heated substance.

Crucibles that are made of high-melting point oxides, high-temperature cracked BN, graphite, refractory metals, etc. can also be used as evaporation vessels. At this time, two methods can be adopted for heating the evaporated substance, namely the ordinary resistance heating method and the high-frequency induction method. The former is heated by wires wrapped around the crucible. The latter uses a water-permeable copper coil as the primary heating induction coil and it heats the evaporated material either in a heated material or in a crucible by creating an inductive current. In the latter case, the substance to be heated or the crucible itself should have a certain electrical conductivity.

2.2.3.2 *Electron beam heating evaporation source*

The aim of the electron beam evaporation method is to place the evaporated material in a water-cooled copper crucible and use electron beam heating directly to make the evaporated material vaporize and then form a film on the substrate surface. The electron beam evaporation method helps overcome many of the disadvantages of the general resistance heating evaporation, such as a high contamination (from crucibles, heating materials, and various supporting components), a low heating power, and a low temperature. So it is an especially suitable method for making high-melting point film materials and high-purity film materials.

(1) **Electron beam heating principle and characteristics**

The electron beam heating principle is that electrons gain kinetic energy to bombard the evaporation material at the anode under the action of an electric field, which heats and vaporizes the evaporation material, and the evaporating coating is realized. Without considering the initial velocity of the emitted electrons, the electronic kinetic energy is equal to the electric power it has, and that is

$$\frac{1}{2}mv^2 = e \cdot U, \tag{2.37}$$

where U is the potential (V) of the electron; m is the mass of the electron (9.1×10^{-28}g); and e is the charge of the electron (1.6×10^{-19}C). So the speed of the electron can be obtained

$$v = 5.93 \times 10^5 \sqrt{U}\,(\text{m/s}). \tag{2.38}$$

If $U = 10\,\text{kV}$, the electronic speed can reach 6×10^4 km/s. Under the action of a certain electromagnetic field, such a high-speed electron flow converges into an electron beam and explodes on the surface of the evaporation material, so that the kinetic energy becomes thermal energy. If the energy of the electron beam is

$$W = ne\,U = IUt \tag{2.39}$$

where n is the electron density; I is the beam current (A) of the electron beam; t is the action time (s) of the beam current, the resulting heat Q is

$$Q = 0.24\text{W}t. \tag{2.40}$$

When the acceleration voltage is high, the thermal energy generated by the electrons given by Equation (2.40) can be sufficient to vaporize the evaporation material, thus becoming a good heat source in vacuum evaporation technology.

The main advantage of the electron beam evaporation source is that the electron beam bombards the heat source with a high beam density, which can obtain a far greater energy density than a resistive heating source. It can reach a power density of 10^4–10^9 W/cm^2 over a large area, so it can evaporate materials with a high melting point (up to 3000°C). The heat can be directly added to the surface of the evaporation material, so the thermal efficiency is high, and the loss of heat conduction and heat radiation is little. Moreover, the evaporated material is placed in the water-cooled crucible, so evaporation of the container material can be avoided, and reaction between the container material and evaporation materials can also be avoided. This is extremely important to improve the purity of the coating material.

One disadvantage of the electron beam heating source is that the primary electrons from the electron gun and the secondary electrons from the evaporation material ionize the evaporated atoms and residual gas molecules, which sometimes affects the film quality. This problem can be solved by designing and selecting electron guns with different structures. Most mixtures will be partially decomposed when subjected to electron bombardment, and residual gas molecules and film material molecules will also be partially ionized by the electrons, which will affect the structure and properties of the thin film. In addition, the structure of the electron beam evaporation device is complicated, so the equipment is expensive.

(2) Electron beam evaporation source structure

According to the different types of electron beam evaporation sources, vacuum evaporation technology that relies on electron beam bombardment evaporation can be divided into ring guns, straight guns (also known as Pierce guns), and e-type guns.

Ring guns rely on ring cathodes to emit the electron beam. After the electron beam is focused and deflected, the crucible material is evaporated in the crucible. Its structure is simple, but its power and efficiency are not high, so it is mostly used in experimental research.

The *straight gun* is an axisymmetric linear acceleration electron gun. The electrons are emitted from the cathode filament and focused into a thin beam. After being accelerated by the anode, they bombard in the crucible to melt and evaporate the evaporation material. The power of the straight gun ranges from several hundred watts to several kilowatts. In addition to obtaining a high energy density, it is easy to control. However, the volume is large and the cost is high, and the vapor deposition material will pollute the gun body. Figure 2.12 is a schematic diagram of a straight gun evaporation source.

The *e-type electron gun* is an electron gun with a deflection of 270°. It helps overcome the shortcomings of the straight gun and is widely used now. Its structure is shown in Figure 2.13. The so-called

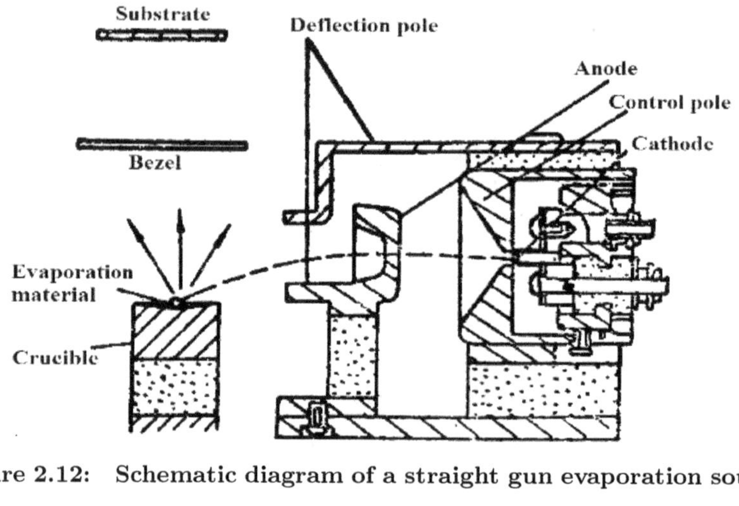

Figure 2.12: Schematic diagram of a straight gun evaporation source.

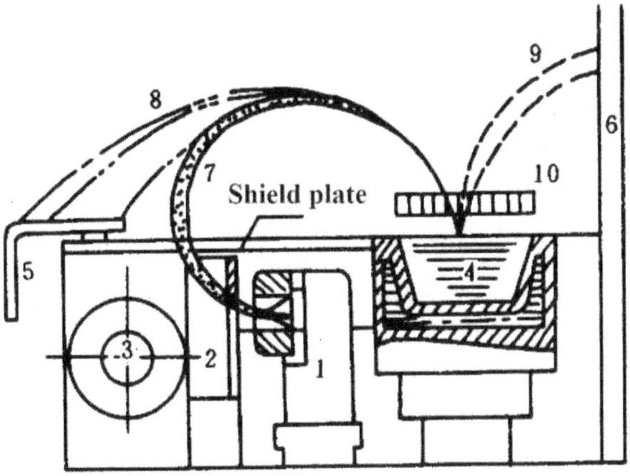

Figure 2.13: Working principle of an e-type electron gun. 1. Emitter 2. Anode 3. Electromagnetic coil 4. Water-cooled crucible 5. Collector 6. Absorption pole 7. Electronic trajectory 8. Positive ion trajectory 9. Scattered electron trajectory 10. Plasma.

e-type is named after the trajectory of the electron. The positive ions released due to the collision of the incident electrons with the evaporated atoms generate a movement in the direction opposite to the incident electrons under the action of the deflection magnetic field. So pollution of the deposited film by the positive ions in the straight gun is avoided. At the same time, the e-gun also greatly reduces the chances of secondary electrons bombarding the substrate.

2.2.3.3 *Arc heating evaporation source*

The arc discharge heating method can avoid contamination of the heating wire or crucible material, and the heating temperature is relatively high; it is particularly suitable for the evaporation deposition of refractory metals with high melting points and a certain conductivity. The equipment used in this method is simpler than the electron beam heating device, so it is a relatively inexpensive evaporation device.

In arc heating, a discharge electrode is made of the material to be evaporated. When the film is deposited, the arc is ignited by adjusting the distance between the electrodes in the vacuum chamber, and the transient high-temperature arc will cause the lower end of the electrode to evaporate for achieving film deposition. Controlling the number of ignitions of the arc can deposit a thin film of a certain thickness. The arc heating method can use either a DC heating method or an AC heating method. The disadvantage of this method is that it is easy to generate micron-sized spatter electrode particles, which will affect the uniformity and quality of the deposited film.

2.2.3.4 *High frequency induction evaporation source*

The purpose of the high-frequency induction evaporation source is to place the crucible containing the evaporation material at the center of the high-frequency (usually radio frequency) spiral coil so that the evaporation material generates a strong eddy current loss or hysteresis loss (for ferromagnets), causing the evaporation material to heat up until the vaporization evaporates. The smaller the volume of the membrane, the higher the frequency of induction. The evaporation source generally consists of a water-cooled high-frequency coil and a graphite or ceramic crucible. Figure 2.14 shows the working principle of a high-frequency induction heating evaporation source.

The characteristics of the high-frequency heating evaporation source are as follows: the evaporation rate is large, which can be

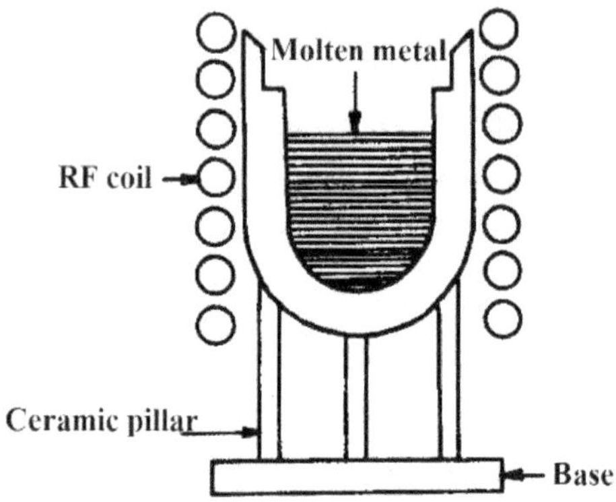

Figure 2.14: Working principle of a high-frequency induction heating evaporation source.

about 10 times larger than the resistance evaporation source; the temperature of the evaporation source is uniform and stable, and it is not easy to generate splashing; the evaporation source is once loaded without the need for a feeding mechanism; the temperature control is easy, and the operation easier. Its main disadvantages are the following: the evaporation device must be shielded, and a more complicated and expensive high-frequency generator is required; in addition, if the pressure near the coil exceeds 10^{-2} Pa, the high-frequency electric field will ionize the residual gas and increase the power consumption.

2.2.3.5 *Laser heating evaporation source*

High-power continuous or pulsed laser beams can also be used as evaporation sources for vacuum evaporation. This method also has the characteristics of high heating temperature, preventing crucible pollution, high evaporation rate of materials, and ease to control the evaporation process. This heating method is particularly suitable for the evaporation of more complex alloys or compound materials. For example, in recent years, this method has been used to study the preparation of complex oxide films such as high-temperature superconducting films and ferroelectric films. This method requires a special window material to introduce the laser beam into the vacuum chamber and requires a lens or a concave mirror to focus the laser beam. For laser beams with different wavelengths, window materials with different spectral transmission characteristics need to be selected.

Some common evaporation substance preparation parameters are listed in Table 2.10, including heating methods, heating temperatures, and applicable crucible materials. The types of substances in Table 2.10 include metals, alloys, oxides, and various compound materials.

2.2.4 *Evaporation of alloys and compounds*

For alloys or compounds composed of two or more elements, attention should be paid to controlling the evaporation rate of each component

Table 2.10: Evaporation process characteristics of common substances.

Substances	Minimum evaporation temperature (*/°C)	Evaporation source status	Crucible material	Deposition rate during electron beam evaporation ($nm \cdot s^{-1}$)
Al	1010	Molten state	BN	2
Al_2O_3	1325	Semi-melted state	—	1
Sb	425	Molten state	BN_4, Al_2O_3	5
As	210	Sublimation	Al_2O_3	10
Be	1000	Molten state	Graphite, BeO	10
BeC	—	Molten state	—	4
B	1800	Molten state	Graphite, WC	1
Cd	180	Molten state	Al_2O_3, Quartz	3
CdS	250	Sublimation	Graphite	1
CaF_2	—	Semi-melted state	—	3
C	2140	Sublimation	—	3
Cr	1157	Sublimation	W	1.5
Co	1200	Molten state	Al_2O_3, B_2O_3	2
Cu	1017	Molten state	Graphite, Al_2O_3	5
Ga	907	Molten state	Graphite, Al_2O_3	—
Ge	1167	Molten state	Graphite	2.5
Au	1132	Molten state	BN, Al_2O_3	3
In	742	Molten state	Al_2O_3	10
Fe	1180	Molten state	Al_2O_3, B_2O_3	5
Pb	497	Molten state	Al_2O_3	3
LiF	1180	Molten state	Mo, W	1
Mg	327	Sublimation	Graphite	10
MgF_2	1540	Semi-melted state	Al_2O_3	3
Mo	2117	Molten state	—	4
Ni	1262	Molten state	Al_2O_3, B_2O_3	2.5
Permalloy	1300	Molten state	Al_2O_3	3
Pt	1747	Molten state	Graphite	2
Si	1337	Molten state	B_2O_3	1.5
SiO_2	850	Semi-melted state	Ta	2

during evaporation to obtain the same membrane as the chemical composition of the evaporated material.

2.2.4.1 *Evaporation of alloys*

Due to the different saturated vapor pressures of the components, the main problem of evaporation of binary and multicomponent alloys and compounds is that the evaporation rates of the components are also different during the vaporization of the evaporation material, which causes the evaporation source to decompose and fractionate and causes deviation of film composition. When vacuum deposition is used to make alloy films of predetermined composition, instantaneous evaporation, dual-source evaporation, and alloy sublimation methods are often used.

(1) **Instantaneous evaporation**

The instantaneous evaporation method is also called the "flashing" evaporation method, and is sometimes called "flash evaporation". The aim of it is to send small alloy particles to a very hot evaporator or crucible one by one so that the particles are completely evaporated instantaneously. If the particle size is small, almost every component can be evaporated at the same time, so the instant evaporation method is often used when the evaporation rates of the elements in the alloy are very different. The advantage of this method is that a thin film with uniform composition can be obtained, and doping evaporation can be performed. The disadvantage is that the evaporation rate is difficult to control.

Figure 2.15 shows the principle diagram of the instantaneous evaporation method. The keys to using this method are to supply the evaporation material to the evaporation source at a uniform rate, and to choose the appropriate powder particle size, evaporation temperature, and ratio of falling dust. A tungsten wire cone basket is a quite good structure for an evaporation source. If an evaporation boat and crucible are used, the powder particles that did not evaporate for a while will remain and become normal evaporation. This evaporation method has been used in the preparation of various

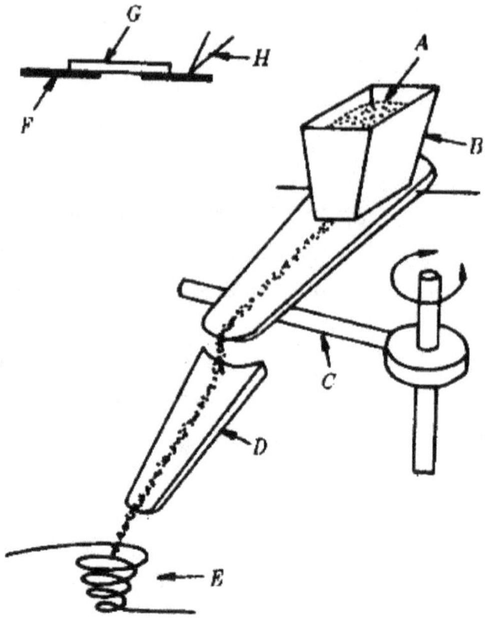

Figure 2.15: Working principle of the instant evaporation method. A. Powdered material; B. funnel; C. funnel support rod; D. chute; E. tungsten wire; F. Bezel; G. substrate; H. thermocouple.

alloy films (such as Ni–Cr alloy films), III–V and II–VI compound semiconductor films. For magnetic metal compounds, films such as MnSb, MnSb–CrSb, CrTe, and $MnSGe_3$ have been successfully prepared.

(2) **Dual-source or multi-source co-evaporation**

The aim of this method is to load each component of the alloy into its own evaporation source, and then the evaporation rate of each evaporation source is controlled independently so that the proportion of atoms reaching the substrate corresponds to the composition of the desired alloy film. To make the film thickness uniform, it is often necessary to rotate the substrate.

Figure 2.16 shows the principle diagram of dual-source co-evaporation. The use of dual-source co-evaporation is helpful in improving the uniformity of the film thickness distribution.

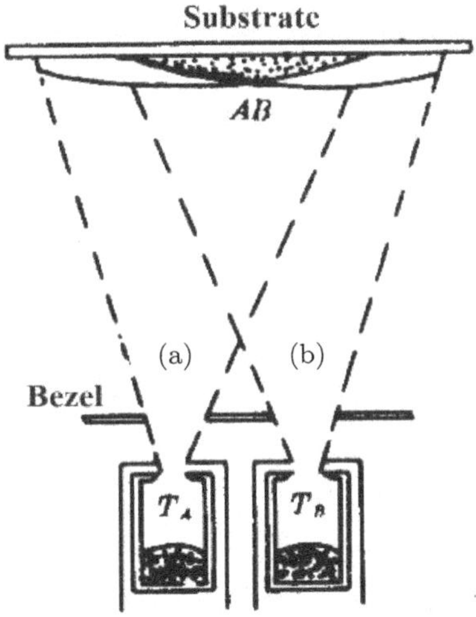

Figure 2.16: Schematic diagram of dual source co-evaporation. T_A, Evaporation temperature of substance A; T_B, evaporation temperature of substance B; (a) vapor stream of substance A; (b) vapor stream of substance B; AB alloy film.

2.2.4.2 *Evaporation of compounds*

There are three methods of compound evaporation: resistance heating, reactive evaporation, dual-source or multi-source co-evaporation, and the three-temperature method and molecular beam epitaxy. The resistance heating method was described earlier. The reactive evaporation method is mainly used to prepare high-melting insulating dielectric films, such as oxides, nitrides, and silicon compounds. The three-temperature method and molecular beam epitaxy are mainly used to make single-crystal semiconductor compound films, especially III–V compound semiconductor films, superlattice films, and various single-crystal epitaxial films. In this section, the reactive evaporation method and the three-temperature method are mainly introduced, and the molecular beam epitaxy method will be specifically introduced later.

(1) **Reactive evaporation**

The aim of the reactive evaporation method is to introduce the active gas into the vacuum chamber so that the atoms, molecules of the active gas and metal atoms escaping from the evaporation source, and low-valent compound molecules react on the substrate surface during the deposition process. So the required high-value compound film can be formed. This method can be used for materials with severe thermal decomposition and low saturated vapor pressure which are difficult to evaporate by resistance heating. The reactive evaporation method is often used to make compound films with high melting points, especially for compound films composed of transition metals and reactive gases such as O_2 and N_2 that are easily desorbed. The reaction evaporation method can be carried out at a lower temperature, and the precipitation or agglomeration in the reaction process is not strong, so it is easy to obtain a uniformly dispersed compound film. In order to accelerate the reaction, the method of evaporating metal and partial active gas discharge can be used to ionize it, thus the active reaction evaporation method is derived, the principle of which is the same as that of active reactive ion plating. The composition and structure of the thin film produced by this method mainly depend on the chemical properties of the reaction material, the stability of the reaction gas, the free energy for forming the compound, the decomposition temperature of the compound, the angle of incidence of the reaction gas on the substrate, the evaporation rate of molecules leaving evaporation sources, and the substrate temperature and other parameters. The principle and device diagram of reactive evaporation are shown in Figure 2.17.

(2) **Three-temperature method**

The three-temperature method is, in principle, a dual evaporation source reactive evaporation method. When the III–V compound semiconductor material is heated and evaporated in a crucible, if the temperature is above the boiling point, the semiconductor material will be thermally decomposed, and the component elements will be fractionated. Therefore, the film layer deposited on the substrate may deviate from the stoichiometric ratio of the compound. Since the vapor pressure of group V elements is much higher than that of group III elements, the three-temperature evaporation method shown in Figure 2.18 has been developed. This method is used to

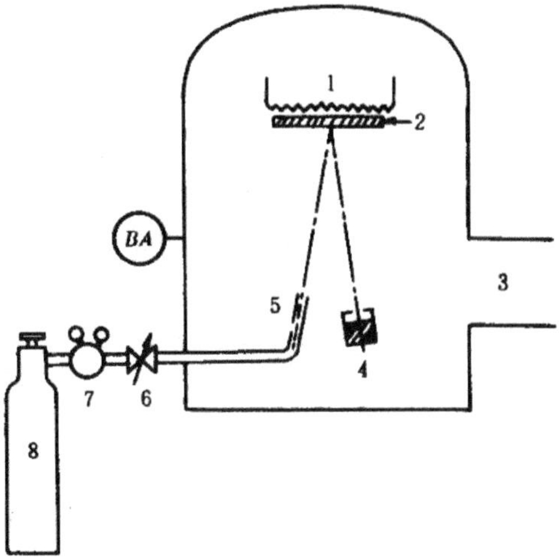

Figure 2.17: Schematic diagram of reactive evaporation coating. 1. Heater; 2. substrate; 3. Exhaust system; 4. evaporation source; 5. gas spout; 6. Adjustable air release valve; 7. pressure reducing valve; 8. reaction gas bottle; BA. vacuum gauge.

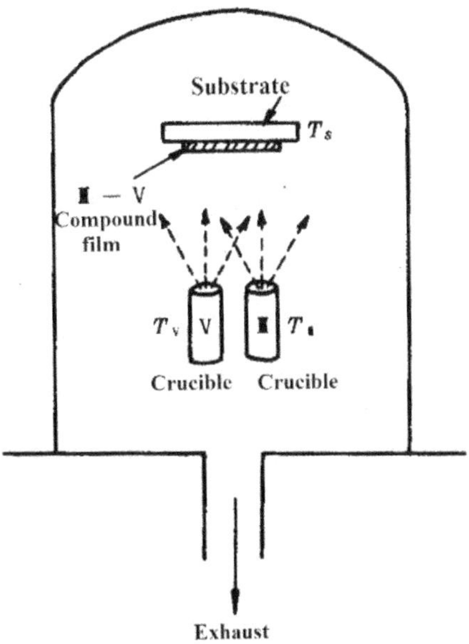

Figure 2.18: Schematic diagram of the three temperature method.

control three temperatures totally, including the evaporation source temperature T_{III} of the low vapor pressure element (group III), the evaporation source temperature T_{V} of the high vapor pressure element (group V), and the substrate temperature T_{S}. It is actually equivalent to evaporating a group III element in an atmosphere of a group V element. Therefore, this definition is also similar to the reactive evaporation method.

2.3 Sputtering Films

2.3.1 *Overview*

The basic principle of sputtering films is that charged ions have a certain amount of kinetic energy after being accelerated in an electric field and then the ions are directed to the target electrode to be sputtered. Under the condition that the ions' energy meets certain conditions, the incident ions will cause the target atoms to be sputtered from the surface during the collision with the target atoms. The sputtered atoms have a certain kinetic energy and will emit toward the substrate in a certain direction, so the deposition of a thin film will be realized on the substrate. Sputtered ions are mostly atomic and are often referred to as sputtered atoms. The incident charged ions that bombard the target can be electrons, ions, or neutral particles. And because ions can be easily accelerated under the action of an electric field to obtain the required kinetic energy, ions are mostly used as bombarded particles, which are also called incident ions. This coating technology is also called ion sputter coating (or deposition). Correspondingly, the effect due to sputtering is equivalent to etching for the target, so etching can also be performed by sputtering. Deposition and etching are two applications of the sputtering process. Sputtered coatings are now widely used for the preparation of various films, such as metal, alloy, semiconductor, oxide, and insulating dielectric films, compound semiconductor films, carbide and nitride films, and high-T_C superconducting films.

Compared with vacuum evaporation coating, sputter coating has the following characteristics:

(1) Every substance can be sputtered. Metals, semiconductors, insulators, compounds and mixtures, or bulk, granular substances can be used as targets as long as they are solid. Since almost no decomposition and fractionation will occur when sputtering oxides and other insulation materials and alloys, sputtering can be used to prepare the following: thin films with a similar composition to the target material, alloy films with uniform composition, and even superconducting films with complex compositions.
(2) The adhesion between the sputtered film and the substrate is good. The energy of the sputtered atoms is 1–2 orders of magnitude higher than the energy of the evaporated atoms, so higher thermal energy will be produced when high-energy particles are deposited on the substrate for energy conversion, which enhances the adhesion between the sputtered atoms and the substrate greatly. Some high-energy sputtered atoms will produce different degrees of injection phenomenon, forming a so-called pseudo-diffusion layer on the substrate so that sputtered atoms and substrate material atoms are "miscible" with each other. In addition, during the bombardment of the sputtered particles, the substrate is always cleaned and activated in the plasma region, which removes the deposited atoms that are not firmly attached and purifies and activates the substrate surface. Therefore, the adhesion between the sputtered film layer and the substrate is greatly enhanced.
(3) A sputtered coating has a high density, fewer pinholes, less pollution, and a high film purity.
(4) It has good controllability and repeatability of film thickness. The discharge current and the target current during sputter coating can be controlled separately, and the film thickness can be controlled by controlling the target current. Therefore, the film thickness controllability of the sputtered coating and the

reproducibility of the film thickness after multiple sputtering are good, and it is relatively effective in plating a thin film of a predetermined thickness.

(5) Sputter coating can deposit thin films with uniform thickness over a large area.

The disadvantages of sputter coating are as follows. The sputtering equipment is complicated and requires a high-voltage device. The film deposition speed of sputtering deposition is low. The substrate temperature is high and is easily affected by impurity gases, etc.

2.3.2 *Glow discharge*

In the sputtering process, the sputtering ions all originate from gas discharge. And different sputtering technologies use different glow discharge methods. DC diode sputtering uses DC glow discharge; tripolar sputtering uses glow discharge supported by a hot cathode; RF sputtering uses RF glow discharge; magnetron sputtering is a glow discharge controlled by a ring-shaped magnetic field. Therefore, glow discharge is the basis of sputtering.

Glow discharge is a gas discharge phenomenon that occurs when a voltage is applied between two electrodes in a thin gas with a vacuum of 10–1 Pa. The principle is shown in Figure 2.19. The target is a material that needs to be sputtered. As the cathode, the target is applied with a voltage of several thousand volts relative to the substrate as the anode. The anode can be grounded, or it can be at a floating potential or at a certain positive or negative potential. After the system is pre-evacuated, an inert gas (generally Ar gas) of appropriate pressure is charged as a carrier for gas discharge, and the pressure is generally in the range of 10^{-1}–10 Pa. Under the action of the high voltage of the positive and negative electrodes, gas atoms between the electrodes will be ionized in large quantities. The ionization process ionizes Ar atoms into Ar^+ and electrons that can move independently. The electrons fly to the anode, while the positively charged Ar^+ fly to the target as a cathode under the acceleration of a high-voltage electric field. The energy is released during the collision, and the atoms on the target surface get out of

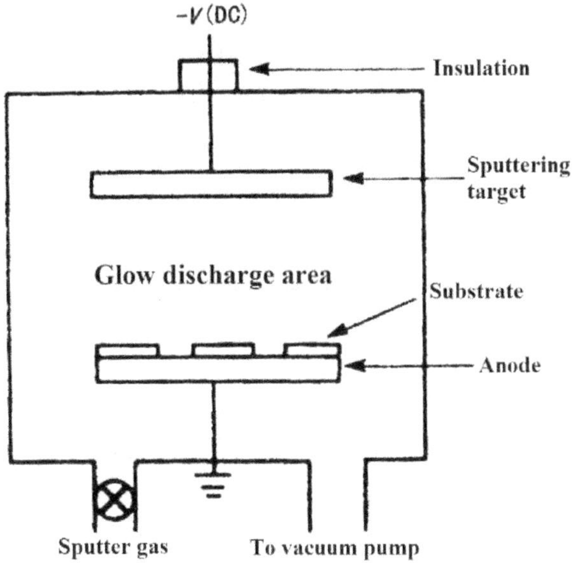

Figure 2.19: Schematic of a DC sputtering deposition device.

the target and fly to the substrate after obtaining a certain amount of energy. In this process, other particles, such as secondary electrons, ions, and photons, may be emitted from the cathode. So relatively speaking, the sputtering process is much more complicated than the evaporation process, and its quantitative description is also much more difficult.

For the DC gas discharge system shown in Figure 2.20(a), a voltage V and a current I are provided between the electrodes by a DC power source with an electromotive force of E, and a resistor R is used as a current limiting resistor. Electrical parameters in the system meet the relationship

$$V = E - I \cdot R. \tag{2.41}$$

Keep the pressure of the gas in the vacuum container at about 1 Pa, and increase the voltage between the two electrodes gradually. In the initial stage, there is almost no current passing between the electrodes, because most of the gas atoms are still in the neutral state at this time. They move in a directional manner under the

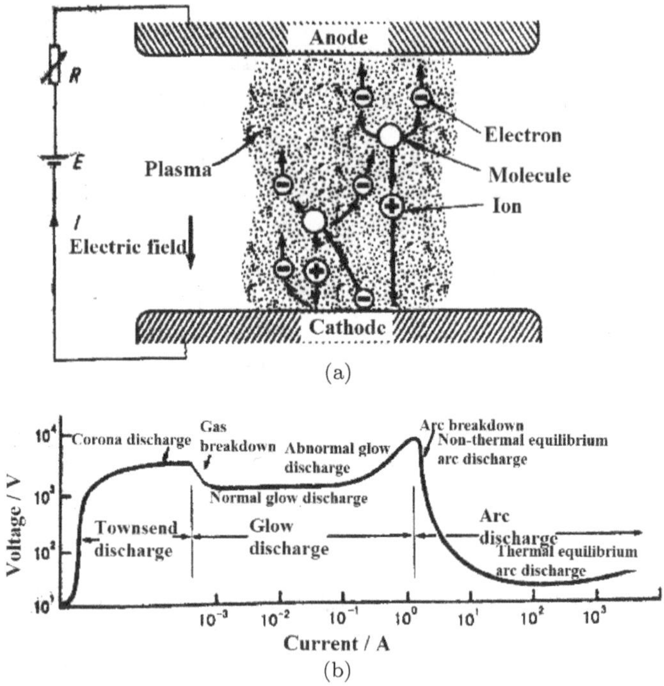

Figure 2.20: DC gas discharge model (a) and volt–ampere characteristic curve of gas discharge (b).

action of an electric field showing a weak current at the macro level, as shown in the beginning of the curve in Figure 2.20(b).

As the voltage gradually increases, the movement of ionized ions also accelerates; that is, the discharge current increases with increasing voltage. When the velocity of this part of the ionized particles reaches saturation, the current no longer increases with the voltage: that is, the current reaches a saturation value, which depends on the number of atoms that have been ionized in the gas.

As the voltage continues to rise, collisions between ions and the cathode and between electrons and gas molecules become important. As collisions become more frequent, the energy transferred from external circuits to electrons and ions is also increasing. The electronic collision begins to cause the gas molecules to ionize, and at the same time, the ions will also cause secondary electron emission when they collide with the cathode. These all lead to the generation of new ions and electrons; that is, the collision process causes the number of ions and electrons to increase avalanche-like. At this time, as the discharge current increases rapidly, the voltage does not change much. This discharge process is called Townsend discharge.

At the later stage of Townsend discharge, the corona discharge begins. At this time, some jumping corona spots begin to appear at the tip of the electrode with higher electric field strength. Therefore, this stage is called Corona discharge.

After Townsend discharge, the gas suddenly experienced a discharge breakdown. The current in the circuit increases significantly while the discharge voltage drops significantly. This is because the gas has been broken down at this time, so the internal resistance of the gas will decrease significantly with the increase of the ionization degree. And the discharge zone changed from the edges and irregularities that had been concentrated only on the cathode to those extending over the entire electrode. At this stage, the number of conductive particles is greatly increased and the energy during the collision is sufficiently high, so a significant glow will be generated.

The continuous increase of the current will make the glow area extend to the entire discharge area. The brightness of the glow will increase, and the voltage will start to rise at the same time as the current increases. This is because increasing the current increases the external voltage accordingly after the discharge has been extended to the entire electrode area.

The above two different glow discharge stages are often referred to as normal glow discharge and abnormal glow discharge. The abnormal glow discharge is a form of gas discharge commonly used in general sputtering methods.

As the current continues to increase, the discharge voltage drops suddenly and sharply again, and the current increases sharply. At this time, the discharge phenomenon begins to enter the arc discharge stage.

During glow discharge, there is obvious discharge glow between the electrodes. The typical area division is shown in Figure 2.21. The entire discharge area from the cathode to the anode can be

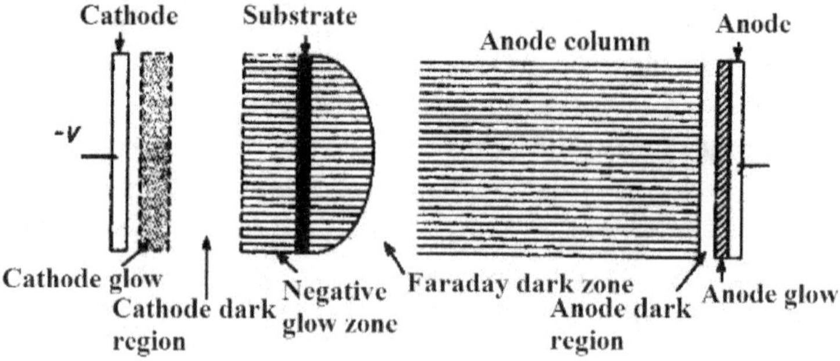

Figure 2.21: The division of the DC glow discharge area.

divided into eight areas with different luminous intensities such as cathode glow area, cathode dark area, negative glow area, Faraday dark area, anode column area, anode dark area, and anode glow area. Among them, the dark region corresponds to the acceleration region where ions and electrons obtain energy from the electric field, and the glow region corresponds to the region where different particles collide, recombine, and ionize.

There are some bright light-emitting layers near the cathode, which are generated by the recombination of positive ions moving towards the cathode and secondary electrons emitted from the anode. These layers are called cathode glow. The cathode dark area is the main acceleration area for secondary electrons and ions, and the voltage in this area accounts for most of the entire discharge voltage. The negative glow area is the area with the strongest light emission. It is the area where the accelerated electrons collide with gas atoms and ionize.

The above division of the discharge area is only a relatively typical case. In fact, the specific discharge situation can be different according to the size of the discharge vessel, the type of gas, the pressure, the arrangement of the electrodes, and the electrode materials. In Figure 2.21, the substrate is actually immersed in the negative glow region because it is closer to the cathode.

The gas, after discharge breakdown, has a certain electrical conductivity. This kind of gas with a certain conductivity is called plasma. It is a substance that consists of ions, electrons, neutral atoms, and atomic groups, and it represents an electrically neutral substance macroscopically. In plasma, there occur electrostatic interactions between various charged particles, so it shows external continuity like a liquid. The glow discharge mentioned above is a type of plasma with a low particle energy, a low particle density, and a high discharge voltage. Its characteristic feature is that the energy of heavy particles (including ions, neutral atoms, and radicals) with larger particle mass in the plasma is much lower than the energy of the electrons. That is, it is a plasma that is in a non-thermal equilibrium state.

For a glow discharge at a pressure of about 1 Pa, the ideal gas law gives the conclusion that the total density of electrons, ions, and neutral particles should be 3×10^{14} particles/cm^3, but among them, the proportion of electrons and ions is only about 10^{-4}. The average kinetic energy E of the electrons in the plasma is about 2 eV, which is equivalent to that the electron temperature is T_e, and $T_e = E/k = 23,000$ K. At the same time, owing to the small electron density, its mass heat capacity and the energy that it can transfer to other particles are extremely limited. Ions and neutral atoms are still in a low-energy state, which is only 1–2% of the energy of the electrons. That is, their temperature is only 300–500 K. The energy of ions is higher than that of neutral atoms, because ions can gain a part of their energy by accelerating in an electric field.

A direct consequence of the different velocities of electrons and ions is the formation of a plasma sheath. That is, any object in the plasma exhibits a negative potential relative to the plasma, and a positive charge accumulation occurs near the surface of the object. This is because every object in the plasma, such as the target and the substrate, will be bombarded by various particles in the plasma. Due to the different speeds of various particles, the density of various particles that bombard the surface of an object is also different. Because the mass of the ions is much larger than that of the electrons, the number of electrons that bombard the surface of the object will be much larger than the number of ions. And the surface of the object will have an excess of negative charges and present a negative potential. The establishment of the potential will repel electrons and adsorb ions, so that the number of electrons reaching the surface of the object decreases and the number of ions increases. Until the number of electrons and ions reaching the surface of the object is equal, the potential on the surface of the object reaches equilibrium. This results in that the substance immersed in the plasma, including the cathode and anode on the outer surface, is at a negative potential without exception relative to the plasma itself. That is, a plasma sheath that repels electrons is formed on its surface, the thickness of which depends on the density and temperature of electrons, and the typical value of the thickness is about 100 μm. Figure 2.22 is a schematic diagram of the plasma potential distribution of a glow discharge, in which the potential of the cathode sheath is enlarged by the superposition of the external electric field and the potential of the anode sheath is reduced by the superposition of the external electric field.

In a glow discharge plasma, the velocity and energy of electrons are higher than those of ions. Therefore, electrons are not only the main carriers in the process of plasma conduction, but also play a very important role in the collision and ionization of particles. In the sheath, the electron density is low, so the probability of collision and ionization is low, which constitutes a dark area. In the entire discharge channel, it is the electrons that play the main role in conduction and collisional ionization.

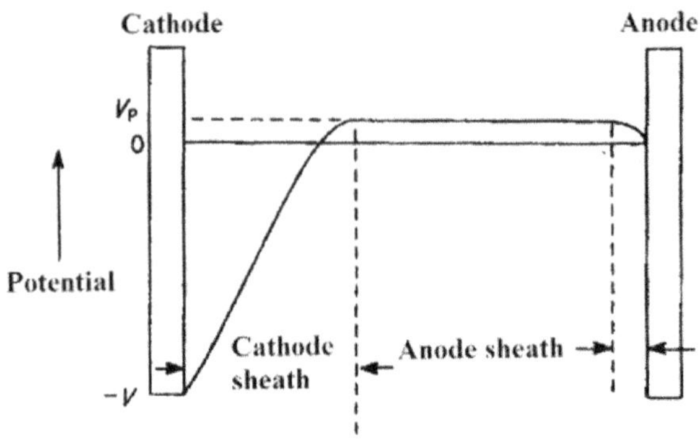

Figure 2.22: Potential distribution of a DC glow discharge and a plasma sheath.

2.3.3 *Basic parameters to characterize sputtering characteristics*

The basic parameters that characterize the sputtering characteristics include mainly the sputtering threshold, the sputtering rate, the speed and energy of the sputtered particles, and so on.

2.3.3.1 *Sputtering threshold*

The so-called sputtering threshold refers to the minimum energy necessary for the incident ions to cause sputtering of target atoms. With the advancement of measurement technology, the sputtering rate below 10^{-5} atoms/ions has been measured. When the incident ions are different, the change in the sputtering threshold is slight, while the change in the sputtering threshold for different targets is more significant. For elements in the same period of the periodic table, the sputtering threshold decreases as the atomic number increases. Table 2.11 lists the sputtering thresholds of some metals.

2.3.3.2 *Sputtering rate*

The sputtering rate represents the average number of atoms that can be sputtered out from the target cathode by each positive ion

Table 2.11: Threshold energy of some metallic elements (eV).

Atomic number	Element	Incident ion				Atomic number	Element	Incident ion			
		Ne^+	Ar^+	Kr^+	Xe^+			Ne^+	Ar^+	Kr^+	Xe^+
4	Ba	12	15	15	15	41	Nb	27	25	26	22
11	Na	5	10	—	30	42	Mo	24	24	28	27
13	Al	13	13	15	18	45	Rh	25	24	25	25
22	Ti	22	20	17	18	46	Pd	20	20	20	15
23	V	21	23	25	28	47	Ag	12	15	15	17
24	Cr	22	15	18	20	51	Sb	—	3	—	—
26	Fe	22	20	25	23	73	Ta	25	26	30	30
27	Co	20	22	22	—	74	W	35	25	30	30
28	Ni	23	21	25	20	75	Re	35	35	25	30
29	Cu	17	17	16	15	78	Pt	27	25	22	22
30	Zn	—	3	—	—	79	Au	20	20	20	18
32	Ge	23	25	22	18	90	Th	20	24	25	25
40	Zr	23	22	18	26	92	U	20	23	25	22

when it strikes the target cathode. The sputtering rate is also called the sputtering yield or the sputtering coefficient, which is usually expressed by S. The sputtering rate is related to the type, energy, and angle of the incident ions, the type of target, the lattice structure, the surface state, and the sublimation heat. If the target is a single-crystal target, the sputtering rate is also related to the crystal orientation.

(1) Relationship between the target material and the sputtering rate

Under the same conditions, bombarding the target materials of different elements by the same ion can give different sputtering rates. It is also found that the sputtering rate changes periodically. Generally, it increases with the increase of the atomic number of target elements, as seen in Figure 2.23. It can be seen from Figure 2.23 that the sputtering rates of copper, silver, and gold are relatively large, while the sputtering rates of elements such as carbon, silicon, titanium, niobium, tantalum, and tungsten are relatively small. When bombarded with Xe^+ with an energy of 400 eV, the

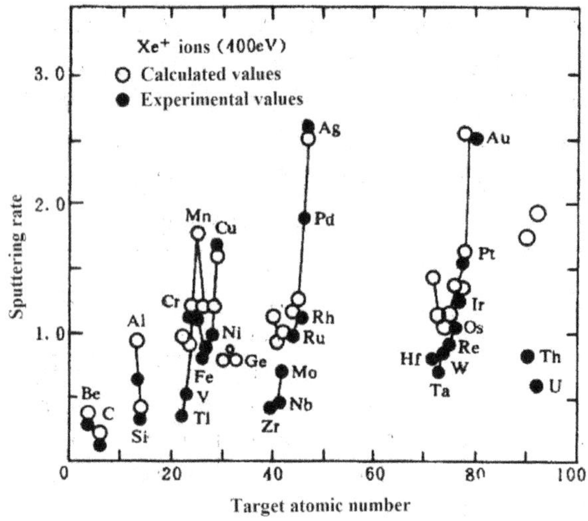

Figure 2.23: The relationship between sputtering rate and atomic number.

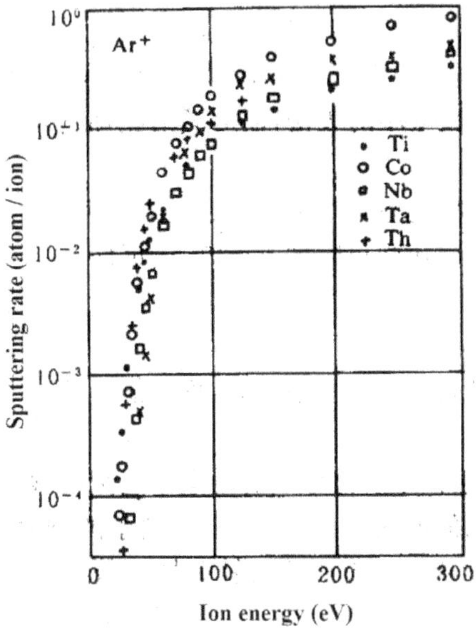

Figure 2.24: Sputtering rate curves of different targets by Ar^+ sputtering.

sputtering rate of silver was the highest and that of carbon was the lowest. Figure 2.24 shows the sputtering rate curves obtained when bombarding several metal targets with Ar^+ with different energies. In addition, metals with a hexagonal lattice structure (such as magnesium, zinc, and titanium) and metals with contaminated surfaces (such as oxide layers) have lower sputtering rates than face-centered cubic metals (such as nickel, platinum, copper, silver, and gold) and clean surface metals. And metals with high sublimation heat have lower sputtering rates than those with low sublimation heat. Table 2.12 shows the sputtering rates of various elements. From the analysis of atomic structure, the above rules are related to the filling degree of the 3d, 4d, and 5d electron shells of the atoms.

(2) **Relationship between incident ion energy and sputtering rate**

The incident ion energy has a significant effect on the sputtering rate. Sputtering occurs only when the incident ion energy is above

Table 2.12: Sputtering rates of some elements.

	Ne$^+$				Ar$^+$			
Target element	100 (eV)	200 (eV)	300 (eV)	400 (eV)	100 (eV)	200 (eV)	300 (eV)	400 (eV)
Be	0.012	0.10	0.26	0.56	0.074	0.18	0.29	0.80
Al	0.031	0.24	0.43	0.83	0.11	0.35	0.65	1.24
Si	0.034	0.13	0.25	0.54	0.07	0.18	0.31	0.53
Ti	0.08	0.22	0.30	0.45	0.081	0.22	0.33	0.58
V	0.06	0.17	0.36	0.55	0.11	0.31	0.41	0.70
Cr	0.18	0.49	0.73	1.05	0.30	0.67	0.87	1.30
Fe	0.18	0.38	0.62	0.97	0.20	0.53	0.76	1.26
Co	0.084	0.41	0.64	0.99	0.15	0.57	0.81	1.36
Ni	0.22	0.46	0.65	1.34	0.28	0.66	0.95	1.52
Cu	0.26	0.84	1.20	2.00	0.48	1.10	1.59	2.30
Ge	0.12	0.32	0.48	0.82	0.22	0.50	0.74	1.22
Zr	0.054	0.17	0.27	0.42	0.12	0.28	0.41	0.75
Nb	0.051	0.16	0.23	0.42	0.008	0.25	0.40	0.65
Mo	0.10	0.24	0.34	0.54	0.13	0.40	0.58	0.93
Ru	0.078	0.26	0.38	0.67	0.14	0.41	0.68	1.30
Rh	0.081	0.36	0.52	0.77	0.19	0.55	0.86	1.46
Pd	0.14	0.59	0.82	1.32	0.42	1.00	1.41	2.39
Ag	0.27	1.00	1.30	1.98	0.63	1.58	2.20	3.40
Hf	0.057	0.15	0.22	0.39	0.16	0.35	0.48	0.83
Ta	0.056	0.13	0.18	0.30	0.10	0.28	0.41	0.62
W	0.038	0.13	0.18	0.32	0.068	0.29	0.40	0.62
Re	0.04	0.15	0.24	0.42	0.10	0.37	0.56	0.91
Os	0.032	0.16	0.24	0.41	0.057	0.36	0.56	0.95
Ir	0.069	0.21	0.30	0.46	0.12	0.43	0.70	1.17
Pt	0.12	0.31	0.44	0.70	0.20	0.63	0.95	1.56
Au	0.20	0.56	0.84	1.18	0.32	1.07	1.65	2.43
								(500)
Th	0.028	0.11	0.17	0.36	0.097	0.27	0.42	0.66
U	0.063	0.20	0.30	0.52	0.14	0.35	0.59	0.97

a certain critical value (i.e., the sputtering threshold). Figure 2.25 shows the typical relationship between the sputtering rate and the incident ion energy.

The curve in Figure 2.25 can be divided into three areas:

(1) $S \propto E^2$ $\quad E_r < E < 500\,\text{eV}$ (E_r is the sputtering threshold);
(2) $S \propto E$ $\quad 500\,\text{eV} < E < 1000\,\text{eV}$;
(3) $S \propto E^{1/2}$ $\quad 100\,\text{eV} < E < 5000\,\text{eV}$.

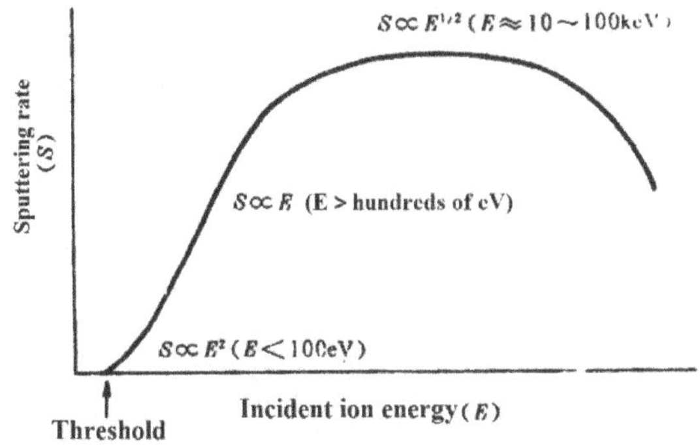

Figure 2.25: Relationship between sputtering rate and incident ion energy.

The above results show that when the incident particle energy is low, the sputtering rate increases exponentially with the increase of incident ion energy, and thereafter, a linear increase region appears with the increase of incident particle energy, and then gradually reaches a flat maximum and becomes saturated. If E is further increased, the value of S starts to decrease due to the ion implantation effect. The relationship between the ionic energy and the sputtering rate during the bombardment of copper with Ar^+ is shown in Figure 2.26. The energy range in Figure 2.26 is expanded to 100 keV. This curve can be divided into three parts. The first part is a low-energy region with almost no sputtering. The energy of the second part ranges from 70 eV to 10 keV; this is the area where the sputtering rate increases with the ion energy and the energy used to deposit the film is mostly in this range. The energy of the third part is more than 30 keV, and the sputtering rate decreases with the increase of ion energy. At this time, the bombarded ions penetrate into the target material, and most of the energy is lost inside the target material instead of being consumed on the target surface.

(3) **Relationship between the incident ion type and the sputtering rate**

The sputtering rate depends on the atomic weight of the incident ions. The larger the atomic weight is, the higher the sputtering rate is. At the same time, it is also related to the atomic number of the

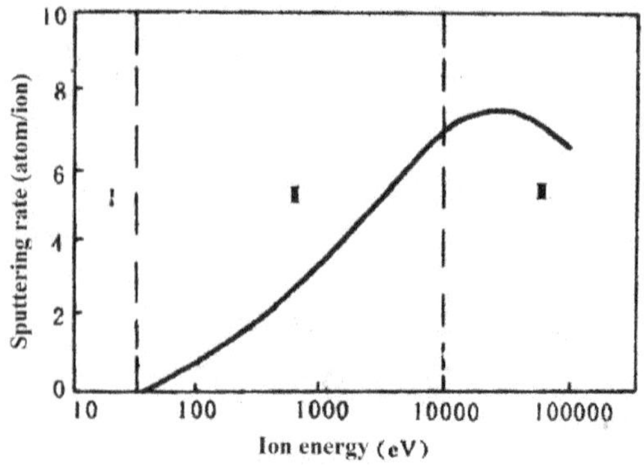

Figure 2.26: Relationship between ion energy and sputtering rate when Ar ions bombard copper.

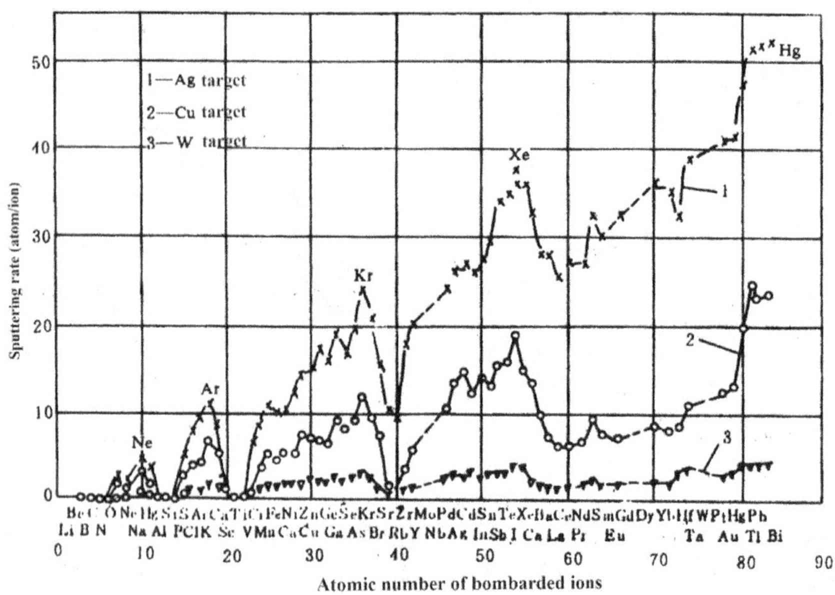

Figure 2.27: Relationship between sputtering rate and atomic number of incident ions.

incident ions, which shows a relationship that changes periodically with the atomic number of the ions. This is similar to the relationship between the sputtering rate and the atomic number of the target, as shown in Figure 2.27.

Figure 2.27 shows that in each row of the periodic table, the element with the electron shell filling has the largest sputtering rate. Therefore, the sputtering rate of the inert gas is the highest. The sputtering rate of the elements located in the middle of each column of the periodic table of elements is the lowest, such as Al, Ti, and Zr. Therefore, in general, most of the incident ions are inert gas ions. In consideration of economy, Ar is usually selected as the working gas. At the same time, the inert gas can avoid a chemical reaction with the target. Experiments show that in the commonly used energy range of incident ions (500–20,000 eV), the sputtering rates of various inert gases are almost the same. Figure 2.28 shows the sputtering rate curves of four inert gas ions (Ne^+, Ar^+, Kr^+, and Xe^+) and mercury ions (Hg^+) bombarding the same tungsten target with different energies, which can be a proof. It can be seen from Figure 2.28 that when the same target is sputtered with different incident ions, the difference in sputtering rates is much smaller than

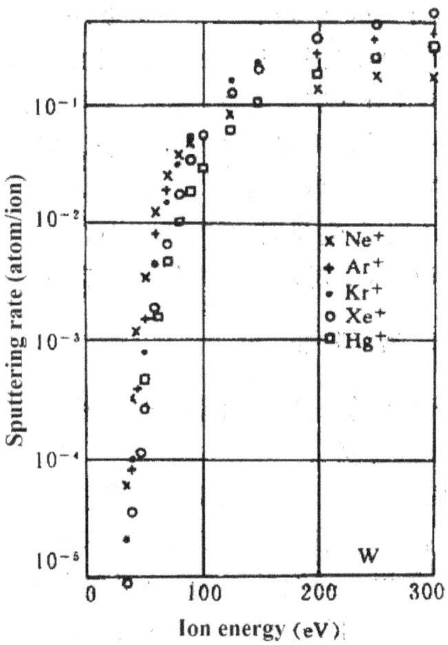

Figure 2.28: Sputtering rate curves of different gas ions bombarding tungsten targets.

the difference in sputtering rates obtained by bombarding different targets with the same ion.

(4) **Relationship between the incidence angle of incident ions and the sputtering rate**

The angle of incidence is the angle between the direction of ion incidence and the normal of the sputtered target surface. Figure 2.29 shows the relationship between the sputtering rate and incident angle of several metals when using Ar^+ to sputter. It can be seen that the sputtering rate gradually increases with the increase of the incident angle, and the relative sputtering rate between 0° and 60° basically obeys the law of $1/\cos\theta$, that is, $S(\theta)/S(0) = 1/\cos\theta$. $S(\theta)$ and $S(0)$ are the sputtering rates when the incident angle is θ and the normal incidence ($\theta = 0$), respectively. The value of S of $\theta = 60°$ is approximately twice as high as that of normal incidence. When the incident angle is 60°–80°, the sputtering rate is the highest. When the incident angle continues to increase, the sputtering rate decreases

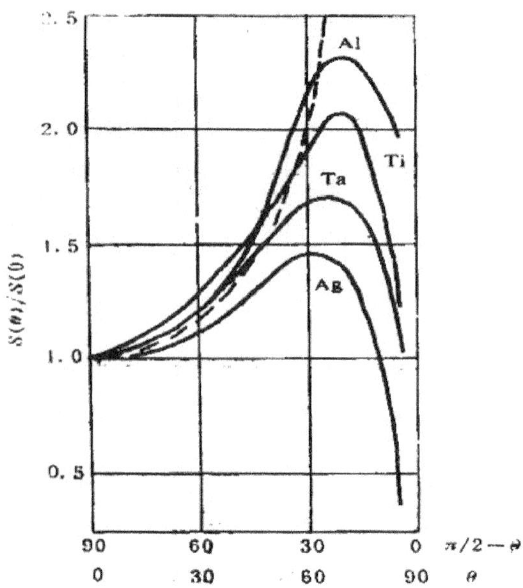

Figure 2.29: Relationship between the incidence angle of Ar^+ and the sputtering rate of several metals.

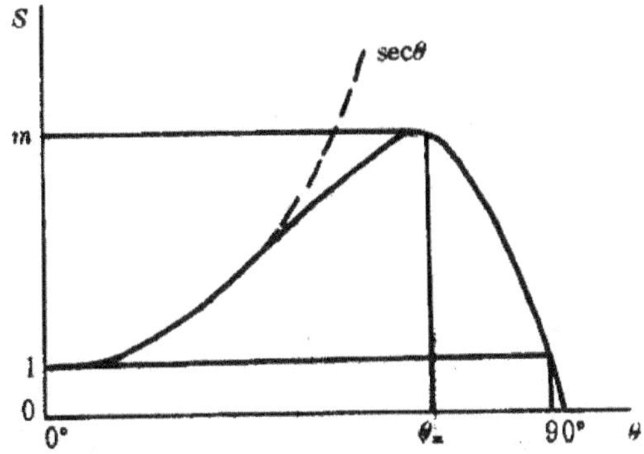

Figure 2.30: Typical relationship between sputtering rate and ion incidence angle.

sharply. When θ is equal to 90°, the sputtering rate is 0. The typical curve of this change is shown in Figure 2.30. That is, for different targets and different incident ions, the S value corresponding to the maximum sputtering rate has an optimal incidence angle θ_m.

(5) Relationship between the target temperature and the sputtering rate

The sputtering rate is also closely related to the target temperature. For a given material, when the target temperature is lower than a certain temperature related to the sublimation energy of the target, the sputtering rate is almost unchanged. However, if this temperature is exceeded, the sputtering rate will increase sharply. Figure 2.31 shows the relationship between the sputtering rate and the target temperature when several targets are bombarded by ions with an energy of 45 keV (Xe^+). Therefore, the target temperature should be controlled during sputtering to prevent a sharp increase in sputtering rates.

Moreover, the sputtering rate is also related to the structure and crystalline orientation of the target, surface morphology, and sputtering pressure. In the process of sputter coating, in order to ensure the quality of the sputtered film and increase the deposition speed of the film, the pressure of the working gas should be reduced as much as possible to increase the sputtering rate of the target.

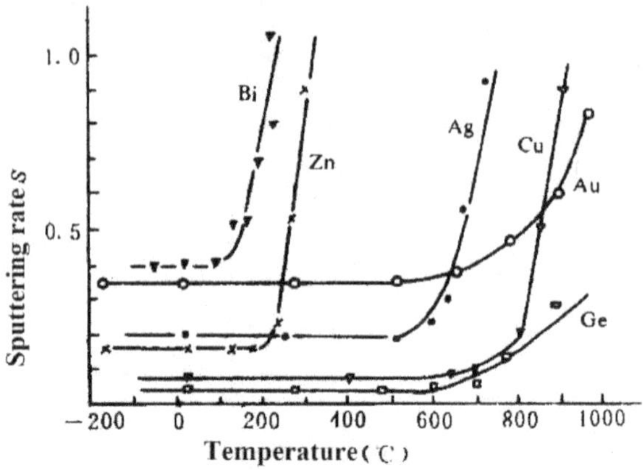

Figure 2.31: Relationship between sputtering rate and target temperature (the target is bombarded with Xe^+ at 45 keV).

2.3.3.3 *Mathematical description of sputtering rate*

As can be seen from the foregoing discussion, there are many factors that affect the sputtering rate. Therefore, it is more complicated to give the expression of sputtering rate in theory. The mathematical expression of the sputtering rate is given in the following three cases, and no specific theoretical derivation is made here.

(1) When the ion energy $E < 1\,\text{keV}$, the sputtering rate S of normal incidence is

$$S = \left(\frac{3}{4\pi^2}\right)\frac{\alpha T_m}{V_0}, \tag{2.42}$$

where

$$T_m = \frac{4m_1 m_2}{(m_1+m_2)^2}E$$

is the maximum energy delivered. For cascade collisions, T_m is also the maximum reflected energy during the sputtering process, V_0 is the barrier height of the target element and also the sublimation energy of the target element, α is the quantity related to m_2/m_1, and m_1 and m_2 are the energy of the target atom and the incident ion. The relationship between the factor α and the mass ratio m_2/m_1 is shown in Figure 2.32.

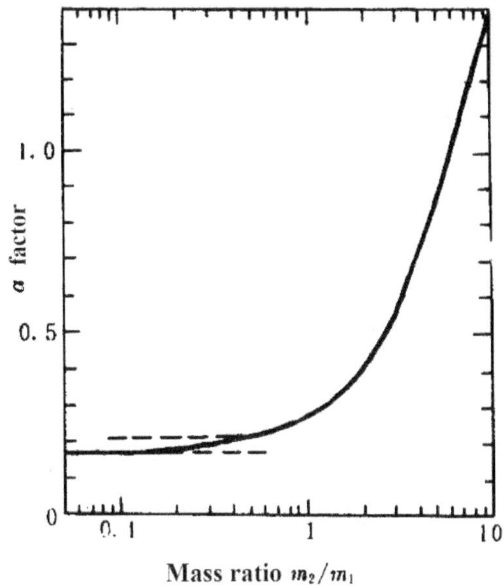

Figure 2.32: The relationship between the factor α and the mass ratio m_2/m_1.

(2) When the ion energy $E > 1\,\text{keV}$, the sputtering rate S of normal incidence is

$$S = 0.042\alpha\, S_n(E)/V_0(\text{Å}^2), \tag{2.43}$$

where Å is Angstrom (1 Å = 0.1 nm), and the definitions of α and V_0 are the same as in Equation (2.42). $S_n(E)$ is given by

$$S_n(E) = 4\pi Z_1 Z_2 e^2 \alpha_{12}\,[m_1/(m_1+m_2)]\, S_n(\varepsilon) \tag{2.44}$$

where

$$\varepsilon = \frac{m_1 E/(m_1+m_2)}{Z_1 Z_2 e^2/\alpha_{12}}. \tag{2.45}$$

In Equation (2.44), $\alpha_{12} = 0.8853\alpha_0\ (Z_1^{2/3} + Z_2^{2/3})^{-1/2}$, and it is called the Thomson–Fermi shielding radius. And $\alpha_0 = 0.0529\,\text{nm}$, and it is the Bohr radius, Z_1 is the atomic number of the bombarded ions, and Z_2 is the atomic number of the target.

Table 2.13: The relationship between ε and $S_n(\varepsilon)$.

ε	$S_n(\varepsilon)$	ε	$S_n(\varepsilon)$
0.002	0.120	0.4	0.405
0.004	0.154	1.0	0.356
0.01	0.211	2.0	0.291
0.02	0.261	4.0	0.214
0.04	0.311	10	0.128
0.1	0.372	10	0.0813
0.2	0.403	40	0.493

ε is a dimensionless parameter called equivalent mass. $S_n(\varepsilon)$ is the called nuclear stop section. The relationship between ε and $S_n(\varepsilon)$ is given in Table 2.13.

(3) In general, the calculation of the sputtering rate can be processed as follows:

$$S = W \times 10^5/mIt, \tag{2.46}$$

where W is the loss amount (g) of the target, m is the atomic weight, I is the ion current (A), and t is the sputtering time (s). W can be calculated as follows:

$$W = RtAd, \tag{2.47}$$

where R is the etching rate (cm/s), A is the sample area (cm^2), and d is the material density (g/cm^3).

$$\text{Ion current} \quad I = JA, \tag{2.48}$$

where J is the ion current density (A/cm^2).
According to the above, it is concluded that the sputtering rate in general is

$$S = \frac{Rd}{mJ} \times 10^5. \tag{2.49}$$

2.3.3.4 *Energy and speed of sputtered atoms*

The energy and velocity of the sputtered atoms are also important parameters to describe the sputtering characteristics. The energy of the sputtering atom is related to the target material, the type and energy of the incident ion, and the directivity of the sputtering atom. Generally, the energy of the atoms evaporated from the evaporation source is about 0.1 eV. In sputtering, the sputtered atoms have a high kinetic energy because the sputtered atoms are sputtered by exchanging energy with high-energy incident ions, and the kinetic energy is about 5–10 eV. Due to this, the sputtering method has many advantages. When different kinds of incident ions bombard different targets, the energy distribution of the escaping atoms is as shown in Figure 2.33. It can be seen from Figure 2.33 that the sputtered atoms have similar energy distribution laws, while the distribution range of energy values is different.

The energy of the sputtering atom is related to the target material, the type and energy of the incident ion, and the directivity of the sputtering atom. After the Hg^+ ions of different energies bombard the Ag single-crystal target, the energy distribution of the escaping Ag atoms is shown in Figure 2.34. Its energy distribution is similar to the Maxwell distribution. The energy of most of the sputtered atoms is lower than 100 eV. The high-energy part has a prolonged tail with an average energy of 10–40 eV. The energy of the bombarding ions increases, and the tail of high energy also

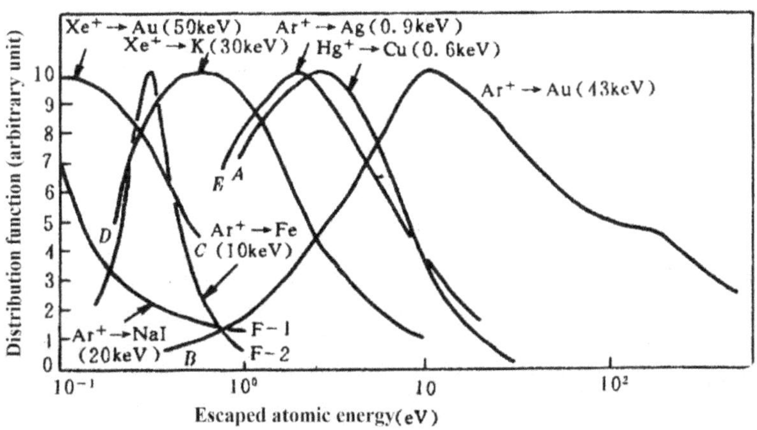

Figure 2.33: Energy distribution of escaped atoms when different incident ions bombard different targets.

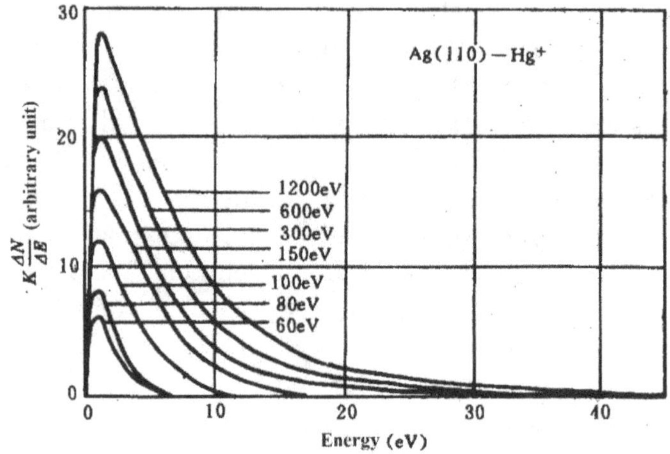

Figure 2.34: Energy distribution of sputtered atoms when Hg^+ ions of different energies bombard Ag.

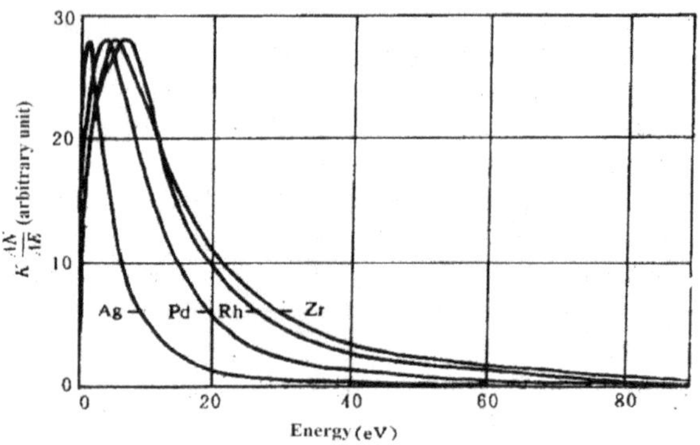

Figure 2.35: Energy distribution of escaping atoms when Kr ions with an energy of 1200 eV bombard different targets.

drags longer. When the incident ion energy is greater than 1000 eV, the average energy of the escaped atoms increases no longer. The energy distribution curve of the escaping sputtered atoms obtained by bombarding the target of different elements with Kr^+ ions with an energy of 1200 eV is shown in Figure 2.35. Rn, Pd, and Ag are adjacent elements in the periodic table, and their atomic weights are almost equal. But their energy distribution curves are quite different.

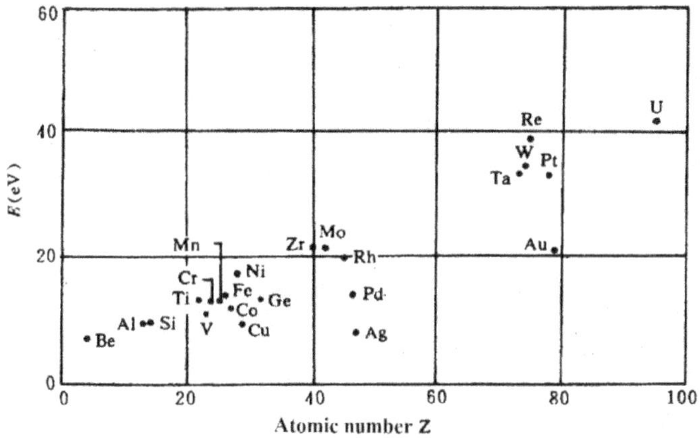

Figure 2.36: Average escape energy of sputtered atoms when Kr ions with an energy of 1200 eV bombard different targets.

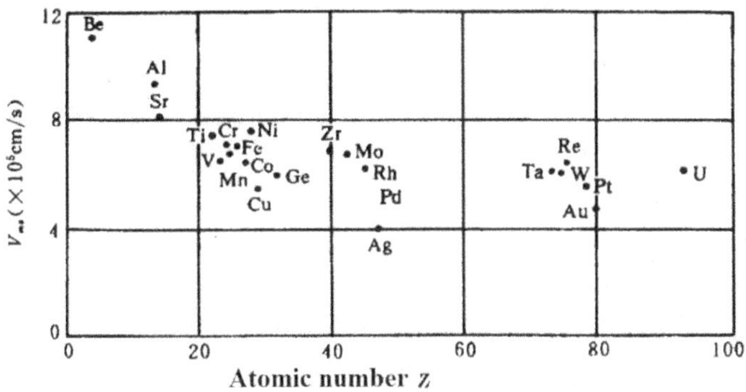

Figure 2.37: The average escape velocity of sputtered atoms when Kr ions with an energy of 1200 eV bombard different targets.

When the same ions bombard different materials, the average escape energy E(eV) and average escape velocity v_{ms} of the sputtered atoms are as shown in Figures 2.36 and 2.37, respectively. Figures 2.36 and 2.37 show that when the atomic number Z > 20, the average escape energy of each element is greatly different, while the difference in the average escape rate is small.

The energy and velocity of sputter atoms in sputter coating have the following characteristics:

(1) Atoms that are sputtered from heavy element targets have a higher escape energy, while light element targets have a higher atomic escape speed.
(2) Different targets have different atomic escape energies. Target orientation and crystal structure have little effect on escape energy, while targets with high sputtering rates usually have lower average atomic escape energies.
(3) At the same bombardment energy, the atomic escape energy increases linearly with the increase of the atomic mass of the incident ions. The atomic escape energy sputtered by the light incident ions is lower (about 10 eV), and the atomic escape energy sputtered by heavy incident ions is larger (above 30 eV on average).
(4) The average escape energy of the sputtered atoms increases with the increase of incident ion energy. When the incident ion energy exceeds 1 keV, the average escape energy gradually approaches a constant value.

2.3.3.5 *Angular distribution of sputtered atoms*

Earlier studies suggested that the angular distribution of the sputtered atoms is in accordance with Knudsen's cosine law. Further studies have found that the distribution of escaped atoms does not follow the cosine distribution law when bombarded with low-energy ions. And the number of atoms escaping perpendicular to the direction of the target surface is obviously less than the number of escaped atoms that should be obtained when cosine distribution is used. The results are shown in Figure 2.38. For different target materials, the deviations of angular distribution and cosine distribution are different. And when the incident angle of the bombarded ions is changed, the number of escaped atoms increases significantly in the positive reflection direction of the incident direction, as shown by the solid line in Figure 2.39.

The main escape direction of the sputtered atoms is related to the crystal structure, which also affects the sputtering rate. For single-crystal target materials, the main escape direction is the direction in which the atoms are arranged most closely, followed by

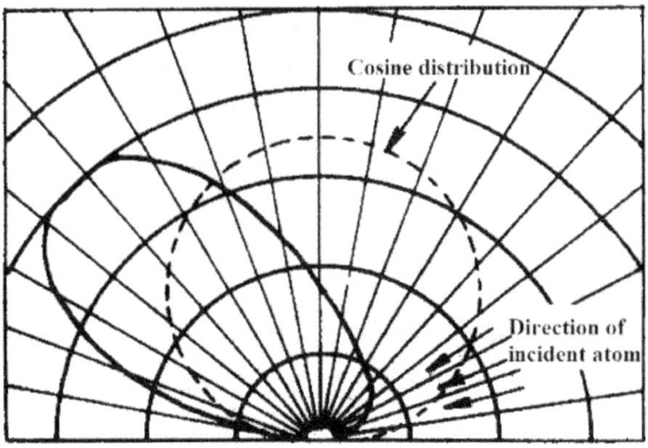

Figure 2.38: Angle distribution of sputtered atoms during oblique bombardment.

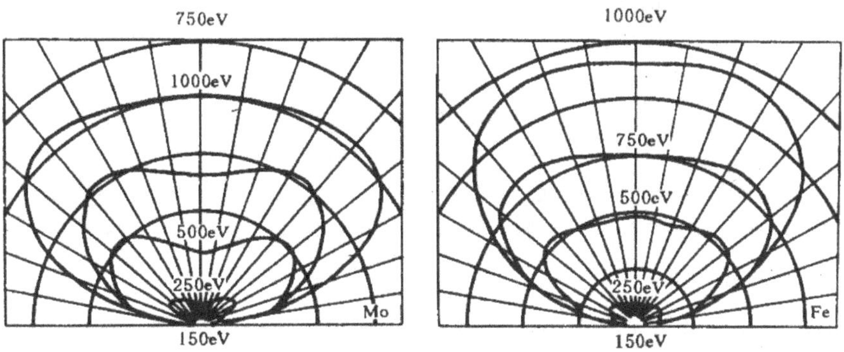

(a) (b)

Figure 2.39: Angular distribution of sputtered atoms of molybdenum (a) and iron (b) when Hg^+ with an energy of 100–1000 eV are incident perpendicularly.

the direction in which they are less dense. For example, for face-centered cubic structure crystals, the main escape direction is [110] crystal orientation, followed by [100], [111] crystal orientation. The angular distribution of the sputtered atoms of the polycrystalline target is significantly different from that of the single-crystal target. For the single-crystal target, an obvious preferred orientation of the sputtered atoms can be observed, while the polycrystalline target shows a cosine distribution basically.

2.3.4 *Sputtering process and sputtering coating*

The sputtering process includes the sputtering of the target, the morphology of the escaping particles, the migration of the sputtered particles to the substrate and the film formation on the substrate.

2.3.4.1 *Sputtering process*

When the incident ions collide with the target, their momentum is transferred to the target atoms. So if the energy the target atoms obtained exceeds their binding energy, the target atoms may be sputtered. This is a process that mainly occurs during the sputtering of a target. In fact, the sputtering process is very complicated. When high-energy incident ions bombard a solid surface, there are many effects. Figure 2.40 shows a schematic view of these effects. For example, the incident ions may be reflected from the target surface, or they may become neutral atoms or molecules after capturing electrons in the bombardment process and then are reflected from the surface. Ions bombardment of the target material causes electrons to escape from the target surface; so-called secondary electrons are generated. Ions penetrate into the target surface to produce an ion implantation effect. In addition, these effects can change the

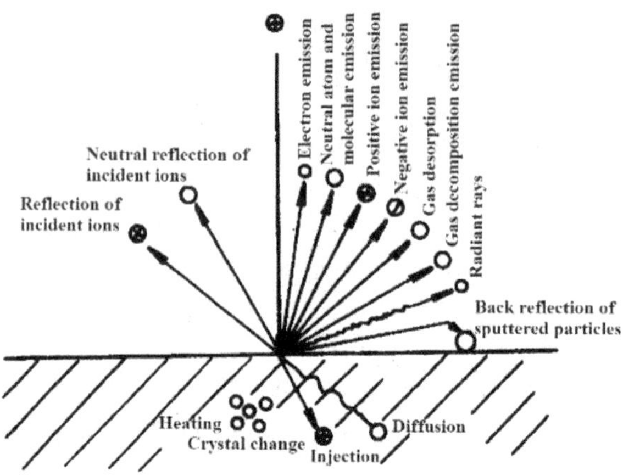

Figure 2.40: Various effects caused by incident ions bombarding a solid surface.

structure and composition of the target surface, desorb the gas adsorbed on the target surface, and generate radiant electrons when high-energy ions are incident, etc. In addition to the neutral particles of the target, that is, atoms or molecules finally deposited into a thin film, some other effects will have a great impact on the growth of the sputtered film layer. Various phenomena produced by ions bombarding the solid surface are related to the type of solid material, and the type and energy of incident ions. Table 2.14 gives an overview of the various effects produced by each incident ion and their probability of occurrence when certain metal surfaces are bombarded with Ar^+ with an energy of 10–100 eV.

It should be noted that the various effects or phenomena shown in Figure 2.40 may also occur on the substrate during the sputtering process. Because in the glow discharge plating process, the substrate's self-bias voltage is the same as that of the ground electrode, and they both will form a negative potential relative to the surrounding environment, therefore, when considering the phenomenon of the substrate during the sputtering process, the substrate can also be regarded as a sputtering target, but the two are very different in degree.

Table 2.14: Various effects of ion bombardment on solid surface and their probability of occurrence.

Effects	Name	Probability of occurrence
Sputtering	Sputtering rate, S	$S = 0.1$–10
Ion sputtering	Primary ion reflection coefficient, ρ	$\rho = 10^{-4}$–10^{-2}
Ion sputtering	Neutralized primary ion reflection coefficient, ρ_m	$\rho_m = 10^{-3}$–10^{-2}
Ion Implantation	Ion implantation coefficient, a	$a = 1$–$(\rho - \rho_m)$
Ion Implantation	Ion implantation depth, d	$d = 1$–$10\,\text{nm}$
Secondary electron emission	Secondary electron emission coefficient, r	$r = 0.1$–1
Secondary electron emission	Secondary electron emission coefficient, k	$k = 10^{-5}$–10^{-4}

2.3.4.2 *Migration process of sputtered particles*

Among the particles that escaped when the target was bombarded, positive ions could not reach the surface of the substrate due to the effect of the reverse electric field, and the remaining particles would migrate to the substrate. The mean free path λ_1 of a large number of neutral atoms or molecules colliding with the working gas molecules during the flight in the discharge space can be expressed by the following formula:

$$\lambda_1 = \overline{c_1}/(v_{11} + v_{12}), \tag{2.50}$$

where $\overline{c_1}$ is the average speed of the sputtered particles, ν_{11} is the average number of collisions between the sputtered particles, and ν_{12} is the average number of collisions between the sputtered particles and the working gas molecules. Generally, the density of the sputtered particles is much lower than the density of the working gas molecules, so $\nu_{11} \ll \nu_{12}$, therefore

$$\lambda_1 \approx \overline{c_1}/v_{12}. \tag{2.51}$$

v_{12} is related to the working gas molecules' density n_2, the average speed $\overline{c_2}$, and the collision area Q_{12} between the sputtered particles and the working gas molecules, and can be expressed as

$$\nu_{12} = n_2 Q_{12}\sqrt{(\overline{c_1})^2 + (\overline{c_2})^2}. \tag{2.52}$$

Since the velocity of the sputtered particles is much faster than the velocity of the gas molecules, it can be considered that v_{12} in the above Equations satisfies $\nu_{12} \approx Q_{12}\overline{c_1}n_2$, and the mean free path of the sputtered particles can be approximately expressed by the following formula:

$$\lambda_1 \approx 1/\pi(r_1 + r_2)^2 n_2. \tag{2.53}$$

The gas pressure of the sputter coating is $10^1 \sim 10^{-1}$ Pa. At this time, the mean free path of the sputtered ions is about 1∼10 cm. Therefore, the distance between the target and the substrate should be approximately equal to this value. Otherwise, the sputtered particles will collide many times during the migration process.

In this way, the kinetic energy of the target atoms is reduced, and the scattering loss of the target is increased.

The sputtered atoms will reduce their energy due to collision with the working gas molecules during the transport process to the substrate, but the energy of these target sputtered atoms is much higher than that of the evaporated atoms. So the energy of the target atoms deposited on the substrate during the sputtering process is still relatively high, even reaching hundreds of times the energy of the evaporated atoms.

2.3.4.3 *Film formation process*

The process of the thin film growth will be introduced in Chapter 5. Here, we mainly introduce several issues that should be considered in the process of film formation when the target particles are incident on the substrate.

(1) **Deposition rate *D***

The deposition rate D refers to the thickness of the substance sputtered from the target material onto the substrate per unit time, and D is proportional to the sputtering speed S.

$$D = CIS, \tag{2.54}$$

where C is a characteristic constant related to the sputtering device, and I is an ion current.

Equation (2.54) shows that for a certain sputtering device (i.e., C is a certain value) and a certain working gas, an effective way to increase the deposition rate is to increase the ion current I. However, without increasing the voltage, increasing the value of I will only increase the pressure of the working gas. Figure 2.41 shows the relationship curve between gas pressure and sputtering rate. It can be seen from Figure 2.41 that when the pressure is increased to a certain value, the sputtering rate will start to decrease significantly. This is caused by the increased back reflection and scattering of the target particles. Therefore, the appropriate gas pressure value should be selected by the sputtering rate, but attention should be paid to the effect of the increased gas pressure on the film quality.

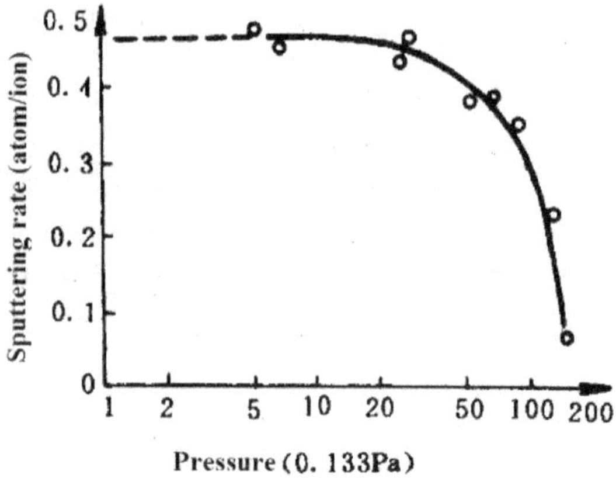

Figure 2.41: Relationship between sputtering rate and Ar pressure.

(2) **Purity of the deposited film**

To improve the purity of the deposited film, the amount of impurities deposited on the substrate must be minimized as much as possible. The impurities referred to here are mainly residual gases. This is because about a few percent of the sputtering gas molecules are usually injected into the film, especially when the substrate is biased. If the volume of the vacuum chamber is V, the partial pressure of the residual gas is P_{C}, the partial pressure of argon is P_{Ar}, the amount of residual gas in the vacuum chamber is Q_{C}, and the amount of argon is Q_{Ar}, then

$$Q_{\mathrm{C}} = P_{\mathrm{C}}V, \quad Q_{Ar} = P_{\mathrm{Ar}}V.$$

That is,

$$P_{\mathrm{C}} = P_{\mathrm{Ar}}Q_{\mathrm{C}}/Q_{\mathrm{Ar}}. \tag{2.55}$$

It can be seen that to reduce the residual gas pressure P_{C} for improving the purity of the film, two effective measures can be taken: increasing the background vacuum and increasing the amount of argon. Generally, the background vacuum should be 10^{-3}–10^{-4} Pa.

(3) **Prevention of pollution during deposition**

Before introducing the sputtering gas, it is necessary to increase the vacuum in the vacuum chamber to a high vacuum region (10^{-4} Pa).

But even so, there may still be many sources of pollution, mainly including: the vacuum chamber walls and other components in the vacuum chamber may adsorb gases, water vapor, and carbon dioxide. These gases may be released again due to the bombardment of electrons and ions in the glow discharge. In order to prevent these pollutions, the surfaces that may come into contact with the glow must be cooled or baked at high temperature during the extraction process. Under the sputtering pressure, the backflow of the diffusion pump oil will also introduce pollution. Generally, this problem can be solved by using a high vacuum valve as a throttle. The impact of particulate matter on the surface of the substrate on the film will also cause pinholes and contamination. Therefore, the substrate should be thoroughly cleaned before deposition to ensure that the substrate is not contaminated or carry micron-like dirt as much as possible.

2.3.4.4 *Control of sputtering conditions*

In sputtering coating, the control of film formation conditions is very important when preparing high-quality thin films. First of all, a kind of gas with a high sputtering rate, inert to the target, inexpensive, and of high purity should be selected as the working gas. Argon is an ideal working gas. Second, attention should be paid to the effects of sputtering voltage and substrate potential on film performance. The sputtering voltage not only affects the deposition rate, but also affects the structure of the film seriously. The substrate potential directly affects the incident electron or ion current. If the substrate is properly biased, it can not only clean the substrate surface and enhance the film's adhesion, but also change the crystal structure of the deposited film. Third, the substrate temperature directly affects the growth and characteristics of the thin film. For example, when depositing a tantalum film, the temperature has little effect on the characteristics of the tantalum film when the substrate temperature is in the range of 200–400°C. However, when the temperature is higher than 700°C, the deposited tantalum film will become a body-centered cubic structure, and it will become a tetragonal lattice when the temperature is lower than 700°C. Fourth, impurities and surface oxides in the target material are important factors that contaminate the film. Attention must be paid to the high purity of the target material and the target surface must be kept clean. Pre-sputtering of the target is usually performed before sputtering deposition, which is an effective method for cleaning the target surface.

Furthermore, during the sputtering process, attention should also be paid to the interactions between parameters such as the electric field, magnetic field, target material, substrate, temperature, geometry, and vacuum degree in the sputtering equipment, because these parameters determine the structure and characteristics of the sputtered thin film.

2.3.5 *Sputtering mechanism*

Sputtering is an extremely complicated physical process, involving many factors. Although a lot of research has been done on the sputtering mechanism for a long time and many theories have been proposed, the sputtering phenomenon cannot be fully explained, and a complete set of theories and models has not been established to systematically elaborate and quantify all experimental results. There are two main theories at present: thermal evaporation theory and momentum transfer theory. And momentum transfer theory is more generally accepted.

2.3.5.1 *Thermal evaporation theory*

In the early days, some people believed that the sputtering phenomenon was caused by positively charged ions of ionized gas, which bombarded the target surface under the acceleration of the electric field and transferred energy to the atoms at the collision. The high temperature causes the target material in this area to melt and then thermal evaporation occurs. Thermal evaporation theory, to some extent, explains certain laws and phenomena of sputtering, such as the relationship between sputtering rate and the evaporation heat of the target material or the energy of bombarded ions, the cosine distribution of the sputtered atoms, and so on. However, this theory cannot explain the relationship between the sputtering rate and the incident angle of the ions, the non-cosine distribution of the angular distribution of the sputtered atoms, and the relationship between the sputtering rate and the mass of the incident ions.

2.3.5.2 *Momentum transfer theory*

For the in-depth study on sputtering characteristics, many experimental results show that sputtering is a momentum transfer process. According to the theory of momentum transfer, when low-energy ions collide with the target, atoms cannot be sputtered directly from the solid surface, but these ions transfer momentum to the atoms being collided, causing chain-like collisions of atoms on the lattice. This collision will occur in all directions of the crystal lattice. At the same time, the collision is the most effective in the lattice direction where the atoms are most closely arranged. As a result, the atoms on the crystal surface get more and more energy from neighboring atoms. If this energy is higher than the binding energy of the atoms, the atoms will be splashed from the solid surface. Momentum transfer theory can well explain problems that cannot be explained by thermal evaporation theory, such as the relationship between sputtering rate and ion incidence angle, and the angular distribution of sputtered atoms.

Essentially, the sputtering process is a series of energy exchange processes by which the incident ions collide with the target. The energy transferred from the incident ions to the sputtered atoms escaping from the target surface is only about 1% of the incident energy, and most of the energy is consumed in the surface layer of the target through cascade collisions and converted into the heat of the crystal lattice vibration. Momentum transfer theory and related theoretical derivation of sputtering will not be introduced in detail.

2.3.6 *Main sputtering methods*

According to the type of electrode structure and the choice of target, there are many types of sputtering coating devices. Targets can be divided into pure metals, alloys, and various compounds. According to the characteristics of the electrode structure, they can also be divided into DC sputtering, RF sputtering, magnetron sputtering, and reactive sputtering. According to the purpose of use, various methods can be subdivided into several types. For example, DC sputtering can be divided into two-pole sputtering, three-pole sputtering or four-pole sputtering, or a method of applying a bias voltage to form bias sputtering. In addition, different methods can be combined to form a new method, such as combining RF sputtering with reactive sputtering to form RF reactive sputtering. In recent years, in order to study the preparation of high-speed and low-temperature equipment for magnetic thin films, a facing target sputtering device has also been developed successfully. Table 2.15 shows several typical sputtering methods.

2.3.6.1 *Diode sputtering*

If the sputtering target (for the cathode) and the film-forming substrate (for the anode) constitute the two electrodes of the sputtering device, it is called two-pole sputtering. Figure 2.42 is a schematic diagram of its structure. When using RF power, it is called RF diode sputtering. When using DC power, it is called DC diode sputtering. Because the sputtering process occurs at the cathode (target), it is also called cathode sputtering. If both of the target and the substrate holder are flat, it is called planar diode sputtering. If the two are coaxially arranged in a cylindrical shape, it is called coaxial diode sputtering.

The cathode target is connected to a negative high voltage. In order to maintain a controllable negative high voltage on the target surface during the glow discharge process, the target must be a conductor. When working, first evacuate the vacuum chamber to a high vacuum (such as 10^{-3} Pa), then pass the working gas argon in, and maintain the pressure in the working room at 1–10 Pa. Turn on the power to generate an abnormal glow discharge between the cathode and anode, and establish a plasma zone, in which positively charged argon ions are accelerated by an electric field and bombard the cathode target, so that the target material is sputtered. In order to increase the deposition rate, the substrate should be as close as possible to the cathode target without affecting the glow discharge.

Table 2.15: Comparison of sputter coating types.*

Serial number	Sputtering method	Sputtering power	Ar pressure (Pa)	Characteristic
1	Diode sputtering	DC 1–7 kV 0.1–1.5 mA/cm^2 RF 0.3–10 kV 1–10 W/cm^2	1	A simple structure, a uniform film can be made on a large area substrate, and the discharge current varies with the pressure and voltage.
2	Bias sputtering	In the range of 0–500 V, make the potential of the substrate to the anode positive or negative	1	During the coating process, light-weight charged particles on the substrate are removed simultaneously, so that the substrate does not contain impurities (H_2O, N_2, etc.).
3	Three- or four-pole sputtering	DC 0–2 kV RF 0–1 kV	6.7×10^{-2}–1.3×10^{-1}	Low pressure and low voltage sputtering can be realized, and the discharge current and ion energy that bombard the target can be controlled independently. Target current can be controlled and RF sputtering can be performed.
4	RF sputtering	RF 0.3–10 kW 0–2 kW	1	Developed for the preparation of insulating films, such as SiO_2, Al_2O_3, glass films, etc., and can also sputter metals.

(*Continued*)

Table 2.15: (*Continued*)

Serial number	Sputtering method	Sputtering power	Ar pressure (Pa)	Characteristic
5	Magnetron sputtering (high-speed sputtering at low temperature)	0.2–1 kV (high temperature and low temperature) 3–30 W/cm^2	$\sim 10^{-1}$	A magnetic field is applied in a direction parallel to the target surface, and the magnetron principle that the electric field and the magnetic field are orthogonal is used to reduce the electron bombardment to the substrate and achieve high-speed sputtering at low temperature.
6	Opposite target sputtering	DC RF	$\sim 10^{-1}$	Two targets are placed opposite to each other, and a magnetic field is applied in a direction perpendicular to the surface of the target to perform high-speed sputtering at low temperature for magnetic materials.
7	Reactive sputtering	DC 1–7 kV RF 0.3–10 kW	Mix an appropriate amount of reactive gas in argon, such as O_2, N_2, etc.	Preparation of compound films of cathode materials, such as TiN, SiC, AlN, Al_2O_3, etc.
8	Ion beam sputtering	DC	$1.3 \times 100.3 \times 10^{-3}$	Under high vacuum, the coating is formed by ion beam sputtering, which is a film formation process in a non-plasma state. Target ground potential is also available.

Note: *DC: direct current; RF: radio frequency.

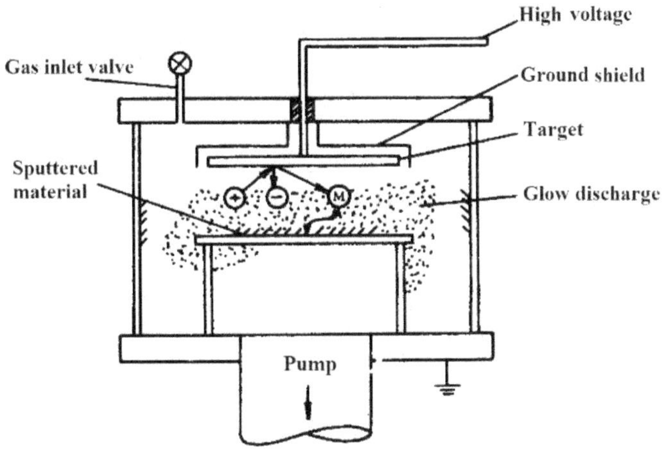

Figure 2.42: Schematic diagram of a two-pole sputtering device.

However, from the perspective of film thickness distribution, the film at the center where the cathode shields the strongest is the thinnest. Therefore, the distance between the cathode target and the substrate is preferably 3–4 times the dark area of the cathode. The deposition rate in Figure 2.44 is determined based on the discharge current (curve A) and the sputtering yield (curve B).

The working parameters of DC diode sputtering mainly include: sputtering power, discharge voltage, gas pressure, and electrode spacing. The main control parameters during sputtering are power, voltage, and gas pressure. When the voltage is constant, the relationship between discharge current and gas pressure is shown in Figure 2.43. Figure 2.44 shows the relationship between the sputtering deposition speed and the working pressure.

The DC diode sputtering device has a simple structure and can obtain a thin film with a large area and a uniform thickness. Its main disadvantages are that it is not easy to independently control the sputtering parameters, and the process repeatability is poor; the residual gas contaminates the film seriously, and the film purity is poor; the substrate temperature is high and the deposition rate is low; the target must be a good conductor. In order to overcome the above disadvantages, the following measures can be

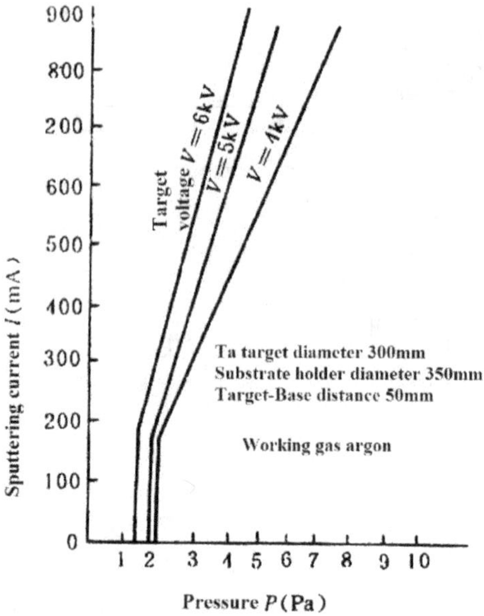

Figure 2.43: Relationship between DC secondary sputtering current and gas pressure.

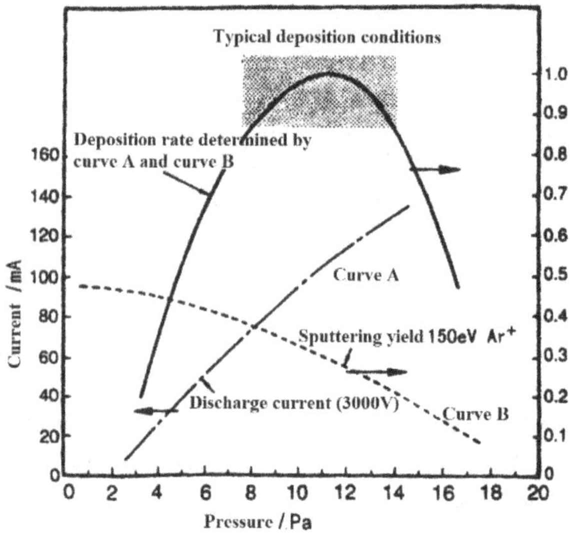

Figure 2.44: Relationship between sputtering deposition rate and working pressure.

taken: generate a glow discharge under a vacuum of more than 10^{-1}Pa and form a high-density plasma that meets the requirements of sputtering; strengthen the cooling of the target and reduce or weaken the substrate bombardment by high-speed electrons emitted by the target; choose appropriate incident ion energy.

2.3.6.2 *Bias sputtering*

Figure 2.45 is a schematic diagram of the principle and structure of DC bias sputtering. The main difference from diode sputtering is that a fixed DC bias is applied to the substrate. If a negative bias is applied, the surface of the substrate will be stably bombarded by gas ions during the film deposition process, thereby removing the gas that may enter the film surface, and the purity of the film can be improved by using it. In addition, the energy of the atoms reaching the substrate can be increased during the film formation, thereby improving the adhesion of the film, and the structure of the deposited film can also be changed by bias sputtering.

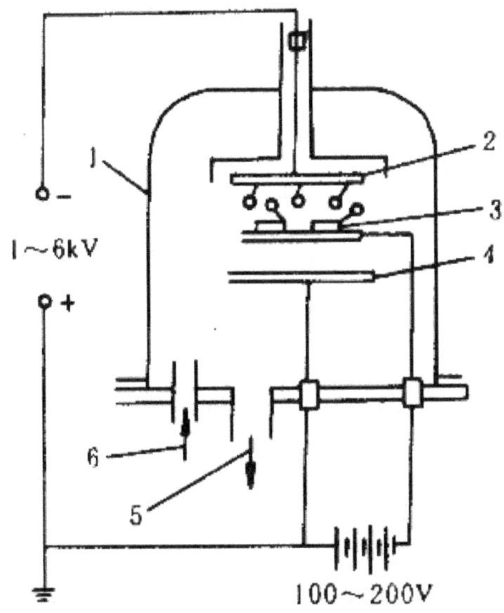

Figure 2.45: Schematic of DC bias sputtering. 1. Sputtering chamber; 2. cathode; 3. substrate; 4. anode; 5. exhaust system; 6. argon gas inlet.

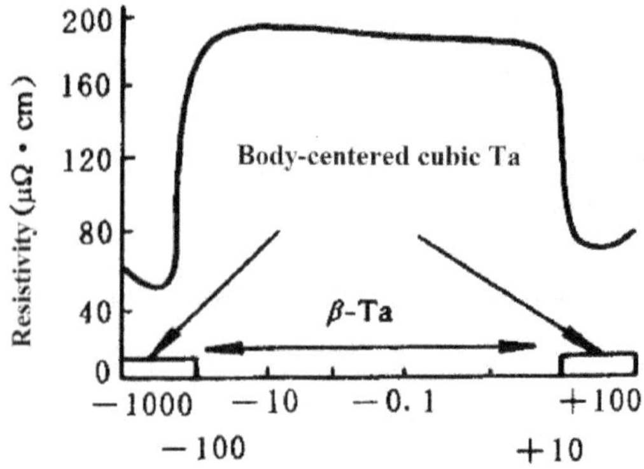

Figure 2.46: Relationship between the resistivity of the tantalum film and the substrate bias.

The application of different bias voltages in bias sputtering has the potential to change the structure of the thin films prepared. Figure 2.46 shows the changes in resistivity of tantalum films prepared by bias sputtering when different bias voltages are applied to the substrate. The bias voltage is in the range of −100 V to +10 V, and the film resistivity is relatively high, which belongs to β-Ta, that is, a tetragonal structure. When the negative bias is greater than 100 V, the resistivity drops rapidly; at this time the tantalum film has changed from β phase to a normal body-centered cubic structure. This situation is probably caused by the positive bias applied to the substrate, which becomes the anode, causing a large number of electrons to flow to the substrate, which causes the substrate to generate heat.

The energy and velocity of sputtered atoms in sputter coating have the following characteristics:

(1) Atoms that are sputtered from heavy element targets have a higher escape energy, while light element targets have a higher atomic escape speed.
(2) Different targets have different atomic escape energies. Target orientation and crystal structure have little effect on escape energy, while targets with high sputtering rates usually have lower average atomic escape energies.

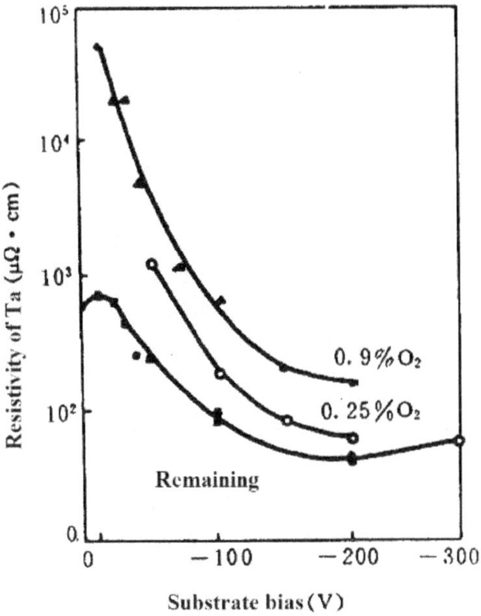

Figure 2.47: Relationship between the resistivity and bias voltage of the tantalum film when the oxygen content in the sputtering gas is different.

(3) At the same bombardment energy, the atomic escape energy increases linearly with the increase of the atomic mass of the incident ions. The atomic escape energy sputtered by the light incident ions is lower (about 10 eV), and the atomic escape energy sputtered by the heavy incident ions is larger (above 30 eV on average).

(4) The average escape energy of the sputtered atoms increases with the increase of incident ion energy. When the incident ion energy exceeds 1 keV, the average escape energy gradually approaches a constant value.

2.3.6.3 *Three- or four-pole sputtering*

DC diode sputtering relies on the secondary electrons emitted when the ion bombards the cathode to maintain the glow discharge. Therefore, it can only work at a higher pressure. Triode sputtering overcomes the disadvantages of diode sputtering. It attaches a hot cathode in the vacuum chamber, which emits electrons and generates a plasma with the anode. At the same time, the target is made to have a negative potential with respect to the plasma, and the target is bombarded with positive ions in the plasma to perform sputtering. In order to introduce hot electrons and stabilize the discharge, a fourth electrode, a stable electrode, is added to form quadrupole sputtering. Figure 2.48 shows the principle of quadrupole sputtering. The function of the stabilizing electrode is to stabilize the discharge.

The three-pole sputtering device can also work at a target voltage of 100 to hundreds of volts. Due to the low target voltage, the sputtering damage to the substrate is small, which is suitable for preparing semiconductor devices and integrated circuits. In addition, because tripolar sputtering no longer depends on the secondary electrons emitted by the cathode, the sputtering rate can be controlled by the emission current of the hot cathode, which improves the controllability of sputtering parameters and the repeatability of the process.

However, tripolar and quadrupole sputtering cannot suppress the bombardment of the substrate by the high-speed electrons generated by the target, so the temperature of the substrate is still relatively

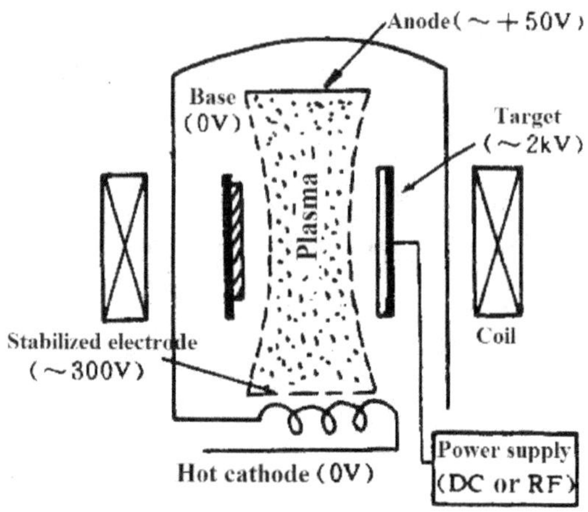

Figure 2.48: Principle of quadrupole sputtering.

high, and the film is contaminated. In addition, this sputtering method is not suitable for reactive sputtering; especially when oxygen is used as a reactive gas, the life of the filament will be significantly shortened.

2.3.6.4 *RF sputtering*

RF sputtering is a sputtering deposition method suitable for various metal and non-metal materials. Figure 2.49 shows the principle of RF sputtering. The reason why it can sputter-coat an insulating target is mainly because a negative bias voltage is established on the surface of the insulating target. This is because, for DC sputtering, if the target is an insulator, the target will be positively charged when positive ions bombard the surface of the target, so that the potential rises, and the accelerated electric field of the ions will decrease accordingly. Glow discharge and sputtering will stop. However, when the RF voltage is applied to the target, when the sputtering target is in the first half of the week, the mass of the electron is much smaller than the mass of the ion, so its mobility is very high. Therefore, the electrons can fly to the target surface in a short time, neutralizing the positive charge on the target surface, which can achieve the sputtering of the insulating material. Another function of electrons flying to the target surface is to quickly accumulate a large number of electrons on the target surface, causing its surface to show a negative potential due to space charges, which also causes the ions to bombard

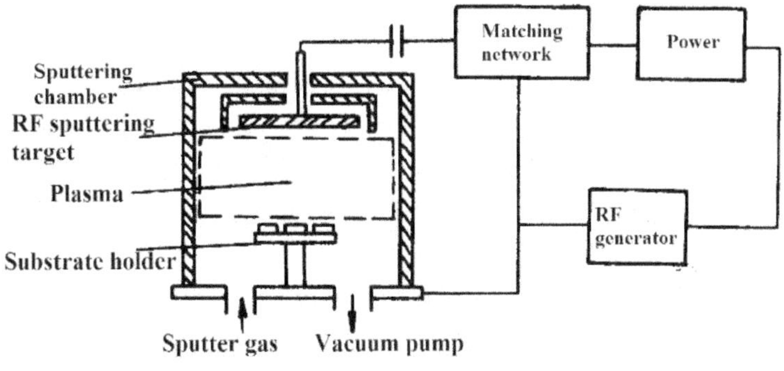

Figure 2.49: Schematic diagram of an RF sputtering device.

the target material in the positive half-cycle of the RF voltage. Thus, sputtering can be generated in both positive and negative half-cycles. RF sputtering can deposit almost all materials, including conductors, semiconductors, and insulators.

The mechanism of RF sputtering can be explained by RF glow discharge. In RF devices, the electrons in the plasma easily absorb energy in the radio frequency field and oscillate in the electric field. Therefore, the probability that the electrons collide with the working gas molecules and ionize them is very large, so that the breakdown voltage and the discharge voltage are significantly reduced, and their values are only about one tenth that of DC sputtering. Due to the existence of gas molecules under the sputtering conditions, the probability of the electrons colliding with the gas molecules during the oscillation process increases, and the direction of its movement also changes from simple harmonic motion to random chaotic motion. Because electrons can continuously absorb energy from the electric field, there is enough energy to ionize gas molecules in constant collisions. Even when the electric field is weak, the electrons can accumulate enough energy to ionize gas molecules, so compared to DC sputtering, RF sputtering can sustain discharge at lower voltages.

RF sputtering does not require subpolar electrons to sustain the discharge. However, when the ion energy is as high as thousands of electron volts, the number of secondary electrons emitted on the insulated target is also quite large, and because the target has a high negative potential, the electrons are accelerated through the glow dark area, which will become a high-energy electron bombarding the substrate, causing the substrate to heat up, charge, and damage the quality of the coating. To this end, the substrate must be placed in a position that is not directly bombarded by secondary electrons, or make sure that the electrons are deviated from the substrate by using a magnetic field.

2.3.6.5 *Magnetron sputtering*

The sputtering deposition method has two disadvantages: first, the deposition rate of the thin film deposited by the sputtering method is low; and second, the working pressure required for sputtering is high,

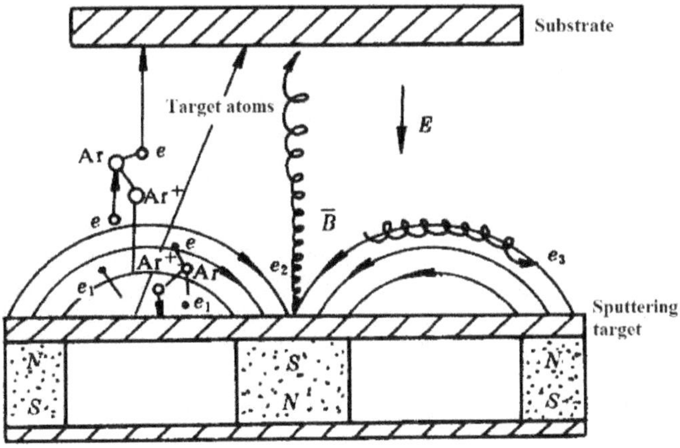

Figure 2.50: Working principle of magnetron sputtering.

thereby increasing the possibility of gas molecules contaminating the film. In order to increase the deposition speed and reduce the working gas pressure, magnetron sputtering technology has been developed.

The working principle of magnetron sputtering can be illustrated by Figure 2.50. Let electrons with velocity v be subjected to the Lorentz force in an electric field E and a magnetic field with magnetic induction intensity B

$$F = -q(E + v \times B). \tag{2.56}$$

Under the action of the electric field E, the electron e collides with the argon atom during its flight to the substrate, causing it to ionize Ar^+ and a new electron, e. The electron flies to the substrate, and Ar^+ accelerates to the cathode target under the action of the electric field and bombards the target surface with high energy, thus sputtering the target material. In the sputtered particles, neutral target atoms or molecules are deposited on the substrate to form a thin film. Once the secondary electron e leaves the target surface, it is simultaneously subjected to the action of an electric field and a magnetic field. It can be approximated that the secondary electrons are only affected by the electric field in the dark area of the cathode, and are only affected by the magnetic field once they enter the negative glow area. Therefore, the secondary electrons emitted from the target surface are first accelerated by the electric field in the dark area of the cathode and fly to the negative glow area. The electrons entering the negative glow region have a certain speed and move perpendicular to the magnetic field lines. In this case, the electrons rotate around the magnetic field lines due to the Lorentz force. After the electron rotates half a circle, it re-enters the dark area of the cathode and is decelerated by the electric field. When electrons approach the target surface, the speed can drop to zero. After that, the electrons again fly off the target surface under the action of the electric field and start a new motion cycle. The electrons repeat in this way, and they drift in the direction of $E \times B$, which is referred to as the $E \times B$ drift, as shown in Figure 2.51.

Under the control of a ring-shaped magnetic field, the secondary electron has a long path of movement and is confined to the plasma region near the target surface. In this region, a large amount of Ar^+ ionization ionizes and bombards the target, causing magnetron sputtering. The deposition rate of the radiation increases. As the number of collisions increases, the energy of the electron e is gradually exhausted, and it gradually drifts away from the target surface, and is finally deposited on the substrate under the action of the electric field. Since the energy of such electrons is very low, the energy transferred to the substrate is also very small, so the temperature rise of the substrate is not great due to the action of the electrons.

It can be seen that the basic principle of magnetron sputtering is to change the direction of electron movement with a magnetic field,

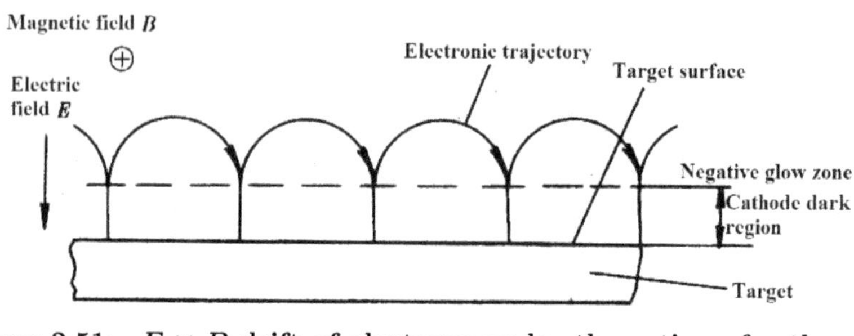

Figure 2.51: $E \times B$ drift of electrons under the action of orthogonal electromagnetic fields.

and constrain and extend the electron trajectory, thereby improving the ionization probability of the electron to the working gas and effectively using the energy of the electron. As a result, the sputtering caused by the bombardment of the target by the positive ions is made more effective; at the same time, the electrons that are thin by the orthogonal electromagnetic field beam can only be deposited on the substrate when their energy is depleted. That is why magnetron sputtering has the two characteristics of "low temperature" and "high speed".

Figure 2.52 shows the three main types of magnetron sputtering sources. Figures 2.52 (a) and 2.52(c) in Figure 2.52 represent columnar magnetron sputtering sources, which are suitable for

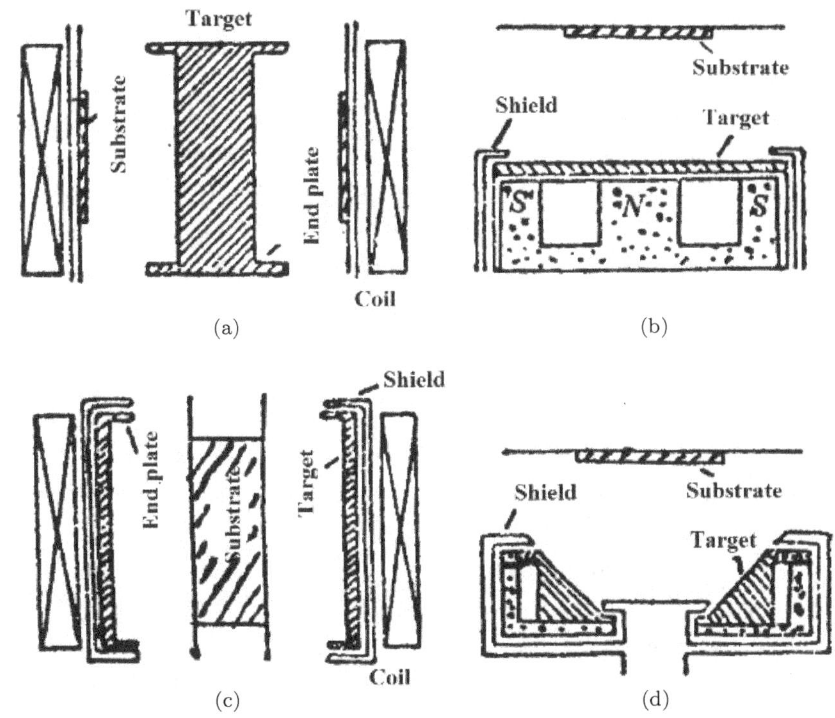

Figure 2.52: Schematic diagrams of several different magnetron sputtering sources: (a) coaxial cylindrical magnetron sputtering source; (b) cylindrical hollow magnetron sputtering source; (c) planar magnetron sputtering source; (d) S-gun sputtering source.

making large-area sputter coatings; Figure 2.52(b) represents planar magnetron sputtering sources, which can be made into small targets, which are suitable for the sputtering of precious metal targets. Figure 2.52(d) represents a sputtering gun (S gun). The structure of the S gun is relatively complicated. It not only has the common working principle of magnetron sputtering and the characteristics of "low temperature" and "high speed", but also shows high target utilization, uniform film thickness distribution, high target power density, and easy replacement of target materials due to its special target shape and cooling method.

Magnetron sputtering not only can obtain a high sputtering rate, but also can avoid secondary electron bombardment and keep the substrate close to the "cold state" when sputtering metals, which is of great significance for the use of single crystal and plastic substrate. The magnetron sputtering power source can work for DC or RF discharge, so it can be used to prepare various materials.

With the demand of industry and the development of surface technology, new magnetron sputtering methods such as high-speed sputtering and self-sputtering have become a new trend in the field of magnetron sputtering. High-speed sputtering can obtain a high rate of deposition of about several μm/min, which can shorten the time of sputtering coating and improve the efficiency of industrial production. It is possible to replace the current environmental pollution from the electroplating process. When the sputtering rate is so high that it can maintain the discharge even when there is no inert gas at all, the steam of the material that separated sputtered can maintain the discharge. This kind of magnetron sputtering is called self-sputtering. The ionization of sputtering materials and the reduction or even elimination of inert gas will obviously affect the mechanism of film formation and strengthen the chemical reactions of alloying and compound formation in the process of film deposition. It is possible to prepare new thin film materials and develop new sputtering techniques, such as self-sputtering of thin films at the bottom of deep holes. High-speed sputtering and self-sputtering are characterized by high target power density $Wt = \text{Pd/S} > 50\,\text{Wcm}^{-2}$ (Pd is the magnetron target power, S is the target surface area).

High-speed sputtering can only maintain high-speed sputtering under special circumstances, such as high enough target source density, high enough target material yield and sputtering gas pressure, and to obtain the maximum gas ionization rate. The cooling of the sputtering target is the biggest problem limiting the deposition of high-speed films. An inherent property of high-rate magnetron sputtering is the generation of a large number of sputtering particles to obtain a high film deposition rate. A high deposition rate means that a high flow of particles flies to the substrate, resulting in a large amount of particle energy being transferred to the growing film during the deposition process, resulting in a significant increase in deposition temperature. Since about 70% of the sputtering ion energy needs to be removed from the cathode cooling water, the maximum sputtering rate of the film will be limited by the sputtering target cooling. Cooling not only depends on adequate cooling water circulation, but also requires good thermal conductivity of the target material and the target thickness of the film. At the same time, the typical target material utilization rate in high-rate magnetron sputtering is only 20–30%, so improving the target material utilization rate is also a problem to be solved.

2.3.6.6 *Opposite target sputtering*

For Fe, Co, Ni, Fe_2O_3, permalloy, and other magnetic materials, low-temperature and high-speed sputter coating and other special requirements are needed. This is because, when these magnetic thin films are prepared by magnetron sputtering, the magnetic resistance of the target is very low, and the magnetic field almost completely passes through the target. Therefore, it's impossible to form a strong magnetic field parallel to the target surface, and it will also increase the temperature of the substrate greatly.

The opposite target sputtering method can be used to achieve low-temperature and high-speed sputter coating on ferromagnetic targets. Figure 2.53 shows a schematic diagram of the sputtering of the opposite target. At this time, the two targets are placed opposite to each other. The applied magnetic field is perpendicular to the target surface, and the magnetic field and the electric field are parallel. The anode is placed in a position perpendicular to the

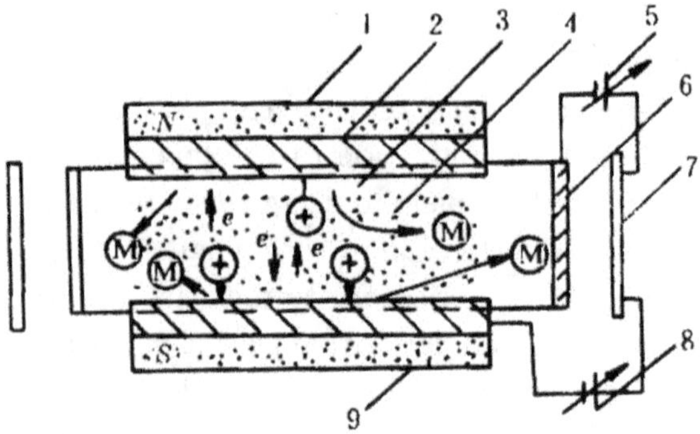

Figure 2.53: Schematic diagram of opposite target sputtering. 1. magnetic field N pole; 2. opposite target cathode; 3. dark cathode area; 4. plasma area; 5. substrate bias power supply; 6. substrate; 7. anode (vacuum chamber); 8. target power supply; 9. magnetic field S pole.

target surface and plays the role of confining the plasma together with the magnetic field. After the secondary electrons fly out of the target surface, they are accelerated by the electric field in the cathode potential drop region of the vertical target. Electrons are subjected to a magnetic field during their movement toward the anode. However, due to the high negative bias applied to the two targets, part of the electrons move almost in a straight line. These electrons are decelerated in the cathode potential drop area of the opposite target, and then accelerated in the opposite direction. Together with the magnetic field, the secondary electrons generated by the target are effectively enclosed between the two targets, forming Columnar plasma. The electrons are reflected back and forth by the two electrodes, which greatly lengthens the path of electron movement, and increases the probability of collision ionization with argon, thereby greatly improving the degree of ionization of the gas between the two targets and increasing the density of argon ions. As a result, the deposition rate is increased.

In addition to being constrained by the magnetic field, the secondary electrons are also subject to strong electrostatic reflections; the plasma is tightly confined between the two target surfaces, thus avoiding the bombardment of the substrate by high-energy electrons, so the substrate temperature does not rise. Therefore, the opposite target sputtering also has the characteristics of a high sputtering rate and a low substrate temperature. This sputtering method can be used to deposit magnetic thin films.

2.3.6.7 *Reactive sputtering*

Reactive sputtering is a commonly used method in the preparation of dielectric thin films; that is, during the sputtering coating, certain reactive gases are introduced to change or control the deposition characteristics to obtain new material films different from the target material. For example, reactive sputtering is performed in O_2 to obtain oxides; reactive sputtering is performed in N_2 or NH_3 to obtain nitrides; nitrogen oxides are obtained in a mixed gas of O_2+N_2.

The necessary condition for the reaction between the reactants is that the reactant molecules must have high enough energy to overcome the intermolecular barriers. The relationship between the potential barrier ε and energy is

$$E_{\mathrm{a}} = N_{\mathrm{A}}\varepsilon, \tag{2.57}$$

where E_a is the activation energy of the reaction and N_A is the Avogadro constant. Like evaporation, the reaction process basically takes place on the substrate surface, and the gas phase reaction is almost negligible. However, the reaction of the target surface during sputtering cannot be ignored. Because the target surface atoms bombarded by ions become active, the target surface temperature will increase to a certain extent. This will greatly increase the reaction speed of the target surface. Therefore, at this time, there are two processes of sputtering and reaction on the target surface. If the sputtering rate is higher than the rate of formation of the compound, it is in a sputtering state; otherwise, the sputtering may be stopped.

In order to ensure a sufficient reaction for forming a compound on a substrate, it is necessary to control the rate of metal atoms and reactive gas molecules incident on the substrate. Under a certain reaction pressure, the larger the sputtering power, the more

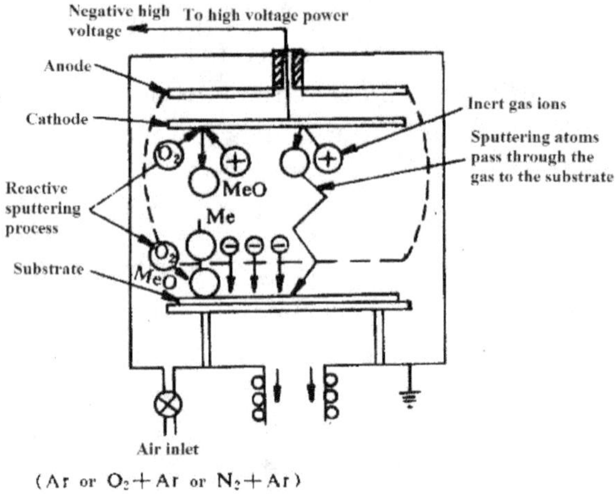

Figure 2.54: Schematic of the reactive sputtering process.

incomplete the reaction may be. By adjusting the sputtering power, or using a constant sputtering power, and adjusting the pressure of the reaction gas, a film of good quality can be obtained.

Figure 2.54 shows a schematic of the reactive sputtering process. Common reaction gases are O_2, N_2, CH_4, CO, H_2S, and so on. As mentioned above, depending on the pressure of the sputtering gas, the reaction process can occur on the substrate or on the cathode target. Generally, the pressure of reactive sputtering is low, and the gas phase reaction is not obvious. However, the current flowing in the plasma is very high, and it plays an important role in the process of decomposition, excitation, and ionization of the reactive gas molecules. Therefore, a particle stream composed of energy-carrying free radicals generated in reactive sputtering, accompanying the sputtered target atoms from the cathode target to the substrate, overcomes the activation energy related to the growth of the thin film and forms a compound on the substrate. This is the main mechanism of reactive sputtering.

In many cases, as long as the ratio of the reactive gas to the inert gas during sputtering can be simply changed, the properties of the thin film can be changed. A large number of experimental results show that almost all the formation of metal compounds occurs on the substrate. The higher the substrate temperature, the faster the film deposition rate.

2.3.6.8 *Ion beam sputtering*

Ion beam sputtering, also called ion beam deposition, is a new film-forming technology developed on the basis of ion beam technology. According to the function of the ion beam used for thin film deposition, it can be divided into two types: primary ion beam deposition and secondary ion beam deposition. The ion beam in the primary ion beam deposition is composed of the ions of the thin film component materials to be deposited, and the ion energy is relatively low. They are deposited into a film after reaching the substrate. Therefore, the primary ion beam deposition is also called low-energy ion beam deposition. The ion beam deposited by the secondary ion beam is composed of ions of an inert gas or a reactive gas. The energy of the ions is high. They hit the target composed of the material to be deposited, causing the target atoms to be sputtered and then deposited on the substrate. So the secondary ion beam deposition is also called ion beam sputtering.

Figure 2.55 shows the principle of ion beam sputtering. The inert gas ions (such as Ar^+, Xe^+, etc.) extracted from the ion source $1^{\#}$ cause their incident on the target to generate a sputtering effect, and the sputtered particles are deposited on the substrate to form a film. The $2^{\#}$ ion source mainly irradiates the formed thin film in order to control the properties of the deposited thin film in a wider range.

Ion beam sputtering (IBS) is a newer thin film preparation technology. Compared with plasma sputtering, it has the following advantages: it is formed in a high vacuum of 10^{-3} Pa and the non-plasma state, and the purity of the film is high; because the substrate is no longer an electrode, the substrate temperature is low; the film formation conditions and parameters can be independently controlled, and the repeatability is good, which is suitable for preparing multi-component dial films and multilayer films, and is also suitable for various powder materials, dielectric materials, metal materials, and compound materials. The main disadvantages of ion

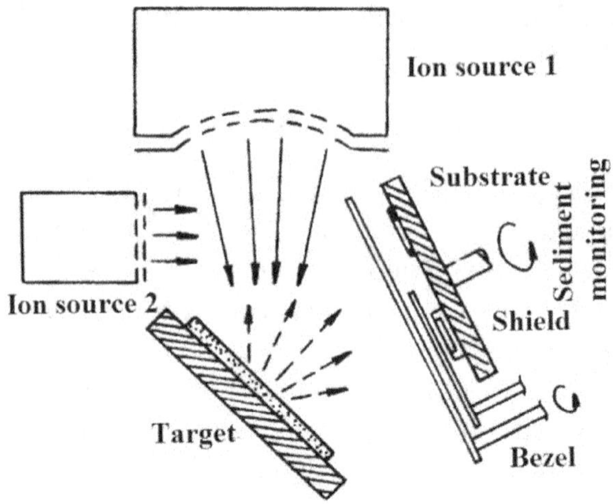

Figure 2.55: Schematic diagram of ion beam sputtering.

beam sputtering are that the device is complicated, expensive, and the film formation rate is low.

In ion beam sputtering, the ion source can be a single source or a multi-source. Although the phenomena involved in this film-forming technology are relatively complicated, by properly selecting the energy and type of the target and ions, it is relatively easy to make a variety of thin films of metals, oxides, and other compounds and it's particularly suitable for making multi-component metal oxide films. At present, this technology has been applied in the preparation of thin films of magnetic materials, superconducting materials and other electronic materials. In addition, due to the strong directivity of the ion beam, the energy of the ion current is easier to control, so it can also be used to study the sputtering process characteristics, such as the bombardment effect of high-energy ions, the sputtering angle distribution of single crystals, and the process of ion implantation and radiation damage.

2.3.6.9 *Other sputtering methods*

In order to improve the quality of the film layer, reduce the residual gas in the film layer (Ar, O_2, N_2, C, etc.), improve the coating rate of the hole bottom, and improve the characteristics of electromobility, collimation sputtering, low-pressure or even zero-pressure sputtering, high-vacuum sputtering, self-sputtering, RF–DC binding bias sputtering, ECR sputtering, and so on have been developed in recent years. Collimator sputtering, for example, is the addition of collimator electrodes between the two sputtering electrodes so that only low-sputtering ions flying perpendicular to the substrate can reach the substrate, effectively improving the coating rate at the bottom of the micropore with a large depth–diameter ratio. Low-pressure sputtering and high-vacuum sputtering are achieved at 10^{-2}–10^{-4} Pa or higher vacuum. Self-sputtering, transforms sputtered atoms into ions. Without the influence of argon atoms, sputtering atoms can fly in a straight line and deposit directly into deep holes. RF–DC combined bias sputtering uses both RF and DC power supply, which can effectively control the target current (DC power supply) and plasma density (RF power supply), respectively, facilitating the purification of the discharge environment and control the sputtering parameters, while ECR sputtering is sputtering or etching by ECR plasma. ECR plasma has high density and purity, and can produce high-quality films. The parameters of ECR source, target, and substrate can be adjusted independently, and the discharge, sputtering, and coating are very stable.

2.3.7 *Analysis of the thickness uniformity of sputter coating*

Film uniformity (including thickness uniformity and performance uniformity) is an important indicator for measuring the film quality and performance of coating devices. In order to improve the uniformity of film thickness, measures such as optimizing the target–base distance, changing the movement mode of the substrate, adding a baffle, and implementing film thickness monitoring can be undertaken. The problem of film thickness uniformity during diode sputtering and magnetron sputtering is mainly discussed below.

2.3.7.1 *Film thickness uniformity of diode sputtering*

The thickness distribution of sputter coating is related to various factors such as the angular distribution of the sputtered particles,

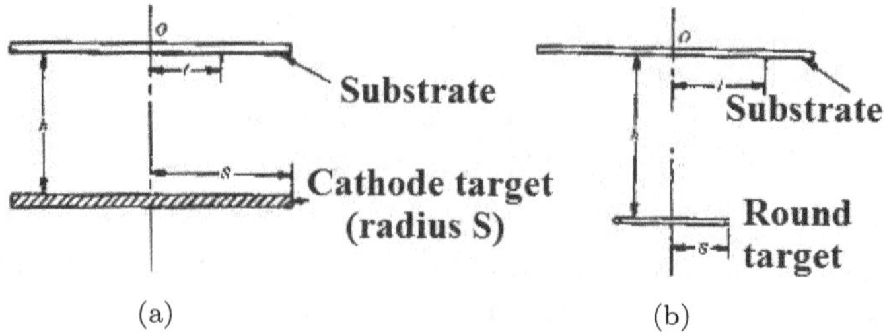

Figure 2.56: Parallel plate target and ring target.

the collision of the sputtered particles with gas molecules, and the configuration of the target. If the angular distribution of the sputtered particles is considered to approximately obey the cosine law, and the collision between the sputtered particles and the gas molecules is ignored, for a planar target as shown in Figure 2.56(a), the distribution of film thickness d/d_0 in the substrate is available. The following formula presents:

$$\frac{d}{d_0} = \frac{(1+R/h)^2}{2(R/h)^2}\left\{1 - \frac{1+(l/h)^2-(R/h)^2}{\sqrt{[1-(l/h)^2+(R/h)^2]^2+4(l/h)^2}}\right\} \tag{2.58}$$

where R is the radius of the target, h is the target–base distance, d_0 is the film thickness at the center of the anode, and d is the film thickness at a distance of l from the center of the anode.

When using a circular target as shown in Figure 2.56(b), the film thickness distribution is

$$\frac{d}{d_0} = \left[1+(R/h)^2\right]^2 \frac{1+(l/h)^2+(R/h)^2}{\left\{[1-(l/h)^2+(R/h)^2]^2+4(l/h)^2\right\}^{3/2}}. \tag{2.59}$$

According to the calculation result of the above equations, the film thickness distribution of different S/h ratios is shown in Figure 2.57.

2.3.7.2 *Film thickness uniformity of magnetron sputtering*

For magnetron sputtering coatings, uneven sputtering and uneven deposition of target atoms will be caused due to the uneven plasma

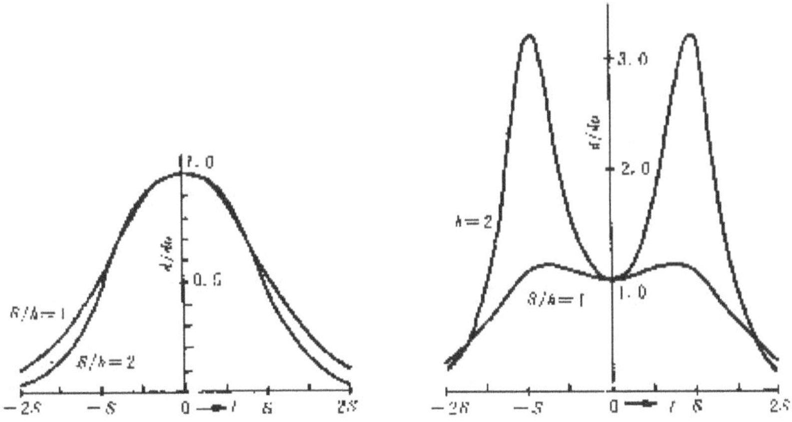

(a) (b)

Figure 2.57: Film thickness distribution curve of a parallel plate target (a) and a ring target (b).

density caused by the non-uniformity of the electromagnetic field (especially the magnetic field). Therefore, the non-uniformity of the film thickness is relatively large.

For the circular planar magnetron target shown in Figure 2.58, because of the etching of the target by magnetron sputtering, its film thickness distribution is

$$d = \frac{2Sh}{\pi\rho_0(R_2^2 - R_1^2)} \times \int_{R_1}^{R_2} \frac{(h^2 + A^2 + R^2)RdR}{[(h^2 + A^2 + R^2 + 2AR)(h^2 + A^2 + R^2 - 2AR)]^{3/2}}, \tag{2.60}$$

where d is the film thickness of P on the substrate; S is the sputtering rate of the magnetron target; ρ_0 is the target density; h is the target–base distance; R_1 and R_2 are the inner and outer radius of the etching area (generally taken as the inner and outer radius of the magnetic pole gap). When $A = 0$, the film thickness d_0 at the center of the substrate can be obtained. By comparing d and d_0, the relative change in film thickness can be obtained, and the optimal target–base distance h can be obtained. Generally $h \approx 2R_2$.

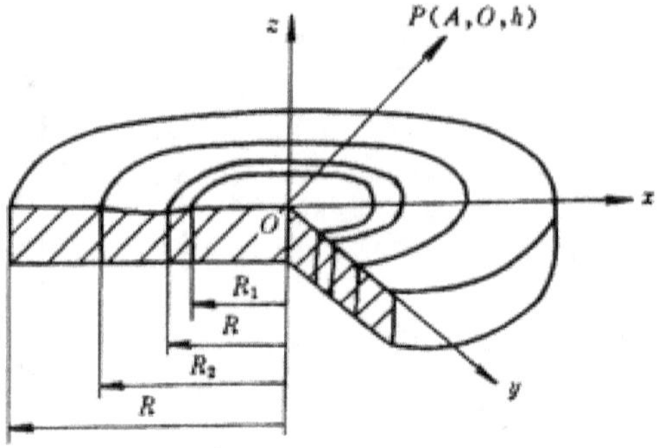

Figure 2.58: Geometric parameters of a circular planar magnetron target.

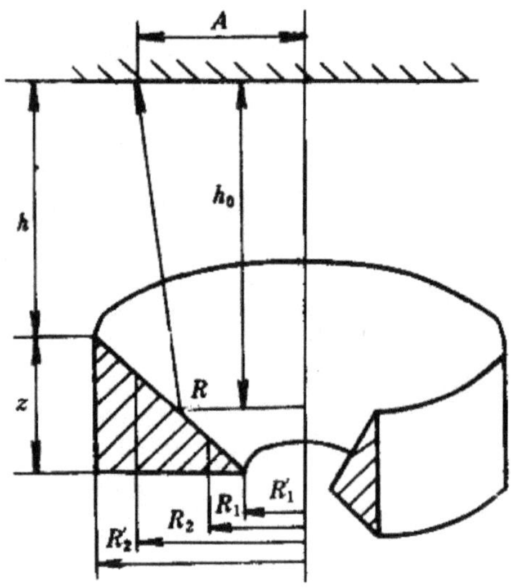

Figure 2.59: Geometric parameters of gun magnetron target.

For the gun magnetron target shown in Figure 2.59, the film thickness distribution should be

$$d = \frac{2S}{\pi \rho_0 (R_2^2 - R_1^2)} \times \int_{R_1}^{R_2} \frac{h_0^2 (h_0^2 + A^2 + R^2) R dR}{\left[(h_0^2 + A^2 + R^2 + 2AR)(h_0^2 + A^2 + R^2 - IAR)\right]^{3/2}}, \tag{2.61}$$

where S is the sputtering rate of the S-gun magnetron target; ρ_0 is the target density; R_1 and R_2 are the inner and outer radius of the etching area of the S-gun target; R is the radius of the filamentary ring source; h_0 is the vertical distance from the filamentary ring source to the substrate. Similar to the discussion of Equation (2.60), when $A = 0$, d_0 can be obtained. Comparing d and d_0, the relative change rate d/d_0 of the film thickness can be obtained, and then the target–base distance h can be obtained. Generally $h \approx 1.5 - 2.0R_2$.

2.3.8 *Comparison of sputter coating and vacuum evaporation coating*

The characteristics of sputtering coating and vacuum evaporation coating are compared as follows:

(1) The sputtering rate increases as the incident ion energy increases; when the ion energy increases to a certain level, the sputtering rate will decrease due to the ion implantation effect.
(2) The size of the sputtering rate is related to the mass of the incident particles.
(3) When the energy of the incident ions is lower than a certain threshold value (threshold value), sputtering will not occur.
(4) The energy of the sputtering atom is many times greater than that of the evaporated atom.
(5) When the energy of the incident ions is relatively low, the angular distribution of the sputtering atoms does not fully comply with the cosine distribution law, and the angular distribution is also related to the direction of the incident ions.
(6) Because the mass of the electrons is small, even when the target is bombarded with electrons with extremely high energy, no sputtering phenomenon occurs.

Sputtering coating and vacuum evaporation coating are two very important coating methods. Therefore, we introduce these two methods in detail in this chapter. The principles of the two methods are different, the coating processes are different, and the application ranges are not exactly the same. Table 2.16 summarizes and compares the two thin film preparation methods, sputter coating

Table 2.16: Comparison of the principles and characteristics of sputtered coating and vacuum evaporation coating.

Sputter coating	Vacuum evaporation coating
Generation process of deposition gas phase	
(1) Mechanism of ion bombardment and collision momentum transfer (2) High sputtering atom energy (2–30 eV) (3) A slightly lower sputtering rate (4) Directional motion of the sputtering atom (5) Guaranteed alloy composition, but some compounds have a tendency to decompose (6) Target purity varies with the type of material	(1) Atomic thermal evaporation mechanism (2) Low atomic kinetic energy (about 0.3 eV at 1200 K) (3) A higher evaporation rate (4) Evaporative atomic motion is directional (5) Element depletion or enrichment occurs during evaporation, and some compounds have a tendency to decompose (6) Higher purity of the evaporation source
Gas phase process	
(1) A slightly higher working pressure (2) The average free path of the atom is less than the distance between the target and the substrate, and multiple collisions are required before atom deposition	(1) A high-vacuum environment (2) Evaporated atoms are deposited directly on the substrate without collision
Thin film deposition process	
(1) Deposited atoms have higher energy (2) Some gas impurities will be introduced during the deposition process (3) High film adhesion (4) Large polycrystalline orientation	(1) Low deposited atomic energy (2) Low gas impurity content (3) Grain size larger than sputter deposited films (4) Facilitates film orientation

and evaporation coating, based on the principle of deposition. In actual work, the reader can choose a suitable film-making method according to different needs.

2.4 Ion Beam Coating

In order to obtain a good single crystal film and a high-purity film for electronic components, it is necessary to form the film in a high-vacuum environment as much as possible. Ion plating mainly includes the ion beam deposition and ion implantation film formation techniques. This technology was first proposed by Sandia in the United States in 1963. Its English name is ion plating, or IP for short. Ion plating is the application of gas discharge to achieve coating; that is, the gas or the substance to be plated is ionized in a vacuum chamber, and under the bombardment of gas ions or the substance to be plated, at the same time, the substance to be plated or its reaction product is plated on the substrate to form a film. This method combines ion plating glow discharge, plasma technology, and vacuum evaporation technology, which not only significantly improves the various properties of the deposited film, but also greatly expands the application range of the coating technology. Compared with evaporation and sputtering, the main advantages of ion plating are a strong film adhesion, a high deposition rate, a fast film formation speed, thicker films can be plated, and a wide range of materials can be plated, which is beneficial to the formation of compound films, etc.

2.4.1 *Principles and characteristics of ion plating*

2.4.1.1 *Basic principles of ion plating*

Figure 2.60 shows a schematic diagram of a DC diode ion plating device. After the vacuum chamber is evacuated to a high vacuum of 10^{-4} Pa, an inert gas (such as argon) is passed in, and a high-voltage power supply is connected, and then a low-pressure gas discharge plasma region is established between the evaporation source and the substrate. Since the substrate is under negative high pressure and surrounded by plasma, it is constantly bombarded by positive ions,

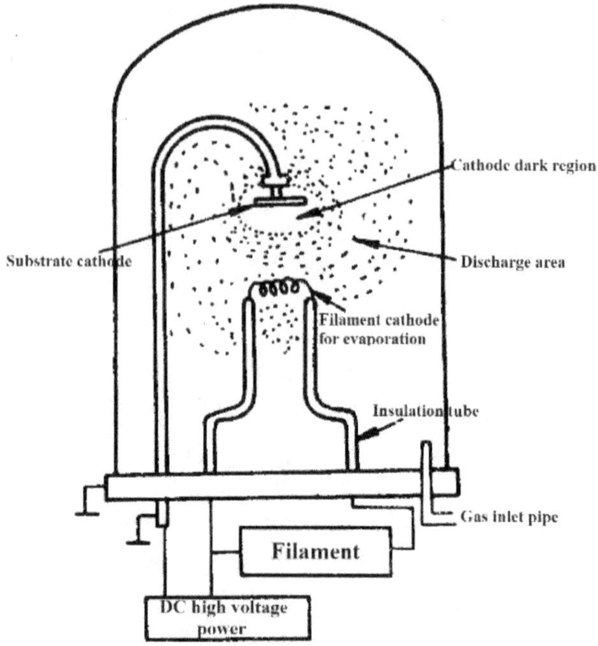

Figure 2.60: Schematic diagram of ion plating.

so the gas and pollutants on the surface of the substrate can be effectively removed, and the surface of the film layer can always be kept clean during the film formation process. At the same time, after the vaporization of the plating material, the evaporated particles enter the plasma area and collide with the positive ions and activated inert gas atoms and electrons in the plasma area.

Some of the evaporated particles are ionized into positive ions. Under the acceleration of a negative high-voltage electric field, these positive ions are deposited on the surface of the substrate to form a film. It can be seen that the energy required for nucleation and growth of the film layer in ion plating is not obtained by heating, but is excited by ion acceleration. The ionized plating material ions and gas ions are accelerated by the electric field and bombard the substrate or the coating surface with quite a high energy. This bombardment effect has been accompanied by the entire process of ion plating. However, because the substrate is a cathode and the evaporation source is an anode, there is also a sputtering effect of positive ions (Ar^+ or ionized evaporated ions) on the substrate while the film material is being deposited. Obviously, film deposition can occur only when the deposition effect exceeds the sputter emission peeling effect.

Since there may be two processes of coating and sputtering in ion plating, the film forming conditions of ion plating are discussed below.

If the glow discharge space contains only metal evaporation materials, and only the deposition of evaporated atoms is considered, the number of metal atoms n deposited on the surface of the unit per unit time can be expressed by the following formula:

$$n = \mu \frac{10^{-4} \rho N_A}{60M}, \tag{2.62}$$

where μ is the deposition rate of the deposited atoms on the substrate surface (μm/min); ρ is the density of the film (g/cm^3); M is the molar mass of the deposited material; $N_A = 6.029 \times 10^{23}$ is the Avogadro constant.

The above analysis has not considered the sputtering peeling effect. If the peeling effect is considered, the sputtering rate should be considered. If the bombarded substrate is a monovalent positive ion (such as Ar^+) and its ion current density is j, the number of ions that bombarded the substrate surface per unit time is n_j.

$$n_j = \frac{10^{-3} j}{1.6 \times 10^{-19}} = 0.63 \times 10^{16} j \ (\mathrm{cm}^2 \cdot \mathrm{S}), \tag{2.63}$$

where 1.6×10^{-19} is the charge amount (C) of the monovalent positive ion; j is the current density (mA/cm^2) formed by the incident ion.

Comparing the above two equations, it is found that in order to obtain a deposited film during the ion plating process, the deposition effect must be better than the sputtering peeling effect: that is, the film formation condition of the ion plating is $n > n_j$ (n_j should also include the number of ions generated by the additional gas).

2.4.1.2 *The main features of ion plating*

Compared with evaporation and sputtering, ion plating has the following characteristics:

(1) During the ion plating process, a large number of high-energy particles generated by glow discharge produce a cathode sputtering effect on the surface of the substrate. Not only can a "pseudo-expanded layer" of the film-based interface be formed at the initial stage of film formation, but also the substrate surface can be cleaned and cleaned, again so the film layer adhesion performance is good.
(2) During the ion plating process, the membrane ions and high-energy neutral atoms reach the substrate with higher energy and can diffuse and migrate on the substrate, so the density of the film layer is high (basically the same as the bulk material).
(3) During the ion plating process, after part of the film material atoms are ionized into positive ions, they will move along the electric line of the electric field. Wherever the electric line is distributed, the film ions can reach, which is known as good diffractive property. In addition, ion plating also has a wide range of materials that can be plated, and it is beneficial to the formation of compound thin films. Ion plating also has characteristics such as high deposition rates, fast film formation speeds, and the ability to plate thicker films.

2.4.2 *Ion bombardment and its role in coatings*

Many characteristics of ion plating different from vacuum evaporation and sputtering are related to the participation of ions and high-speed neutral particles in the coating process. Moreover, there is ion bombardment throughout the entire process of ion plating.

2.4.2.1 *Ionization rate*

The ionization rate is an example of the percentage of the number of atoms being ionized to the total number of evaporated atoms. It is an important index to measure the characteristics of ion plating, especially in reactive ion plating, because it is the main parameter to measure the degree of activation. The degree of ionization of the vaporized atoms and the reaction gas can directly affect the various properties of the film.

(1) Neutral particle energy W_ν
W_ν mainly depends on the level of evaporation temperature, and its value is

$$W_v = n_v E_v \tag{2.64}$$

where n_ν is the number of particles deposited on a unit area per unit time; E_ν is the kinetic energy of evaporated particles, $E_\nu = 3kT_\nu/2$, k is the Boltzmann constant, and T_ν is the temperature of the evaporated substance.

(2) Ion energy W_i
W_i is mainly determined by the cathode acceleration voltage,

$$W_i = N_i E_i \tag{2.65}$$

where N_i is the number of ions bombarded per unit area per time; E_i is the average energy of the ions, $\mathrm{E}_i \approx \mathrm{eV}_i$, V_i is the average acceleration voltage of the deposited ions.

(3) Active energy coefficient ε
ε can be roughly calculated from the following formula:

$$\varepsilon = (W_i + W_v)/W_v = (n_i E_i + n_v E_v)/n_v E_v. \tag{2.66}$$

when $n_\nu E_v \ll N_i V_i$, we can obtain

$$\varepsilon = \frac{n_i E_i}{n_v E_v} = \frac{eV_i}{3kT_v/2}\left(\frac{n_i}{n_v}\right) = C\frac{V_i}{T_v}\left(\frac{n_i}{n_v}\right), \tag{2.67}$$

where n_i/n_ν is the ionization rate; C is an adjustable parameter.

As can be seen from the above equations, due to the presence of the substrate acceleration voltage V_i during the ion plating process, even the ionization rate that is very low will affect the energy activity

Table 2.17: Surface energy activity coefficient of different coating processes.

Coating processes	Energy activity coefficient ε	Parameters	
Vacuum evaporation	1	Energy of evaporation particles $E_\nu \approx 0.2\,\text{eV}$	
Spattering	5–10	Energy of sputtered particles $E_S \approx 1\text{–}10\,\text{eV}$	
		Ionization rate n_i/n_ν	Average acceleration voltage U
Ion plating	1.2	10^{-3}	
	3.5	10^{-2}–10^{-4}	50–5000 V
	25	10^{-1}–10^{-3}	50–5000 V
	250	10^{-1}–10^{-2}	500–5000 V
	2500	10^{-1}–10^{-2}	500–5000 V

coefficient of ion plating. The energy of bombarding ions in ion plating depends on the substrate acceleration voltage. The neutral atoms produced by sputtering also have a certain energy distribution, and their average energy is about several electron volts. In ordinary electron beam evaporation, if the evaporation temperature is 2000 K, the average energy of the evaporated atoms is 0.2 eV. The values of the energy activity coefficients ε achieved by various coating methods are shown in Table 2.17. From this table, it can be seen that in ion plating, the value can be increased by 2–3 orders of magnitude by changing V_i and N_i/N_ν. If the average acceleration voltage of ions $U_i = 500\text{V}$ and the ionization rate n_i/n_v is 3×10^{-3}, the energy activity coefficient of ion plating is the same as that of sputtering. Therefore, the ionization rate is very important in the ion plating process. Figure 2.61 is the relationship between the energy activity coefficient and the ionization rates n_i/n_v and U_i at a typical evaporation temperature $T_\nu = 1800\,\text{K}$. It can be seen from Figure 2.61 that the relationship between the energy activity coefficient and the acceleration voltage is largely limited by the ionization rate.

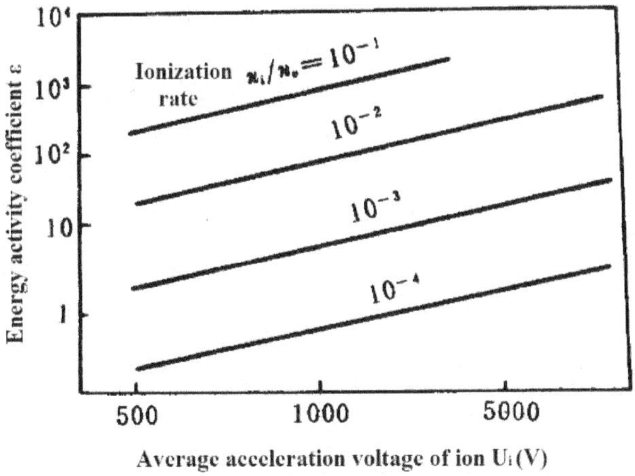

Figure 2.61: The relationship between the energy activity coefficient ε and the ionization rate n_i/n_ν and the average acceleration voltage U_i of the ion ($T_\nu = 1800\,\mathrm{K}$).

2.4.2.2 *Sputter cleaning*

The effects of ion bombardment on the surface of the substrate before the film began to deposit mainly include:

(1) Sputter cleaning effect. It can effectively remove impurities such as gases and oxides adsorbed on the substrate surface.
(2) Defects and dislocations are generated. When the energy delivered by the incident ions to the target atoms exceeds the minimum energy at which the target atoms are ionized, the lattice particles will migrate to the interstitial space of the lattice, thereby forming vacancies, interstitial atoms, and thermal excitation. Certain defects can also migrate and aggregate into dislocation networks.
(3) Destroy the crystal structure on the surface. If the defects produced by ion bombardment are stable, the crystal structure on the surface will be destroyed and become amorphous.
(4) Gas incorporation. Low-energy ion bombardment can cause gases to be incorporated into surfaces and deposited films. Bombardment heating can release gas.

(5) The surface composition changes. Because the sputtering rate of each component in the system is different, the surface composition and the overall composition will be different, and the surface diffusion will have a significant effect on the composition.
(6) The surface morphology changes. After the surface is bombarded with ions, the surface morphology of both crystalline and amorphous substrates will change, increasing the surface roughness and changing the sputtering rate.
(7) The temperature rises, because most of the energy that bombards the ions is converted, into thermal energy.

2.4.3 *Effect of particle bombardment on film growth*

During ion plating, on the one hand, particles of the plating material are deposited on the substrate, and on the other hand, high-energy ions bombard the surface, causing some ions to be sputtered out. The former is faster than the latter, and the film increases. This special process of deposition and sputtering gives the film-based interface many characteristics.

First, on the basis of mixed sputtering and deposition, a "pseudo-diffused layer" was formed on the base surface of the film due to the continuous thickening of the evaporated particles. This is a kind of physical mixing phenomenon of substrate elements and evaporated film elements at the film-based interface: that is, a component transition layer with a certain thickness is formed at the interface between the substrate and the film, which can disperse the mismatch of substrate and film material in a wide thickness area. It is very beneficial for improving the adhesion strength of the film-substrate interface.

Second, the surface morphology of the ion bombardment is damaged, which may provide more nucleation sites than the unbroken surface, increase the nucleation density, and provide good nuclear growth conditions for the deposited particles. Higher nucleation density is also beneficial for reducing the gap between the substrate and the film layer, which can also improve the adhesion of the film substrate. The bombardment of ions on the film also affects the morphology and composition of the film. In addition, the internal stress is also significantly affected by ion bombardment. Internal stresses are generated by atoms that are not yet in the lowest energy state. On the one hand, the bombardment of particles forces a part of the atoms to leave an equilibrium position in a higher energy state, thereby causing an increase in internal stress. On the other hand, the self-heating effect produced by ion bombardment on the surface of the substrate is beneficial to the diffusion of atoms. Therefore, proper use of the thermal effects of bombardment or proper external heating can reduce internal stress, on the one hand, and also be beneficial for improving the crystalline properties of the thin film.

Generally speaking, the films produced by evaporation have tensile stress, the films deposited by sputtering have compressive stress, and the ion-plated films also have compressive stress.

2.4.4 *Types and characteristics of ion plating*

According to different gasification and ionization methods of membrane materials, different types of ion plating devices can be constructed. The gasification methods of the membrane include resistance heating, electron beam heating, plasma electron beam heating, high-frequency induction heating, and cathode arc discharge heating. Ionization and activation of gas molecules or atoms are of the following types: glow discharge type, electron beam type, thermionic type, plasma beam type, multi-arc type and high-vacuum arc discharge type, and various types of ion sources, as shown in Figure 2.62. Different evaporation sources and different ionization methods can have many different combinations.

Table 2.18 lists the types of ion plating currently used, their evaporation, ionization methods, gas pressures, and their advantages and disadvantages.

2.4.4.1 *DC diode ion plating*

The previous Figure 2.60 is a schematic diagram of the structure of DC diode type ion plating. It has the characteristic that ions are generated by a glow discharge between two electrodes, and

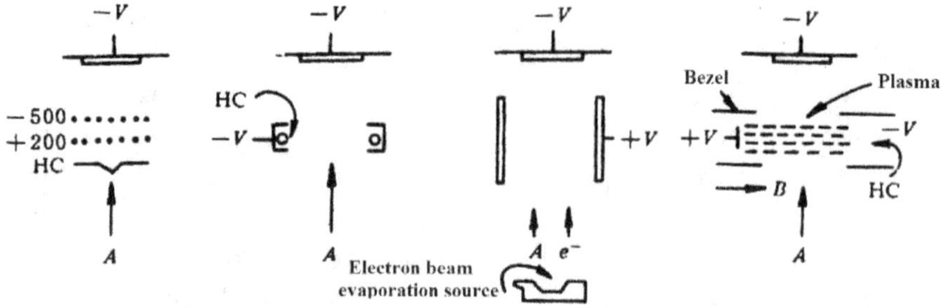

Figure 2.62: Various ionization methods during ion plating.

the ions are accelerated by a negative bias voltage applied on the substrate. According to the theory of gas discharge, the pressure of glow discharge can only be maintained at 6.67×10^{-1}–1Pa. Due to the high working pressure, resistance heating evaporation sources are mostly used for metals with a melting point below 1400°C, such as Au, Ag, Cu, and Cr. The discharge space charge density of DC diode type ion plating is low, and the cathode current density is only 0.25–0.4 mA/cm^2, so the ionization rate is low. The coated layer is uniform and it has good adhesion and strong particle diffraction. The equipment is relatively simple, and the coating process is easy to implement, and it can be modified by ordinary coating machines. The disadvantage is that the energy of the bombarded particles is large, which has a peeling effect on the formed film layer; at the same time, it will cause the substrate temperature to rise, and the surface of the film layer will be rough. In addition, the process parameters are difficult to control and the film purity is not high.

2.4.4.2 *Three-pole or multi-cathode ion plating*

These are two improved methods of the bipolar type, as shown in Figures 2.63 and 2.64. In DC discharge ion plating, introducing low-energy electrons into the ion region and increasing the average free path of the electrons in the plasma region can significantly improve the ionization effect of the deposited ions. In Figure 2.63, a hot cathode is used to emit a large number of thermionic electrons, and the ionized ion current collides across the evaporated ion current

Table 2.18: Comparison of various ion plating equipment.

No.	Type	Evaporation source	Filled gas	Degree of vacuum (Pa)	Ionic mode	Ionization acceleration method	Substrate temperature rise	Other features	Application
1	DC secondary type (DCIP)	Resistance heating or electron beam heating	Ar. But can also be filled with a small amount of reaction gas	6.67×10^{-1}–1	The substrate to be plated is a cathode, and a high-voltage DC glow discharge is used.	Acceleration is performed at voltages of hundreds to thousands of volts. Ionization and ion acceleration proceed together.	Big	Diffraction is good. Good adhesion. Substrate temperature is easy to rise. Film structure and morphology are poor. If using electron beam heating, differential pressure plate must be used.	Corrosion-resistant lubrication mechanical products.
2	Multi-cathode type	Resistance heating or electron beam heating	Vacuum inert gas or reaction gas	10^{-4}–1	Relying on hot electrons, cathode-emitted electrons, and glow discharge	0 to several thousand volts of acceleration voltage. Ionization and ion acceleration can be performed independently.	Small. Sometimes need to heat the substrate.	Low-energy electrons have high ionization efficiency, and the film quality can be controlled.	Precision machinery, electronics, devices, decorations.

(*Continued*)

Table 2.18: (*Continued*)

No.	Type	Evaporation source	Filled gas	Degree of vacuum (Pa)	Ionic mode	Ionization acceleration method	Substrate temperature rise	Other features	Application
3	Active reaction evaporation (ARE)	Electron beam heating	Reaction gases O_2N_2, C_2H_2, CH_4,etc.	10^{-4}–10^{-1}	Rely on the low-pressure plasma glow discharge between the forward-biased probe and the electron beam. Secondary electron.	No acceleration voltage. There are also AREs that apply 0 to thousands of volts of acceleration voltage to the substrate.	Small. The substrate needs to be heated.	High evaporation efficiency. It can obtain thin films such as Al_2O_3, TiN, TiC.	Mechanical products, electronic devices, decorations
4	Hollow cathode ion plating (HCD)	Plasma electron beam	Ar. and other reactive gases	10^{-2}–1	Electron beam collision using a low voltage and a high current.	0 to several hundred volts of acceleration voltage. Ionization and ion acceleration are performed independently.	Small. The substrate needs to be heated.	High ionization rate, large electron beam shift. It can be coated with metal film, dielectric film, compound film.	Decorative coating, wear-resistant coating, mechanical products.
5	RF ion plating (RFIP)	Resistance heating or electron beam heating	Vacuum, Ar. and other inert gases, reaction gases, O_2, N_2, C_2H_2, CH_4, etc.	10^{-3}–10^{-1}	RF Plasma Discharge (13.5 MHz).	0 to several thousand volts of acceleration voltage. Ionization and ion acceleration are performed independently.	Small	Less impure gas, good film formation, suitable for compound film plating. Difficult to match.	Optics, semiconductors, devices, decorations, automotive parts.

Table 2.18: *(Continued)*

No.	Type	Evaporation source	Filled gas	Degree of vacuum (Pa)	Ionic mode	Ionization acceleration method	Substrate temperature rise	Other features	Application
6	Enhanced ARE	Electron beam heating	Ar. and other inert gas reaction gases, O_2, N_2, CH_4, etc.	10^{-2}–10^{-1}	In addition to the primary and secondary electrons that attract the electron beam, the probe enhances the low-energy electrons emitted by the electrode to promote ionization.	No acceleration voltage. There are also enhanced ARE that apply 0 to thousands of volts of acceleration voltage to the substrate.	Small. The substrate needs to be heated.	Easy to dissociate, the required power and discharge power of the substrate can be independently adjusted, and the film quality and thickness can be easily controlled.	Mechanical products, electronic devices, decorations, electronic devices.
7	Low-pressure plasma ion plating (LPPD)	Electron beam heating	Reactive gas	10^{-2}–10^{-1}	Plasma.	DC or AC, 50V	Small. The substrate needs to be heated.	Simple structure. It can obtain thin films such as Al_2O_3, TiN, TiC.	Mechanical products, electronic devices, decorations
8	Electric field evapo-ration	Electron beam heating	—	10^{-4}–10^{-2}	Metal plasma formed by the electron beam is used.	Accelerated voltage of hundreds to thousands of volts. Ionization and acceleration are linked.	Small. The substrate needs to be heated.	Vacuum evaporation with electric field. Good coating quality.	Electronic devices, audio devices.
9	Induction heat-ing ion plating	High-frequency induction heating	Inert gas. Reaction gas	10^{-4}–10^{-2}	Inductive magnetic leakage.	DC1–5 kV	Small.	Ability to obtain compound coatings.	Mechanical products, decorations, electronic devices.

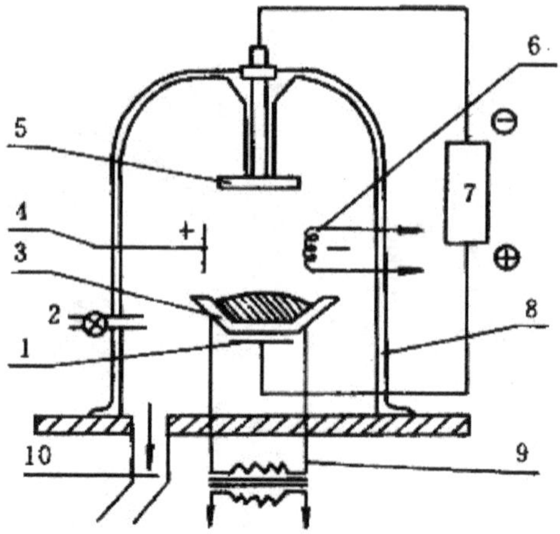

Figure 2.63: Three-pole ion plating. 1. Anode, 2. air inlet, 3. evaporation source, 4. electron collector, 5. substrate, 6. electron launch gun, 7. DC power supply, 8. vacuum chamber, 9. evaporation power supply, 10. exhaust system.

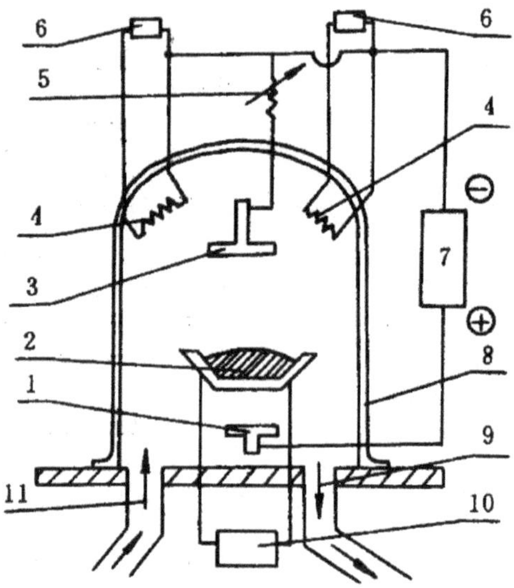

Figure 2.64: Multi-cathode ion plating. 1. Anode, 2. evaporation source, 3. Substrate, 4. cathode filament, 5. variable resistance, 6. filament power, 7. negative high voltage power supply, 8. vacuum chamber, 9. extraction system, 10. evaporation power supply, 11. air inlet.

under the action of collector 4. Compared with the bipolar type, the ionization rate of the tripolar type can be significantly improved, and the substrate current density can be increased by 10–20 times.

The multi-cathode type uses a substrate to be plated as the main cathode, and arranges several hot cathodes next to it. The electrons emitted by the hot cathode promote gas ionization.

In fact, the discharge is maintained under the voltage of the hot cathode and the anode. Because the discharge can be maintained under low pressure in this way, ion plating can be achieved under low pressure.

Because the voltage applied to the main cathode to maintain the glow discharge is not high and the multi-cathode filament is located around the substrate so that the cathode region is enlarged, we find that the diffraction property is improved and the bombardment of the substrate by high-energy ions is reduced. Therefore, the characteristics of severe sputtering, rough film formation, rapid temperature rise, and difficulty to control the bipolar ion plating are avoided. In addition, by adjusting the negative potential of the multi-cathode filament, the amount of ions received by the multi-cathode filament can be adjusted, thereby adjusting the number of ions reaching the substrate, which is also beneficial for controlling the temperature of the substrate. For insulator substrates, the accumulation of charge on the substrate can be eliminated by electron emission from the multi-cathode filament, allowing the process to proceed. Therefore, the multi-cathode type expands the applications of ion plating.

2.4.4.3 *Reactive ion plating*

In the ion plating process, a gas that can react with metal vapor, such as O_2, N_2, etc., can be introduced into the vacuum chamber instead of Ar or mixed in Ar gas, and the molecules and atoms of the metal vapor and the reactive gas are activated and ionized by various discharge methods, and the chemical reaction between them is promoted, so a compound film can be obtained on the surface of the substrate. This method is called activated reactive evaporation (ARE), and has a wide range of practical values.

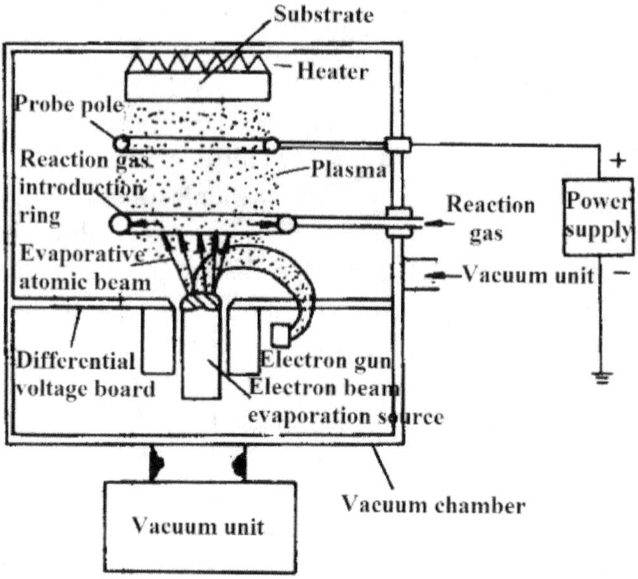

Figure 2.65: Schematic illustration of a typical ARE equipment.

A typical ARE equipment is shown in Figure 2.65. The evaporation source is usually an "e" gun. The vacuum structure is divided into two upper and lower chambers. The upper part is the evaporation chamber, and the lower part is the hot wire emission chamber of the electron beam source. There is no pressure difference hole between the two chambers. The film was allowed to evaporate. The use of such a gun can both heat and evaporate high-melting-point metals, and provide electrons for activating metal vapor particles, providing a good heat source for the preparation of high-melting-point metal compounds.

Different compound gases can be obtained by selecting different reaction gases. The main characteristics of active reactive ion plating are the low heating temperature of the substrate; the high deposition rate; the film can be formed on any substrate; by adjusting or changing the evaporation rate and the pressure of the reaction gas, it is very easy to prepare similar kinds of compounds with different proportions and different properties.

2.4.4.4 *Radio frequency ion plating*

This method was first proposed by Sanyo Mura of Japan in 1973, and its basic principle is shown in Figure 2.66. The ionic plating discharge is stable and can be coated under high vacuum. Because of the use of radio frequency excitation, the ionization rate of vaporized molecules can reach about 10%, and the working pressure is relatively low. Its main characteristics are that evaporation and acceleration can be controlled independently; high ionization rate and good film quality; easy for reactive ion coating, low substrate temperature, and easy to control. Its main disadvantage is that the working vacuum is high, so the coating diffraction is poor, and RF is harmful to the human body, and should be shielded. In addition to the above four, there are magnetron sputtering ion plating, vacuum arc ion plating, hollow cathode discharge ion plating, and so on. Ion plating

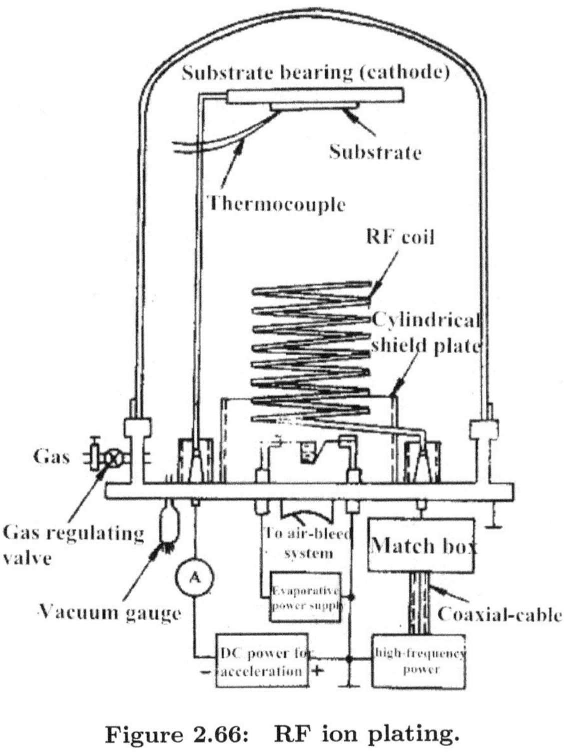

Figure 2.66: RF ion plating.

technology has been widely used in manufacturing sensitive, heat-resistant, corrosion-resistant, and decorative films.

2.5 Molecular Beam Epitaxy

2.5.1 *Basic concept of MBE*

Epitaxy is a new technique for preparing single-crystal thin films. It is a membrane-making method to grow a new single-crystal layer with a complete lattice structure along the original crystal axis of the substrate on a substrate that meets certain conditions and under appropriate conditions. This process is called epitaxy, and the new, long single-crystal layer is called epitaxy. The epitaxial layer is called homoepitaxy if it is identical in structure and properties to the substrate material, such as the epitaxial silicon layer on a silicon substrate. If the two are different in structure and nature, it is called heteroepitaxy, such as the sapphire growth epitaxy silicon layer. The typical epitaxy methods include vapor phase epitaxy (VPE), liquid phase epitaxy (LPE), and MBE.

The VPE method is also known as chemical vapor deposition (CVD). Because its reaction source is supplied from the outside, it is easy to control, and can be used for crystal growth of multiple elements (such as $GaAs_{1-x}P_x$, $In_xGa_{1-x}P$), and because it can make the lattice constants of the substrate and the epitaxial gradually converge, it is possible to grow the mixed crystal in a gradual change. LPE is a method in which the solute is put into the solvent to form a uniform solution at a certain temperature, and then the solution is slowly cooled through the saturation point (liquidus). This method can produce epitaxial layers with high purity and excellent crystallization, and can grow multilayer crystalline films continuously.

MBE is a new epitaxial membrane preparation method and a special vacuum coating process. It is a technique for ejecting a molecular beam of thin-film components directly onto the substrate surface under ultra-high vacuum conditions to form an epitaxial layer on it. The main advantage of MBE is the ability to grow extremely thin monocrystalline film layers and the ability to accurately control the film thickness and composition and doping. It is suitable for making microwave, photoelectric, and multilayer structure devices, which provides a powerful means for making integrated optics and very large-scale integrated circuits.

2.5.2 *Principle and device of MBE*

The MBE device is shown in Figure 2.67. It mainly consists of a studio, a molecular beam jet source, and a variety of monitors. In order to ensure the quality of the epitaxial layer and reduce defects, the pressure in the workshop should not be higher than 10^{-8}Pa, and the pressure for the generation of molecular beams should also be 10^{-6}Pa. Therefore, the studio should generally be vented by an oil-free ultra-high vacuum system with a high pumping rate. For more complex epitaxial films with growth components or for doping, multiple injection sources are required, as shown in Figure 2.68. There is a liquid oxygen shield around the injection source and a special auxiliary exhaust system behind the injection source to reduce

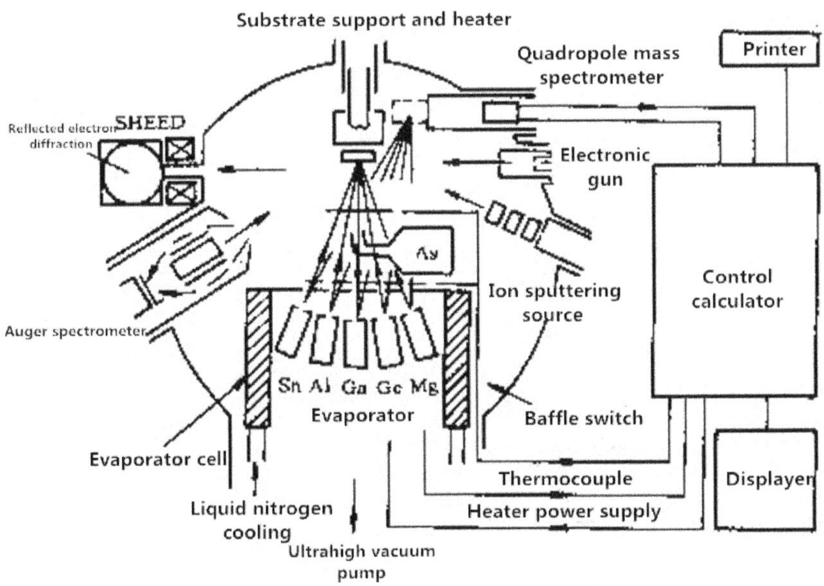

Figure 2.67: Schematic diagram of an MBE device.

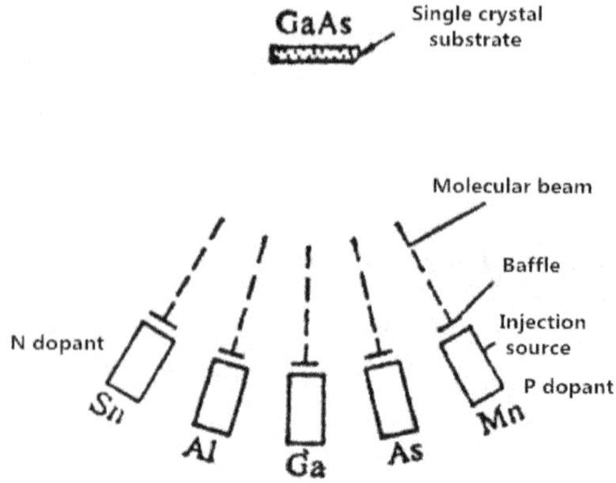

Figure 2.68: Schematic diagram of the MBE injection principle.

the pressure of the gas around the injection source. An Auger electron spectrometer and an electron diffractometer are used to monitor and study the growth process of epitaxial layer single crystals, evaluate the crystal quality, and analyze the components. A quadrupole mass spectrometer is used to detect molecular beam flow. The electron beam emitted by the electron gun is diffracted through the thin film and then received by the detector. As the number of deposited atomic layers increases, the diffraction intensity changes periodically, and the deposition thickness can be obtained by measuring the electron diffraction signal.

When the vapor molecules are spaced a few atoms apart from the surface of the substrate during the jet process, they are adsorbed to the substrate surface due to the surface tension and can migrate further along the surface, then release latent heat at appropriate locations to form crystal nuclei or graft to the lattice nodes. But it's also possible to get back into the gas phase because it's so energetic. Therefore, adsorption and desorption are in dynamic equilibrium at a certain temperature. The ratio of the number of molecules chemically adsorbed to the number of molecules incident on the surface is usually expressed by the adhesion coefficient. Because adsorption is usually an exothermic process, it will increase the substrate temperature and is not conducive to adsorption, resulting in the decline of the adhesion coefficient.

Substrate material is an important factor that affects the quality of the epitaxial layer. In addition to requiring the substrate material to be stable at the deposition temperature, not to undergo thermal decomposition, not to be eroded by the epitaxial material and its vapor, and to have a similar expansion coefficient to the epitaxial material, the most critical thing is requiring the lattice type and the lattice constant to be as close as possible to the epitaxial material. The crystal growth and orientation behavior of the epitaxial layer mainly depend on the matching of the lattice structure and atomic spacing between the epitaxial material and the substrate material. In general, a single epitaxial relationship can be maintained when the lattice mismatch reaches 7%, but when the lattice structure is different or the mismatch is very large ($\geq$10%), the deposition may show different epitaxial relationships, or the epitaxial single-crystal film may not be available at all. In addition, the ultra-high vacuum environment in the deposition process, the suitable substrate temperature, the heating temperature of each jet furnace, and the strength and type of molecular beam are also the conditions to control the epitaxial film growth.

Doping in MBE involves loading impurity elements into a jet source for doping during crystal growth. For example, the n-type doping used in GaAs is Sn, Ge and Si. P-type impurities include Mn, Mg, Be, Zn, etc. In MBE, as long as the substrate temperature is appropriately selected, the epitaxial layer with a growth rate of 0.1–1 nm/h and a doping concentration of 1013–1019/cm^3 can be obtained by controlling the strength of the molecular beam.

2.5.3 *The characteristics of MBE*

MBE is a method of depositing atoms one by one directly on the substrate, so it has the following characteristics compared with the usual CVD epitaxy and vacuum evaporation:

(1) Although MBE is also an evaporation process based on the molecular theory of gas, it does not take the evaporation temperature as the control parameter. Instead, it uses the quadrupole mass spectrometer, atomic absorption spectroscopy and other

modern analytical instruments in the system to accurately monitor the type and intensity of molecular beam, so as to strictly control the growth process and growth rate.

(2) MBE is a physical deposition process of ultra-high vacuum, which does not need to consider intermediate chemical reactions and is not affected by mass transfer, and instantaneous control of growth and interruption (switch baffle) is also carried out. Therefore, the composition and doping concentration of the film can be adjusted rapidly with the change of evaporation source.

(3) The substrate temperature of MBE is low, so the lattice mismatch effect introduced by thermal expansion at the interface and the influence of substrate impurities on self-doping diffusion of the epitaxial layer are reduced.

(4) MBE is a kinetic process in which the incident neutral particles (atoms or molecules) are stacked one by one on the substrate for growth, rather than a thermodynamic process, so it can grow thin films that are difficult to grow by normal thermal equilibrium growth methods.

(5) Another notable feature of MBE is its slow growth rate, about 1mm/h, which is equivalent to the growth of a single atomic layer per second. Therefore, it is beneficial to achieve accurate control of film thickness, structure and composition as well as the formation of heterojunctions. Therefore, MBE is particularly suitable for growing superlattice materials.

(6) In the technology for obtaining monocrystalline thin films, the substrate temperature of MBE is the lowest, which is conducive for reducing self-doping.

(7) MBE is conducted in an ultra-high vacuum environment, and the substrate and molecular beam source are far apart, so a variety of surface analysis instruments can be used to observe the composition, structure, and growth process of the growth surface in real time, which is conducive to scientific research.

In addition, MBE can effectively utilize the planar technology to produce Schottky barrier characteristics equal to or exceeding those of CVD.

Thus, MBE is particularly suitable for growing thin films or ultra-thin film epitaxial layers with complex profiles. MBE has been widely used in the research on solid-state microwave devices, photoelectric devices, multilayer periodic structure devices, and monolayer thin films.

2.6 Pulsed Laser Deposition Technology

2.6.1 *Overview*

Pulsed laser deposition (PLD) membrane fabrication technology utilizes a high-energy laser beam as a heat source. It is a new technique for bombarding materials to be vaporized and then evaporating a thin film on a substrate. The laser source can be excimer excited laser, CO_2 laser, Ar laser, neodymium glass laser, ruby laser, yttrium–aluminum garnet laser, or other high-power lasers, and is placed outside the vacuum chamber. The high-energy laser beam enters the vacuum chamber through the window, and is focused through a prism or concave mirror, after which it is illuminated to the steam. The material is heated (or ablated) and vaporized. The power density of the focused laser beam is very high, up to 10^6W/cm. In 1987, Bell Labs's Venkatesan and his collaborators first used PLD technology to grow high-quality YBCO films. PLD technology has received extensive attention. In the past five years, the PLD method has been used to prepare films. The number of compounds is approaching 200. So far, PLD has been used to deposit diamond-like films, high-temperature superconducting thin films, various nitride films, complex multi-component oxide thin laser observation window CCD films, ferroelectric films, nonlinear waveguide films, synthetic nanocrystalline measuring target tube point films, etc. Figure 2.69 shows the basic structure diagram of the pulsed laser deposition device. A laser that is generated from a laser passes through a mirror to change its direction and then through a convex lens to focus the laser beam onto the target surface. The focused laser produces an electric field large enough to remove electrons from the atom through a nonlinear process that falls within the range of light absorption depths. For a nanosecond

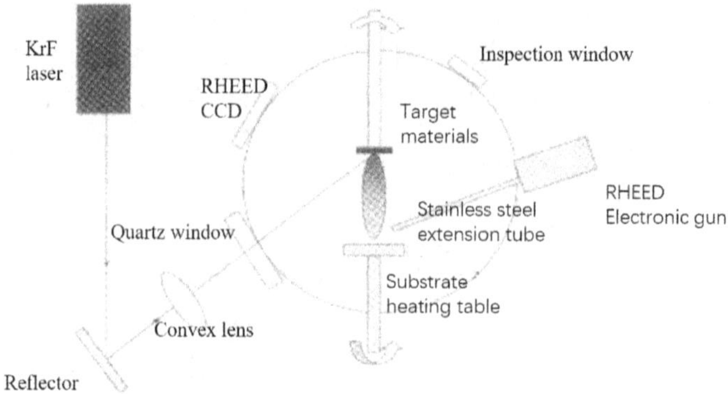

Figure 2.69: Basic structure diagram of a pulsed laser deposition device.

laser pulse, the process is used in a very short time, about 10 ps. For most of the material, the interaction of laser and target range, the depth of the light absorption is within about 10 nm. For $1\,\text{mm}^2$ of focused spot, there will be about 10^{15} outer electrons. These electrons oscillate in the electromagnetic field generated by the laser beam while colliding with nearby atoms or ions, so part of the lattice energy is transferred to the target surface. Once enough free electrons are generated, inverse duction radiation will become the main process. When the laser pulse is stopped, because of the coulomb repulsion and recoil effect on the surface of the target by high-energy ions, electrons, atoms, and molecules composed of plasma form steam into the target of the near adiabatic expansion and spread quickly, with the background gas generating a complex gas phase reaction. The expansion from the plasma of the visible light spectrum is the plume. Their temperature is close to 10,000 K or higher. The time required for these high-energy particles to get desorbed to deposit on the substrate is about a few μs, during which they are adsorbed on the substrate surface with a certain probability after reaching the substrate surface. Although the PLD film deposition process and the principle are simple, but it is still difficult to get high-quality film. This is because in the thin film deposition process, the control of thin film growth is by many parameters, such as the type of substrate, substrate temperature, atmospheric pressure

Table 2.19: The main laser parameters used in the PLD technique.

Laser type	Wavelength (nm)	Pulse energy (J)	Impulse frequency (Hz)	Pulse width (s)
CO_2 TEA laser	10600	7.0	10	$(2–3) \times 10^{-6}$
Nd:YAG laser	1064	1.0	20	$(7–9) \times 10^{-9}$
Second harmonic laser	532	0.5	20	$(5–7) \times 10^{-9}$
Third harmonic laser	355	0.24	20	$(4–6) \times 10^{-9}$
XeCl excimer laser	308	2.3	20	40×10^{-9}
ArF excimer laser	193	1	50	$(1–4) \times 10^{-9}$
KrF excimer laser	248	1	50	$(1–4) \times 10^{-8}$

of the laser energy density, background, target distance, deposition rate, annealing temperature and annealing pressure, etc., and the interaction between these parameters are also very complex. With the development of laser technology, the main parameters used in laser PLD technology are shown in Table 2.19.

2.6.2 *Characteristics of PLD*

Compared with other continuous physical vapor deposition methods, such as sputtering, molecular beam epitaxy, co-evaporation reaction, and so on, the most important characteristic of PLD is the discontinuity of particle supply, namely from the evaporation of the target material, supply is discontinuous, so the surface of the substrate in PLD is discontinuous supersaturation. At the same time, of because the interaction between laser and target is strong, in the pulsed laser deposition method, supersaturation is higher than that in the average continuous deposition method by about four orders of magnitude. Immediately after the laser is stopped, supersaturation decreases to zero. At the beginning, the supersaturation is high, leading to critical nucleation volume decreases, and increasing the nucleation density. Subsequently, the low supply of supersaturation and particle suspension contributes to surface diffusion of the adsorbed particles. The nucleation of high density and high particle mobility films obtained from two-dimensional layer epitaxial growth is very important, and PLD with supersaturation discontinuity can meet the above requirements simultaneously. However, the plume in the duration of the substrate is shorter, only about a few μs; therefore, the instantaneous deposition rate can be as high as 10^4 nm/s, which is two orders of magnitude higher than that of other thin film deposition technologies. This is a significant feature of PLD deposition films.

In the general process of thermal deposition, the evaporation rates of different elements from the source surface are different (Ba is easier to evaporate than Cu for YBCO, while Y is the most difficult to evaporate), which leads to a membrane component and the components of the target material deviation. But, unlike the general thermal deposition technology, PLD is a kind of nonthermal thin film deposition technique, namely the deposit to the substrate surface of particles is produced by the thermal process. This characteristic makes it possible to easily realize thin film deposition with the same components. Generally speaking, the "nonthermal" refers to all the particles (atoms, ions, or free radicals) physical and chemical processes, such as fracture and formation of chemical bonds. It has one or more energy degrees of freedom (translational degree of freedom, rotational degree of freedom, freedom degree of vibration or electricity). However, these energy degrees of freedom do not follow the Maxwell–Boltzmann statistical distribution, so their characteristics and laws cannot be described by the simple physical quantity of temperature. In experiments, two methods are usually used to obtain non-thermal conditions: discharge or condensed matter and a strong light field are interacting with each other. PLD is using a pulsed laser to achieve the thermal conditions. Although the traditional deposition methods, such as CVD, can be applied through the thermal conditions of the plasma enhanced CVD (PECVD) technique and laser enhanced CVD (LECVD), the application of non-thermal conditions does not change the nature that the deposited particles are produced by a thermal process, and only uses the non-thermal conditions to adjust the particle. PLD deposition methods are in essence the opposite of traditional sedimentation methods. At the same time, in the non-thermal deposition technology, deposition to the thin film on the surface of the particle energy is about 0.1 eV, lower than the surface atoms bond energy by one or two orders of magnitude, so on the surface of the atomic diffusion the contribution of energy transfer between the films is essentially trivial. However, in the non-thermal deposition technique, the energy of incident particles as the binding energy of the material is basically the same order of magnitude or higher. When they adhere to the film on the surface, the surface of the adjacent lattice energy transfer can be used to change their surrounding environment. This means that the moment of interaction can strengthen the surface of the atomic diffusion. In this way, the energy required for crystal growth is mainly provided by the deposition of particles rather than by the heating of the film and substrate. So high-quality films can usually be deposited on the substrate at a lower temperature, which is also an important characteristic of the PLD deposition technology. Sankur *et al.* found that Ge heteroepitaxy could be achieved using PLD at 300°C. The temperature required to achieve the above process by using MBE is more than 700°C. Therefore, high-energy particles generated by non-thermal processes play an extremely important role in the epitaxial growth of the film.

As early as 1987, Venkatesan and his collaborators realized that the rate of laser interaction with the target made it impossible to separate from the irradiated target, thus "magically" replicating the chemical and crystallographic properties of the target onto a thin film. It can achieve thin film deposition with the same components. This is not only the main characteristic of PLD deposition films, but also the biggest advantage of PLD deposition films. In 2000, Willmott *et al.* reported on Si (001) substrate heteroepitaxy growth of mixed Nd and Cr of $Gd_6Sc_2Ga_6O_{12}$; the materials not only include six different elements, but also contain 160 atoms in crystal cells, which give full play to the advantages of the PLD with component deposition films.

PLD mainly has the following characteristics:

(1) Laser heating can reach extremely high temperature, evaporate any high-melting point material, and obtain a very high evaporation rate.

(2) Because of the non-contact heating, the laser can be installed outside the vacuum, which avoids the pollution from the evaporation source and simplifies the vacuum chamber. It is very suitable for preparing high-purity films under ultra-high vacuum.
(3) Laser beam heating can cause "scintillation evaporation" of some compounds or alloys, which is conducive for ensuring the chemical ratio of membrane components or preventing decomposition of materials to be evaporated; and because the material gasification time is short, there is not enough time to make all around the material reach the evaporation temperature, so laser evaporation is not prone to fractionation. Therefore, laser flash deposition is a good method for deposition of dielectric films, semiconductor films and inorganic compound films.
(4) The background air pressure has a wide range (from high vacuum to several kPa), and can achieve reactive deposition in the atmosphere, forming high-quality multi-component films;
(5) High-energy atoms or ions in the plume contribute to chemical reactions in the gas phase or substrate.

PLD deposition films also have their own disadvantages, such as the formation of small particles on the surface of the film and the uneven thickness of the film, which limits its application to some extent. Methods such as baffle method, off-axis deposition and velocity filter are usually used to reduce the particles on the surface of the film.

For the chemical structure and components of complex transition metal oxides, such as high-temperature superconducting copper oxide and pang magnetoresistance materials, the material is closely related to the chemical composition, and the properties of any component of the tiny deviation will lead to a significant change in material properties. Therefore, the growth of the thin film must be controlled while any properties of the material can be changed flexibly and in random, so that its component and target material compositions are exactly the same. PLD is just the kind of method that can realize the goal of research on thin film deposition technology.

2.6.3 *Improvement of PLD*

There are particles of different sizes deposited on the surface of the film by pulsed laser deposition. However, commercial application requires less than one large particle per cm^2, which is one of the main reasons why this technology is difficult to commercialize. In order to overcome these serious shortcomings, a lot of research and improvement has been made on the mechanism of film formation and experimental methods. The optimization of experimental parameters and the use of new ultra-short picosecond or femtosecond lasers are the key.

The optimization of experimental parameters is the key to the preparation of high-quality films. The optimum configuration of the main parameters, such as laser wavelength, laser energy intensity, pulse repetition frequency, substrate temperature, atmosphere type, air pressure, ion beam auxiliary voltage and current, target–base distance, etc., is the prerequisite for the preparation of ideal films. In addition, whether the target material matches the substrate lattice, the surface polishing and cleaning of the substrate all affect the strength of the film–substrate bonding force and the smoothness of the film surface. A particle beam discharge assist can screen the particle orientation deposited to the substrate and increase the smoothness of the film surface. A laser energy intensity of suitable size, the target–base distance, the substrate rotation method, or a chopper that can filter slow and large-mass particles can all play the role of a smooth surface. In order to prepare large-area uniform films by the PLD method, Keyi Caiyomh used laser circular scanning and laser composite scanning to deposit films so that the laser beam can rotate according to a certain trajectory. The rotating laser beam is shot into the vacuum system to peel off the target, and the plasma cloud acts on the substrate rotating at a certain angular speed to form a film. After parameter optimization, large-area films with uniformity greater than 98% and diameter greater than 50mm can be obtained.

It is also instructive to optimize experimental parameters through computer simulation methods; the main simulation methods are numerical analysis and Monte Carlo simulation. Among them, the Monte Carlo method was first adopted by Bird to calculate the single gas relaxation problem. The essence of this is to use the appropriate number of simulations instead of a large number of real gas molecules, i.e. using computer simulation. Due to the motion of the gas molecules that collide, momentum and energy transport, heat exchange, the aerodynamic force and the pneumatic macro physical process, they can be more real than numerical analysis methods to the real situation of the simulation experiment. Itina used the ideas of Bird on the pulsed laser deposition film process; simulation methods are compared for a series of successful applications. It considered in detail the deposition, diffusion, nucleation, growth and diffusion of atoms and evaporation, and background gas and air pressure differences. The effect of different mass numbers of particles on the thin film deposition rate allowed many successful estimates. For example, a particle with a small mass number (less than 27) at 25 Pa and a large mass number at 40Pa can achieve the best deposition uniformity.

Using femtosecond laser as a laser light source can reduce the generation of particles on the surface film. Because at the instant of the interaction between the laser and the target material, all the energy of the photons before transfer to the lattice has been transferred to all melting steamed materials in the electronics, the melting efficiency of steam is closer to unity. Steaming and melting zones around have little damage, so as to avoid the splash of small particles from the target surface. But femtosecond laser is expensive and the inconvenient operation also limits the use of it.

2.6.4 *Development of PLD*

2.6.4.1 *Ultra-fast PLD*

With the development of laser technology, people have developed the use of picosecond and femtosecond pulse lasers to produce thin films.

In 1997, Gamaly *et al.* from Australia first proposed and designed a femtosecond pulsed laser film deposition device. Then picosecond and femtosecond pulsed laser deposition technology emerged in the United States, Europe, and Australia. This technique uses the methods of low pulse energy and high repetition frequency to achieve the purpose of high-speed deposition of high-quality films. The main reasons are as follows:

(1) Each ultra-short pulse laser with low energy can only vaporize a few atoms, so the generation of large particles can be prevented correspondingly. Repetition frequencies up to tens of megahertz allow the resulting vapor to interact with the substrate, compensating for the low evaporation rate per pulse and resulting in a very high deposition rate overall. At the same time, it can also effectively prevent the large particles generated by the inhomogeneity of the target material, the fluctuation of the laser beam, and other irregularities in the deposition process of traditional PLD technology, which is another good method to improve the surface quality of the film in addition to the mechanical filtration method to prevent the large particles from reaching the substrate. Therefore, ultra-fast PLD technology is very effective in overcoming the shortcomings of traditional PLD in preparing films with large surface particles.

(2) As the repetition frequency reaches tens of megahertz, each pulse is very close to each other in space. In this way, continuous films of complex components can be made by scanning the laser beam on the target material and rapidly and continuously evaporating multiple targets with different components. Ultra-fast PLD can be used to produce multilayer films, mixed-component films, and single-atom films efficiently. In 1997, the Australian national university designed and manufactured the first femtosecond pulsed laser deposition facility. The results show that the microroughness of the diamond-like film is only within the range of atomic thickness, which is greatly improved compared with the traditional PLD method. This is because the pulse width of the laser is shorter than 1ps, so there is no energy exchange between electrons and ions during the pulse

duration. The mechanism of the nanosecond pulse width laser in the traditional PLD method is that the vapor is generated first and then ionized after the vapor energy exceeds the energy threshold under the action of subsequent pulses. Ultra-fast PLD, by contrast, produces no hot shock, and the absorbed laser energy is efficiently transferred to the stripped particle.

Ultra-fast PLD technology has the following features:

(1) Low mono-pulse energy is used to inhibit the generation of large particles;
(2) The repetition frequency is high enough to quickly sweep multiple targets for obtaining continuous films of complex components at a high film production efficiency;
(3) The deposition rate is 100 times that of the traditional PLD method. There are still many ununderstood phenomena in the interaction between ultra-fast pulsed lasers and solids in Dan that need to be studied and explained to accelerate the process of practical application of PLD of thin films.

2.6.4.2 *Pulsed laser vacuum arc technology*

Pulsed laser vacuum arc is a combination of pulsed laser deposition and vacuum arc deposition, and its schematic diagram is shown in Figure 2.70. The basic principle is as follows: in a high-vacuum environment, a high voltage is applied between the target and the electrode, and the laser is introduced from the outside and focused on the surface of the target to vaporize, thus triggering a pulse arc between the electrode and the target. The arc acts as a secondary excitation source to re-excite the surface of the target material, so that the matrix surface pulse can form the required film. Before the arc combustion point of the cathode fully develops into random motion, the arc is cut off by a predesigned pulse circuit. The lifetime of the arc and the size of the cathode's combustion zone near the combustion point depend on the duration of the pulse generated by the external current supply.

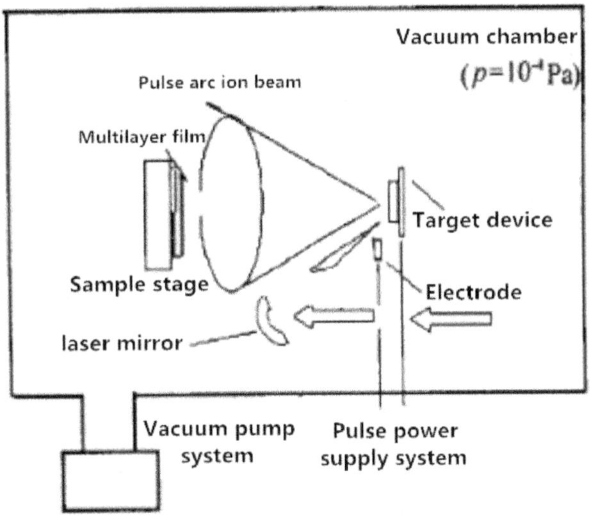

Figure 2.70: Schematic diagram of a laser vacuum arc deposition device.

By moving the target or moving the laser beam, the laser can scan the whole surface of the target. Because of the high repetition rate and the high pulse current, the method can achieve a high deposition rate. Combining the controllability of the former with the high efficiency of the latter, a controllable pulsed laser-excited plasma source can be obtained. It can be used for the preparation of films with large area and scale and the deposition of high-precision multilayer films with complex structure. This technology has shown its unique advantages in some experimental studies and practical applications; especially in the preparation of hard films and solid lubricating material films it will have a very wide range of applications.

Dr. Siemroth in 1990 first prepared diamond films using laser–arc technology, and then adjusted the parameters, from diamond, and graphite to the class of different types of carbon membranes in the glassy state, using the technology in alloy steel, hard metal, copper, aluminum and brass substrate surface for the preparation of high-hardness, low-coefficient of friction, and high-wear resistance diamond films, thereby the technology of diamond films can reach the standard of optical applications. Laser–arc technology has been used in industries such as drill bits, cutters, shank milling cutters, rough-cut hobs, and spherical ring flow switches. The good controllability and the uniform and effective excitation on the surface of the cathode material make it suitable for the deposition of complex and high-precision multi-layer films. After Ti/ TiC multilayer films, experiments on Al/C, Ti/C, Fe/Ti, Al/Cu/Fe and other nanoscale multilayer and monolayer films have been successful. The obtained multilayer film is well combined with the membrane base, and the monolayer film is smooth and dense.

2.6.4.3 *Dual beam pulsed laser deposition technique*

Double beam pulsed laser deposition (DBPLD) technology is using two laser beams for spectroscopic methods, at two different target materials at the same time, and through the control of the two laser beams focuses on the power density, thickness, and chemical composition to design the ideal functionally gradient films, which can accelerate the metal doped thin films such as complex compound films' speed of development of new materials. The device diagram is shown in Figure 2.71.

In 1997, the DBPLD method was first used to prepare Bi/Te thin films with gradually graded components on glass in Japan. In other words, a beam of light was divided into two beams at the temperature of (200–350°C) and Bi and Te targets were bombarded simultaneously. When the target distance is 30 mm for film distribution of components in the horizontal plane of 10 mm distance as Bi:Te = 1:1.1–1:1.5, the electric heating coefficient and resistance coefficient are about 170 μV/K and $2 \times 10^{-3} \Omega \cdot$ cm. In Singapore, Ong *et al.* used DBPLD technology to simultaneously react on YBCO and Ag targets, and realized *in-situ* doping by precisely controlling the intensity of the two beams of light. A long columnar Ag structure of 150 μm was observed on the membrane for the first time, which is of practical significance for preparing conventional superconductors and metallic superconducting Josephson junctions. Schenck and Kaiser of Germany used DBPLD technology to prepare BST series ceramic

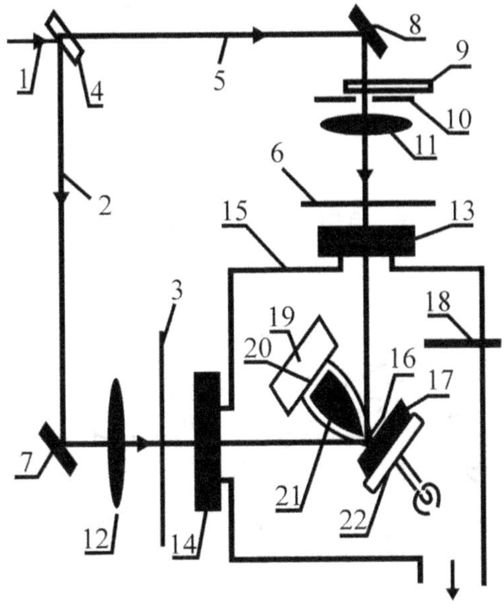

Figure 2.71: Schematic diagram of double beam pulsed laser deposition system. 1: Incident laser beam; 2: laser beam 1; 3: guided wave plate 1; 4: beam splitter; 5: laser beam 2; 6: guided wave plate 2; 7: reflecting mirror 1; 8: reflecting mirror 2; 9: beam energy controller; 10: diaphragm; 11: focus mirror 1; 12: focus mirror 2; 13: laser window 1; 14: laser window 2; 15: PLD deposition chamber; 16: doping target; 17: target; 18: ventilation pipe; 19: substrate heater; 20: substrate; 21: plasma plume; 22: target platform.

films using $BaTiO_3$ and $SrTiO_3$ as target materials. By controlling the energy intensity and action time of each beam, doped gradient films with gradually different components can be prepared. If the temperature, atmosphere type, pressure, and light intensity can be successfully supplemented by a device that can rotate multiple target seats, it is expected to solve the problem of repeatedly preparing targets with different components, such as $YBCO_x$, $Ba_x\ Sr_{1-x}\ TiO_3$, etc. The efficiency of preparing films can be greatly improved.

2.6.4.4 *Laser molecular beam epitaxy*

Laser molecular beam epitaxy (LMBE) is a new technology for thin film preparation developed in recent years. It combines the

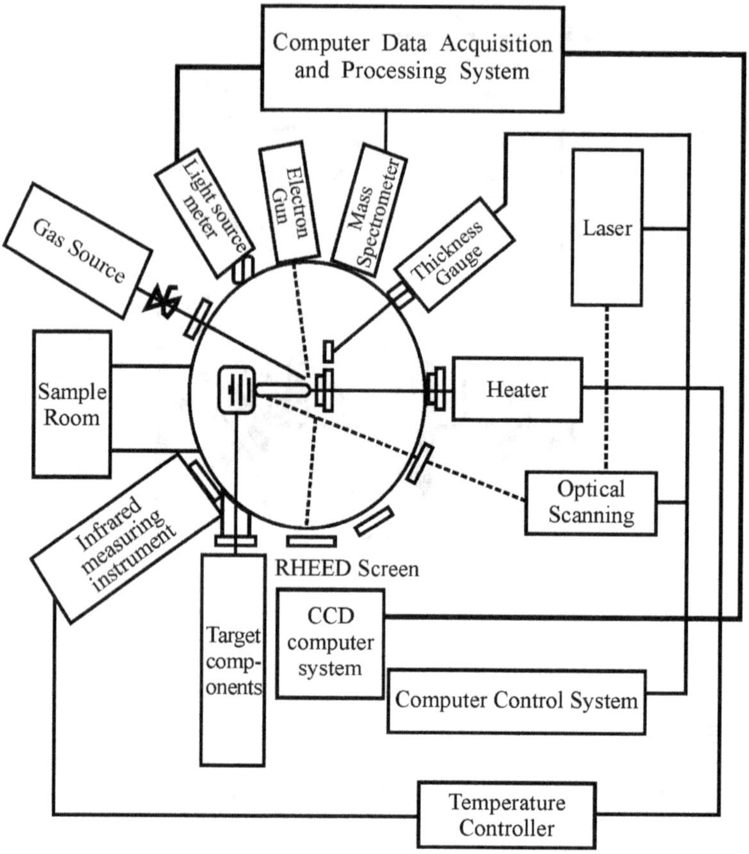

Figure 2.72: Device for deposition of thin films by a laser molecular beam.

advantages of ultra-high vacuum, *in situ* monitoring of MBE, and easy control of chemical components and a wide range of uses of PLD. It is an effective method to prepare high-quality epitaxial films, especially multi-layer and superlattice films. Its device is shown in Figure 2.72.

The laser molecular beam epitaxy equipment is mainly composed of the following four parts:

(1) **Laser system:** A high-power UV pulse laser source usually using an excimer laser (e.g., KrF, XeCl or ArF). The laser pulse width

is about 20–40 ns, the repetition rate is 2–30 Hz, and the pulse energy is greater than 200 mJ.

(2) **Vacuum deposition system:** Consists of an injection chamber, a growth chamber, a turbine a molecular pump, an ion pump, a sublimation pump, etc. The injection chamber is equipped with a sample transfer device, and the growth chamber is equipped with a rotatable target bracket and a substrate heater. The vacuum degree of the injection chamber is 6.65×10^{-4} Pa; the limit vacuum degree of the epitaxial growth chamber is 6.65×10^{-8} Pa. There are 4–12 target boxes on the target bracket, which can be changed at any time as needed. The sample rack can achieve 3D movement and rotation; the heater can make the surface temperature of the substrate reach up to 850–900°C and can work normally under the condition of higher gas partial pressure (e.g., 200 mTorr).

(3) ***In situ* real-time monitoring system:** Equipped with a reflective high-energy electron diffractometer (RHEED), a thin film thickness measuring instrument, a quadrupole mass spectrometer (QMS), a grating spectrometer, X-ray photoelectron spectrometer (XPS), etc.

(4) **Computer precision control, real-time data acquisition and data processing system:** The laser beam is reflected and focused on the target through a quartz window, and the mirror is rotated by a computer to allow the beam to be scanned in two dimensions on the target.

The basic process of laser beam epitaxial growth of thin films or superlattices is to focus a strong pulsed ultraviolet laser beam into the target through a quartz window into the growth chamber and make the target partially heated and evaporated instantaneously, resulting in plasma plumes containing the components of the target. The material in the plume reaches the surface of the substrate opposite the target and is deposited into a membrane. The epitaxial growth of the membrane layer is controlled in real time by the precision of the monoatomic or protocell layers changing the target alternately. And repeating the above process can periodically deposit multilayer films or superlattices on the same substrate. Usually, the laser beam after focusing is incident to the target surface at a 45° angle with an energy density of 1–5 J/cm^2. The local temperature on the target surface can be as high as 2000–3000 K; thus heating the target surface can steam out atoms, molecules, or molecular clusters. Further absorption of laser energy by these atoms and molecules near the target surface immediately transforms them into plasma and precipitates along the target surface normal direction towards the substrate at an extremely fast speed ($\sim 10^5$ cm/s). The distance between the substrate and the target is adjustable between 3 and 10 cm. For different substrates and target materials, the heating temperature of the substrate surface during the deposition process is also different, varying in the range of about 600–900°C. In general, after the film deposition is completed, it is also subjected to proper annealing to achieve the desired crystalline phase and improve the lattice integrity. The best energy density, reaction gas pressure, temperature of substrate, and the distance between substrate and target can be controlled by proper selection of laser wavelength, pulse repetition frequency, and the appropriate deposition rate, and if the best film-forming conditions can be obtained, then a high-quality film or superlattice can be prepared.

When the film grows in a two-dimensional atomic layer manner, the *in situ* monitored RHEED spectrum oscillates periodically with the increase of the film at atomic scale. When the new film starts to grow, the RHEED spectral strength is always at the maximum, and the film thickness corresponding to its oscillation period is the thickness of each new epitaxial layer. In addition, emission spectroscopy, mass spectrometer, high-speed CCD photography, a photoelectron spectrometer, a quartz crystal oscillator film thickness monitor, etc. are also commonly used for medium ion diagnosis and real-time monitoring and the analysis of the structure and composition in the film making process.

Laser molecular beam epitaxy, which concentrates the main advantages of traditional MBE and PLD methods and overcomes their shortcomings, has remarkable characteristics:

(1) According to different needs, it is possible to design manually and tailor different structures with special functional multilayer films (e.g., $YBa_2Cu_3O_7/BaTiO_3/YBa_2Cu_3O_7$) or superlattices, and use RHEED and film thickness gauges to monitor the film growth process *in situ* and realize the extension of atomic and molecular levels, which is conducive to the development of new thin film materials.

(2) It can grow films *in situ* with the same stoichiometric ratio as the target composition, even if the target composition is complex and contains five, six or more elements, and high-quality films can be made as long as it can form a dense target. For example, a single poly compound target can be used to grow films with the same stoichiometric ratio of protocell scale deposition to target composition, or several pure elemental targets can be used to grow the multicomponent film with monoatomic layer epitaxy.

(3) It is widely used and has a high deposition rate (up to 10–20 nm/min). Because of the strong directionality of the laser plume, the matter in the plume causes little pollution to the system and is easy to clean, so films from many kinds of materials can be prepared on the same equipment, such as various superconducting films, optical films, ferroelectric films, ferromagnetic films, metal films, semiconductor films, piezoelectric films, insulator films, and even organic polymer films, etc. Because it can operate under high reactive gas partial pressure, it is especially beneficial for preparing thin films with a complex oxide structure.

(4) It is convenient for studying in depth basic physical problems such as the interaction kinetics between the laser and the material target and the mechanism of film formation under different technological conditions, so as to select the best conditions for film formation, guide the preparation of high-quality films, and develop new film materials. For example, by means of a quadrupole mass spectrometer and spectrometer, the composition of the product after laser heating the target, the energy state, and the rate distribution of the atoms and molecules in the plasma plume can be analyzed and studied;

the film deposition rate, surface smoothness, crystal structure, and lattice re-structure mechanical processes can be observed *in situ* with RHEED, film thickness gauge and XPS.

(5) Because it can control the growth of thin films by atomic layer scale, it enables people to study microscopically the basic physical properties of thin films and related materials, such as the concentration of diffusion components between layers, the selective substitution of ions, the number of protocell layers, the coupling effect between layers, and the proximity effects on the physical origin and structural properties of materials.

2.6.4.5 *Pulsed laser liquid epitaxy*

Laser liquid phase epitaxy mainly uses laser pulse to radiate the target surface to produce plasma plume containing neutral atoms, molecules, active groups, a large number of ions, and electrons. Plasma plume absorbs laser energy in a laser pulse cycle and becomes a high-temperature and high-density plasma mass. In this process, the active particles in the system collide with each other, and the product is related to the laser energy and reaction conditions. Pulsed laser liquid phase epitaxy not only has the characteristic of pulsed laser deposition, but also has the advantage of liquid phase epitaxy. But the main drawback of this method is that the smooth surface and complete film can be grown only when the dislocation degree of the lattice is less than 1%. The composition of the epitaxial layer is related to not only the composition of the liquid phase, but also the partition coefficient of the elements. The inhibition of large droplets by this method is not ideal. In 1999, Ni–Pd–P nanofilms were prepared by this method on a silicon base by Zhenglie Gong *et al.* In 2004, Jie Zhu *et al.* used the pulsed laser liquid phase method to prepare nanoparticles with relatively good concentration and uniformity.

2.6.4.6 *Pulsed laser-induced crystallization*

Pulsed laser-induced crystallization is often used to crystallize large-area and amorphous silicon films; the crystallization process can achieve transient processing, and the crystalline film has a low defect density of states. In 2003, Xin Li *et al.* treated inorganic luminescent films prepared by the traditional plasma-enhanced chemical vapor deposition method with pulsed laser-induced crystallization, first inducing a laser with an energy density of 779 mJ/cm^2, and then treating with conventional thermal annealing at 900°C for 30 min to produce a high-density, size-controllable nanoscale silicon quantum dot film. In 2004, Feng Qiao *et al.* reported a nanosilica multilayer film prepared by the pulsed laser induction method. The results showed that a precise and controllable size of nanosilica particles could be obtained by changing the thickness of the original deposited α-Si silicon sublayer, and the crystallization ratio was high. The fluorescence emission spectrum shows that the peak intensity of the luminescence at 650 nm increases with the increase of the energy density of the laser irradiation.

2.6.4.7 *DC-discharge assisted pulsed laser deposition*

The basic principle of DC-discharge assisted pulsed laser deposition is that the pulsed laser burns the target material to form a plasma, and then the plasma is further pushed and heated by a strong current discharged on the plasma motion path, and the neutral particles are filtered out by a magnetic field filter. DC-discharge assisted pulsed laser deposition breaks the limitation of the traditional pulsed laser deposition method, and improves the average kinetic energy of plasma through DC discharge, and effectively reduces the output energy of the pulsed laser. Jian Guo *et al.* reported in 2002 that crystalline carbon films were deposited on monocrystalline silicon and K9 glass substrates by pulsed laser deposition under different negative DC substrate biases. The silicon-based GaN films prepared by DC-discharging-assisted pulsed laser deposition were studied in detail in 2004 by Xinglin Tong *et al.* And the results show that increasing the deposition pressure is beneficial for improving the crystallization quality of GaN films. The surface structure of the GaN film is improved by the increase of the incident laser pulse energy in the range of 150–220 mJ. Under the optimized conditions of a substrate temperature of 700°C, a deposition pressure of 20 Pa, and an incident laser pulse energy of 220 mJ, the deposited grown GaN film has good quality.

2.7 Exercises and Questions

(1) Design an experiment to determine the experimental value of the vaporization heat of metals by means of conventional thin film deposition and characterization equipment.

(2) If iron is evaporated on a surface with 1550°C and an area of $1\,cm^2$, the desired evaporation rate can be achieved by increasing the temperature by 100°C, but this will burn out the surface material. In addition to increasing the temperature, what factors strengthen the surface and can make the evaporation rate meet the requirements?

(3) There are two Al and As evaporation sources with an area of $4\,cm^2$ separated from each other on a molecular beam epitaxy device with an evaporation source distance to (100)GaAs substrate of 10 cm. The Al source was heated to 1000°C and the As source was heated to 300°C. What is the rate of AlAs growth in Å/s? [*Note*: AlAs has the same crystal structure and lattice constant (5.661 Å) as GaAs].

(4) If Al films are deposited at a rate of $1\mu m/min$ in a vacuum at 25°C and the film is estimated to contain 10^{-3} oxygen, what is the partial pressure of oxygen in the system?

(5) The deposition of YBa_2Cu_3 alloy films requires the formation of Y, Ba, and Cu in three dot-like evaporation sources. These three points are placed on three vertices of an equilateral triangle with a side length of 20 cm. Place a small substrate directly above the centroid of the triangle in parallel to the face at which the triangle is located; the entire deposition system forms a regular tetrahedron with a side length of 20 cm.

   (a) On heating the Y evaporation source to 1740 K, it can produce a vapor pressure of 10^{-3}Torr. Then to what temperature does the Cu evaporation source need to be heated for maintaining the chemical composition of the film?

(b) If the point evaporation source of Cu is replaced by a surface evaporation source, what is the temperature of the point evaporation source of Cu to ensure the chemical composition of the film?

(c) How much partial pressure of oxygen does it require to deposit $YBa_2Cu_3O_7$ superconducting films? Atomic mass: Y = 89, Cu = 63.5, Ba = 137, O = 16.

(6) A method of depositing a specific thickness of metal film: the metal is made into an evaporating source and heated to dry (that is, the metal is placed in a crucible and evaporated to no residue), and 5000 Å Au film is now required on the inner surface of a hemispherical shell with a diameter of 30 cm.

(a) Please design two different types of evaporation source configurations (type of evaporation source and location of placement) to produce uniform films.

(b) How much each of the two approaches requires? Assume complete evaporation.

(7) Let the process of electron shock ionization and the generation of secondary emission electrons by ions control the current in a sputtering system denoted as J. According to the Townsend equation:

$$J = \frac{J_0 \exp \alpha d}{1 - \gamma[\exp(\alpha d) - 1]},$$

where J_0 is the Main current density from external sources, α is the Number of ions per unit length generated by electrons, γ is the Number of secondary electrons emitted per incident ion, d is the Space between the electrodes.

(a) Calculate the ratio coefficient of Cu if the rate of sputtering deposition is proportional to the production of JS. The ion deposition rate of Ar was 200 Å/min at 0.5 keV. Assume that $\alpha = 0.1$ ion/cm, $\gamma = 0.08$ electron/ion, $d = 10$ cm, $J_0 = 100\,\text{mA/cm}^2$.

(b) What is the deposition rate of 1 keV Ar when $\alpha = 0.5$ ion/cm and $\gamma = 0.1$ electron per ion?

(8) A direct current planar magnetron system operates at a voltage of 1000 V, with a distance of 10 cm between the two poles. How much magnetic field intensity is required to capture electrons in the range of 1 cm on the target?

(9) What rate of deposition of In on the Si substrate will cause the film to melt within 1 min? The melting point of In is 155°C.

(10) (a) When sputtering gold at a voltage of 1 keV, it is assumed that the Au and Ar atoms will have two collisions before they can be deposited. So what is the energy of the Au atoms in the deposition?

(b) The possibility of gas phase atoms moving x away without collision is $\exp(-x/\lambda)$ and λ is the average free path between two collisions. λ of Au is assumed to be 5 cm of 1 mTorr pressure. If the distance between the target and the anode is 12 cm, then at what working pressure will 99% of the Au atoms collide before deposition?

(11) A continuous coating of 2 micron-thick Al is required on a 1-m wide steel strip. A row of electron beam evaporators along the y direction on the x–y surface of the steel strip maintains a uniform coating thickness in the wide direction of the steel strip. If the evaporator can evaporate 20 g of Al per second, how long will it take to send the steel strip over the surface evaporation source? Assuming that Equation (3.18) of Chapter 3 determines the coating thickness in the x direction, the distance between the evaporation source and the steel strip is 30 cm.

(12) Select appropriate thin film deposition processes for the following applications (evaporation, sputtering, etc., source, target, etc.):

(a) Plate a layer of Rh on the lens of a large telescope.
(b) Plate a layer of Al film on a bag of potato chips.
(c) Deposit Al-Cu-Si thin-film internal connections on integrated circuits.
(d) Deposit a TiO_2-SiO_2 multilayer on artificial gemstones to enhance their color and reflectivity.

(13) The theory points out that when neutral atoms are sputtered out of the surface, their kinetic energy and angular distribution can be given as follows:

$$F(E,\theta) = CS\frac{E}{(E+U)^3}\cos\theta,$$

where U is the binding energy of the surface atoms, C is the constant, θ is the Angle between sputtered atoms and normal surfaces.

(14) To indicate better the agglomeration nucleation in the ICB process,

(a) how much supersaturated vapor pressure is required, if the surface tension is 1000 ergs/cm^2, to form an agglomeration of 1000 Au atoms?

(b) if such an agglomeration is ionized and accelerated to an energy of 10 keV, how much energy will each agglomerate atom give to the substrate?

Chapter 3

The Chemical Synthesis Technology of Thin Films

3.1 Overview

Among the chemical depositing and processing technologies of thin films, chemical vapor deposition (CVD) is a very important synthesis method that is widely used, which is the basis of the chemical synthesis technology of thin films. According to the pressure in the reaction vessel, the depositing temperature, the temperature of the reaction vessel wall, and the activation method of the depositing reaction, CVD can be divided into several categories. According to the pressure in the reaction vessel, it can be divided into normal pressure CVD and low pressure CVD; According to the depositing temperature, it can be divided into low temperature (200–500°C) CVD, moderate temperature (500–1000°C) CVD, and high temperature (1000–1300°C) CVD; According to the temperature of the reaction vessel wall, it can be divided into hot wall CVD and cold wall CVD. According to the activation methods of the reaction, it can be divided into thermal activated CVD and plasma activated CVD.

Another technology for depositing thin films using chemical deposition is the chemical solution method. It is a process technology for depositing thin films on the surface of a substrate using a solution related with reagents and adopting a chemical reaction or an electrochemical reaction. This technology does not require a vacuum environment; since the equipment is cheap and the process is also easy to master, it has a great value for the application. For the deposition of the film in a chemical solution, the soft solution processing (SSP) technology developed in recent years has attracted wide attention, especially the hydrothermal–electrochemical technology of SSP. It may be developed into a depositing technology of thin films with almost zero pollution and zero emission. In addition, the LB technology improved from the Langmur–Blodgett method has also been widely used in recent years.

3.2 Chemical Vapor Deposition

3.2.1 *Introduction*

CVD is a way to deposit thin films through chemical reactions. Its basic feature is that the reactants (precursors) in the chemical reaction are gases and one of the products (usually being the expected thin film) is solid. Chemical reactions in CVD include many reactions, such as thermal decomposition reactions, reduction and displacement reactions, oxidation or nitridation reactions, hydrolysis reactions, disproportionation reactions, and the like. In most cases, these reactions rely on the thermal excitation, therefore the reaction temperature, especially the substrate temperature, needs to be strictly controlled. In general, the substrate temperature is high, in a certain temperature range which is conducive to the smooth progress of the reaction.

The process of film deposition via CVD can be divided into four main stages:

(1) The reaction gas diffuses to the substrate surface;
(2) The reaction gas is adsorbed on the substrate surface;
(3) The reaction gas reacts on the substrate surface to form the required solid film;
(4) The gases generated from the gas reaction on the substrate surface are separated from the substrate surface, diffused out of the substrate or sucked away by a vacuum pump; meanwhile, there is a non-volatile solid product film left by the reaction.

Obviously, in order to enable the process of depositing films via CVD to proceed smoothly, the reaction gases could be generated by suitable precursors; the reaction gases related to the film composition ratio required would be guided to the substrate surface.

Then specified chemical reactions could occur at a certain substrate temperature. As mentioned above, the four stages of depositing film via CVD would be performed in sequence, where the slowest process would limit the rate of depositing film by type of chemical reaction.

The process of depositing film via CVD involves many reactions with multidisciplinary fields such as reaction chemistry, reaction thermodynamics, reaction kinetics, transport process, film growth mechanism, and reactor engineering. In order to research the process of depositing film via CVD in depth and get the proper conditions for deposited film, we have to carefully analyze the process of thermodynamics of the deposition to find the factors in the chemical reaction process showing the expected direction and the maximum solid phase yield (or conversion efficiency) during dynamic equilibrium.

3.2.2 *Principle of CVD*

The CVD process is a complex process involving thermodynamics and kinetics related to chemical reactions. According to the principle of thermodynamics, the free energy changes (ΔG_r) of a chemical reaction can be calculated using the standard free energy G_f of the reaction product.

$$\Delta G_r = \Sigma\Delta G_f \text{ (product)} - \Sigma\Delta G_f \text{ (reactant)}. \tag{3.1}$$

The relationship between ΔG_r and K_ρ (the equilibrium constant related to each partial pressure in the reaction system) is as follows:

$$\Delta G_r = -2.3RT\log(K_\rho). \tag{3.2}$$

Actually, the kinetic process in chemical reactions includes many processes, such as the diffusion of the reactive gas on the substrate surface, adsorption of the reactive gas on the substrate surface, chemical reactions on the substrate surface, and desorption and diffusion of by-products from the substrate surface. When the substrate temperature is low, the reaction rate τ can change with the temperature according to the following exponential law:

$$\tau = Ae^{-\Delta E/RT}, \tag{3.3}$$

where A is the frequency factor of effective collision, and E is the activation energy. When the substrate temperature is high, the diffusion rate of the reactants and by-products becomes the main factor determining the reaction rate. At this time, the relationship between the reaction rate τ and temperature T can be expressed as $\tau \sim T^{1.5-2.0}$. CVD technology mainly involves the following seven types of chemical reactions.

3.2.2.1 *Thermal decomposition reaction*

Hydrides, hydroxy compounds, and organometallic compounds in many elements can exist in a gaseous state, which under appropriate conditions can undergo thermal decomposition reactions on the substrate surface to form the thin film. The thermal decomposition method generally involves using a simple high-temperature furnace to heat the substrate to acquire a desired substrate temperature under the protection of a vacuum or an inert gas, then introducing a reactant gas to cause thermal decomposition, and finally depositing the thin film on the substrate surface. The general formula of the thermal decomposition reaction is

$$\mathrm{AB(g)} \rightarrow \mathrm{A(s)} + \mathrm{B(g)}.$$

In this reaction formula, A and B refer to chemical elements (or compounds), respectively; s represents a solid phase; and g represents a gas phase. A typical example is polysilicon and amorphous silicon are deposited via the pyrolytic reaction of SiH_4

$$\mathrm{SiH_4(g)} \xrightarrow{700-1100^\circ\mathrm{C}} \mathrm{Si} + 2\mathrm{H_2}\uparrow.$$

In addition, the reaction of thermal decomposition of used nickel carbonyl in a traditional nickel purification process to generate metal Ni also belongs to the thermal decomposition reaction:

$$\mathrm{NI(CO)_4(g)} \xrightarrow{180^\circ\mathrm{C}} \mathrm{Ni(s)} + 4\mathrm{CO(g)}.$$

The thermal decomposition reaction is the simplest deposition reaction, and many types of films have been deposited by this method.

3.2.2.2 *Chemical synthesis reaction*

Most processes of the depositing film involve the chemical reaction between two or more gaseous reactants on a heated substrate. This type of reaction is called the chemical synthesis reaction. It is more common that hydrogen reduction halides using the method can deposit various metal and semiconductor films on the substrate surface. In addition, involving the reaction in the deposition process of the insulating film using a suitable hydride, halide, or metal organic compound is also called the chemical synthesis reaction, such as

$$SiCl_4 + 2H_2 \xrightarrow{1150\sim1200^\circ C} Si + 4HCl,$$

$$SiCl_4 + CH_4 \xrightarrow{1400^\circ C} SiC + 4HCl.$$

The chemical synthesis reaction route is more widely used than the thermal decomposition route. Although there are not many compounds that can be deposited by thermal decomposition, any kind of inorganic material can be synthesized in principle by suitable chemical reactions.

3.2.2.3 *Reduction reaction*

Although halides, hydroxy compounds, and oxyhalides in many elements can exist in the gaseous state, their thermal stabilities are considerably poor. Therefore, they have to be replaced with a suitable reducing agent, such as depositing single-crystal silicon on the epitaxial layer using H_2 to reduce the $SiCl_4$ reaction

$$SiCl_4(g) + 2H_2(g) \xrightarrow{1200^\circ C} Si(s) + 4HCl(g),$$

and the preparation of various refractory metal films such as W, Mo, and other metals:

$$WF_6(g) + 3H_2(g) \xrightarrow{300^\circ C} W(s) + 6HF(g),$$

where the gaseous substances suitable for use as reducing agents are relatively easy to obtain, the one that's used the most is H_2.

3.2.2.4 *Oxidation reaction*

In contrast to the reduction reaction, SiO_2 film can be deposited by the oxidation reaction of SiH_4 using O_2 as an oxidant; the reaction formula is

$$SiH_4(g) + O_2(g) \xrightarrow{450^\circ C} SiO_2(s) + 2H_2(g).$$

Alternatively, SiO_2 film can also be prepared by using the reaction among SiH_4, O_2, and H_2.

$$SiCl_4(g) + 2H_2(g) + O_2(g) \xrightarrow{1500^\circ C} SiO_2(s) + 4HCl(g).$$

The two methods are suitable for the deposition of semiconductor insulating layers and the deposition of optical fiber raw materials, respectively. The former requires a low deposition temperature, and the latter requires a high deposition temperature, but both require a faster deposition rate.

3.2.2.5 *Disproportionation reaction*

Some elements have multiple gaseous compounds with different stabilities, and the changes in external conditions can often promote the conversion of one compound into another compound with higher stability. This is the basic meaning of the disproportionation reaction. In depositing thin films via CVD, the disproportionation reactions can be used to achieve the deposition of thin film, such as

$$2GeI_2(g) \xrightarrow{300\text{–}600^\circ C} Ge(s) + GeI_4(g).$$

Some metal halides have such characteristics, namely the metal elements in them can often form compounds of different valences. Increasing the temperature is conducive to increasing the stability of lower valence compounds. Therefore, the transfer and deposition of Ge can be achieved by adjusting the temperature of the CVD reaction chamber.

3.2.2.6 *Reversible reaction*

Similar to the disproportionation reaction described above, the characteristic that the relative stability of the same compound in certain elements can change with temperature can also be used to achieve the transfer and deposition of the substances, such as

$$As_4(g) + As_2(g) + 6GaCl(g) + 3H_2(g) \xrightarrow{750\sim850^\circ C} 6GaAs(s) + 6HCl(g).$$

The above formula tends to turn left at high temperatures (850°C); whereas it turns to the right at relatively low temperatures (750°C). Using this characteristic, GaCl can be loaded with $AsCl_3$ gas and it can react with As_4 vapor at an appropriate temperature, thereby GaAs is deposited.

3.2.2.7 *Chemical transport reaction*

In general, the substances of depositing film are used as the source materials (non-volatile materials) and react with the appropriate gas medium to form gaseous compounds. Subsequently, this gaseous compound is transported to a deposition area which is different from the source area by chemical migration or by a carrier (carrier gas), and a reverse reaction may occur on the substrate, so that the source material can be deposited again on the substrate. This reaction process is called a chemical transport reaction, where the gaseous medium is called a transport agent. This method was first used for the purification of rare metals, such as

$$Ge(s) + I_2(g) \underset{T_2}{\overset{T_1}{\rightleftarrows}} GeI_2$$

The transmission reaction (to the right) would occur in the source area (temperature T_1), and the source material Ge would interact with I_2 to generate gaseous GeI_2; after the gaseous products were transported to the deposition area, the deposition reaction (proceed to the left) would occur in the deposition area (temperature T_2). Then Ge would be re-deposited. There are many similar reactions,

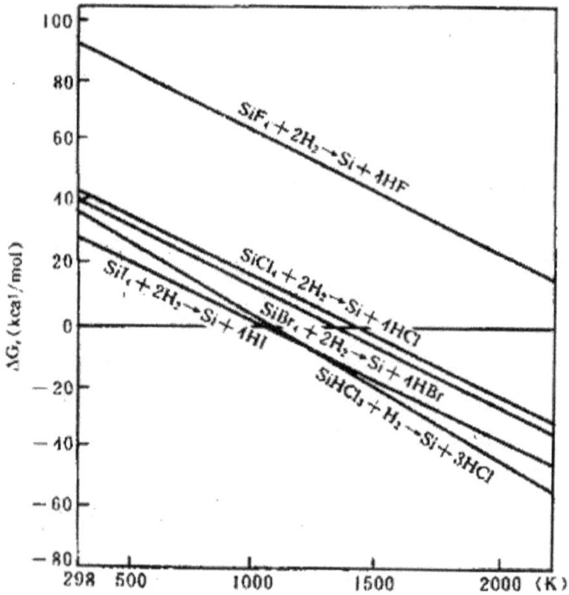

Figure 3.1: Relationship curve between the free energy in hydrogen reduction reaction of silicon halides and temperature.

such as

$$2\text{CdTe(s)} \underset{\text{T}_2}{\overset{\text{T}_1}{\rightleftarrows}} 2\text{Cd(g)} + \text{Te}_2\text{(g)}.$$

In most cases, the above-mentioned various types of reactions rely on thermal excitation. Therefore, temperature is an important factor for the CVD method, which has limited the choice of substrate materials.

Figure 3.1 shows the relationship between the free energy in the hydrogen reduction reaction of silicon halides by using CVD (see Equation (3.1)) and the temperature. It can be seen from Figure 3.1 that a reaction can proceed smoothly, because the change value of free energy ($\Delta G_{\rm r}$) in the process of the reaction must be negative, and the value of $\Delta G_{\rm r}$ would decrease as the temperature increases. Therefore, increasing the temperature is beneficial for proceeding smoothly with the spontaneous reaction. Table 3.1 lists the film materials of some metals, alloys, oxides, nitrides, carbides, silicides, borides, etc. that can be deposited by CVD.

Table 3.1: Films prepared by the CVD method.

Thin film		Source materials: Raw material name	Source materials: Gasification temperature (°C)	Reaction temperature (°C)	Transport or reaction gas
Metal	Cu	CuCl, $CuCl_3$, CuI	500–700	500–1000	H_2 or Ar
	Be	$BeCl_3$	290–340	500–800	H_2
	Al	$AlCl_3$	125–135	800–1000	H_2
		$AlCH_2$ $CH(CH_3)_2$	−38	93–100	H_2 or Ar
	Ti	$TiCl_4$	20–80	1100–1400	H_2 or Ar
	Zr	$ZrCl_4$	200–250	800–1000	H_2 or Ar
	Ge	GeI_2	250	450–900	H_2
	Sn	$SnCl_4$	25–35	400–550	H_2
	V	VCl_4	50	80–1000	H_2 or Ar
	Ta	$TaCl_5$	250–300	600–1400	H_2 or Ar
	Sb	$SbCl_3$	80–110	500–600	H_2
	Bi	$BiCl_3$	240	240	H_2
	Mo	$MoCl_5$	130–150	500–1000	H_2
		$Mo(CO)_6$		150–160	H_2 or Ar
	Co	$CoCl_3$	60–150	370–450	H_2
	W	WCl_6	165–230	600–700	H_2
		$W(CO)_6$	50	350–600	H_2 or Ar
	Cr	CrI_2	100–130	1100–1200	H_2
		$Cr[C_5H_4CH(CH_3)_2]_3$		–400	H_2 or Ar
	Nb	NbI_3	200	1800	H_2
	Fe	$FeCl_3$	317	650–1100	H_2
		$Fe(CO)_4$	102	140	H_2 or Ar
	Si	$SiCl_4$	20–80	770–1200	H_2
		SiH_2Cl_2	20–80	1000–1200	H_2
	B	BCl_3	–30–0	1200–1500	H_2
	Ni	$Ni(CO)_4$	43	180–200	H_2 or Ar
	Pt	$Pt(CO)_2Cl_2$	100–120	600	H_2 or Ar
	Pb	$Pb(C_2H_5)_4$	94	200–300	H_2 or Ar
Alloy	Ta/Nb	$TaCl_5$+$NbCl_5$	250	1300–1700	H_2 or Ar
	Ti–Ta	$TiCl_4$+$TaCl_5$	250	1300–1400	H_2 or Ar
	Mo–W	$MoCl_6$+WCl_6	130–230	1100–1500	H_2 or Ar
	Cr–Al	$CrCl_3$+$AlCl_3$	95–125	1200–1500	H_2 or Ar
Oxide	Al_2O_3	$AlCl_3$	130–160	800–1000	H_2+H_2O
	SiO_2	$SiCl_4$	20–80	800–1100	H_2+H_2O
		SiH_4+O_2		400–1000	H_2+H_2O
	Fe_2O_3	$Fe(CO)_5$		100–300	N_2+O_2
	ZrO_2	$ZrCl_4$	290	800–1000	H_2+H_2O

(*Continued*)

Table 3.1: (*Continued*)

Thin film		Raw material name (Source materials)	Gasification temperature (°C) (Source materials)	Reaction temperature (°C)	Transport or reaction gas
Nitride	BN	BCl_3	–30–0	1200–1500	N_2+H_2
	TiN	$TiCl_4$	20–80	1100–1200	N_2+H_2
	ZrN	$ZrCl_4$	30–35	1150–1200	N_2+H_2
	HfN	$HfCl_4$	30–35	900–1300	N_2+H_2
	VN	VCl_4	20–50	1100–1300	N_2+H_2
	TaN	$TaCl_5$	25–30	800–1500	N_2+H_2
	AlN	$AlCl_3$	100–130	1200–1600	N_2+H_2
	Si_3N_4	$SiCl_4$	–40–0	–900	N_2+H_2
		SiH_4+NH_3		550–1150	Ar or H_2
	Th_3N_4	$ThCl_4$	60–70	1200–1600	N_2+H_2
Carbide	BeC	$BeCl_3+C_6H_5CH_3$	290–340	1300–1400	Ar or H_2
	SiC	$SiCl_4+CH_4$	20–80	1900–2000	Ar or H_2
	TiC	$TiCl_4+C_6H_5CH_3$	20–140	1100–1200	H_2
		$TiCl_4+CH_4$	20–140	900–1100	H_2
		$TiCl_4+CCl_4$	20–140	900–1100	H_2
	ZrC	$ZrCl_4+C_6H_5$	250–300	1200–1300	H_2
	WC	$WCl_5+C_6H_5CH_3$	160	1000–1500	H_2
Silicide	MoSi	$MoCl_5+SiCl_4$	–50–13	1000–1800	H_2
	TiSi	$TiCl_4+SiCl_4$	–50–20	800–1200	H_2
	ZrSi	$ZrCl_4+SiCl_4$	–50–20	800–1000	H_2
	VSi	VCl_4+SiCl_4	–50–50	800–1100	H_2
Boride	AlB	$AlCl_3+BCl_3$	–20–125	1100–1300	H_2
	SiB	$SiCl_4+BCl_3$	–20–0	1100–1300	H_2
	TiB_2	$TiCl_4+BCl_3$	20–80	1100–1300	H_2
	ZrB_2	$ZrCl_4+BBr_3$	20–30	1000–1500	H_2
	HfB_2	$HfCl_4+BBr_3$	20–30	1000–1700	H_2
	VB_2	VCl_4+BBr_3	20–75	900–1300	H_2
	TaB_2	$TaCl_5+BBr_3$	20–100	1300–1700	H_2
	WB	WCl_6+BBr_3	20–35	1400–1600	H_2

3.2.3 *The main features of the CVD method*

Compared to other thin-film fabrication technologies, CVD holds the below characteristics:

(1) Both metal and non-metallic films can be made, and CVD can be used to make multi-component alloy films by adjusting the flow of different air sources accordingly. And it can also be used to make mixed crystals and complex crystals.
(2) The rate of film formation is fast and can reach several μm per minute or even hundreds of μm. By regulating the flow of various gases, the composition of the products can be controlled within a larger range in the same furnace, i.e. the CVD method can be used to prepare thin films on a large substrate, or to place large quantities of substrates in the same furnace.
(3) The CVD reaction is carried out in a normal atmospheric or a low vacuum. And the diffraction of the precursor reaction gas is fantastic. Even the complex surface or a deep hole and a fine hole of the workpiece can be evenly coated.
(4) The film coating can be obtained with high purity, good density, small residual stress, good crystallization and smoothness. Because the film grows at a much lower temperature than the melting point of the membrane material, the film layer can be obtained with high purity and complete crystallization. In the process of film deposition, the reaction gas, reaction products, and collective surface atoms will be mutually dispersed, and the adhesion of the film is relatively good.
(5) The surface of the film deposited is relatively smooth. The surface roughness is small, and the film's radiation damage is low due to the adoption of gas phase reaction deposition.

The main disadvantage of using CVD technology is that the temperature of the reaction is high, even reaching 1000°C. And many matrix materials cannot be subjected to such high temperatures, thus limiting their use. When using metal–organic chemical vapor deposition (MOCVD), the appropriate reaction precursor is selected. Sometimes the precursor may not be available in the market, and needs to be synthesized by oneself, and in the process of preparation of products and disposing of waste gas, it is necessary to be careful not to cause harm to the environment and the experimenter.

In recent years, CVD technology has developed rapidly and has gained wide attention, becoming a very important method in film preparation that has numerous applications in microelectronics, photoelectronic, mechanical, aerospace materials, etc.

3.2.4 *Brief introduction to several major CVD techniques*

According to the pressure of the reaction vessel in the CVD technology, the deposition temperature of the film, the temperature of the reaction vessel wall and the activation method of reaction, chemical vapor deposition is divided into several classes to meet the needs of different kinds of applications. The choice of the CVD reaction and reaction vessel depends on many factors, mainly the properties, quality, cost, equipment size, convenience of operation, purity, and cost of raw materials, and safety and reliability. But the reaction system used should satisfy all the following three conditions:

(1) At the deposition temperature, the reactant must have a high enough vapor pressure to ensure that the reactants are introduced into the reaction chamber at an appropriate speed.
(2) The other reactant must be volatile, except for the sediment that is needed as the solid membrane.
(3) The vapor pressure of the deposited film itself must be low enough to ensure that the deposited film stays on the hot substrate during the whole deposition process. The vapor pressure of the substrate material at the deposition temperature must also be low enough.

As a result, CVD devices often include the following basic components:

(1) A supply of the reaction gas and the carrier gas and the corresponding gas metering devices.

(2) The necessary heating and cooling systems.
(3) A device for removal of the reaction gas by-product from the product.

3.2.4.1 *Normal pressure CVD*

This is the most commonly used type in the CVD reaction chamber. This kind of reaction chamber is usually operated under normal pressure, so that it is convenient to load and unload. The whole unit includes a gas purification system, a gas measurement and control system, a reactor, an exhaust gas disposing system, a vacuum system, etc., as shown in Figure 3.2.

When you choose the normal pressure CVD technique, the material doesn't have to be all gas at room temperature. If a liquid material is used, it is heated to produce steam, and then carried by the carrier gas into the reaction chamber. If a solid material is used, the steam generated by the heating is carried by the carrier gas into the reaction chamber. Usually it is expected that these reagents don't react with each other before they reach the sediment zone. Therefore, substances that react with each other at low temperatures should be separated before entering the sedimentary zone.

The process feature of atmospheric CVD is that it can provide a continuous air supply and exhaust, and the transportation of material is generally realized by an inert gas, which does not participate in the reaction. Since there is at least one reaction product that can be discharged continuously from the reaction area, the reaction will always be in a non-equilibrium state, which is beneficial to the formation of the thin film layer. In addition, the deposition process of atmospheric CVD is easy to control, and the process

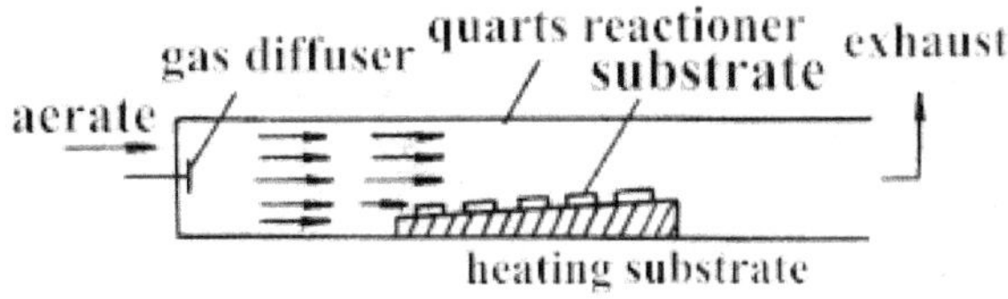

Figure 3.2: Schematic diagram of an atmospheric CVD reactor.

repeatability is better, and the workpiece is easy to take, and the same reactor configuration can be used many times.

3.2.4.2 *Low pressure CVD*

The CVD technique corresponding to atmospheric CVD is the low pressure CVD technique. CVD devices with working pressure below 0.1 MPa are low pressure CVD (LPCVD) devices. Reducing the pressure of the reaction chamber can increase the diffusion capability of the reaction gas and reaction products through the boundary layer. Meanwhile, in order to partially offset the effect of the pressure reduction on the film preparation, the concentration ratio of the gas in the total gas flow can be increased. Compared to normal pressure CVD devices, the pressure of low pressure CVD devices is often reduced to 100 Pa, which increases the diffusion of the gas.

The schematic of a typical low-voltage CVD device is shown in Figure 3.3. The main difference between the low pressure CVD device and the normal atmospheric CVD device is that the low pressure CVD device requires a vacuum pump system to keep the whole system working at a lower pressure.

3.2.4.3 *High temperature and low temperature CVD*

Two most important physical quantities of the preparation of films using CVD technology are: one is the degree of supersaturation of the gas reactant and the other is the deposition temperature. These two

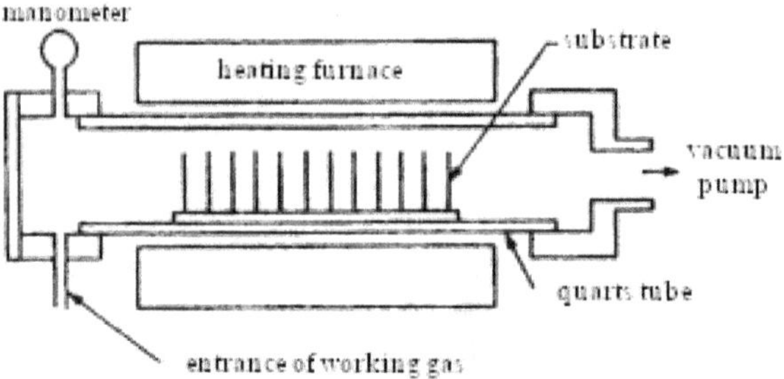

Figure 3.3: Schematic diagram of a low pressure CVD device reaction chamber.

quantities determine the nucleation rate of the film during deposition, the deposition rate of the film, and the integrity of the thin film microstructure. By adjusting these two parameters, the resulting deposition can be in a single crystal state, a polycrystalline state, or even a non-crystalline state.

In order to achieve a highly complete single crystal deposition, the general condition is that the degree of supersaturation of the gas phase is low and the deposition temperature is high, while the opposite condition promotes the formation of polycrystalline and even amorphous materials. Therefore, if the integrity of materials is emphasized, a high temperature CVD system is mostly used in the application, and the low temperature CVD system should be used with special attention to the low temperature preparation conditions of thin film material.

The high-temperature CVD system is widely used in the preparation of semiconductor epitaxial films to ensure the quality of the material. These types of systems can be divided into two types: hot wall and cold wall type, as shown in Figure 3.4. The characteristic of such devices is to heat the reaction chamber to a higher temperature using an external heater.

The first two devices, shown in Figure 3.4, are cold wall systems, which are characterized by the use of an induction heating device to heat a sample table with a certain conductivity, while the wall of the reaction chamber is made from materials with poor conductivity and cooled to a lower temperature by the cooling system. Figure 3.4(c) gives a schematic diagram of the hot wall CVD device. In this device, a resistance heater is used to heat the reaction chamber.

In the semiconductor industry, low-temperature CVD devices are widely used for various dielectric films, such as SiO_2, Si_3N_4, etc. In addition to the epitaxial growth of semiconductor materials, high temperature CVD devices are widely used in the preparation of wear resistant coating of metal parts.

3.2.4.4 *Plasma enhanced CVD*

In the process of low pressure chemical vapor deposition, the technique of using glow discharge plasma to exert a certain influence

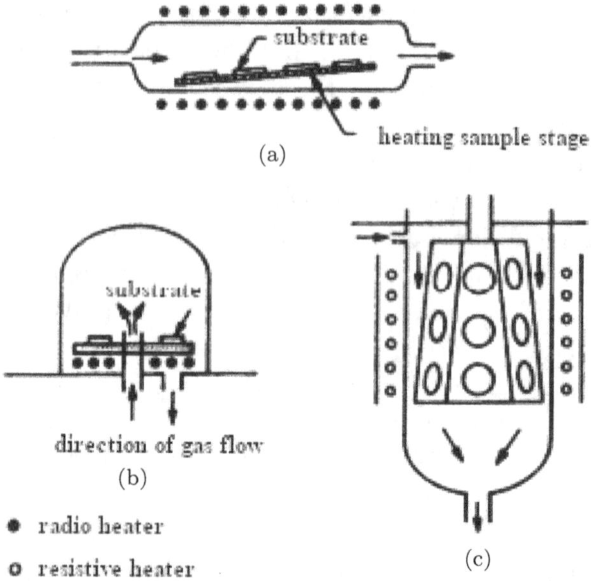

Figure 3.4: Several schematic diagrams of CVD device reaction chamber heating. (a) and (b) Cold wall CVD; (c) hot-wall CVD.

on the reaction process is called plasma enhanced CVD (PECVD). In a PECVD device, the work pressure is about 5–500 Pa, the electron and ion density can reach 10^9–10^{12}cm^{-3}, and the average electron energy can reach 1–10 eV.

The difference between the PECVD method and other CVD methods is that the presence of plasma can promote the decomposition, combination, excitation, and ionization of gas molecules and also promote the formation of reactive groups, thereby significantly reducing the temperature range of the reaction deposition, so that some of the reactions that need to be performed at high temperature can be achieved at low temperature.

Because the main application of the PECVD method is in the low-temperature deposition of insulating dielectric films, the RF method is usually adopted to generate the plasma of these films. Rf fields can be coupled in two different ways, namely inductive coupling and capacitance coupling. Figure 3.5 is a schematic of the typical structure of an RF PECVD device with capacitance coupling.

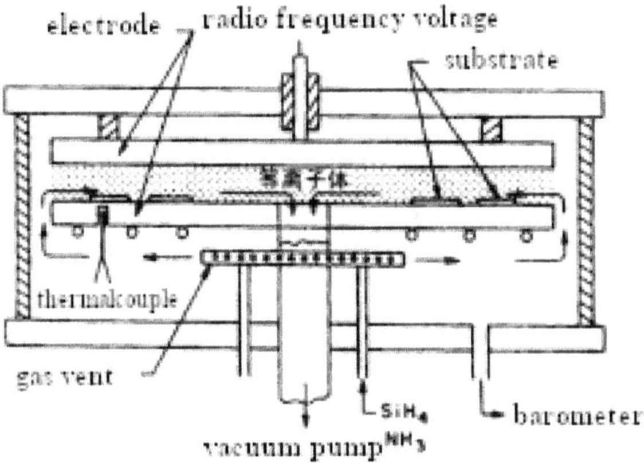

Figure 3.5: Capacitance coupled RF PECVD device.

In this device, an RF voltage is imposed on two relatively placed plate electrodes, through which the reaction gas passes and then the corresponding plasma is produced. The deposition of the film on the substrate is carried out by the active functional groups of the plasma. The CVD process for Si_3N_4 generated by SiH_4 and NH_3, for example, is performed at about 900°C in the atmospheric CVD device, while it is performed in the low pressure CVD unit at around 750°C. The application of the PECVD device can help achieve a uniform and large-area deposition of Si_3N_4 dielectric film at a low temperature of 300°C. At the same time, because the device works also under lower pressure, the diffusion capability of the active group can be enhanced, and the film can grow at the rate of tens of micrometers per minute.

A schematic of the inductive coupling of the PECVD device is given in Figure 3.6. The high frequency coils in the device are placed outside the reaction vessel, and the alternating magnetic field generated by it induces an alternating current in the reaction chamber, thus forming the electrode free discharge of the gas. Because of the electrode free nature of the plasma discharge, it is possible to avoid the contamination of electrode discharge. Table 3.2 lists some typical examples of the application of the PECVD system to deposit films.

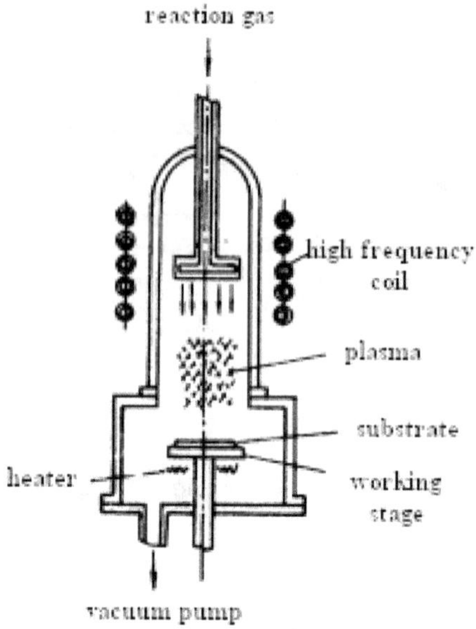

Figure 3.6: Schematic diagram of an inductive coupling radiofrequency PECVD device.

It is important to note that the use of the PECVD technique for depositing films is a very complex process involving both chemical vapor deposition processes, glow discharge processes, and plasma enhancement of film growth. In other words, there exist extremely complex plasma chemical reactions. The plasma that is used to excite CVD includes radiofrequency plasm, DC plasma, pulsed plasma, microwave plasma, electron cyclotron resonance plasma, and so on. These different types of plasma are generated by the glow discharge of a rare gas excited by radiofrequency, DC pressure, pulsed high voltage, microwave, and electron cyclotron resonance. The pressure of the discharge gas in the reaction chamber is usually 1–600 Pa.

3.2.4.5 *Electron cyclotron resonance plasma enhanced CVD*

Another common improvement of PECVD is shown in Figure 3.7. Such a PECVD device, known as an electronic cyclotron resonance (ECR) PECVD device, uses a power source of microwave frequency to generate the plasma. The microwave energy is coupled by a

Table 3.2: Typical applications of PECVD.

Applications	Constituent	Gas source
Insulating and passive films	SiO_2	$SiCl_4+O_2$ $Si(OC_2H_5)_4$
	Si_3N_4	SiH_4+NH_3
	Al_2O_3	$2AlCl_3+3CO_2+3H_2$
Amorphous silicon solar cell, electronic photographic photocopying	$\alpha-Si$	$SiH_4(SiH_2Cl_2)+B_2H_6$ (pH_3) or mixed gas $SiH_4+SiF_4+Si_2H_6$
Plasma polymerization	Organic compound	—
Wear-resistant film	TiC	$TiCl_4+CH_4$
	TiN	$TiCl_4+N_2$
	TiC_xN_{1-x}	$TiCl_4+CH_4+N_2$
Other films	SiC	$SiH_4+C_2H_2$ Si_4+CH_4 (or CF_4)
	Al_2O_3	$AlCl_3+O_2$
	BN	$B_2H_6+NH_3$
	P_3N_5	$P+N_2$

microwave waveguide into the reaction vessel and causes the gas to produce plasma breakdown and discharge. In order to promote the absorption of energy from the microwave field by the electrons in the plasma, a magnetic field coil is set up in the device to produce a magnetic field with a certain divergence. A cyclotron resonance occurs when an electron moves in the microwave field and the magnetic field.

The ECR method has a high degree of vacuum, typically 10^{-1}–10^{-3} Pa. Therefore, the ECR method gives a plasma ionization degree one to three orders of magnitude higher than the normal PECVD method. This means that the plasma is highly reactive, and can even be thought of as an additional ion source. In addition, ECR type PECVD has other advantages, such as low-pressure and low-temperature deposition, high deposition rate, zero electrode pollution, and good plasma controllability, which is why ECR technology is widely used in film deposition and film etching.

3.2.4.6 *Laser assisted CVD*

Laser assisted CVD technology is a thin film deposition technique that promotes or controls the CVD process by using the laser as an auxiliary excitation method. As a light source of high intensity, monochromaticity and directionality, the laser plays an important role in the CVD process, with two main effects:

(1) **Thermal effect:** The heating of the substrate by the laser can promote the chemical reaction on the substrate surface, so that the CVD can be achieved when the substrate is not too hot.
(2) **Light effect:** High-energy photons can directly facilitate the decomposition of the gas molecules in the reactants. Since the decomposition of many common reactant molecules (such as SiH_4, CH_4, etc.) requires the wavelength of the photons to be less than 220 nm, the effect usually only exists in the ultraviolet band of the excimer laser. Figure 3.8 shows the two mechanisms of laser beam action in laser assisted CVD technology.

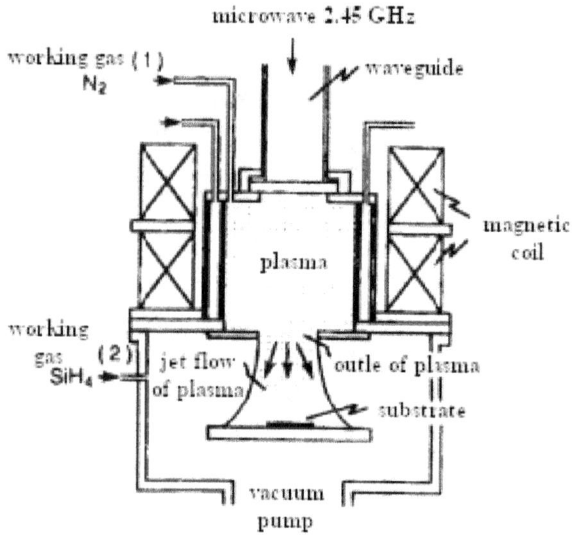

Figure 3.7: Schematic diagram of ECR PECVD deposition.

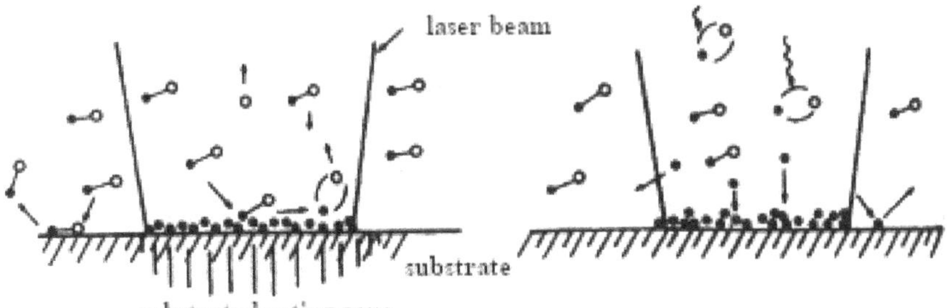

Figure 3.8: Two mechanisms of action of a laser beam on the surface of the CVD substrate. (a) Pyrolysis, (b) Photoactivation.

Another feature of the laser assisted CVD technology is that the selective film deposition of the reactant on the substrate can be achieved by using the above thermal effect and the photon activation effect, which means that the laser beam can be applied to the substrate surface only when it needs to be deposited, thus obtaining the desired deposition pattern. In addition, laser assisted CVD technology can effectively reduce the base temperature of the CVD process. For example, SiO_2 film deposition can be achieved at the base temperature of 50°C, thanks to a laser assisted approach.

Laser assisted CVD technology has been widely used in the deposition of metal and insulating dielectric films. The former directly exploits the decomposition of some metals under light radiation, while the latter exploits the chemical reactions of a variety of gases under the promotion of photons.

3.2.4.7 *Metal–organic compound CVD*

MOCVD is a CVD technique that utilizes the thermal decomposition and subsequent chemical synthesis of metal organic compounds to epitaxially grow thin films in the gas phase. This method has been widely used in the gas phase growth of compound semiconductor thin films. The principle is the same as that of silicon epitaxial growth by SiH_4 thermal decomposition, such as

$$(CH_3)_3Ga + AsH_3 \xrightarrow{630\sim675^\circ C} GaAs + 3CH_4.$$

In MOCVD, raw material compounds containing semiconductor compound elements must meet the following conditions: stable at room temperature and easy to handle. Meanwhile, the by-products should not interfere with the crystal growth, do not contaminate the growth layer, and have a vapor pressure around room temperature.

MOCVD is a new epitaxy technology developed rapidly in recent decades, which has been successfully used in the preparation of superlattice structures, ultra-high-speed devices and quantum well lasers. It has low deposition temperature and can be widely applied in the range of almost all compounds and alloy semiconductor growth. In MOCVD technology it is relatively easy to control the deposition rate, which is beneficial in growing films that have a big change of component in the thickness direction, and can also deposit very thin film layers of different composition repeatedly. Thus, it can be used to produce superlattice materials and epitaxially grow various III–V and II–VI compound semiconductor heterojunction films. Its recent application includes the preparation of high temperature superconducting ceramic films. Figure 3.9 is the schematic diagram of an MOCVD device growing $Ga_{1-x}Al_xAs$.

However, the traditional MOCVD process adopts the precursor of gaseous metal–organic matter, and for the corresponding gases

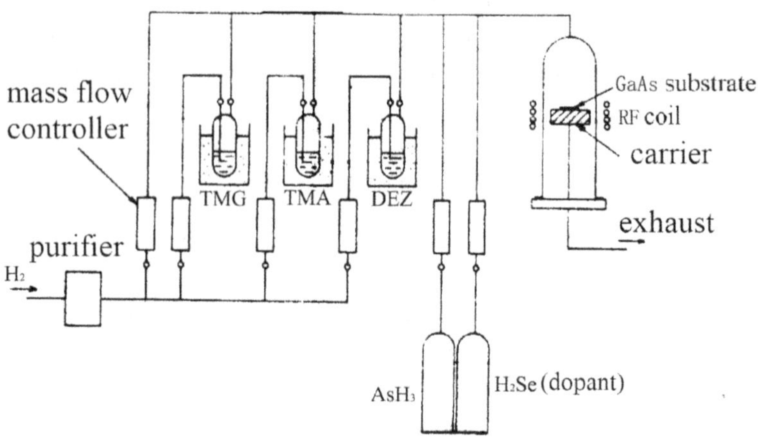

Figure 3.9: MOCVD device schematic diagram for $Ga_{1-x}Al_xAs$ film preparation.

the gaseous precursor can't be obtained for some complex metal–oxide films, or the gaseous precursor is too expensive. And given that many of the metallic–organic compounds in the membrane materials are readily available or obtained, source MOCVD (liqued source MOCVD, or LSMOCVD) has been developed in recent years. LSMOCVD has been used as a new method for the preparation of multi-component metal–oxide films. It is a good way to avoid the complexities of multi-source delivery. The various sources are dissolved into organic solvents, and a well mixed precursor solution is obtained, which is then fed into the evaporation chamber to obtain the gaseous source material, and then through the flow control is fed into the reaction chamber, or directly into the reaction chamber, and vaporized and deposited in the reaction chamber. The advantage of this method is the simplification of the delivery of source materials, and the requirement for source materials is reduced, so it is convenient to realize the alternating deposition of various films in order to obtain the superlattice structure. Various MOCVD technologies have been used to prepare excellent superconducting films, ferroelectric films, semiconductor superlattices, and multilayer films.

3.2.4.8 *Hot wire chemical vapor deposition*

Hot wire chemical vapor deposition (HWCVD) is a newly developed film preparation method. It uses a high temperature hot wire to decompose the precursor gas, and obtains a large area of high quality sedimentary film by adjusting the precursor component contrast and the hot wire temperature. The hot wire chemical vapor deposition method has advantages such as simple installation, low deposition temperature, and no introduction of plasma. A good crystalline cubic silicon carbide film can be prepared under the lower substrate temperature (400°C) under the Si (100) substrate with the HWCVD method.

In recent years, HWCVD technology has developed rapidly. For example, the μc–SiC film was prepared using the HWCVD process. The conductivity of the n-type material is 5 s/cm, and the conductivity of the p-type material can be increased to 1×10^{-2} s/cm.

The absorption spectrum of the μc–GeC film is shifted to the higher photon energy side (compared with the crystal Ge). The data suggest that μc–SiC and μc–GeC are promising next-generation thin-film solar cell materials.

3.2.4.9 *Atomic layer deposition*

Atomic layer deposition (ALD) was also called atomic layer epitaxy (ALE), atomic layer growth (ALG), atomic layer CVD (ALCVD), and molecular layer epitaxy (MLE). ALD is a method of depositing substances in the form of a single atomic film layer by layer on the substrate by selective pulse of precursor and reactive gases into the chamber and to the substrate at a controlled rate, resulting in physical and chemical adsorption or surface saturation reactions on the substrate surface. The typical ALD process is shown in Figure 3.10. The most distinctive feature of ALD is that it has a self-limiting growth nature; each time only one atomic or molecular layer can grow. This self-limiting reaction is repeated until a desired thickness of the film is prepared. Therefore, ALD offers the best method for controlling the film thickness and surface smoothness in truly nanometer or sub-nanometer range.

An ALD deposition cycle can be divided into four steps:

(1) A chemical adsorption or reaction occurs between the first precursor and the substrate surface.

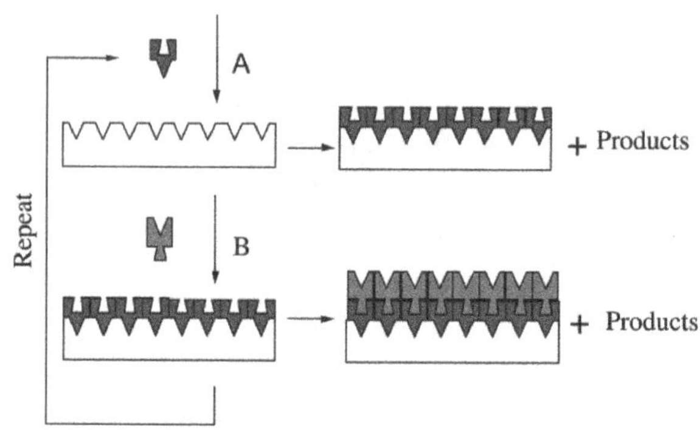

Figure 3.10: Schematic representation of ALD using self-limiting surface chemistry and an AB binary reaction sequence.

(2) The excess precursors and byproducts are removed from the reaction chamber by inert gas.
(3) The second precursor reacts chemically with the first precursor on the surface of the substrate to form a thin film.
(4) After the reaction is complete, the excess precursors and by-products are cleared out of the reaction chamber with inert gas.

The choice of proper precursors is the key issue in the successful design of an ALD process. A variety of precursors have been used in ALD. For example, elemental zinc and sulfur were used in the first ALD experiments for the growth of ZnS. Metal chlorides were studied soon after the first demonstrations of ALD. Metalloragnic compounds including both organometallic compounds and metal alkoxides are widely used. For non-metals, the simple hydrides have mostly been used: H_2O, H_2O_2, H_2S, H_2Se, H_2Te, NH, N_2H_4, PH_3, AsH_3, SbH_3 and HF.

Since the emergence of ALD technology, a series of new ALD technologies have been developed, such as plasma enhanced ALD, electrochemical ALD, strong oxidant-assisted ALD, magnetic plasma-assisted ALD technology, and so on.

ALD technology has been widely used in many fields because of its excellent shape preserving property and uniformity, high step coverage rate, and rate controllability. Thin films of various materials including various oxides, nitrides, florides, elements, II–VI and III–V compounds in epitaxial, polycrystalline or amorphous form deposited by ALD are summarized in Table 5.2 of Chapter 5.

In addition to the above commonly used CVD methods, there appear many other CVD technologies, which will not be described here.

3.3 Chemical Solution Preparation Technique for Thin Films

The chemical solution preparation technique refers to a technique of depositing thin films on the substrate surface by chemical reaction or electrochemical reaction in solution. It includes chemical reaction deposition, sol–gel method, anodic oxidation, electroplating, and so on. This kind of technique does not need a vacuum environment and too many equipment. It can deposit the film in every kind of matrix surface, and the raw material is relatively easy to obtain. Therefore, it has been widely used in the field of electronic components, surface plating, decoration, etc.

3.3.1 *Chemical reaction plating*

Chemical reaction plating is one of the most important kinds of solution film making. It is mainly used to deposit thin films on substrates by various chemical reactions (such as redox, displacement, and hydrolysis).

3.3.1.1 *Electroless plating*

Electroless plating, commonly referred to as non-power supplied electroplating, is a method of using reductants to reduce metal ions in metal salts to atomic states and deposit them on the surface of the substrate to obtain a film. It belongs to the method of chemical reaction deposition without electrification, together with chemical sedimentation. The difference between the two is that the reduction of chemical plating must be conducted under the action of catalyst, and the deposition reaction occurs only on the surface of plated parts. Otherwise, the chemical deposition reduction reaction is uniform in the whole solution, only part of the metal layer is deposited on the plating piece, and most of it is deposited in the solution as metal powder. That is to say, electroless plating is a redox process that occurs on the plating (substrate) under catalytic conditions. In this plating process, metal ions in the solution are reduced continuously and deposit on the substrate surface, catalyzed by the growing plating surface. Therefore, catalysis of the surface of substrate material is quite important, and metal elements of group V and group III in the periodic table have the catalytic effect that is needed in the process of electroless plating.

The catalysts mentioned above mainly refer to sensitizers and activators, which can promote the electroless plating process to occur on the surface of the plating with catalytic activity. If the metal to be plated itself does not catalyze automatically, the deposition process will stop automatically after the active surface of the plating is covered by the deposited metal. In contrast, metals like Ni, Co, Fe, Cu, Cr, etc., which themselves catalyze the reduction reaction, and allow the plating process to continue until the film thickness reaches the requirement. Such electroless plating, which relies on the autocatalytic action of the metal to be plated, is called autocatalytic electroless plating. Most of the normal electroless plating is this kind of autocatalytic electroless plating. Autocatalytic electroless plating has the following characteristics:

(1) Can form a uniform plating on a complex plating surface;
(2) Low porosity of the film layer;
(3) Can directly plate on plastics, ceramics, glass and other non-conductors;
(4) No power supply, no electrodes;
(5) The plating has special physical and chemical properties.

In electroless plating, the ionization potential of the reductant used must be lower than that of the deposited metal electrode, but the potential difference between the two should not be too large. Common reductants are hypophosphate and formaldehyde, the former for Ni plating, and the latter for Cu plating. The reductants must provide the electrons needed for the reduction of metal ions. That is,

$$\mathrm{M}^{+n} + n\mathrm{e}^- \rightarrow \mathrm{M}.$$

This reaction can only be carried out to obtain the plating on the surface of the plating with catalytic properties, and once the deposition begins, the deposited metal must continue this catalytic function so that the deposition process can continue and the plating can be thickened. Therefore, in this sense, electroless plating must be a controlled autocatalytic chemical reduction process. Electroless nickel plating is a process in which nickel ions are reduced to metallic nickel by the nickel salt solution under the action of a strong reducing agent, hypophosphate, and phosphorus is precipitated by hypophosphate decomposition. A film of Ni–P alloy is obtained on the substrate with catalytic surface. The reaction process is

$$H_2PO_2^- + H_2O \rightarrow HPO_3^{2-} + H^+ + 2H,$$
$$Ni^{2+} + 2H \rightarrow Ni + 2H^+,$$
$$2H \rightarrow H_2 \uparrow,$$
$$H_2PO_2^- + H \rightarrow H_2O + OH^- + P.$$

Thus, the equation for oxidation of hypophosphate and reduction of nickel ion is obtained as follows:

$$Ni^{2+} + H_2PO_2^- + H_2O \rightarrow HPO_3^{2-} + 3H^+ + Ni.$$

3.3.1.2 *Displacement deposition plating*

This method is also known as immersion plating. The principle is to deposit the metal on the substrate by chemical replacement in the solution of the salts to be plated, without the need for an external power supply. For example, when the Fe substrate with low potential is immersed in the copper salt solution with high potential, the copper will be precipitated on the surface of the substrate because of the potential difference and the formation of a microcurrent. The reaction is as follows:

$$Fe + Cu^{2+} \rightarrow Cu + Fe^{2+}.$$

This reaction is conventionally called a displacement reaction. Under this condition, the precipitated metal is attached to the surface of the substrate metal to deposit a plating.

Displacement deposition is essentially a process of exchanging metal atoms and ions between solid and liquid phases at the interface. It does not need to add extra reductant, because the substrate itself is a reductant. In order to improve the porosity and poor bonding of the film, additives or complexing agents can be added to improve the binding force of the film.

3.3.1.3 *Solution hydrolysis plating*

The essence of this method is to synthesize hydrocarbon oxygen compounds by compounding group IV elements with group III and group V elements in the periodic table, and using some inorganic salts, such as oxides, nitrates and acetate as plating materials. These film-forming substances, dissolved in some organic solvents, such as acetic acid or cupric, will become a plating solution, which is put on the surface of the rotating flat-glass plating in the plating bath, hydrolyzed to form a colloidal film, and then dehydrated, finally obtaining the oxide film of the element. For example, the preparation of $\lambda/4$-$\lambda/2$-$\lambda/4$ three-layer wide band anti-permeability films and other optical films by ethyl titansate [$Ti(OC_2H_5)_4$] and ethyl silicate [$Si(OC_2H_5)_4$] is achieved using this method, which is close to the sol–gel method described below.

3.3.2 *Sol–gel method*

The required oxide films can also be obtained by hydrolysis with appropriate solution of metal organic compounds.

The essence of this coating method by solution hydrolysis is to synthesize hydrocarbon oxygen compounds by using the elements in group III, IV, and V of the periodic table, and dissolve them in some organic solvents (such as propyl acetate copper or other organic solvent) together with some inorganic oxides such as nitrate, and acetic acid salt to get the sol, then to plate the sol on the substrate surface by using the methods of impregnation and centrifugal glue. The sol forms a colloid film on hydrolysis, and then on dehydration it solidifies into a solid film. The thickness of the film depends on the concentration of the metal–organic compounds in the solution, the temperature and viscosity of the sol, the rotational speed and angle of the substrate, and the ambient temperature.

The main steps of sol–gel process are as follows: first dissolve metal alcohols in organic solvents, then dehydrate and add other components (in the form of inorganic salts, so long as they can be miscible after addition) to make a homogeneous solution, and then hydrolyze and polymerize at a certain temperature to form a gel.

The main processes include:

Hydrolysis reaction:

$$M(OR)_n + H_2O \rightarrow (RO)_{n-1}M\text{-}OH + ROH.$$

Polymerization reaction:

$$(RO)_{n-1}M\text{-}O + RO\text{-}M(OR)_{n-1} \rightarrow (RO)_{n-1}M\text{-}O\text{-}M(OR)_{n-1} + ROH,$$

where M is a metallic element, such as titanium (Ti), zirconium (Zr), etc. and R is an alkoxy group.

For example, the reaction process of preparing TiO_2 film with ethyl titanate is as follows:

$$Ti(OC_2H_5)_4 + 4H_2O \rightarrow H_4TiO_4 + 4C_2H_5OH,$$

$$H_4TiO_4 \xrightarrow{120^\circ C} TiO_2 + 2H_2O \uparrow .$$

A film prepared by the sol–gel method has the advantages of uniform mixing of multiple components, easy control of components, uniform film formation, low cost, short cycle, and easy industrial production. It is used to prepare TiO_2, Al_2O_3, SiO_2, $BaTiO_3$, $PbTiO_3$, PZT, PLZT, and other thin films at present.

3.3.3 *Anodic oxidation method*

The anodic oxidation method is to put some metals or alloys (such as aluminum, tantalum, titanium, vanadium, etc.) in the corresponding electrolyte as the anode, together with the cathode of graphite (or the metal itself) and apply a certain dc voltage. Due to the electrochemical reaction, the metal–oxide film will be formed on the surface of these metals; this process is called the anode oxidation, and the film-forming method is called anodic oxidation method. Like other electrolytic processes, it follows Faraday's law, which strictly quantificationally converts a certain amount of electricity into an oxide of the metal. However, during the anodic oxidation process, a certain amount of oxide is dissolved in the electrolyte, so the effective mass of the oxide film on the metal surface is lower than the theoretical value. In other words, there are two opposite processes of metal dissolution and film formation in the anodic oxidation process, and the film formation is a combination of the two. Therefore, the formation of oxidation film is a typical non-uniform reaction. The following reactions occur in the plating:

Oxidation of metal M

$$\mathrm{M} + n\mathrm{H_2O} \rightarrow \mathrm{MO}_n + 2n\mathrm{H}^+ + 2ne^-.$$

Dissolution of metal:

$$\mathrm{M} \rightarrow \mathrm{M}^{2n+} + 2ne^-.$$

Dissolution of MO_n oxide:

$$\mathrm{MO}_n + 2n\mathrm{H}^+ \rightarrow \mathrm{M}^{2n+} + n\mathrm{H_2O}.$$

The above simultaneous reactions were used to produce an anodic oxide film. In the early stage of film formation, the film formation reaction and the metal dissolution reaction occur simultaneously. The solution reaction produces hydrated metal ions and colloidal precipitated oxides of hydroxyl compounds or oxides. The metal activation and dissolution will stop when the oxide covers the surface. The continuous oxidation reaction makes the metal ions and electrons pass through the insulating metal oxide and continue to form oxides on the surface of the film. In order to maintain the movement of ions and ensure the growth of oxide films, a certain electric field is required. The composition and structure of the anodic oxide film are related to the type and concentration of electrolyte, the process parameters, and some other factors.

3.3.4 *Electroplate method*

The chemical reaction caused by passing a current through an electrolytic salt solution is called electrolysis, and the process of plating on the substrate in the cathode is called electroplating. Since electroplating and electrolysis are carried out in aqueous solution, while vacuum evaporation, ion plating and sputtering are carried out in vacuum, the former can also be called wet plating technology, while the latter is called dry plating technology. With the development of plating technology, electroplating can also be carried out in non-aqueous solution (such as molten salt).

Electroplating is a process to get metal thin films in which positive ions are discharged on the surface of the cathode by applying a direct current in an aqueous solution containing metallized ions. Electroplating mainly refers to electroplating in aqueous solution and has been widely used. In the electroplating process, the cathode surface is plated with a layer of metal film by applying an external direct current field to reduce the potential of the cathode and reach the precipitation potential. At the same time, the anode potential must be increased. Only when the applied potential is much larger than the anode potential can the anode metal be continuously dissolved and the dissolution rate exceed the deposition rate of the cathode, so as to ensure a normal plating process. The electroplating process follows two basic rules proposed by Faraday: (1) the amount of chemical reaction is proportional to the passing current; (2) under the same electric flux, the amount of different substances deposited on the cathode or decomposed from the anode is proportional to their molar mass.

The electrolytic solution used in electroplating is called the electroplating solution. Generally, the salts that are used to plate the metals can be divided into single salts and complex salts. Single salts like chloride, sulfate, etc. and complex salts like cyanide are contained in the electroplating solution. A single salt is safe and cheap, but the film quality is poor, and relatively rough. A complex salt is expensive and toxic, but the film surface is bright. Different kinds of solutions can be selected according to different requirements. Usually a single salt solution is used to plate nickel, platinum, and a complex salt solution is used to plate copper, gold, etc. In the electroplating process, the basic requirements for electroplating are: fine crystal, smooth and firm plating, no pinhole, etc. Because electroplating is carried out at room temperature, plating has the advantages of being compact, smooth, pinhole free and easy to control the thickness, so it has been widely used in the electronic industry.

In addition, the chemical solution method includes electrophoretic deposition, etc., which will not be detailed here due to the limited space.

3.3.5 *Spray pyrolysis*

3.3.5.1 *Fundamentals of spray pyrolysis*

The MOCVD method is used to prepare thin film at low deposition temperature and high deposition rate, and can obtain homogeneous film and multiplex film, which is compatible with the semiconductor process, and is widely used in the preparation of semiconductor, ferroelectric, superconductive, and other thin film materials. However, because the source of metal–organic materials (MO source) is gas or liquid with high vapor pressure, it is hard to prepare and purify.

The variety of preparation and the easy hydrolysis, and the inconvenience during transport and usage to some degree limit the development of MOCVD. In order to overcome the shortage of MOCVD, a new method for preparing the film was developed, namely spraying pyrolysis (spray pyrolysis, or SP). The spray pyrolysis method, which combines the advantages of the liquid phase method and the gas phase method, shows wider application prospects.

Spray pyrolysis uses organic solution or aqueous solution analogous to the those in the metal–organic decomposition (MOD) or Sol–gel method (sol–gel) to atomize the liquid phase into droplets, and uses a method similar to CVD to send the droplets into the reaction chamber, and to react on the heating substrate to deposit films. According to the different atomization methods, the technology of spray pyrolysis can be divided into pressure atomization, deposition, ultrasonic atomization, and electrostatic atomization deposition. Although different in atomization mode, each system includes the transport of the liquid precursor, the atomization of the precursor, the transport of the atomized droplet, the heating of the substrate, and the film deposition.

Since Chamberlin and Skarman used SP to fabricate the CdS membrane for solar cells in 1966, SP has been widely used in the preparation of mono oxide, spinel oxide, perovskite type oxides, as well as sulfides and selenides. The SP method is also used in the preparation of ceramic powder of multi-chemical composition.

There are several advantages of the SP system in thin film technology:

(1) Simple equipment, no need of high vacuum, and can work under normal pressure;
(2) Deposition of large-scale film, deposition on the surface of a solid, high deposition speed, and easy to industrialize.
(3) Many choices of precursor, easy to control the stoichiometry of the films, and easy to dope and change the concentration of components in the precursor solution for fabricating multi-layer films or component gradient films.
(4) Overcomes the difficulty of the sol–gel method in preparing thick film by adjusting the atomization parameters to control the thickness of the films.
(5) The deposition temperature is mostly below 600°C, relatively low.

Whatever be the method of atomization, the film prepared by the SP method is made of droplets or fine particles of fine powder. The surface of the film is not as smooth as that of the film made by MOCVD, and the porosity of the film is also higher. Moreover, the size and distribution of the atomized droplets, the properties of the solution, and the temperature of the substrate are very important to the surface morphology of the films prepared by SP. Therefore, spray pyrolysis cannot prepare smooth and compact films easily and it is easy to introduce impurities during the deposition, and SP is mainly limited to the preparation of oxides, sulfides and other materials.

3.3.5.2 *Improved techniques for spray pyrolysis*

In order to improve the surface morphology and properties of the thin film, the main research is focused on three aspects: (1) improving the atomization technology, such as pulse intermittent spray and nano jet, (2) choice of source material and solvent, and improvement of the preparation of the precursor, and (3) a new method for heating deposition, such as microwave heating, thermal plasma separation, cold plasma atmosphere deposition, etc.

The type of plasma enhanced spray pyrolysis is based on the way the plasma is produced: (1) corona discharge spray pyrolysis, (2) microwave discharge plasma spray pyrolysis, and (3) radiofrequency plasma spray pyrolysis.

(1) Corona discharge spray pyrolysis

The corona discharge plasma is weakly ionized, in which neutral particles are much more than ions (10^4–10^7:1), the electrons temperature is lower, and the temperature of ions is much lower. Because of the low efficiency of the traditional pressure spray pyrolysis, the corona discharge is used to charge the droplets and make it deposit onto the substrate, improving the deposition efficiency. If the ultrasonic atomization technology is used to atomize the precursor, using nitrogen as the carrier gas, the corona discharge plasma is generated by applying a high voltage of 20–60 kV to the blade fixed above the flow of air mist, which causes the atomized droplets to be charged, and the charged droplets are deposited on the ground substrate in a fixed direction, and the deposition efficiency is increased to 80%.

(2) Microwave plasma spray pyrolysis

Microwave discharge is a way of discharge by converting the energy of microwave into internal energy of the gas molecules, which excites and ionizes to give a plasma discharge. Usually a frequency of 2.45 GHz is adopted, and it has many similarities to radiofrequency discharge. Microwave discharge does not require us to set up the electrode in the discharge space, and the power can be localized, thus obtaining a high-density plasma. In the process of preparing thin film by microwave plasma spray, the mass of droplets is much larger than the mass of charged particles, so the microwave energy transferred to the droplets by collision is rather small, which makes it difficult to vaporize the source material particles. Studies show that microwave coupling to the droplets of salt solution can't work because the microwave coupling of 2.45 GHz is not strong enough. Using microwave plasma spray pyrolysis, ceramic powders of alumina and zirconium oxide were synthesized. The morphology was compared with the powder prepared by pyrolysis, and it was found that there was no obvious difference between the two, proving that the process of plasma pyrolysis is a pure thermal process, and the plasma enhanced chemical reaction has little effect on the surface morphology of the powder.

(3) **Radio frequency inductive plasma spray pyrolysis**

Because the radiofrequency inductively coupled plasma (Spray-ICP) is a hot plasma with a temperature of more than 5000 K, it can vaporize and ionize almost all kinds of source materials and use them to deposit the films. Commonly used RF power sources work at 13.56 MHz, using metal coils to couple electricity energy to the plasma. The Spray-ICP technology atomizes the precursor solution into droplets, which are introduced into the plasma torches with the carrier gas, and the droplets undergo the process of drying, gasification, decomposition, and ionization, and finally the particles are deposited on the substrate. The technology is mainly used in the preparation of metal–oxide powder, oxide film, and epitaxial thick film. However, because the substrate is heated by the wake flame of the plasma torch to a high temperature, and there exists strong electromagnetic interference, the temperature of the substrate is difficult to control and measure accurately. Thus, the stoichiometry of each component of the film is more difficult to control because of the impacts from the temperature of the substrate and the position of the substrate.

3.4 Soft Solution Preparation Technology of Thin Films

3.4.1 *Fundamentals of SSP*

In order to develop a large number of advanced functional materials in an environmentally friendly way and realize the sustainable development of economy, many researchers have put forward new technological concepts and technological processes, and this should

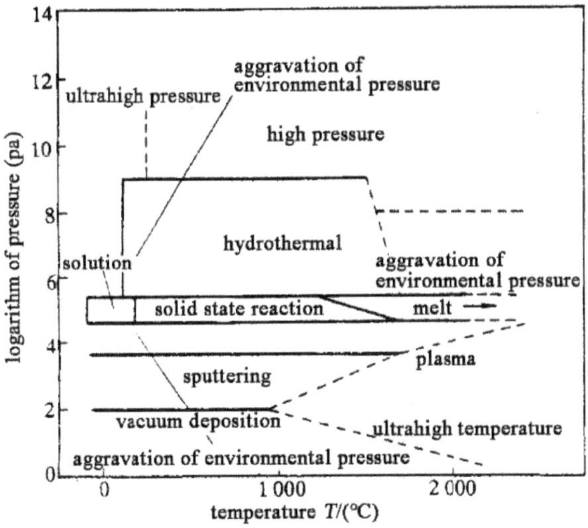

Figure 3.11: Temperature and pressure diagram of different preparation processes of inorganic ceramic materials (P–T).

become the strategic direction of material workers. Based on this, Masahiro Yoshimura of Tokyo Institute of Technology proposed a new technology for preparing advanced inorganic materials based on years of research on ceramic film preparation technology — SSP. This technology points out a new and environmentally friendly development direction for the preparation of advanced inorganic powders, crystals, and films.

From the viewpoint of the Earth's ecological material cycle and energy cycle, the solution process route is the most environmentally friendly. Figure 3.11 shows the relationship between temperature and pressure (P–T) in different preparation processes of inorganic ceramic materials. Figure 3.12 shows the relationship between free energy and temperature (G–T) of several preparation processes of ceramic materials in the unit system.

It can be seen from Figure 3.10 that the range and characteristics of aqueous solution processes in the P–T diagram exactly meet the conditions of life activities on the Earth, while other processes are related to higher temperature and higher (or lower) pressure, so they are out of harmony with the environment. From Figure 3.11, the

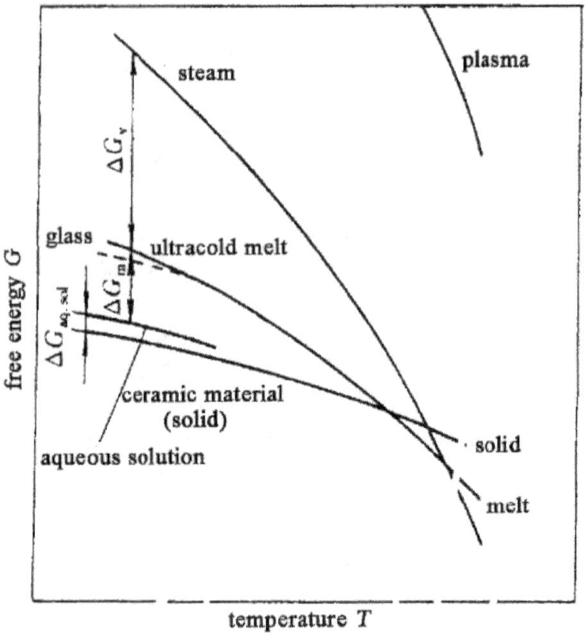

Figure 3.12: Free energy and temperature (G–T) diagram of several ceramic preparation processes in the unit system.

aqueous solution system has the lowest total energy consumption among all the process routes. The results in Figures 3.10 and 3.11 show that the basic starting point of the concept of SSP is based on the theoretical results of thermodynamic analysis and research on problems related to material processes.

The concept of SSP is similar to "Soft Chemistry" or "Mild Chemistry" or "Biomimetic Process"; the main advantages of this technology are as follows:

(1) One-step (or direct) molding or directional deposition in the preparation of ceramic materials;
(2) Low energy consumption;
(3) Materials with any shape and size are expected to be prepared according to the preparation technology;
(4) The whole process is carried out in a closed system, so it is easy to load, separate, and recycle;

(5) Relatively high yield;
(6) Multifunctional products are expected to be produced;

There are many kinds of SSP technology; some of the technologies mentioned above, such as hydrothermal technology and electrochemical technology, belong to the soft solution technology.

3.4.2 *Hydrothermal electrochemical*

Hydrothermal processes require higher temperature and pressure, and because of poor airtightness of a single electrochemical technology, some materials, especially perovskite-type dielectric ceramic films, are difficult to prepare and the film-forming rate is low. So it is possible to combine the hydrothermal method and the electrochemical method to form the hydrothermal electrochemical method.

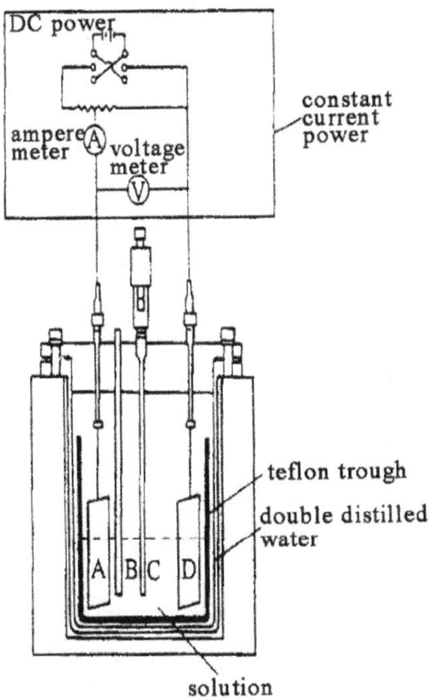

Figure 3.13: Schematic diagram of a hydrothermal electrochemical reaction device.

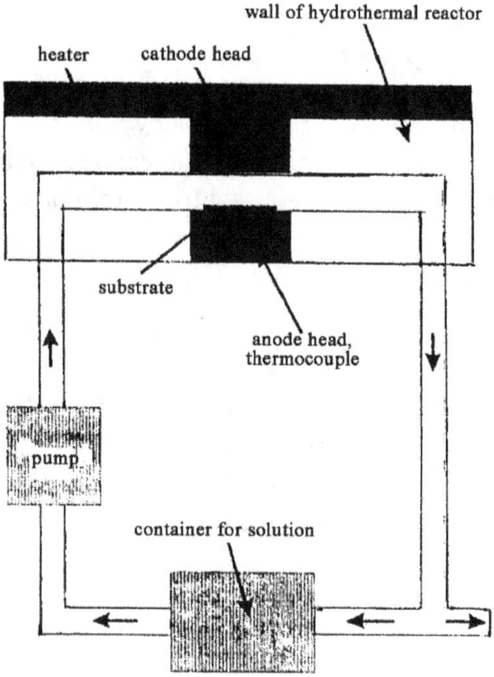

Figure 3.14: Schematic diagram of a hydrothermal electrochemical device of the flow system.

At present, the method is successfully used in the preparation of perovskite (ABO3) ferroelectric ceramic films. Figure 3.13 shows a schematic diagram of the device for the hydrothermal electrochemical reaction. By adding a dc electric field to the hydrothermal reactor, a ceramic film is prepared under hydrothermal conditions and an electric field. Figure 3.14 shows a schematic diagram of a hydrothermal electrochemical device of the flow system.

The basic principle of hydrothermal electrochemical technology: in a special closed reaction vessel, use an aqueous solution as the reaction medium, create a high-temperature and high-pressure reaction environment by heating the reaction vessel, so that the insoluble or hardly dissoluble substance can be dissolved or its solubility can be increased. On the one hand, the viscosity, dielectric coefficient, and expansion coefficient of water will change under hydrothermal conditions. In the rarefied-gas state, the viscosity of water increases with higher temperature, but decreases with higher temperature when compressed into a dense state. The viscosity of the hydrothermal solution is about two orders of magnitude lower than that of the solution at room temperature and room pressure. Because the diffusion is proportional to the viscosity of the solution, there is a very effective diffusion in the hydrothermal solution; thus the growth rate of the crystal nucleus and grains is higher than that of other aqueous solution systems and the diffusion region is narrower near the interface and so on, which creates a condition for the preparation of film materials.

3.5 LB Technology for Ultrathin Organic Thin Films

3.5.1 *Overview*

An Langmuir–Blodgett (LB) film is a kind of ultra-thin organic film, and the LB technology is the kind of film-making technology that can accurately control the thickness and molecular structure of the film. This technology is the monolayer preparation technology established in the 1930s by American scientist I. Langmuir and his student K. Blodgett. In this technology, insoluble molecules are arranged closely and orderly to form a monolayer, and then it is transferred to the surface of a solid. Thus, an LB film is an ordered organic molecular film.

Some organic materials have the ability to form a film with molecular thickness on the interface of air and water when put into the water surface, and the most typical example of organic compounds is stearic acid. These materials are amphiphilic molecules. One end of the molecule is a hydrophilic group, which is soluble in water; The other end of the molecule is hydrophobic, which is insoluble in water but lipophilic.

To obtain a monomolecular film, the solid material must be first dissolved in a suitable solvent, which must be able to dissolve an appropriate amount of the monomolecular material. The solvent is used to extend the monolayer, so it does not react chemically with the monolayer, and of course it does not dissolve in the base phase solution, and the solvent eventually evaporates at the right time without remaining in the monomolecular compression layer. Examples of common solvents include trichloromethane, n-hexane, and so on.

Usually, the dissolved solution is dripped onto the appropriate liquid base phase (usually ultrapure water). At this point, the solution dropped on the surface of the base phase immediately expands outward, and the organic solvent volatilizes during this period, leaving molecules of dissolved condensed material to distribute disorderly on the surface of the base phase. Monomolecular thin films can be obtained by ordering the disordered molecules of such a condensed matter and transferring them to the required substrates (or other substrates).

3.5.2 *Langmuir–Blodgett technology*

A floating monomolecular film in condensed matter can be transferred to solid substrates in a variety of ways, the most common of which is LB deposition. This technology allows a solid substrate, such as a silicon chip or a glass sheet, to pass back and forth through the interface between the monolayer and water at an appropriate rate in the case of constant pressure of the surface of monolayer, while maintaining the pressure on the surface of the monolayer, and the molecular film is transferred to the surface of the solid substrate layer by layer.

Figure 3.15 shows a schematic diagram of the device used to prepare LB films. The water container is called a Langmuir trough, which consists of the following parts:

(1) There is a trough spreading on the monolayer, and the surface of the tank is usually coated with inert materials (such as polyvinyl fluoride), which is convenient to clean and avoid being polluted.
(2) There is an independent movable belt, whose purpose is to fix the monolayer within a certain area. The belt can be driven by a motor to change the pressure and area of the surface of the liquid.

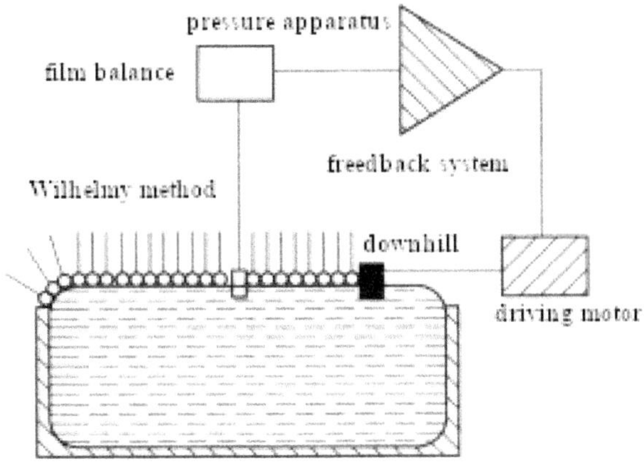

Figure 3.15: Schematic diagram of an LB film preparation device.

(3) There is a pressure sensor to monitor the surface pressure at any time.

(4) There is a deposition pulling mechanism used to fix the substrate sheet. The speed is adjustable, and it can reciprocate smoothly through the interface of monolayer film and water. The typical film-pulling speed is (0.5–10) mm/min.

(5) There is a circuit control feedback system, which connects the surface pressure test, the drive motor of the belt, and the pulling mechanism to realize the constant pressure of the surface and automatic adjustment of the area of the floating monolayer during the film-pulling process; if two alternative films of different materials are prepared, two separate surface regions should be designed in the tank to control the surface pressure, respectively.

Figure 3.16 is a schematic of the process of fabrication of a single molecule membrane. Firstly, select a suitable solvent to dissolve the membrane material (amphipathic molecule) and make into a spread solution with a specific concentration (step one). Then, fill the LB sink with the subphase solution, use a microinjector (10–100 μl, and clean in turn with chloroform and ethanol) to transfer a specific volume of spreading solution and drop it onto the surface of subphase solution. Pay attention to avoid the overconcentration

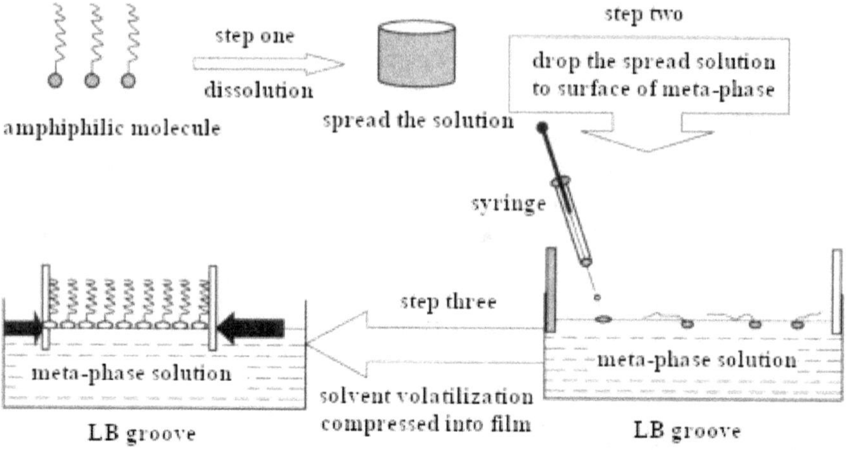

Figure 3.16: LB films with different transfer types.

of the spreading solution, in case it submerges to the bottom of the subphase (step two). After the spreading solution volatizes completely (15–30 min), a single molecule film can be formed when you move the surface of the sliding barrier compression subphase slowly.

According to the order of the monolayers, LB films are divided into X, Y and Z type films. Usually the substrate is chemically treated to make its surface hydrophobic or hydrophilic to ensure a stronger binding force during the transfer of the first monolayer. For example, when a hydrophilic treatment substrate passes through the floating layer upward, the hydrophilic end is deposited on the substrate, along with the hydrophobic end and the whole molecule. The orientation is perpendicular to the substrate plane. The substrate then passes down through the monolayer and the second layer is deposited. The hydrophobic end of the second layer is connected to the first layer. Then, the third layer is deposited upward, and by analogy, many layers can be transferred. This kind of transfer structure is called Y-type, as shown in Figure 3.17(a). If it is transferred only when the substrate passes through the monolayer downward, this transfer structure is called X-type; if it is transferred only when substrate

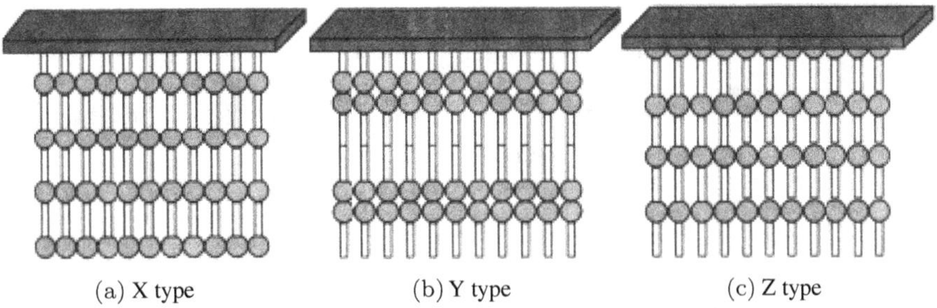

Figure 3.17: LB films with different types of transfer (the sphere represents the hydrophilic part and the rod represents the hydrophobic part).

passes through the monolayer upward, the transfer structure is called X-type, as shown in respectively Figures 3.17(b) and 3.17(c).

3.6 Exercises and Questions

(1) (a) Please give the chemical reaction equation for the Al_2O_3 film formation by CVD from a mixing gas composed of $AlCl_3$+CO_2+H_2.
(b) If a film thickness of 2 μm with a diameter of 2 cm is to be deposited, and the substrate is placed on a tubular reactor with a diameter of 5 cm and a length of 50 cm, calculate the minimum amount of $AlCl_3$ to be used.
(c) If you are fabricating a VC film with VCl_4+$C_6H_5CH_3$+H_2, please recalculate the above two problems.

(2) Consider the following CVD reaction:

$$A_g \underset{T_2}{\overset{T_1}{\rightleftarrows}} B_s + C_g (T_2 > T_1).$$

Under the pressure of 1 atm (P_A+ P_C = 1), the free energy of the reaction is $\Delta G^0 = \Delta H^0 – T\Delta S^0$.

(a) Transform $\Delta P_A = P_A(T_1) - P_A(T_2)$ to a function of T_1, ΔH, ΔS.
(b) Make a plot of ΔP_A–ΔH

(c) Use the function of H to describe the movement direction and magnitude of the gas.

(3) In the process of fabrication of epitaxial Ge film, the thermodynamics data are as follows:

$$I_{2(g)} = 2I_{(g)} \qquad \Delta G^\circ = -38.4T \text{ cal/mol}$$
$$Ge_{(s)} + I_{2(g)} = GeI_{2(g)} \qquad \Delta G^\circ = -1990 - 11.2T \text{ cal/mol}$$
$$Ge_{(s)} + GeI_{4(g)} = 2GeI_{2(g)} \qquad \Delta G^\circ = 36300 - 57.5T \text{ cal/mol}$$

(a) What's the value of ΔG° in $Ge_{(s)} + 2I_{2(g)} = GeI_{4(g)}$?
(b) Devise reaction apparatus and point out which part is hotter and which part is colder?
(c) Make a rough estimation of the working temperature of the reaction apparatus.
(d) How to change the design of the reaction apparatus for the fabrication of polycrystalline films.

(4) (a) The growth rate of Si films fabricated by various different gases under 1200°C is listed in the table below.

Gas	Growth rate (μm/min)
SiH_4	1
SiH_2Cl_2	0.5
$SiHCl_3$	0.3
$SiCl_4$	0.15

(b) The observed density, of the crystal nucleus during the formation of polycrystalline silicon film by different kinds of gases under 1000°C are as follows: SiH_4 — $10^{10}\,\text{cm}^{-2}$, SiH_2Cl_2 — $5 \times 10^{-7}\,\text{cm}^{-2}$, $SiHCl_3$ — $3 \times 10^{6}\,\text{cm}^{-2}$. Are the results from (1) and (2) consistent? How to explain the two discoveries?

(5) Supposing you have participated in a program of fabrication of ZnS/CdS infrared coating. The thermodynamics data

show that

$$H_2S_{(g)} + Zn_{(g)} \rightarrow ZnS_{(s)} + H_{2(g)}$$
$$\Delta G = -76,400 + 82.1T - 5.9T \ln T(\text{cal/mol})$$
$$H_2S_{(g)} + Cd_{(g)} \rightarrow CdS_{(s)} + H_{2(g)}$$
$$\Delta G = -50,000 + 85.2T - 6.64T \ln T \ (\text{cal/mol})$$

(a) Are these reactions endothermic or exothermic?
(b) Actually, reactions 1 and 2 happen at 680°C and 600°C respectively. Supposing it is under equilibrium, make an estimation of the value of P_{H_2}/P_{H_2s}, according to the vapor pressure of Zn and Cd at the corresponding temperature.
(c) Make a design of the reaction apparatus that could be used to fabricate ZnS and CdS; the way to add the reactants and heating of the substrate should be included.

(6) The following phenomena are observed when WF_6 pass through a substrate containing Si and SiO_2:

(a) W will deposit on Si instead of SiO_2.
(b) When the thickness of W film reaches some value (like 100–150 Å), the reaction will be suppressed automatically, and W cannot be deposited after that.

Come up with a method so that we can fabricate a thicker W film.

(7) Si + $SiCl_4$ = $2SiCl_2$ ($\Delta G^\circ = 83,000 + 3.64T \log T - 89.4T$ (cal/mol)) is a reversible reaction that happens inside a tubular atmosphere reactor with a diameter of 15 cm. Si are deposited on the substrate with a temperature of 750°C and a distance of 25 cm from the reaction source. The reaction temperature is 900°C. The substrate and the reaction source are in thermodynamic equilibrium; calculate the flux of $SiCl_2$. The viscocity of the gas is 0.08cP. [Hint: see problem 2.]

(8) Find out the chemical reaction equation for the following films:

(a) Fabrication of silicon nitride through PECVD; there are 20 at% H, and the ratio of Si and N is 1.2.
(b) Fabrication of silicon nitride through LPCVD; there are 6% H and the ratio of Si and N is 0.8.
(c) Fabrication of silicon oxide through LPCVD with a density of $2.2\,g/cm^3$ and an H content of 3×10^{21} atoms/cm^3.

(9) Hydrogen flux with 0.5% $SiCl_4$ is flushed to a tubular CVD reactor with a diameter of 12 cm. There is a flat plate, above which there is placed a silicon wafer. If the gas viscosity at 1200°C is 0.03 cP,

(a) What's the Reynolds number of the airflow?
(b) Make an estimation of the thickness of the edge layer 50 cm below the plate.
(c) If the epitaxial silicon film is deposited at the rate of 1 μm/min, make an estimation of the diffusion rate of Si to the edge layer.

(10) At 540°C, polycrystalline silicon is deposited at a rate of 30 Å/min. If the activation energy of the deposition of film is 1.65 eV, what's the deposition rate at 625°C?

(11) When the first chemical reaction happens inside a long tube-type CVD reactor, the stationary diffusion and convection are maintained. Supposing the concentration $C(x)$ satisfies the following equation:

$$D\frac{d^2C}{dx^2} - v\frac{dc}{dx} - KC = 0,$$

K is the chemical rate constant, and x is the distance inside the reactor:

(a) If $C(x = 0) = 1$, $C(x = 1\,\text{m}) = 0$, derive the expression for $C(x)$.
(b) If $C(x = 0) = 1$, $\text{d}C/\text{d}x(x = 1\,\text{m}) = 0$, derive the expression for $C(x)$.

(c) If $D = 1000\,\text{cm}^2/\text{s}$, $v = 100\,\text{cm/s}$, $K = 1\,\text{s}^{-1}$, derive the expression for the concentration.

(12) Choose any kind of film material (e.g. semiconductor, oxide, nitride, carbide, alloy, etc.), and make thin films with PVD and CVD, respectively. Compare the structure, chemical composition, and properties of the two methods, and write a report.

Chapter 4

Related Technologies of Film Preparation

Besides deposition technology, there are many special technologies we need to master and use skillfully in the preparation process of films so as to obtain films and devices with good performances.

4.1 Substrates

Films with thin thickness cannot support themselves and need carriers. Besides having enough adhesive force to support films, the ideal substrates should have no interaction with films. In addition, the substrates should be appropriate for the deposition process and the following preparation and application processes. The cost is another important concern. Hence, the desired performances of an ideal substrate are shown in Table 4.1.

Because the substrates need to have high mechanical strength, high resistivity and good thermal stability, the substrates are mostly glass, ceramics and single crystals. The generic metal, organic plastics, and semiconductor materials can be used only for particular conditions.

4.1.1 *Properties of various substrates*

4.1.1.1 *Glass*

As a kind of transparent, smooth-surfaced and stable material, glass can be used below 500°C. The thermal and chemical properties of glass vary obviously with components. Transparent quartz glass, with total content of impurities less than 100 ppm, presents transparency in the wavelength range of 0.18–4 μm and has a small coefficient of

Table 4.1: The desired performances of ideal substrates.

Required performances	Reasons
Surface roughness of the order of an atom	Obtaining high homogeneity
Completely flat	Obtaining high definition
No pores	Avoiding overmuch degasification process
Mechanical strength	Avoiding splintering
Thermal coefficient of expansion being equal to deposited films	Avoiding stress
High thermal conductivity	Avoiding overheating of circuit components
Thermal shock resistance	Avoiding being destroyed in technological processes
Heat stability	Permission for being heated in technological processes
Chemical stability	Using chemical reagents without restriction in technological processes
High resistance	Making the circuit components insulated
Low cost	Allowing a large number of applications

thermal expansion of 5.4×10^{-7}°C^{-1} in the temperature range of 20–1000°C.

Table 4.2 lists the composition (wt.%) of typical glass substrates. The precise compositions of glass substrates are different for different manufacturers.

Quartz glass has excellent chemical durability, heat resistance, and thermal shock resistance. The common glass pane and microscope glass lenses belong to soda lime glass, which are easy to fuse and model, but with a large expansion factor. Through the substitution of B_2O_3 with Na_2O, the expansion factor will be decreased.

Quartz glass can be fabricated via the fused quartz method, which is transparent with crystals as raw materials and non-transparent with quartz sand and quartz as raw materials. Except for different transparencies, there is hardly any difference in physical and chemistry properties between these two kinds of glass.

Table 4.2: The composition (wt.%) of glass.

Types	SiO_2	Na_2O	K_2O	CaO	MgO	B_2O_3	Ai_2O_3	Remarks
1. Transparent quartz glass	99.9							Fused quartz
2. 96% quartz glass	>96	<0.2	<0.2			2.9	0.4	Vycor (brand)
3. Borosilicate glass with low expansion factor	80.5	3.8	0.4			12.9	2.2	Pyrex (brand)
4. Aluminosilicate glass	55	0.6	0.4	4.7	8.5	4	22.9	Jena supermax CGW 1270#
5. Aluminoborosilicate glass	74.7	6.4	0.5	0.9	BaO 2.2	9.6	5.6	for physical and chemical tests
6. Low alkali glass	49.2	<0.2	<0.1		BaO 25.0	14.5	10.9	CGW 7059#
7. Common glass panes	71–73	13–15		8–12	1–3		1–2	

Table 4.3: The purities of quartz glass (ppm).

Materials	Al_2O_3	Fe_2O_3	TiO_2	Na_2O	K_2O
$SiCl_4$	<3	<1	<1	<1	<1
Crystal	30–100	1–5	1–10	1–6	1–5
Quartz	200–500	20–100	10–20	20–50	10–30
Quartz sand	500–3000	100–1000	10–100	20–100	20–100

Table 4.3 lists the purity of all kinds of quartz glass. The quartz glass fabricated with the Verneuil flame melting method has the highest purity. As shown in Table 4.3, purity is determined by mineral contents. When iron oxide content is higher than 100 ppm, Fe^{3+}, Fe^{2+} can absorb, respectively, near ultraviolet of 0.38 μm and near-infrared waves of 1.1 μm. The fracture surfaces of plate-type samples are green. Melted glass materials in a strong oxidizing atmosphere have reduced content of Fe^{2+}, which makes fracture surfaces present hazel. Impurity-free glass is pure white.

Table 4.4 lists the physical properties of transparent quartz glass. The density, Poisson ratio, specific heat capacity, etc., are smaller

Table 4.4: The properties of transparent quartz glass.

Parameter	Data	Parameter	Data
Density (g/cm^3)	2.203	specific resistance($\Omega \cdot$cm)	
Strength of extension (kgf/cm^2)	490	(20°C)	10^{19}
		(500°C)	10^8
Young's modulus (kgf/cm^2)	7.78×10^5	dielectric constant (1 MHz)	3.6
Stiffness coefficient (kgf/cm^2)	3.41×10^5	refractive index ($\lambda = 5891$)	1.458
Poisson's ratio	0.14	Abbe number	67.62
Coefficient of linear expansion °C^{-1} (less than 1000°C)	5.4×10^{-7}	visible-light transmissivity of panes	93.3%
Specific heat capacity (kcal/g°C)	0.251	visible-light reflectivity of panes	6.7%

than the ones of common glass panes. Besides the influence of temperature, the specific resistance is susceptible to water content and impurity. Quartz glass is often used under 1100°C. However, when the water content in glass is high, the operating temperature needs to decrease by 50–100°C for safe use. The critical temperatures used at high temperatures are approximately the warpage deformation temperature of viscosity equal to $10^{13.5}\rho$.

4.1.1.2 *Ceramics*

The insulating ceramics, such as alumina and forsterite, have excellent insulation resistance, withstand voltage and dielectric loss even at service conditions of high frequency electric and high temperature. Main applications and properties of ceramic insulating materials are listed in Tables 4.5 and 4.6, respectively.

The properties of ceramics can be affected by many factors, such as crystalline types of the main components, compositions of non-main components, impurities, vitreous intermediates, amount and distribution of pores, composition deviation, and structural defects formed during the crystallization process, and so on. High-precision ceramic materials can meet the demands of modern film fabrication

Table 4.5: The applications of ceramic insulating materials.

Applications	Components	Materials
Hybrid integrated circuit	Substrates for thick films, substrates for thin films, sapphire substrates	A, F
Large-scale semiconductor integrated circuits	Multi-layer ceramic package, dual-in-line ceramic package amplifier, multilayer ceramic micromodule, sapphire substrates	A
Semiconductors	Multilayer substrates, transistor bases, diode bases, power transistor bases	A A, S B, A
Electronic tubes	Magnetron shell, controlled silicon rectifier shell, electronic gun unit, insulator inside the tubes, cathode ray tube shell	A A, F F

Note: A: alumina; F: forsterite; S: talc high frequency insulation material; B: beryllium oxide.

Table 4.6: The properties of ceramic insulating materials.

	Alumina 92%	Alumina 96%	Alumina 99%	Sapphire	Forsterite
Relative density	3.6	3.7	3.8	4.0	2.8
Bending strength (MPa)	320	280	310	1300	150
Coefficient of thermal expansion ($\times 10^{-6°}C^{-1}$) (40–400°C)	6.5	6.7	6.8	7	8.8–0.6
Thermal conductivity coefficient (cal/(cm · s · °C))	0.04	0.05	10.06	0.1	0.01
Dielectric strength (kV/mm)	10	10	10	48	10
Resistivity (Ω · cm)					
(20°C)	$>10^{14}$	$>10^{14}$	$>10^{14}$	10^{16}	10^{14}
(300°C)	10^{13}	10^{14}	10^{14}	10^{14}	10^{13}
(500°C)	10^{10}	10^{11}	10^{11}	10^{11}	10^{10}
Dielectric constant (1 MHz)	8.5	9.4	9.7	10	6.5
Dielectric loss ($\times 10^{-4}$) (1 MHz)	3	2	2	1	3

and applications in terms of functions, material features, complex machining for metal connecting, dimensional accuracy, and shape precision.

The manufacturing technology of alumina develops rapidly and alumina ceramics with purity 92–99% has been widely used. The membrane circuits mainly use grinding alumina substrates, glazed substrates fabricated by coating glass glaze on alumina substrates, glabrous and high-purity alumina substrates, and grinding sapphire substrates.

Sapphire substrates have excellent electrical insulation property, chemical thermal stability, heat-conducting property, and surface roughness for application at present. Sapphire substrates are used in a wide range of applications, such as electronic materials, optical materials, machine part materials and decorative materials. In the past, such substrates were fabricated using the Verneuil oxy-hydrogen flame fusion method and the Czochralski single crystal method. After 1975, the Kyocera Company (Japan) used the edge-defined film-fed growth (EFG) method to produce film substrates, silicon-on-sapphire (SOS) substrates and silicon–sapphire wafers.

Due to the low dielectric loss and large thermal conductivity of the sapphire substrate in the frequency range of GHz, it is very suitable for microwave substrates. In addition, the sapphire surface is highly polished to be used as hybrid integrated circuit substrates for micro-pattern fabrication and-as precise film resistor substrates. Table 4.7 lists the dimensions of mass-produced sapphires used as substrates and other shaped sapphires.

In order to meet the need of a high density integrated circuit, multi-chip micromodules with high density, high reliability and high speed can be fabricated on multilayer ceramic substrates. Multilayer ceramic substrates often adopt green sheet lamination, thick-film lamination, and thin-film lamination methods, as shown in Figure 4.1.

Forsterite ($2MgO \cdot SiO_2$) has the features of low dielectric loss, and large insulation resistance under high frequency. It is easy to obtain a smooth surface to use as substrates or cores for metal film resistor, carbon resistor, and winding resistance, and as transistor

Table 4.7: The standard shape and dimensions of sapphires made by the EFG method.

Shape	Dimensions	Manufacturable size (mm)
	Width	75 (maximum)
	Length	150 (maximum)
Plate	Thickness	0.25–3.0
	Crystal plane	R plane, A plane, C plane
	Surface roughness	Double-sided mirror; one side is a mirror and the other side a grinding surface
rod	Diameter length	1.0–10 1000 (maximum)

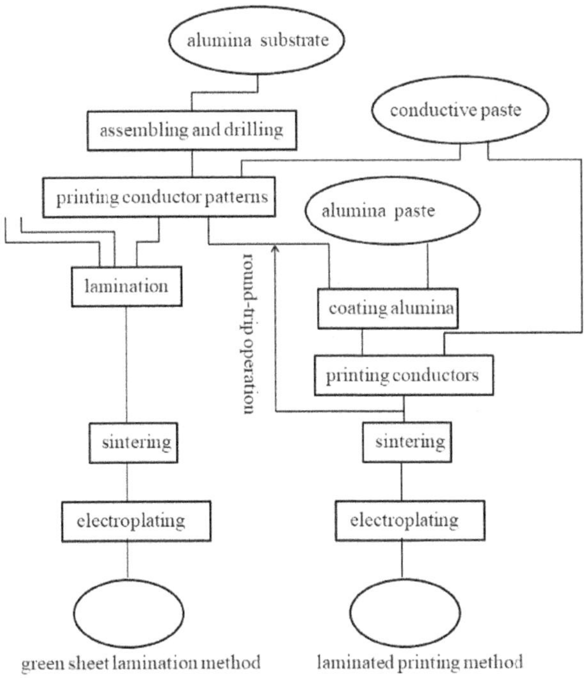

Figure 4.1: The manufacturing process of ceramic multilayer wiring substrates.

base as well as integrated circuit substrates. The dielectric constant of forsterite is smaller than the one of alumina, so it has a short delay time for signal transmission. The dilatation coefficient of forsterite is close to the ones of glass panes and most metals, and changes with composition, which is different from alumina and easy to choose matched gas sealing materials.

Although alumina ceramics have higher thermal conductivity, it has a high cost and its powder and steam are harmful to the body, restricting its applications.

High-thermal conductive insulating silicon carbide has high thermal conductivity (0.65(cal/cm · s°C) at 25°C] and high electrical resistivity ($10^{13}\Omega \cdot$cm at 25°C) at the same time. In addition, it has large bending strength and elastic coefficient and its coefficient of thermal expansion is close to the one of monocrystalline silicon (3.7×10^{-6}/°C in the range of 25–400°C), which applies to loading large components. The dielectric constant of silicon carbide is about 40. Because the signal delay time is in direct proportion to the square root of dielectric constant, the signal delay time of silicon carbide is double of alumina's, which is the disadvantage. The metallization of silicon carbide with Cu, Ni can expand its application areas, such as integrated circuits, packaging, and so on.

4.1.1.3 *Single-crystal substrates*

Single-crystal substrates are important to the formation of epitaxial layers. Many properties of various epitaxial layers are decided by single-crystal substrates. Table 4.8 lists basic properties of common single-crystal substrates. In order to grow epitaxial films on high temperature substrates, a good understanding of thermal properties of single-crystal substrates is necessary. The anisotropy of crystal substrates can produce cracks, and the big difference in coefficient of thermal expansion between substrates and films will lead to large residual stress in films, which decreases durability of films to a large extent. Hence, there are some announcements when we use single-crystal substrates; they are as follows:

(1) **Thermal expansion ratio $\triangle L/L$**

Replacing the commonly used coefficient of thermal expansion α (°C^{-1}) with the thermal expansion ratio $\triangle L/L$(%) has the following benefits.

Table 4.8: Properties of common single crystal substrates.

Materials		Sapphire			Rutile	
Chemical name		$\alpha - Al_2O_3$			TiO_2	
Space group		Hexagonal			Tetragonal	
Lattice constant (Å)		$\alpha_0 = 4.758 \quad c_0 = 12.991$			$\alpha_0 = 4.584 \quad c_0 = 2.953$	
Density (kg/m^3)		3970			4240	
Melting point (K)		2326			2143	
Cleavage plane		—			(110)	
Elastic constant ($\times 10^{10} N/m^2$)		C_{11}49.6 C_{12}10.9 C_{33}50.2 C_{13}4.8 C_{44}20.6 C_{14}3.8			—	
Dielectric constant (298 K, 10^6 Hz)		9.34⊥c,11.54∥c			—	
Thermal expansion ratio $\Delta L/L$ (%)	T (K)	$\Delta L/L \parallel c$	$\Delta L/L \perp c$	T (K)	$\Delta L/L \parallel$c	$L/L\Delta L \perp c$
	100	–0.078	–0.060	100	–0.155	–0.120
	293	0	0	500	0.197	0.150
	1000	0.593	0.552	1000	0.740	0.570
	1900	1.555	1.440	1400	1.200	0.927
Thermal conductivity λ (W/(m.K))	T(K)	$\lambda \parallel c$	$\lambda \perp c$	T (K)	$\lambda \parallel c$	$\lambda \perp c$
	5	410		1	(2.6)	(2.3)
	15	8700		5	313	265
	30	20700^M		12	2060^M	1700^M
	40	12000		20	1000	690
	1000	10.5		50	66	45
Specific heat capacity C_p(cal/(kg.K))	T (K)	C_p		T (K)	C_p	
	533	263				
	672	272				
	1088	297				
	1227	306				

For various materials, $\triangle L/L$ (%) can be expressed as the following smooth curve:

$$\triangle L/L = C_0 + C_1T + C_2T^2 + C_3T^3, \tag{4.1}$$

where, C_0, C_1, C_2, C_3 are material constants to control the testing accuracy of $\triangle L/L$ under $\pm 5\%$. Using the four $\triangle L/L$ values presented in Table 4.8 we can obtain thermal expansion ratio with Equation (4.1). Meanwhile, using the formula $\alpha = (\triangle L/L) \triangle T$ we can obtain thermal expansion coefficient α values.

(2) **Thermal conductivity λ**

In the x direction of the crystal, there exists the temperature gradient dT/dX. Suppose thermal flux density Q_x produced by the temperature gradient, which using $\lambda = -Qx/(dT/dX)$ enables to define thermal conductivity λ. For ideal insulating crystals, the general thermal conductivity curve is shown in Figure 4.2, which has a maximum.

In Table 4.8, the materials fitting the shape of this curve are C, Si, Ge, MgO, LiF, NaCl, $MgAlO_4$, Al_2O_3, TiO_2, quartz, and so on. Top-down values in Table 4.8 represent the conductivity values at the A, B, M, C, D points in Figure 4.2. In addition, for materials which are unmatched with the curve in Figure 4.2, such as InAs, InSb, quartz

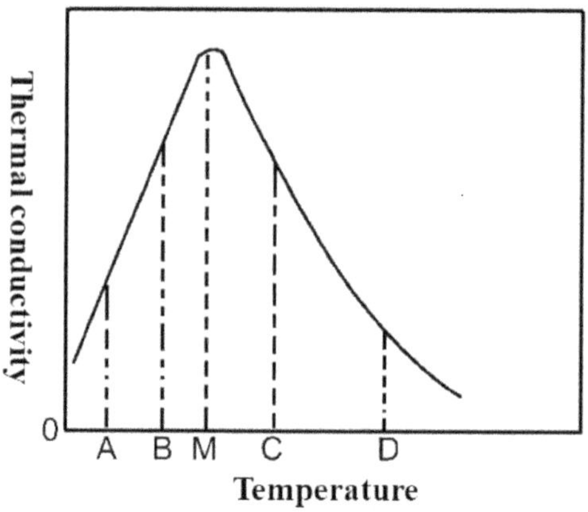

Figure 4.2: The general curve of thermal conductivity for insulating crystals.

glass and mica, this table presents two or three conductivity values at middle temperature besides the highest and lowest measured temperatures.

(3) **Specific heat capacity** c

The specific heat capacity is the quantity of heat required to raise the temperature by one degree Celsius, which generally changes with temperature. Specific heat capacity can be divided into constant-pressure specific heat capacity c and constant-volume specific heat capacity c_v. For the solid and liquid, the difference in specific heat capacity is about 3–10% of c_v.

4.1.2 *Substrate conditions*

4.1.2.1 *Principles*

Any substrates must be thoroughly cleaned before use. The used cleaning methods are relevant to the substrate properties, the nature of contamination and the required cleanliness. The contamination includes the contamination in the manufacturing process, human contact (protein), flying dust in the air, fibers, and oil molecules. Cleaning means to destroy the adsorption bonds between substrates and contaminants but not damage the substrate itself. The amount of energy to clean is provided directly by heating or ion bombardment, chemical reaction or dissolution, or mechanical scrubbing.

The detailed methods of cleaning substrates for deposited films are mainly decided by the growth method and the purpose of the films. The surface condition will affect seriously the structures and physical properties of prepared films. Substrate cleaning methods are generally divided into methods of removing physical attachment or chemical attachment on the surface.

The general cleaning method is as follows: mechanical scrubbing in a solution at first, chemical reaction and dissolution, then heating or ion bombardment (generally carried out in vacuum or by adopting vacuumed cleavage).

The examination of surfaces after cleaning with different methods can be done by low energy electron diffraction (LEED), reflection high-energy electron diffraction (RHEED), and Auger electron spectroscopy (AES). In addition, the various performances of films can also be used to judge whether the film surface meets the requirements.

4.1.2.2 *Cleaning methods used in the laboratory*

The general cleaning methods used in the laboratory roughly include chemical cleaning, ultrasonic cleaning, ion bombardment cleaning and heating cleaning. In actual cleaning processes, some cleaning methods can also be used in combination.

(1) **Detergent cleaning**

The method of removing oily components on the substrate surface is as follows. Firstly, steep substrates in boiled detergent for about 10 min, and rinse them well with moving water, then dry them using a dryer after soaking in ethanol. After cleaning, we need to use bamboo tweezers to hold substrates to avoid human hand grease. A simple cleaning method is to use a gauze soaked with a detergent to scrub the surface adequately, and then dry the substrates as above.

(2) **Chemicals and solvents cleaning**

In order to remove oily grease on glass surfaces, the cleaning method used in the laboratory is to steep glass in chromic acid and sulfuric acid mixture for several hours. But to dispose of cleaning waste is troublesome and contaminated. At present, the cleaning solvent, especially for cleaning the semiconductor surface, will adopt strong alkali solution.

When adopting acetone as solvent, we generally use the cleaning order in method (1). In addition, the solvent vapor can also be used to degrease and isopropanol solvent is extremely effective. The solvent vapor cleaning method is superior to the ultrasonic cleaning one.

(3) **Ultrasonic cleaning**

Ultrasonic cleaning makes use of cavitation generated by the transmitting of ultrasonic waves in a liquid medium to clean the substrates. For different cleaning purposes, solvent, washing liquid, and distilled water, or a proper combination of these liquids can be used as a liquid cleaning medium. Some generally use the cleaning order in method (1).

(4) **Ion bombardment cleaning**

Ion bombardment cleaning use accelerated positive ions to strike the substrate surface to remove the contamination and absorbed matter. In the sample studio of vacuum degree 10^{-1}–10 Torr, the electrode located in front of the substrate is applied a voltage of 500–1000 V to produce low-energy glow discharge. As noted above, excess energy ions will make the substrate surface spurt and partially decompose oil vapor in the chamber, and then the decomposition products will contaminate the substrate surface.

(5) **Heating cleaning**

If the substrates are thermal stable, then the substrates can be heated to about 300°C in vacuum with a vacuum degree as high as possible to remove effectively the adsorbed things on the surface, such as water molecules. At such time, it's better not to use oil in the vacuum exhaust system to avoid the adsorption of decomposed products of oil.

4.1.2.3 *Cleanliness test*

After cleaning substrates, the cleanliness needs inspection and assessment. Cleanliness testing has many methods, such as the exhalation imaging test, droplet test, static friction test, and so on. However, the testing methods must meet the requirements of cleanliness level.

(1) **Exhalation imaging test**

Exhaling to the glass surface, water will attach to the surface. Through flame cleaning, the glass surface is in a clean state. After, exhaling, the surface will, form a uniform water film, and a black exhalation image without diffuse reflection of light. The glass surface without cleaning will not to wet by water and form a grey exhalation image. This kind of method is the simplest test.

Cover the sample on clean hot steam, and observe the adhesion and evaporation state of water, which is called the steam test.

Table 4.9: The steam test results and contact angle AES analysis.

Steam test results (appearance level status of cleanliness)	States of attached water on ground quartz	Contact angle (°)	Number of monolayers of contamination
Very clear stripes	Forming a uniform rainbow interference fringe during the attachment and evaporation processes	4	< 0.1
Clear stripes	Forming a uniform interference fringe during attachment process and uneven interference fringe during the evaporation process	4	< 0.1
Unclear stripes	Forming uneven interference fringe during the attachment process	4	≤ 0.1
Orange peel shape	Colorless, visible large water droplets and transparent	5–10	0.1–1
Opacitas	Many small droplets and translucent	> 10	> 1

As shown in Table 4.9, the steam test can do a semi-quantitative inspection of trace contamination on the quartz plate surface.

(2) **Droplet test**

The droplet test is to place water or ethanol on the substrate surface and judge the cleanliness by diffusion degree, infiltration, and contact size of the liquid. The contact angle parameter for testing surface cleaning can achieve a quantitative inspection and can be used for non-stick water pollution assessment. Contact angle test can effectively check the pollution of monolayer level. The relationship between glass cleaning methods and droplet contact angle is shown in Figure 4.3.

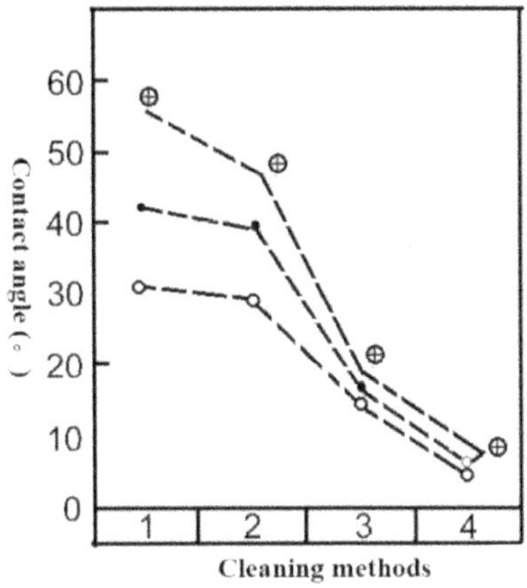

Figure 4.3: Relationship between cleaning methods of glass substrate and contact angle with water. 1: Grinding and drying; 2: organic solvent cleaning; 3: chromic acid cleaning; 4: glow discharge treatment.

(3) **Static friction coefficient test**

The sticking grease on the surface will decrease static friction coefficient. In the actual inspection, place the loaded glass on the surface to be tested, pull the glass horizontally and get the pulling force when the glass starts moving. Or slowly tilt the substrate and get the tilt angle when the small plate starts sliding. The static friction coefficients of the glass surface measured with different cleaning methods are listed in Table 4.10.

(4) **Radioisotope method**

This test method uses the radioisotope tracer method to count pollutants. The advantages are that this method has no connection with smoothness of the substrate surface and inspects the trace contaminants on a large scale. For example, ^{14}C is used to trace stearic acid contaminants, and the result turns out that general organic solvents and ultrasonic cleaning cannot remove stearic acid completely, leaving 0.1 molecular layer.

(5) **Surface analysis method**

Surface analysis methods, such as X-ray photoelectron spectroscopy (XPS) or electron spectroscopy for chemical analysis (ESCA), X-ray

Table 4.10: Relationship between the cleaning methods of glass substrates and static friction coefficients.

Cleaning methods	Static friction coefficients (μ_s)
(1) Isopropanol vapor cleaning	0.5–0.64
(2) trichloroethylene vapor cleaning	0.39
(3) Teepol anionic surfactant detergent cleaning and drying with cloth	0.07
(4) Teepol anionic surfactant detergent cleaning and glow discharge treatment	0.8
(5) Teepol anionic surfactant detergent cleaning, ethanol cleaning and drying with kapok	0.33
After (6) and (5) processes and glow discharge treatment	0.8
After (7) and (1) processes and glow discharge treatment	0.8
After (8) and (5) processes and flame heating treatment	0.11

microanalysis (XMA), Auger electron spectroscopy (AES) and ion microprobe mass spectrometry (IMA), are used to carry out a microscopic analysis of contaminated surfaces.

4.1.3 *Ultra-clean surface*

4.1.3.1 *Definition of ultra-clean surface*

In experiments on films with high performance, the impurities on the substrate surface are required to be controlled within a few percent of the number of atoms in the first layer of the surface, and the substrate surface should be a smooth structure with few defects. Such ultra-clean surfaces are also commonly known as atomic-level cleaning surfaces.

4.1.3.2 *Preparation of ultra-clean surface*

There are mainly two methods for actually obtaining an ultra-clean surface: one is to prepare a new surface that is not contaminated, and the other is to remove impurities remaining on the surface by a cleaning method. The two methods are explained below.

(1) Methods of forming a new surface: Broken; cleavage; forming films.
(2) Methods of cleaning impurities: mechanical grinding and chemical cleaning; ion bombardment; electric field evaporation; high temperature heating; laser irradiation.

The clean surfaces obtained through the above methods are also quickly contaminated by the diffusion of adsorbed gas molecules and impurity atoms to the surface.

For repeated experiments, the surfaces obtained by method (1) cannot be used again after the first use, while the surfaces obtained by method (2) can be used again through cleaning a few times after the first use, which makes the number of uses higher than that of method (1). However, method (2) is easy to change the surface structure and element composition, which causes that the surfaces cleaned by this method are not suitable for the study of oxides. The various preparation methods, characteristics and related problems to be considered for surface cleaning are shown in Table 4.11.

As can be seen from Table 4.11, each method has its own advantages and disadvantages, the most widely used one being ion bombardment. Along with the progress of molecular beam epitaxy (MBE) technology, the use of thin films to form clean surfaces has been well applied in recent years. Recently, a method to prepare a clean surface was designed: to prepare a single crystal by the molecular beam epitaxy method and clean the surface. Among various methods for preparing an ultra-clean surface, ion bombardment has a utilization rate of about 50%, and the utilization rates of the cleavage method and the high-temperature heating method are 25% and 15%, respectively. The utilization rate of the remaining methods is a few percent. The crushing method, mechanical grinding method, and electric field evaporation method have special properties, and hence, their applications are restricted.

4.1.3.3 *Main preparation technologies for ultra-clean surfaces*

At present, the cleavage method, ion bombardment method, high temperature heating method and laser irradiation method are the most important methods for preparing ultra-clean surfaces.

Table 4.11: The various preparation methods, and characteristics for getting ultra-clean surfaces.

Methods	Advantages	Disadvantages
(1) Composition methods of new surfaces		The contaminated surface cannot form a new surface.
(a) Crushing methods The sample is pulverized (thousands of times) with an electromagnetic hammer to form a sample mainly composed of cleavage surfaces.	Getting a large surface area ($\approx 1\text{m}^2/\text{g}$).	The surface orientation is not clear, and there are more defects. It is difficult to get a very regular surface, and it cannot be used repeatedly.
(b) Cleavage methods Use a hard material such as WC to pry, align the blade to the top and bottom, or support it with a copper plate. The cutting edge is angled at 30°, and the supporting blade is at an angle of 90° and 120°.	Short time; accurate cleaning and elemental composition.	Being applicable to cleavable materials, and limited to single crystals required for cleavage planes of a certain size. Cleavage repetitions are up to 4–5 times.
(c) Forming films	Getting a large surface area ($\approx 10^2\text{cm}^2$); no restrictions on material types.	It is necessary to have a knowledge of the fabrication and properties of films. In order not to reduce vacuum degree, deflate time is needed.
(2) Methods of removing impurities		It is easy to change the compound compositions.
(a) Mechanical grinding Use rotary cutters and boring tools to grind the surface	For soft materials with low melting point, this method can be used when other methods are invalid.	It is difficult to achieve finish machining of mirrors. It is easy to cause defects. Impurities from the grinding tool may invade the surface.

Table 4.11: (*Continued*)

Methods	Advantages	Disadvantages
(b) Ion bombardment $A_r^+(\approx 500V, 5\mu A/cm^2)$ is sputtered for a few minutes after which defects are repaired. Then annealing is used to release Ar, which is bombarded into the surface. The annealing temperature is about 2/3 of the melting point.	It is used widely with a few exceptions.	It is necessary to anneal due to the formed defects. The selective sputtering of compound results in inconsistent elemental compositions, and requires the charging of rarefied gas and exhaust time.
(c) Electric field evaporation	It can observe the peeling of dirt.	It is only for tubular specimens and needs high voltage ($\approx 10^4$ V).
(d) high temperature heating Different samples require different heating temperatures. Heated temperature (1000–2500 K), holding time (10 min to several hundred hours)	It is easy to operate. It is carried out in oxygen, and hydrogen, and is good to remove dirt.	It is not suitable for low melting point materials, but not always effective for high melting point materials. Sometimes it produces small facet deflation. The pumping process for exemption from recontamination takes a long time (a few days to a week).
(e) Laser irradiation (pulse laser) ruby laser, Nd:YAG laser, etc. Generally pulsed laser with a width of dozens of nanoseconds is used and the output power is 1–2 J/cm^2.	It can be cleaned in a very short time, and the risk of contamination is small.	Heating and cooling time is short, and it is easy to cause deformation. It is not suitable for light-transmitting materials. Cost of the light source is high. Sometimes there will remain a few percent of impurities.

(1) **Cleavage method**

The cleavage method is the most ideal method for the preparation of ultra-clean solid surfaces, but it is only suitable for those materials that can be applied on crystal surfaces. It is mainly used for the preparation of compound cleansing surfaces. A good cleavage plane is like a mirror surface. There is a step line along the cleavage direction in the cleavage plane, and it is difficult to obtain a flat surface with a large atom. Even for the cleavage surface, it is necessary to carry out cleavage operation in a vacuum in order to maintain its ultra-clean state. Crystals with an ultra-clean surface obtained by the cleavage method are shown in Table 4.12. A schematic diagram of various methods to prepare cleavage planes is shown in Figure 4.4. Among them, Figure 4.4(a) shows a standard method, in which a wedge is inserted into the cleavage plane of the crystal to cause a strain greater than the fracture limit.

Figure 4.4(c) presents a method that was conceived to obtain a cleavage plane using few steps, which is called the well-known Gobeli–Allen method. The samples with L shape are fixed on the test bench. Before cleavage, a grinding hole has to be machined first on the surface of the sample, then an impact bar is used to impact the sample to crack it. The step plane expands in all directions from the crack, but does not spread to the corners, leading to fewer steps at corners. In Figure 4.4(d), the blade is pressed in the middle of the sample, so that cracking occurs from the inner face. At the same time, the two chucks clamping the sample are rotated by 90°, so that the two cleavage planes are in the same direction. Figure 4.4(e) is a device for cracking mica and other layered materials. The cutting of layered material does not use a blade, which eliminates the contamination caused by the contact of the blade with the sample.

(2) Ion bombardment, high-temperature heating and the laser irradiation method

When there is a little dust on the surface, rinsing with water can give a smooth surface. To remove dirt that has firmly adhered to

Table 4.12: The materials with an ultra-clean surface obtained by the cleavage method.

Crystal structure		Crystal	Cleavage plane
Monomer	Close-packed hexagonal structure	Zn	(0001)
	diamond structure	Si $(2 \times 1^{(1)})$, Ge $(2 \times 1^{(1)})$	(111)
	Graphite structure	C	(0001)
	Arsenic structure	As, Sb	(111)
	Tellurium structure	Te $(2 \times 1^{(1)})$	$(10\bar{1}0)$
Diatomic compound	NaCl structure	(1) LiF, NaF, NaCl, KCl; LiI (2) MgO, CaO, MnO, CoO, NiO, EuO (3) PbS, PbSe, PbTe (4) TiC, UC	(001)
	sphalerite structure	(1) $ZnS^{(2)}$, $ZnSe^{(2)}$, ZnTe, CdTe (2) HgTe, $Cd_xHg_{1-x}Te$ $(x = 0.2, 0.31, 0.39)$ (3) GaP, GaAs, GaSb (4) InP, InAs, InSb	(011)
	Wurtzite structure	ZnO, $CdS^{(3)}$, CdSe	(0001), $(10\bar{1}0)$ $(11\bar{2}0)$
		PbO	(001)
		GaSe	(0001)
Polyatomic compound		MoS_2(0001) $BaTiO_3$(001) VO_5(001) $Bi_2(Te_{1-x}S_x)_3$(0001) ReO_3(011) mica$^{(4)}$(001)	

Notes: (1) Periodic arrangement is different from bulk crystals (refer to LEED-related contents).

(2) Wurtzite structure also exists.

(3) Sphalerite structure also exists.

(4) Polymorphic substance (Multiple crystalline states have been confirmed).

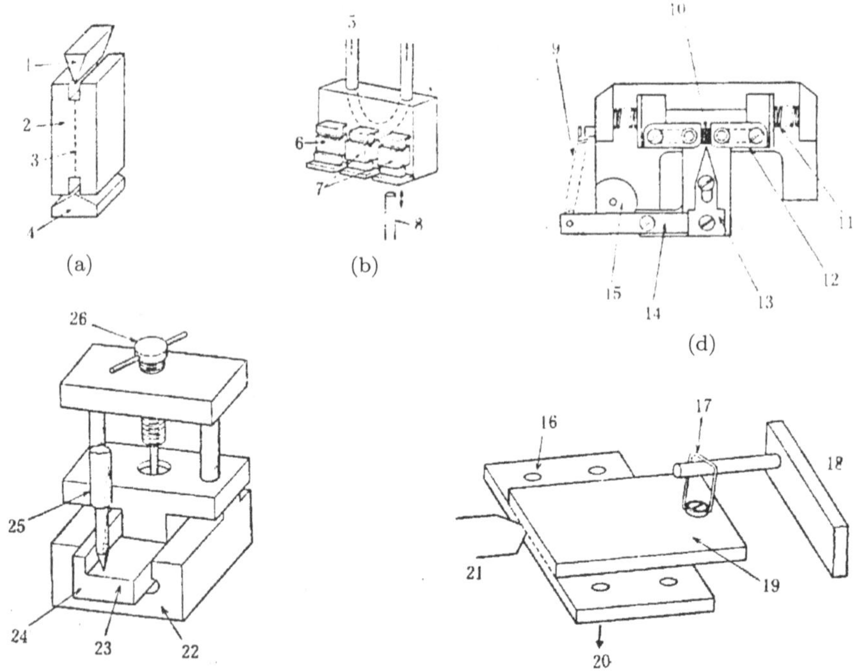

Figure 4.4: Various devices for preparing cleavage surfaces. (a) Standard method; (b) device capable of splitting three samples at low temperature; (c) method for preparing a cleavage plane with fewer steps;(d) device causing the cleavage plane to be in the same direction; (e) device using open layered material without blade. (1) 30° wedge; (2) sample; (3) expected cleavage plane; (4) 120° wedge (or 90° wedge or oxygen free copper plate); (5) liquid nitrogen; (6) cleavage plane; (7) pre-cleavage sample; (8) impact bar; (9) spring; (10) sample; (11) spring; (12) sample chuck; (13) wedge; (14) wedge drive rod; (15) hammer guide; (16) fixing hole of support desk; (17) hook; (18) support bar; (19) sample; (20) support desk; (21) expected cleavage plane; (22) support frame; (23) parts with fewer steps; (24) sample; (25) impact bar; (26) fixing bolts.

the surface for a long time, the surface layer must be peeled off. High-temperature heating or laser irradiation is suitable for removing a little dust and dirt. Ion bombardment is suitable for removing strongly adhered dirt, but damage to substrates is inevitable on peeling off the dirt on the surface. In the cleaning process of peeling off rust by ion bombardment, brown rust is prone to form on the substrate surface, so a better way to prevent this is to ionize inert gas, accelerate ion velocity, and bombard the substrate surface. After the bombardment, there is a moon-shaped large undulating surface, which makes the lattice arrangement disordered. Therefore, the substrate needs to be heat-annealed to make the crystal lattice neat. It can be seen that heat annealing treatment is indispensable for ion bombardment cleaning. The ion bombardment method can be effectively used for the cleaning treatment of not easily cleaved surfaces of metal materials and non-cleavable surfaces of compounds. The choice of ion accelerating voltage and ion current should be considered to minimize the damage to surface and not affect the composition ratio of the surface.

The high temperature heating method is suitable for high-melting point materials that can withstand a high heating temperature and for materials that can form a volatile compound on the surface even at a low temperature through a suitable pretreatment. Sometimes high temperature heating performed in an oxygen or hydrogen environment is better than that performed only in vacuum.

The laser irradiation cleaning method is a newly developed method in recent years, which has the advantage of being able to locally heat the surface. Because the laser can be used in an intermittent manner, the parts that do not require heating will not emit gas due to the temperature rise, so this application has considerable prospect. The disadvantages of this method are that the sample surface temperature is difficult to measure and control, and there is a large strain on the surface due to the instantaneous action of temperature. Therefore, more in-depth research is needed. The lattice arrangement period of the laser-irradiated surface is mostly different from that of other methods, and the surface properties also have many problems to be studied.

4.2 Film Thickness Measurement and Monitoring

Almost all film properties are related to film thickness, so film thickness directly affects various properties of films. We need to not only monitor the film thickness in real time during film formation, but also accurately measure the thickness of the produced film.

The measurement of the film thickness is mainly carried out by measuring the changes in certain physical properties due to the changes in films. Film thickness measurement methods commonly include mechanical methods, electrical methods, and optical methods.

4.2.1 *Mechanical methods*

4.2.1.1 *Quartz crystal method*

Quartz crystal resonance method is one of the most commonly used methods of film thickness measurement. It is also a dynamic real-time measurement method which has been widely used for film thickness measurement and detection of vacuum evaporation. The basic principles are shown in Figures 4.5 and 4.6.

We use the piezoelectric effect of a quartz crystal; when the source material is evaporated onto the substrate, it is also plated on a quartz plate connected to the oscillating circuit. The thickness of the film is calculated according to the new change in the resonant frequency of the quartz crystal.

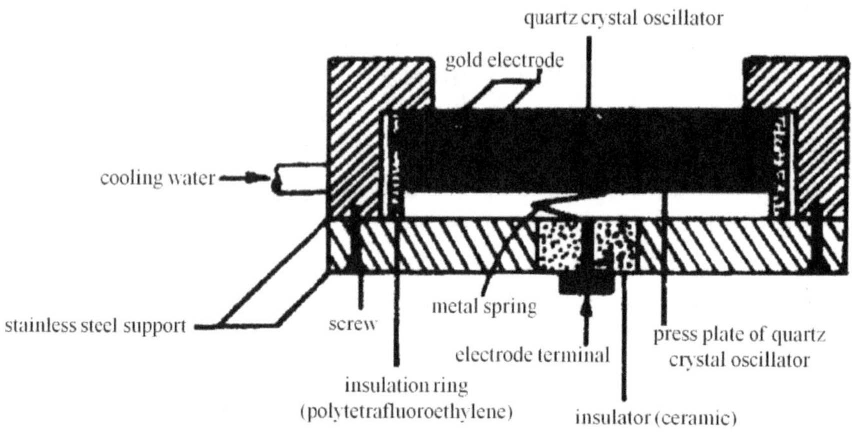

Figure 4.5: Measuring element of the quartz crystal oscillation method.

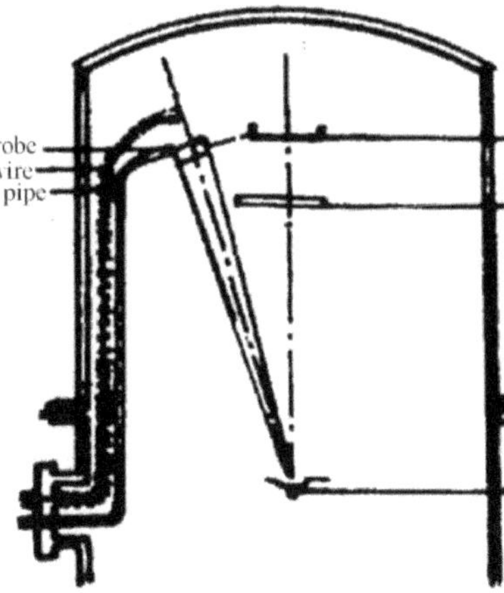

Figure 4.6: Location of the quartz crystal oscillator probe installed in the vacuum chamber.

The relationship between elastic modulus E_0 of the quartz crystal, its mass M_0, and the inherent characteristic frequency is

$$f_0 = N\left(\frac{E_0}{M_0}\right)^{1/2}, \tag{4.2}$$

where N is a proportional constant related to the cutting pattern and size of the quartz crystal piece. If other substances are deposited on the wafer surface, m_0 in Equation (4.2) will increase, leading to a change of the resonant frequency of the quartz crystal, namely

$$df = -\frac{1}{2}NE_0^{1/2}\left(\frac{1}{M_0}\right)^{3/2} dm. \tag{4.3}$$

The negative sign in Equation (4.2) indicates that the resonant frequency f of the quartz crystal decreases as the deposition material increases, and is obtained from Equations (4.1) and (4.2)

$$df = \left[-\frac{1}{2}\frac{f_0}{m_0}\right] dm = k \cdot dm. \tag{4.4}$$

Set up a quartz crystal with working area S, and deposit a film with density ρ and thickness d, then $dm = \rho S d$,

$$df = K\rho S d. \tag{4.5}$$

Therefore, when the amount of frequency shift is measured, the film thickness d can be obtained from Equation (4.5)

$$d = \frac{1}{kS}\frac{1}{\rho}df. \tag{4.6}$$

Of course, since the density ρ of the film is generally slightly smaller than that of the corresponding block, an error occurs when the bulk density is used instead of film density ρ.

For the generation of a larger frequency shift df, there is no longer a linear relationship between df and dm in Equation (4.3), which can be corrected by the following equation:

$$(f_0 - df)^2 \approx f_0(f_0 - 2df). \tag{4.7}$$

The quartz crystal method is characterized by simplicity and sensitivity, and can measure the thickness of the metal, the semiconductor, and the dielectric films. The main disadvantage is that this test method is an indirect measurement method, and the quartz probe is prone to electric shock when the df is too large.

4.2.1.2 *Microbalance method*

The masses of the substrate are weighed by a microbalance with a thickness sensitivity of 10^{-8} g/m^2 before and after film formation, and the film mass m on a given area can also be determined. The film thickness can be calculated from the bulk density of film

$$d = m/(S\rho). \tag{4.8}$$

Also, using bulk density instead of film density leads to a smaller calculation result calculated by Equation (4.6). The measurement accuracy of this method depends on the accuracy of using the balance.

4.2.2 *Electrical method*

4.2.2.1 *Resistance measurement*

Resistance measurement is a simple method to measure the film thickness of metals (including semiconductor) and uses a principle in which the resistance value has a certain relationship with shape to measure the film thickness.

According to the measured sheet resistance R_s of the film and the resistivity ρ of the film-forming substance, the film thickness can be calculated as

$$d = \rho/R_s. \tag{4.9}$$

The resistivity of bulk material used for film formation has a definite value, which is independent of material shape. However, the film resistivity is not a certain value. It is related to object shape, and its value varies greatly depending on the film thickness.

This is because film structure is different from that of bulk material. It has many effects on resistance, such as the kinds of lattice defects, the scattering effect of film interface, adhesion and absorption of residual gases, so that using the resistivity of block material to calculate film thickness can cause large errors.

When film thickness is measured by the electric resistance method, in order to obtain a more realistic film thickness, a thicker layer (about 200 nm) of the same material as the film being coated can be deposited on the substrate in advance, and then evaporate the film to be measured on the surface of the thicker layer.

Because the electric resistance of a metal-conductive film decreases as film thickness increases, the film thickness can be determined according to the amount of resistance reduction. The initial sheet resistance of the film is R_s, and after evaporating the film, the sheet resistance changes to ΔR_s, and the corresponding film thickness changes to Δd. After differentiation, Equation (4.8) changes to

$$\Delta d = \Delta R_s(-\rho/R_s^2). \tag{4.10}$$

This is the basic formula for measuring film thickness with resistance.

Resistance measurement can also be used to monitor film thickness during film crystallization. As shown in Figure 4.7, a monitoring piece is used as an integral part of the Wheatstone bridge and is placed in the vacuum coating chamber. Both ends of the monitoring piece stock conductive bands with a thickness of about 200 nm for the electric contactor. When deposited film is evaporated, the current

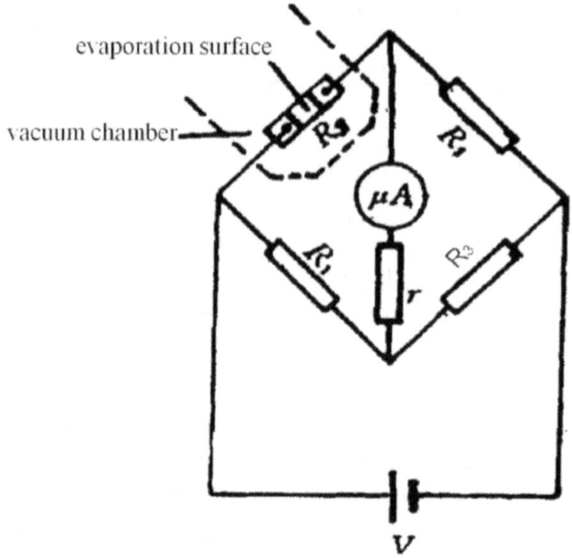

Figure 4.7: Bridge circuit for measuring resistance to measure film thickness.

flowing through the ammeter can be obtained from the imbalance of the bridge.

$$1 = \alpha \frac{R_s - r}{R_s + \rho}, \tag{4.11}$$

where

$$\alpha = V \frac{R_3}{R_1R_2 + R_2R_3 + R_3R_1 + \gamma(R_3 + R_1)},$$

$$\rho = R_2 \frac{R_3R_1 + \gamma(R_3 + R_1)}{R_1R_2 + R_2R_3 + R_3R_1 + \gamma(R_3 + R_1)},$$

$$\gamma = \frac{R_1R_2}{R_3}.$$

If it is initially balanced, then $r = R_s$, and after differentiation, Equation (4.11) can change to

$$\Delta R_s = \frac{(R_s + \beta)^2}{\alpha(\beta + \gamma)} \Delta I = \frac{R_s + \beta}{\alpha} \Delta I, \tag{4.12}$$

Therefore,

$$\frac{\Delta R_s}{R_s} = \left(1 + \frac{\rho}{R_{s0}}\right)\frac{\Delta I}{\alpha}. \tag{4.13}$$

The change of current in the ammeter is the corresponding change of sheet resistance. Therefore, when the resistance changes to the desired value, evaporation can be stopped to obtain a film of desired thickness.

Resistance measurement is suitable for measuring film thicknesses in a wide range, especially for evaporation coatings with high deposition rates and low residual gas pressure.

4.2.2.2 *Capacitance measurement*

The thickness of a dielectric film can be determined by measuring its capacitance, and Figure 4.8 shows the measuring device using this as the test principle. Evaporation of the film onto the electrode causes a change in measured capacitance.

If the dielectric constant ε and the electrode area S of the films are known, then the film thickness is

$$d = \frac{\varepsilon S}{3 \cdot 6\pi C}(\text{cm}). \tag{4.14}$$

The unit of capacitance C is pF, and the unit of S is cm ^{2}.

When depositing dielectric films, it can also be monitored by the capacitance method: to measure film thickness by measuring the change in capacitance of a specially designed plate capacitor.

4.2.2.3 *Quality factor (Q value) change measurement*

As shown in Figure 4.9, when a coil with radius r and an alternating current as power is placed at a distance h from a metal film with thickness d, the eddy current induced in metal film will lose a part of electrical energy in the coil. Therefore, both the Q value of the coil and the resonant frequency change.

Based on the change of Q values, the C value and the frequency value of the used capacitor, the resistance value of the film can be calculated, and film thickness can be obtained by Equation (4.8).

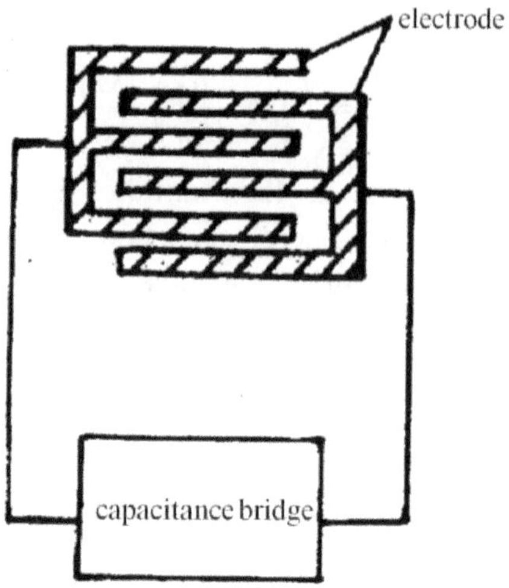

Figure 4.8: Schematic diagram of measuring capacitance to monitor film thickness.

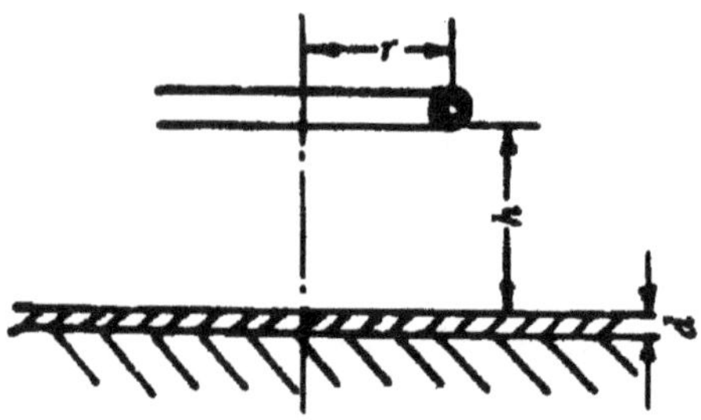

Figure 4.9: When the film thickness is monitored by the change of Q value, the position of the coil above the film.

The advantage of Q value change measurement is that it is mainly non-destructive measurement, therefore it has received the attention of people. It can also be used to monitor the evaporation process.

4.2.2.4 *Ionization method*

The ionization method is based on measuring the ion current generated by the vapor of evaporating materials, and it is assumed that all particles of the reaching substrate are agglomerated on the surface. By measuring vapor flow using the ionization method, film thickness can be calculated. The most common measurement system

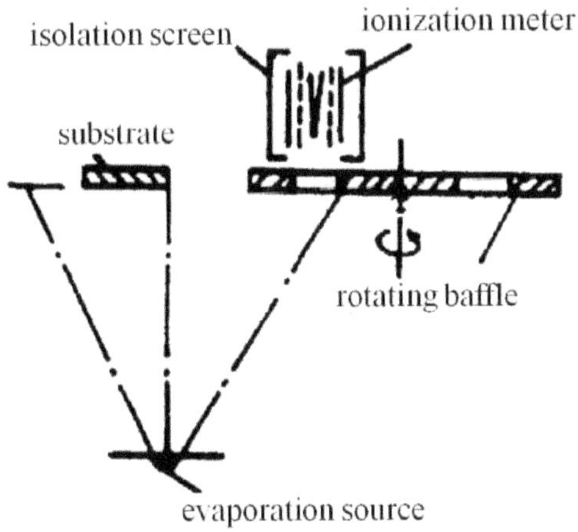

Figure 4.10: Film thickness monitoring with an ionization gauge.

is an ionization gauge placed in the vapor stream of evaporating materials, as shown in Figure 4.10. Electrons emitted from the cathode are accelerated by the positive grid and collide with vapor molecular flow to ionize them. The ions are collected by a collector whose potential is negative relative to the cathode, and then ion current can be measured. However, when vapor flow passes through the ionization chamber, residual gas is also ionized at the same time, so it is necessary to differentiate the ions of the evaporation material from the ions of residual gas. A vibrating baffle and a rotating baffle can be used to modulate vapor flow, and then ion flow has two components, with one being the direct current component produced by residual gas and the other being the modulation component produced by measured vapor. The latter is recorded by a frequency-selective amplifier, which is a linear function of deposition rate. This linear relationship can reflect the value of this approach.

The relationship between the ion current I_+ generated by vapor and the vapor molecular density (vapor pressure P) is

$$I_+ = SpI, \tag{4.15}$$

where S is sensitivity (1/Torr) of the ionization gauge, and I is electron current (A), and the film thickness can be determined by Equation (4.13) and the related evaporation rate formula.

4.2.3 *Optical method*

4.2.3.1 *Light absorption method*

The intensity attenuation of the plane wave after passing through a material can be described by Lambert's law

$$I = I_0 e^{-\alpha d}, \tag{4.16}$$

where I_0 is initial light intensity, I is the light intensity after passing through the substance with thickness d, and α is the absorption coefficient.

Let light with intensity I_0 be incident perpendicularly from the air to the film sample with thickness d, and the transmitted light intensity after passing through the sample is I_t, then the transmission coefficient T_t can be defined as the ratio of I_t to I_0

$$T_t = I_t/I_0. \tag{4.17}$$

If the absorption coefficient of the sample is α and the reflection coefficient is R, the light intensity reflected at the first interface becomes RI_0, and the light that passes through the first interface to enter sample is $(1 - R)I_0$. Due to the absorption of the sample, the light intensity changes to $(1 - R)I_0 \exp(-\alpha d)$ when reaching the second interface with film thickness d.

The intensity of light reflected back into the sample at the second interface is $R(1-R)^2 I_0 \exp(-\alpha d)$, which is reflected inside the sample to the first interface and absorbed part. And the transmission and reflection process will occur again on the first interface. Multiple reflections in the sample and transmission to the outside of the sample occur as many times as this. When the number of reflections is large, you can get

$$I_t = \frac{(1 - R)^2 \exp(-\alpha d)}{1 - R^2 \exp(-2\alpha d)}. \tag{4.18}$$

When the value of αd is sufficiently large or R^2 is small so as to satisfy $R^2 \exp(-2\alpha d) \ll 1$, the second term of the denominator of

Equation (4.17) can be omitted; thus

$$I_t \approx (1 - R)^2 \exp(-\alpha d). \tag{4.19}$$

From this,

$$I_t \approx I_0(1 - R)^2 \exp(-\alpha d). \tag{4.20}$$

Therefore, for a given light irradiation, if α and R are known, I_t and I_0 can be measured to calculate film thickness d. This method is also suitable for controlling the evaporation process and monitoring the deposition rate. However, the magnitude of the absorption coefficient is related to the used light wave. The light wavelength is different, and the difference in absorption coefficient can be up to several orders of magnitude.

4.2.3.2 *Optical interference method*

If a film is irradiated with light, the light reflected directly from the film surface and the light passing through the film reflected from the substrate surface have an optical path difference due to the difference in refractive index between air and film, so that the two beams produce interference of light. At the same time, light waves entering the substrate also interfere.

Using the optical interference of the film to measure the film thickness, a complete method has been established, such as the interferometric film thickness method and the equal thickness interference fringe method, that is, the multi-beam interferometry (MBI) method. It is suitable for measuring film thicknesses from 1/8 to 4 or 5 wavelengths.

(1) **Interferometric film thickness method**

When light pass perpendicularly into the film, and the optical thickness of the film nd (the product of refractive index n and film thickness d) is an odd multiple of $\lambda/4$, then reflected light intensity exhibits an extreme value. It is assumed that the refractive indices of air and substrate are n_0 and n_c, respectively, and if $n_0\langle n\rangle n_c$, the light intensity of reflection exhibits a minimum value. When the

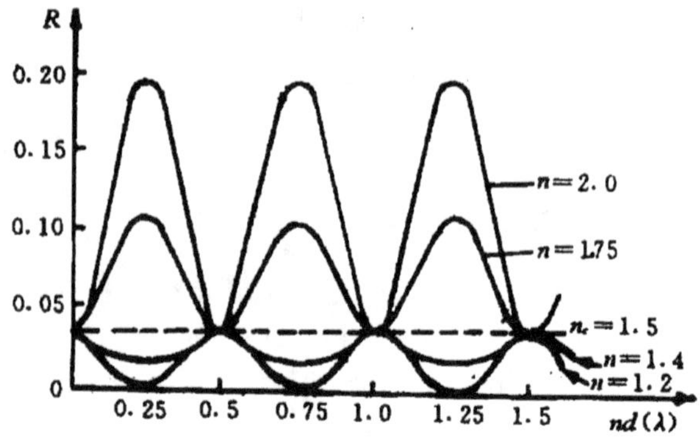

Figure 4.11: The reflectivity R varies with optical thickness nd(in wavelength) for several materials with different reflectivities.

optical thickness of the film is an odd multiple of $\lambda/2$, the intensity of reflected light is independent of film refractive index; that is, it is the same as that before coating. Transmitted light shows the opposite situation. If reflected light is strong, transmitted light is weak, and when reflected light is weak, transmitted light is strong. Therefore, both reflected light and transmitted light periodically change with the optical thickness of the film, exhibiting a series of maxima and minima. The light intensity can be measured optically. Figure 4.11 shows the periodicity of reflectance of films with various refractive indices deposited on a glass substrate having a refractive index of $n_c = 1.5$ as a function of its optical thickness.

The optical path difference formed between reflected light on the film surface and reflected light at the interface of film and substrate is

$$\delta = 2d(n^2 - n_0 \sin^2 \varphi)^{1/2}, \tag{4.21}$$

where d and n are the thickness and the refractive index of the film, respectively, and n_0 is the refractive index of the air above the film, and its value is 1 and φ is incident angle when light enters the film surface. According to the maximum condition of light interference, Equation (4.21) can be expressed as

$$\delta = 2d(n^2 - \sin^2 \varphi)^{1/2} - k\lambda, \tag{4.22}$$

where k is an integer to describe the interference order and λ is the wavelength.

When light is incident perpendicularly ($\varphi = 0$), if the kth maximum value appears at the wavelength λ_1 and the (k+1)th maximum value appears at the wavelength λ_2, then

$$2nd = k\lambda_1, \tag{4.23}$$

$$2nd = (k+1)\lambda_2. \tag{4.24}$$

Multiply λ_2 and λ_1 by the above two equations, and then subtract the two equations to obtain

$$2nd = \frac{\lambda_1\lambda_2}{\lambda_1+\lambda_2}. \tag{4.25}$$

If the refractive index of the film is known, the film thickness d can be obtained by measuring λ_1 and λ_2 values according to the spectrometer.

(2) **Equal-thickness interference fringe method**

The equal-thickness interference fringe method measures film thickness according to the wall-tip interference principle, and the parallel monochromatic light is vertically irradiated onto the film, and the interference fringes are generated by multiple reflection interferences, and then according to the deviation of stripes to obtain film thickness. This is also one of the methods commonly used in film thickness measurement. Measuring film thickness by this method requires the film to be stepped; for this purpose, a stepped film is formed with a high reflectivity material on the target surface as a comparative film for film thickness measurement, and the comparative sheet is placed near the substrate to be film-formed so that the shallow film is formed under exactly the same film-forming conditions, and a portion of the comparative sheet is covered with a plate having a sharp edge. If monochromatic light is irradiated on the comparative sheet after film formation, parallel streaks between light and dark are generated due to interference, as shown in Figure 4.12. At this time, the light interference on the comparative sheet can be regarded as the film interference of the tip shape. From the

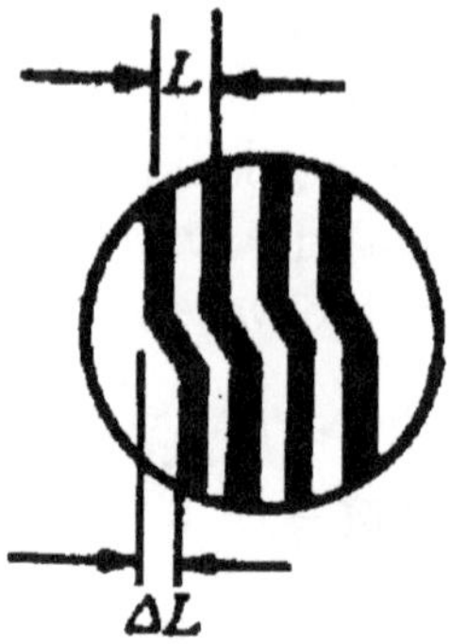

Figure 4.12: Displacement of film step by the equal-thickness interference fringe method.

stripe pitch L and the displacement ΔL of the stripe at the film steps and monochromatic light wavelength λ, with reference to the relevant conclusion of tip interference, the film thickness d and the relationship between them are obtained.

So,

$$d = (\Delta L\lambda)/(2L). \tag{4.26}$$

That is to say, the ratio of the strip offset to the stripe spacing multiplied by half wavelength is the step height, which is film thickness.

Measuring the film thickness by the interference method has the advantages of being simple, fast and without damaging the film layer. However, it requires that the film must have a highly reflective and flat surface, otherwise it will affect the test precision. The minimum film thickness measurable by the interferometry method is 2.5 nm ± 0.5 nm, the maximum film thickness is 2000 nm, and the film thickness measured by the equal-thickness interference fringe method is 20 nm.

4.2.3.3 *Ellipsometry method*

The film thickness can be measured within a certain range by the light interference phenomenon, but the precision is low. The optimal film thickness measurement is a method of measuring the film

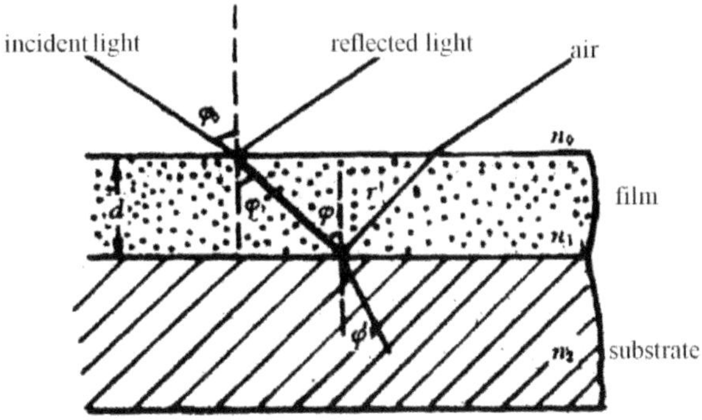

Figure 4.13: Basic principle diagram of measuring the film thickness by the ellipsometry method.

thickness by the polarization effect, and is also called the ellipsometry method.

The basic principle of the ellipsometry method is shown in Figure 4.13. When the incident angle is relatively large, the ratio and difference in phase shift between reflected light amplitude polarized along the parallel incident plane and reflected light amplitude polarized along the vertical incident plane. During the measuring process, a single-wavelength laser beam is converted into elliptically polarized light through a polarizer and a quarter wave plate, and is incident on the sample surface at an angle. After reflection, the polarization state changes, and then relative amplitude attenuation ($tg\varphi$), and the difference in phase rotation are measured. Using the relationship between φ, Δ and refractive index n, the thickness d, the film refractive index, and thickness can be determined based on light source wavelength and incident angle.

The ellipsometry method is also referred to as the polarization analysis method. In this method, the elliptically polarized light reflected by the film is analyzed to determine the film thickness and refractive index. The ellipsometry method is a vibrating surface segmentation type interferometry method with the following advantages:

(1) Light path is easy to adjust;
(2) Stable vibration intensity, and high precision measurement;
(3) No need of a coherent light source.

Consider that the sample has an incident linear polarized light at an angle of 45° to the sample surface, and the amplitudes of the parallel p-polarized component and the vertical s-polarized component on the incident surface are equal, and set to 1. Reflecting relative to each component, the amplitude reflection coefficients of light are equal to r_p and r_s, respectively. The r_p, s of a single layer film is generally expressed in plural

$$r_{p,s} = \rho_{p,s} \exp(-i\delta_{p,s}). \tag{4.27}$$

Because of $\delta_p \neq \delta_s$, the front end of electric field vectors usually formed with T_p and T_s, their in-plane scanning trajectory perpendicular to the light is an ellipse. The ratio of the two amplitudes of the elliptical pattern can be determined by

$$\frac{T_p}{T_s} = \frac{\rho_p}{\rho_s} \exp\{-i(\delta_p - \delta_s)\} = \tan\varphi \exp(i\Delta). \tag{4.28}$$

Polarized light analysis is a method of measuring the ellipse shape and evaluates the refractive index and film thickness of a sample very quickly.

By processing the data with a computer, the film thickness and optical constant can be quickly determined, and it is also the only feasible route to measure the thickness and optical constant of an ultra-thin transparent film deposited on a metal surface.

4.2.4 *Other film thickness monitoring methods*

4.2.4.1 *Photoelectric method*

The photoelectric method to monitor the thickness of an optical film is the most suitable and relatively intuitive. The main methods of monitoring are the monochromatic method, wavelength scanning method, and two-color method.

If the film does not absorb incident light, and the incident light intensity I is constant, the sum of reflected light intensity I_r of the film and transmitted light intensity I_t is equal to the incident light intensity; that is, $I_r + I_t = I$. The relationship between the light intensity and thickness of the optical film obeys the law of sinusoidal variation. Take the transmitted light intensity I_t as an example; there is

$$I_t = \frac{(1 - R^2)^2 I}{1 + R^4 - 2R^2 \cos 4\pi \frac{d}{\lambda}}, \tag{4.29}$$

where R is the reflection coefficient of the film, λ is the light wavelength, and d is the thickness. As can be seen from the above equations, the light transmission intensity I_t is extremely weak when the film thickness d is an odd multiple of a quarter wavelength, but extremely strong when d is an even multiple of a quarter wavelength, so it can be used as a monitoring point.

4.2.4.2 *Stylus method*

This method is to use the method of measuring the surface finish to measure the film thickness directly, as shown in Figure 4.14. The stylus of the surface finish meter is scanned vertically on surface of the film to be tested. As the needle moves up and down the surface, the electrical signal induced by the induction coil changes, and the film thickness is measured after amplification.

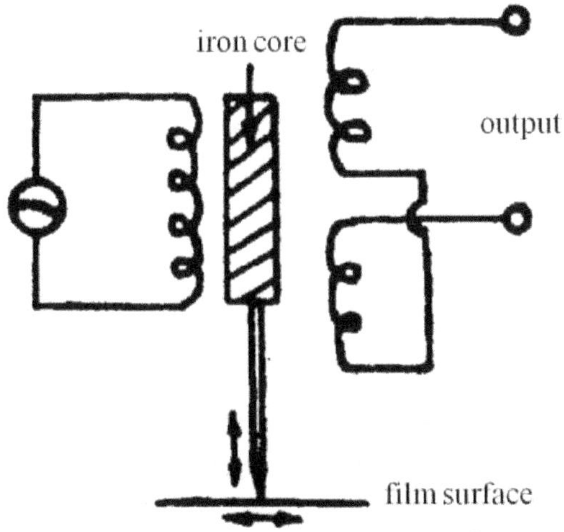

Figure 4.14: The stylus part of the differential transformer and the finish meter.

The stylus is a diamond fine needle with a diameter of 0.7–2μm, and is applied to the film surface with a pressure of 50 MPa during use. This method can quickly determine the thickness distribution on the surface and surface structure, and shows considerable precision. Moreover, damage to the film surface is small, and the minimum thickness difference recorded is about 2.5 nm.

The results obtained by the stylus method are in good agreement with the measurement results of the interference method, but the stylus method does not require coating an additional film layer on the sample surface. In addition, the stylus method cannot record narrow cracks and breaks on the film surface.

In addition to the above methods, there are other special film thickness monitoring methods, such as the radiation-absorption method, radiation-emission method, work function changes, etc. They will not be introduced one by one here.

4.3 Thin Film Graphic Manufacturing Technology

The preparation of electronic components and circuits using thin films requires the manufacture of certain geometric patterns on film. When depositing films, the film is deposited only at a certain place on the substrate by the masking method. Or else coat with film the whole substrate first, and then selectively etch it to leave the desired film pattern. With both methods, a pattern can be produced. Both methods use a mask with a pattern to transfer the pattern onto the substrate.

4.3.1 *The main methods of film pattern processing*

There are roughly three methods of forming a desired film pattern on the substrate surface:

(1) Use screen printing or a photosensitive resin (photoresist) to form a negative image on the substrate surface, then a full surface coating is performed by vacuum evaporation, sputtering, CVD, and other methods. Then immerse the substrate in solvent to

dissolve the negative image substance. Thus, while the negative-forming substance is swollen and dissolved, the film plated on the substrate is removed, and finally the desired positive film pattern is left on the substrate surface.

(2) After coating on the surface of the entire substrate by the above-described evaporation methods or the like, use screen printing or photoresist to form a positive image. Then the film, equivalent to the negative portion (exposed portion), is removed by chemical (wet) etching or dry etching. The ink for screen printing or photoresist remaining on the positive image is removed by the corresponding solvent or drying treatment, and finally a desired film positive image is formed on the substrate.

(3) A mask with a negative image is attached to the substrate, and then the film is plated on the entire surface by the above-described evaporation methods or the like, and the desired film positive image is obtained by removing the mask.

Each of the above three methods has its advantages and disadvantages, as shown in Table 4.13. It can be flexibly mastered in the actual use process. The method with minimum process in forming the thin film pattern is the above-described third method, the so-called "mask method". This simplest method is now mostly used in the semiconductor industry, but the evaporation mask used in this method needs to be produced by a special method.

The mask for evaporation is preferably of a material showing a small stretchability when the evaporation mask is heated (usually 300°C); therefore, metals such as molybdenum or cobalt, graphite, glass, or the like is often used. In addition, in order to obtain a clear film pattern, the plated mask should be tightly attached to the substrate surface so the vaporized vapor does not enter the mask, and the evaporation mask is required to be flat and smooth.

Methods for effectively improving dimensional accuracy of thin film patterns include chemical etching (wet etching), dry etching, and electroforming based on electrospray. Since chemical etching and dry etching employ visible light or ultraviolet light exposure at the beginning of the plate making stage, that is, at the beginning of

Table 4.13: Main methods of film pattern formation.

Methods	Advantages	Disadvantages
Reverse etching (etching method)	(1) It can be used for chemical etching (wet etching) but is difficult for dry etching. (2) It has a wide range of applications.	(1) Line width and line spacing are minimum 0.08 mm when screen printing is used. (2) Low sharpness of grooves. (3) Line width and line spacing are minimum 0.03 mm when using photoresist. (4) There are multi-residue inclusions on the surface.
General photolithography method	(1) It has a wide range of general purposes. (2) It is easy to realize clean operation. (3) It can use micro-machining, and processing accuracy is also determined by flatness of substrate, etc. The line width, line spacing can be as small as 0.005 mm. (4) It can make island (isolated) graphics.	(1) Multi-layer film processing is complicated. (2) There is ion collision damage in substrate materials and thin film materials such as semiconductors during dry etching. (3) It is difficult to choose a photoresist with small abrasiveness during dry etching and the line width is difficult to control with excessive photoresist.
Mask method	(1) It is suitable for film materials which are difficult to form using chemical etching (wet) and dry etching. (2) The evaporation mask can maintain a consistent accuracy in the same process. (3) Thin film graphics process is simple. (4) It has a wide range of applications. (5) The accurate repeatability of film pattern size is good.	(1) There is a need to develop a technique for reducing the gap between substrate and evaporation mask. (2) It is difficult to improve the accuracy (alignment) of dimensional position of evaporation masks, and it is difficult to automate. (3) It takes time to make island-like (isolated) graphics.

patterning, these methods are generally referred to as photolithography. Photolithography and electroforming can be performed separately or in proper combination in a series of processes to finish micromachining, which is collectively referred to as photo processing. In the process of photo processing to fabricate the evaporation mask, it is necessary to first produce a desired original film pattern. In order to make the original plate, the pattern is printed larger (original picture) on the film with small temperature and humidity deformation characteristics, and then the original image is reduced by the micro camera according to processing size, and finally the back image is made on the photo-drying plate or a photosensitive material such as a film to make the original film pattern. This basic original pattern is copied multiple times using a step repetition camera. In addition, the coordinates (data) of the base map can be read, and the coordinate data is input into an electronic computer. After being processed by the electronic computer, the original plate is output according to the required size, and the plate is automatically drawn by an automatic plotter.

The photoelectric forming method is essentially an electroforming method in which a conductor such as a metal or a conductive nonmetal is selected as the support, and then the support surface is cleanly and uniformly coated with photoresist and dried.

Subsequent exposure and development are carried out in accordance with the aforementioned photolithography process, except that the image of photoresist formed on the bottom surface of the conductive support needs to be a negative image. Secondly, in order to improve plating resistance of the bottom support surface, it should be heated and the conductive support is used as a cathode for electroplating. After evenly electrolyzing the required thickness, an evaporation mask is obtained by peeling off the electrodeposited metal layer from the bottom surface of support. The bottom surface of support can be used multiple times without damaging the photoresist image, and a plurality of evaporation masks having the same circular shape can be manufactured.

4.3.2 *Photolithography method*

Photolithography is a surface microfabrication technique that combines copying of images with chemical etching. The process is to use photocopying first, and accurately print the pattern of the photolithographic mask on a photoresist-coated film or metal foil surface. Then the corresponding pattern is gotten by etching the film or metal foil using selective protection of photoresist.

4.3.2.1 *Lithography process*

The lithography process generally includes processes such as gluing, pre-baking, exposing, developing, hardening, etching, and de-gluing. Figure 4.15 is a flow chart of the photolithography process.

First, the surface to be processed of the part is coated with photosensitive photoresist, and after appropriate drying, a mask made of a light-blocking substance with a desired pattern is masked

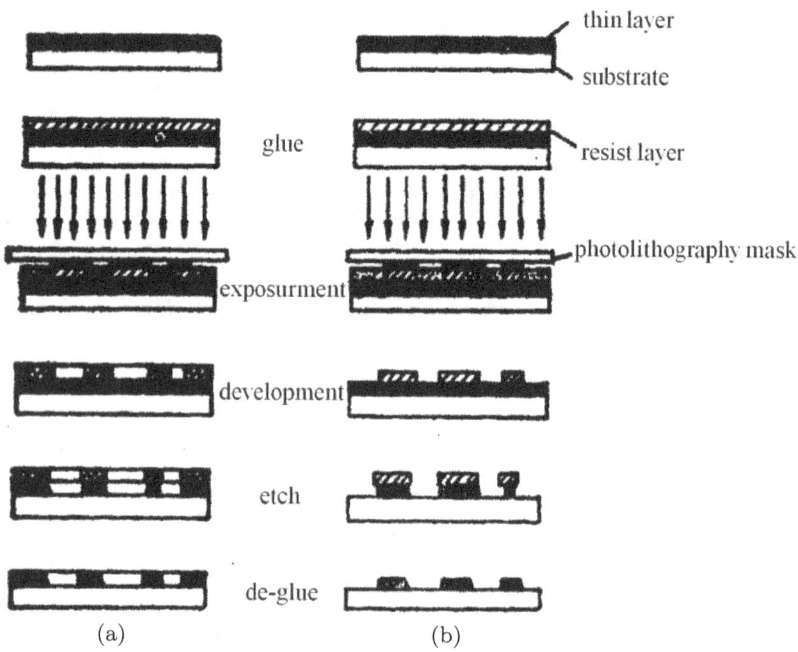

Figure 4.15: General process of lithography.

on the resist, and then exposed. The resist coating corresponding to the light-transmitting region of mask undergoes a photochemical reaction, and the portion covered by the opaque region of the mask remains unchanged. By using the apparent difference in the solubility of photosensitive and non-photosensitive portions of photoresist, depending on photochemical reaction characteristics of the used photoresist, an appropriate solution is used to remove the unsensitized portion (see Figure 4.15(a)), or dissolve the photosensitive portion (see Figure 4.15(b)). Thereafter, the obtained resist coating with a certain pattern is subjected to appropriate heat curing to perform protective corrosion. After completely etching, the resist coating completes its action and can be removed to obtain a desired pattern on the film or foil.

4.3.2.2 *Photoresist*

The photoresist is prepared by using a photosensitive resin material as the main component and adding sensitizer and solvent as additional materials, and is substantially a polymer solution which can undergo photochemical reaction such as crosslinking decomposition or polymerization after light irradiation. According to the characteristics of the photochemical reaction mechanism and the change of properties during exposure, photoresist can be divided into positive and negative types. The former obtains the same anticorrosion coating on the photoresist-coated substrate as the light-blocking pattern on the mask plate; while the latter is the opposite, and the obtained resist coating is exactly the same as the unblocked portion of the mask plate.

The lithography process requirements for photoresist are as follows: an accurate circular shape should be formed as much as possible, and it should have sufficient stability for post-treatment, there is no deformation with changes in heat and mechanical conditions, and there is no factor harmful to the electrical characteristics and reliability of components; in the final de-glue treatment it can be removed cleanly without leaving a residue. These requirements should be met as much as possible, otherwise it will adversely affect the etching. Since the negative photoresist has the characteristics of fewer pinholes, corrosion resistance, high sensitivity, and stability, it has become the mainstream of current photoresists. Table 4.14 lists some of the necessary conditions for photoresist.

4.3.2.3 *Exposure types*

Exposure is the selective exposure of a substrate coated with photoresist, and causes a photochemical reaction in the illuminated film after absorbing energy from the light, finally forming a "latent image" of a desired pattern. The method of exposure has a great influence on the precision of the lithography process, especially the resolution of the pattern. Currently used exposure types include ultraviolet exposure, electron beam exposure, and X-ray exposure. Figure 4.16 shows a comparison of these three exposure techniques. Ultraviolet exposure is easy to operate, economical, and commonly used, but there is diffraction, reflection, and interference of the mask–resist–substrate interface, so the resolution is limited, and the method is suitable for making micron-scale lines. Electron beam exposure has a short wavelength, a large energy (about 5–20 eV), a small beam diameter of 0.01–0.5 μm of focusable size, and easy deflection, so it can produce patterns with an accuracy of 1 μm or lower. However, it requires a long exposure time, a large secondary effect, and high equipment cost. The X-ray exposure requires a wavelength of about a few tenth of a nanometer, which neglects the diffraction, reflection and interference of light. It has strong transmission and is less affected by dust, which is beneficial to improve the yield of components. At the same time, because the secondary effect is small, a clear resist pattern can be obtained. But it is not as easy to focus and deflect as an electron beam.

In addition, ion beam exposure has appeared in recent years. Compared with electron beam exposure, it has higher resolution and sensitivity, and has almost no proximity effect. The effect on resist is almost 100 times that of the electron beam, which makes exposure time shorter and obtains circuit patterns with actual resolution less than 0.5 μm or even smaller. Thus, it has received great attention.

Characteristic	Serial number	The necessary conditions for photoresist
Optical properties		Small dispersion of irradiation energy
	1	High absorption rate
	2	
Photochemical properties		High sensitivity
	3	Large γ values
		Not easily affected by the surrounding atmosphere
	4	
	5	
Graphic forming characteristics		It has high resolution even if the film is thick
	6	The edge linearity of graphics is good
		Good repeatability
	7	No residual base film and edge sawtooth after development treatment
	8	
	9	
Drug characteristics	10	For films, there are few defects such as pinholes
	11	Strong corrosion resistance to various corrosive agents
	12	Invariant after treating with various drugs
	13	Easy to peel off without leaving a residue
Mechanical properties	14	Small brittleness
	15	Good adhesion to the substrate
	16	Poor adhesion on the surface of the coated film
Thermal characteristics	17	In pre-baking, thermal base film is small
	18	There is no dimensional change in the lower layer during post-baking
Mass production characteristics	19	Stable in material quality and production technology, no deviation
	20	Wide process tolerance
Impurity	21	It does not contain impurities that have a bad influence on component characteristics

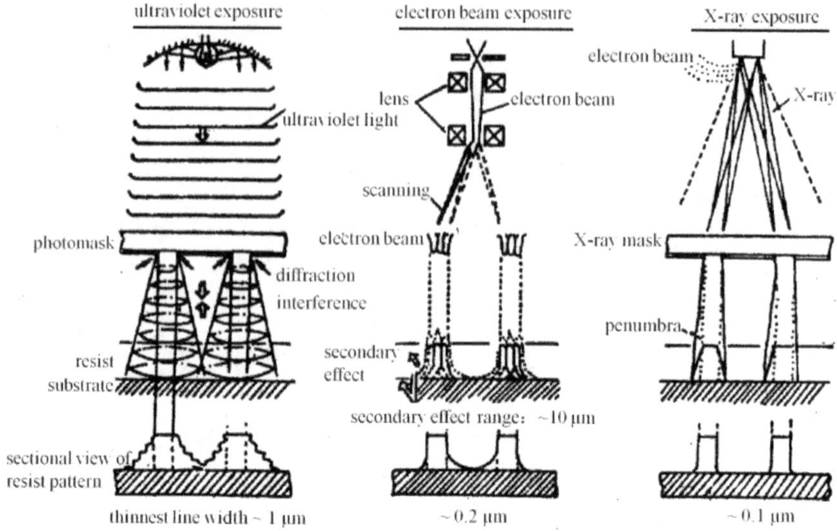

Figure 4.16: Comparison of various exposure techniques.

Of course, the resists used are different due to the difference in exposure form. In order to obtain higher resolution and sensitivity, photoresist and electron beam resist may be respectively used depending on exposure form. The main factors affecting the precision of lithography are the material quality and precision of the photomask, photoresist characteristics, graphic formation method and device accuracy, position alignment method during exposure and device accuracy, and etching method.

4.3.2.4 *Corrosion*

After the above exposure and development, a resist-protected pattern can be obtained on the substrate, and the process of removing the film at the exposed portion of the substrate is called etching. Corrosion is a very important part in the lithography process, and its quality directly affects the resolution and accuracy of the graphics.

(1) **Wet etching**

A widely used method in production is chemical etching using an etching solution, that is, wet etching. This method is simple, and suitable for mass production and operation, and it is relatively easy to produce patterns of about 3 ± 0.3 μm, so the method is commonly used. For example, the SiO_2 film uses HF acid as etching solution, and the reaction is

$$SiO_2 + 4HF = SiF_4 \uparrow + 2H_2O.$$

Thereby, the film exposed to etching liquid can be removed. Some commonly used membranes and corrosive fluids used are shown in Table 4.15.

(2) **Dry etching**

Since wet etching is prone to erosion, it is very difficult to make small-sized, high-precision patterns. At the same time, a large amount of waste solution needs to be processed, so dry etching techniques without corrosive liquid are used, including ion etching, plasma etching, and reactive ion etching. This is known as dry etching.

(a) *Ion etching*

Ion etching includes ion beam etching and sputtering etching. These are all methods of stripping corrosion by sputtering, which occurs

Table 4.15: Examples of wet etching.

Etched material	Etched solution
Si	HF–HNO_3–CH_3COOH; KOH; N_2H_4+$CH_3CHOHCH_3$
Al	H_3PO_4–HNO_3–CH_3COOH; KOH–K_3[Fe(CN)6]; HCl; H_3PO_4
Mo	H_3PO_4–HNO_3
Ti	HF; H_3PO_4; H_3PSO_4; CH_3–COOH(I_2)–HNO_3–HF
Ta	HNO_3–HF
W, Pt	HNO_3–HCl
Au	I_2–KI
Ag	$Fe(NO_3)_3$–ethylene glycol
Cu	$FeCl_3$
SiO_2	HF–NH_4F with buffer
PSG	HF
BSG	HF–HNO_3
SiN_4	H_3PO_4; HF; HF–CH_3COOH
Al_2O_3	H_3PO_4; H_2SO_4 $\rightarrow$BHF

when an inert gas (such as Ar gas) is bombarded on the substrate surface, and are basically a physical process. Ion beam etching is substantially the same as the sputtering of ion beam described above.

Using a hot cathode electron impact ion source to generate argon plasma, Ar^+ ions are extracted from the ion chamber and accelerated into a large-diameter ion beam and enter a high-vacuum sample chamber. The ion beam then neutralizes electrons from the cathode and becomes neutral. The atomic beam bombards the sample, and the sample is sputtered and peeled off to achieve the purpose of corrosion. The apparatus for sputter etching is basically a flat electrode system in which the substrate is placed on the cathode, and ions formed by gas discharge directly bombard the substrate under the bias of the cathode to cause etching.

In ion etching, it is necessary to use a mask with a certain pattern and small etching speed, place it on a solid sample, and then carry out uniformly ion beam irradiation of the sample to achieve selective sputter etching on the sample surface. The etching rate is related to factors such as ion acceleration voltage, incident angle, and sample types, but the etching rate is low. Since the bombardment of the ion or ion beam has a certain direction, the etching has obvious directivity and does not cause undercutting, and can form fine patterns of the order of micrometers or submicron.

(b) *Plasma etching*

Plasma etching uses a chemical reaction caused by plasma in gas glow discharge to achieve the purpose of etching. It uses low temperature plasma, and Figure 4.17 shows a schematic of plasma etching. Suitable gas is introduced into the reaction chamber, a certain pressure is maintained, and a high frequency power is applied to obtain plasma. The low-temperature plasma has a small degree of ionization, and the gas particles are mostly neutral atoms and molecules, and a considerable part of them are atoms and molecules in an excited state.

The atoms and molecules in the excited state have strong chemical activity. Therefore, when a solid sample is placed in gas

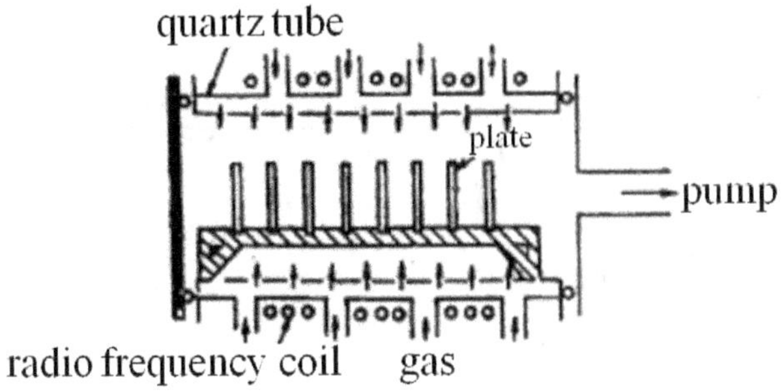

Figure 4.17: Device schematic of plasma etching.

plasma, the atoms on the surface of the solid sample react chemically with excited atoms and molecules in plasma. As a result of reaction, volatile substances are generated; that is, the atoms are etched away from the solid sample surface, which is the principle of plasma etching.

Since the plasma etching effect is mainly caused by chemical reaction of the active atomic group generated by discharge, it is generally impossible to control the direction in which the activating group acts on the substrate, and thus this etching is similar to wet etching, and is isotropic. Then it will cause more serious undercut. However, this etch provides greater selectivity by selecting different gas and gas mixtures.

(c) *Reactive ion etching*

The devices for reactive ion etching and sputter ion etching are basically parallel plate electrode systems. The main difference between reactive ion etching and sputter etching is that an ion beam of an inert gas is sputter-etched, and a reactive gas such as a carbon halide compound is used for reactive ion etching. The main difference between reactive ion etching and flat-plate plasma etching is that the coupling mode of the radio frequency power source and electrode is different from the pressure of the reaction gas. Reactive ion etching not only has the physical effect of ion bombardment, but also the chemical reaction of plasma active particles. However, the ion bombardment effect referred to here is different from the

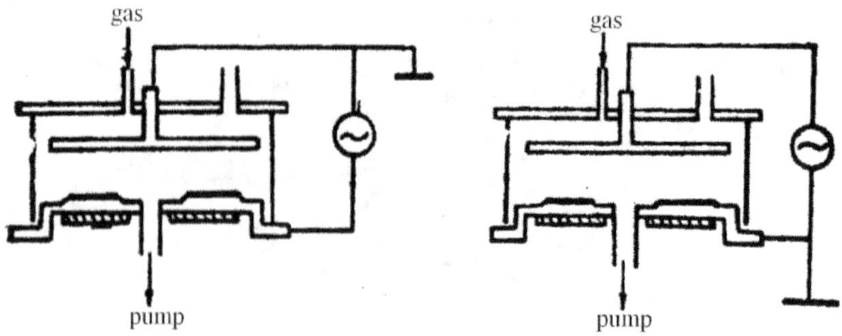

Figure 4.18: Schematic diagram of a device for reactive ion etching.

purely physical process in ion etching, which will have a significant enhancement effect on the chemical reaction. Thus some call it enhanced plasma etching. Figure 4.18 is a schematic diagram of a device for reactive ion etching.

4.3.2.5 *Comparison of various corrosion methods*

Table 4.16 lists the basic mechanisms of ion etching, reactive ion etching, plasma etching, and wet etching. Table 4.17 compares the various characteristics of these corrosion methods.

Compared with dry etching, wet etching has advantages of micro-machining, high precision, less damage, and high process cleanliness. Both wet etching and plasma etching are chemical etching. However, in the case of wet etching, there is a problem that the resist penetrates into the interface between resist and substrate. In order to suppress the incorporation of etching liquid, it is baked to form a hardened film, which often causes deformation of resist, and is disadvantageous for microfabrication. Ion etching is almost purely a physical effect. Because of the small lateral corrosion, it is a powerful means for microfabrication. The biggest disadvantage of this method is that the difference in etching rate between different materials is small, thus limiting mask materials, and at the same time easily causing the layer below the film to be etched corrosion. Reactive ion etching combines physical and chemical corrosion with good overall performance.

Table 4.16: Corrosion method and its basic properties.

Reaction mechanism	Corrosion method	Process
Physical chemistry	Ion etching	Dry
	Reactive ion etching	
	Plasma etching	
	Wet etching	Wet

Table 4.17: Comparison of the characteristics of three corrosion methods.

Project	Corrosion method		
	Ion etching	Plasma etching	Wet etching
Working environment	Good		Not good
Waste treatment	Easy		Difficult
Cleanliness	High		Low
Automation	Slightly difficult	Easy	Difficult
Control	High	Medium	Low
Device	Large	Slightly larger	Easy
Sample selectivity	Small	Large	
Corrosion rate	Small	Large	
Undercut/side erosion	Small	Medium	Large
Corrosive penetration	No		Large
Problems caused by stickiness	No		Yes
Damage	Large (atomic level)	Medium (electronic level)	No
Temperature	Large		Small

Therefore, in the specific process, various factors and requirements should be considered to properly select the corrosion method for corrosion processing.

4.4 Environment of Preparation Films

4.4.1 *Dust and pinholes*

When coating, the air bubbles, moisture and dust in the environment are the main problems affecting the quality of films. In general, dust

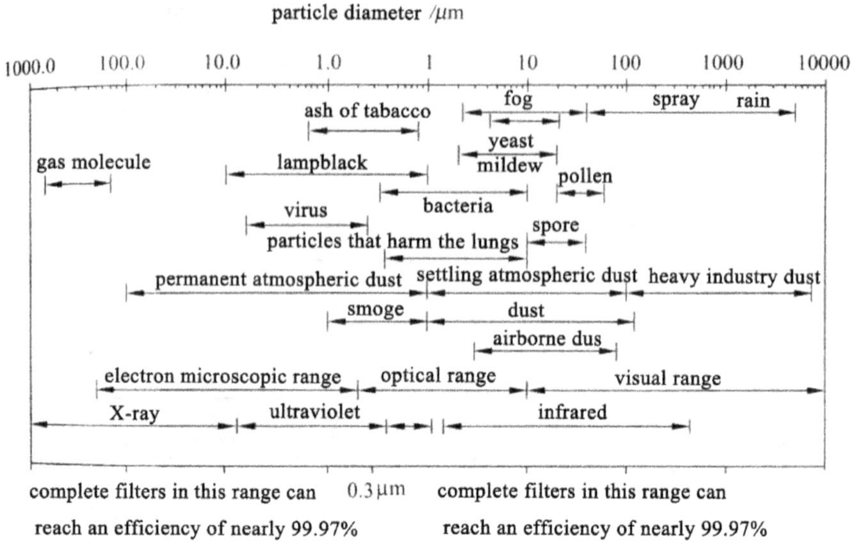

Figure 4.19: The size of dust particles in the atmosphere.

above 0.5 μm reaches 177,000 L even under relatively clean conditions in an average city. The size range of the dust in the atmosphere is shown in Figure 4.19. Obviously, if the film is prepared directly in such an environment, the quality of the film will be very poor. Therefore, removing dust and keeping clean are very important to prepare fine products and components.

4.4.2 *Standards and levels of an ultra-clean chamber*

The preparation of thin films in a clean environment is one of the key technologies to ensure the quality of thin films. The clean room was originally defined as Federal Standard 209, developed in the United States. Later, the definition of the clean room level was developed based on the number of particles above a certain diameter per unit volume. The metric version of the clean room level is shown in Table 4.18. The vertical coordinate of Figure 4.19 represents the number of particles per unit volume, with the right

Table 4.18: Federal standard 209E metric definition of particulate purification levels in the air.

Level	Number of particles per cubic meter				
	0.1 μm	0.2 μm	0.3 μm	0.5 μm	5 μm
M1	3.5×10^2	7.57×10^1	3.09×10^1	1.00×10^1	
M1.5	1.24×10^3	2.65×10^2	1.06×10^2	3.53×10^1	
M2	3.5×10^3	7.57×10^2	3.09×10^2	1.00×10^2	
M2.5	1.24×10^4	2.65×10^3	1.06×10^3	3.53×10^2	
M3	3.5×10^4	7.57×10^3	3.09×10^3	1.00×10^3	
M3.5		2.65×10^4	1.06×10^4	3.53×10^3	
M4		7.57×10^4	3.09×10^4	1.00×10^4	
M4.5				3.53×10^4	2.47×10^2
M5				1.00×10^5	6.18×10^2
M5.5				3.53×10^5	2.47×10^3
M6				1.00×10^6	6.18×10^3
M6.5				3.53×10^6	2.47×10^4
M7				1.00×10^7	6.18×10^4

side corresponding to the British system and the left side to the metric system. According to Federal Standard 209A, the number of particles above 0.5-μm per cubic foot are 100, 10,000, and 100,000, corresponding to FS100, FS10000, and FS100000, respectively. The levels 100, 10,000 and 100,000 are the ones we often use.

As the requirements for a clean room increase, the standards become more detailed. In recent years, it is stipulated that the number of particles above 0.1 μm per cubic meter corresponds to new level 1 (10), new level 2 (100), new level 3 (1000), new level 4 (10,000) and so on, as shown in Figure 4.20 on the left.

In the preparation of films, the most suitable environment depends on the specific situation. For thin films of semiconductors, integrated circuits, etc., substrate cleaning should be carried out in a clean room of new class 5 (FS100) to new class 6 (FS1000), and coating in a clean environment superior to new class 3 (FS1). Therefore, the cleanliness and design of the workshop must be determined from the purpose of preparing the film.

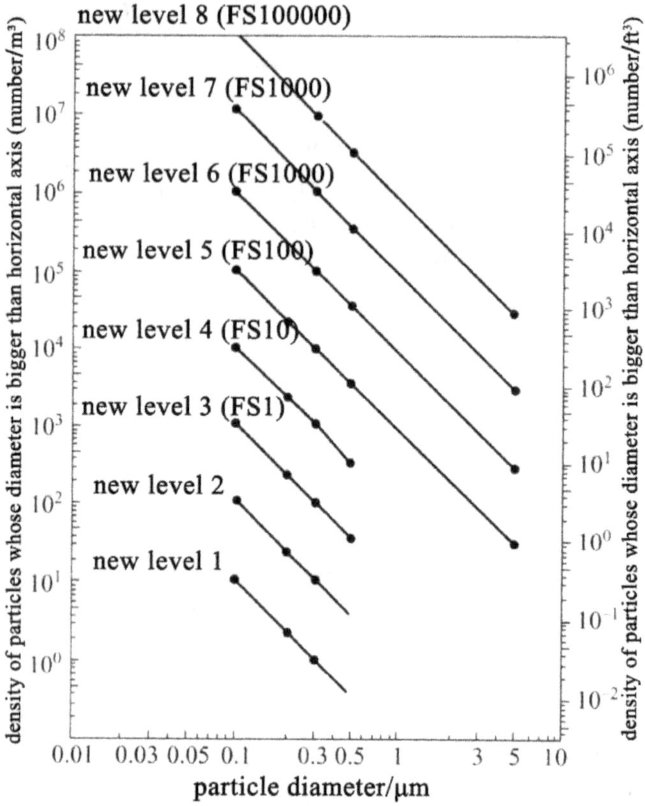

Figure 4.20: Relationship between the cleanliness of the air and the diameter of dust particles.

4.5 Exercises and Questions

(1) Metal films with same thickness are plated on both sides of a thin substrate, and residual tension is found in the films by experiment. Use the re-sputtering method to remove thin film on one side. Please describe qualitatively how the remaining film–substrate will deform.

(2) Matrix deformation occurs due to the stress field around a fault.

   (a) In a film, a row of edge dislocations in the same direction are generated near the substrate interface and parallel to the substrate interface. Indicate the internal stress of the film.

(b) If these dislocations are screw dislocations, how should the stresses of the film be different?
(c) By annealing, some dislocations will climb and some will disappear. What effect will they have on internal stress?

(3) At 500°C, a GaSb epitaxial film was grown on an AlSb substrate.

(a) What are the expected point defects at 500°C?
(b) If $E_{\mathrm{GaSb}} = 91.6$ GPa and $V_{\mathrm{GaSb}} = 0.3$, What is the thermal stress of the film at 20°C?

(4) The properties of stress (σ_f) and thickness (d) of a thin film are described as functions of d (K and n are both constants). Compare the stresses of various films and instantaneous stresses on d.
(5) The internal stress of SiO_2 is 10^{10} dynes/cm^2. In the range of 900–1500°C, if the viscosity coefficient of SiO_2 film $\eta(T) = 1.5 \times 10^{-8} \exp E_v/RT(E_v = 137\,\mathrm{kcal/mole})$. At 1000°C, how long does it take for the film to reach half of its final stress? Assume that E= 6.6×10^{11}dynes/cm^2.
(6) An engineer wants to make sure that at 20°C, silicon wafers are coated with SiO_2 films one micron thick or Si_3N_4 films one micron thick, which has a greater curvature. Both films were deposited on a silicon substrate at 500°C. At this deposition temperature, the internal stress of the two materials is SiO_2: -3×10^9dynes/cm^2, Si_3N_4: -6×10^9dynes/cm^2. If the respective modulus of the two materials is $E_{\mathrm{SiO2}} = 7.3 \times 10^{11}$dynes/cm^2, $E_{\mathrm{Si3N4}} = 15 \times 10^{11}$dynes/cm^2, and the thermal expansion coefficient is $\alpha_{\mathrm{SiO2}} = 0.55 \times 10^{-6°}\mathrm{C}^{-1}$, $\alpha_{\mathrm{Si3N4}} = 3 \times 10^{-6°}\mathrm{C}^{-1}$. Calculate the curvature of the two wafers respectively [*Note*: $E_{\mathrm{Si}} = 16\times10^{11}$dynes/cm^2, $\alpha_{\mathrm{Si}} = 4\times10^{-6°}\mathrm{C}^{-1}$]. If the Poisson's Ratio of the film to the substrate is 0.3, what is the radius of curvature of a wafer with a diameter of 15 cm? What is the difference between the height of the center and edge of the wafer?

(7) The thickness of films deposited sequentially is very thin compared to the substrate; each film is introduced into the bending moment and curvature independently.

(a) $\frac{1}{R_1} + \frac{1}{R_2} + \cdots + \frac{1}{R_n} = \frac{1-\nu_s}{E_s}\frac{6}{d_s^2}(\sigma_1 d_1 + \sigma_2 d_2 + \cdots + \sigma_n d_n)$, 1, 2, ..., n represents the number of layers of thin films, and σ_n and d_n represent stress and thickness.

(b) A layer of 5000 Å Al thin film is deposited on a silicon wafer with a diameter of 12.5 cm and a thickness of 0.5 mm. The deposition temperature is 250°C, so stress relaxation does not occur when cooled to 20°C. Heat the Al–Si system to 500°C and the Al will relax completely. A Si_3N_4 thin film with a thickness of 2 μm is doposited on the substrate with an internal compressive stress of 700 MPa. After cooling to 20°C, what is the final radius of curvature?

	Si	Al	Si_3N_4
E/GPa	160	66	150
α/°C^{-1}	4×10^{-6}	23×10^{-6}	3×10^{-6}

Assume $v = 0.3$ for all materials.

(8) For adjacent films 1 and 2, the corresponding film thickness, modulus, and untensioned lattice constants are d_1, E_1, a_0 (1) and d_2, E_2, a_0 (2). The lattice constant on the film interface is $\bar{a}_0$.

(a) What is the strain of each film?
(b) What are the corresponding stresses?
(c) If the forces reach equilibrium, verify that

$$\bar{a}_0 = a_0(1)a_0(2)\left(1 + \frac{E_2}{E_1}\frac{d_2}{d_1}\right)/a_0(2) + \frac{E_2 d_2}{E_1 d_1}a_0(1).$$

(9) Consider a substrate with a thickness of d_s, which consists of a thin film with a thickness of d_f and has a homogeneous tension σ_f. It is assumed that the elastic coefficients of the film and the substrate are equal.

(a) Measure the stress on the substrate, assuming it is in equilibrium.

(b) Show that under the above conditions, the net torque on the center axis of the substrate disappears.

(c) After complete annealing, the stress of the layer disappears without affecting the substrate and other films. So what is the net force? What is the combined torque?

(d) After the external inhibition disappears, the film–substrate will undergo elastic deformation to achieve the new stress balance and torque balance. A force or torque is required to compensate for the mechanical imbalance in (c). How much does this force affect the stress on the remaining films? What is the maximum effect of this torque on the remaining films?

(e) What is the maximum stress in the remaining films?

(10) Holes and gaps are often observed at faults, compounds, diffusion, and interfaces between thin film and substrate. Identify the source of these defects on this interface. Which interface is prone to microcrack? Why is that?

Chapter 5

Formation and Growth of Thin Films

The formation and growth process of thin films can be roughly divided into the stages of condensation, diffusion, nucleation, grain growth, combination, and connection of atoms or particle clusters into thin films on substrates. The formation and growth of thin films are not only related to the film materials and substrate materials, but also affected by the state and energy of atoms and ions, deposition rate, substrate temperature, impurities, and other factors. The growth process of the films will determine the microstructure and structure of the films, and the physical and chemical properties of the films are directly determined by the microstructure and structure of the films. Therefore, it is necessary to study the formation and growth of thin films before discussing their structure and properties.

Although the preparation methods of the film are various, the formation mechanisms of the film are different. However, in many respects, different preparation methods of films, film formation and growth processes still have many of the same characteristics. In this chapter, we mainly discuss vacuum filming as an example.

5.1 Condensation Process and Surface Diffusion Process

The condensation process refers to a phase transition process from the vapor phase to adsorption phase and then to condensate phase after the atoms, ions or molecules evaporated or sputtered from evaporation source or sputtered target are incident on the substrate surface. It is the first stage of film formation.

5.1.1 *Adsorption process*

A major difference in crystal structure between a solid surface and a solid interior is the interruption of chemical bonds between atoms or molecules. The interruption bonds formed by atoms or molecules on solid surfaces are called unsaturated bonds or suspended bonds. This bond has the ability to attract foreign atoms or molecules. The phenomenon that gaseous atoms incident on the substrate surface are attracted by the suspended bond is called adsorption. It is called physical adsorption if the adsorption is only acted by the van der Waals force between the electric dipole moments of atoms, and chemical adsorption if the adsorption is acted on by the bonding force of chemical bonds. The special state of solid surface makes it have a special energy, which is surface free energy. The surface free energy decreases due to the adsorption phenomenon. The energy released with the occurrence of the adsorption phenomenon is called adsorption energy. The removal of vapor atoms adsorbed on solid surfaces is called desorption, and the removal of energy adsorbed by vapor atoms is called desorption energy.

The gaseous atoms incident on the substrate surface from the evaporation source or the sputtering target have a certain amount of energy. Three phenomena can occur after they reach the substrate surface:

(1) The gaseous phase atoms are adsorbed by energy exchange with the surface atoms of the substrate.
(2) After adsorption, the gaseous atoms still have a large desorption energy, and then are desorbed and evaporated (re-evaporation or secondary evaporation) after a short stay on the substrate surface.
(3) The gaseous atoms do not exchange energy with the substrate surface, and are incident on the substrate surface and are immediately reflected back.

The energy curve of the adsorption process is shown in Figure 5.1. When the kinetic energy of gaseous atoms incident on the substrate surface is small, they are in the state of physical adsorption, and their

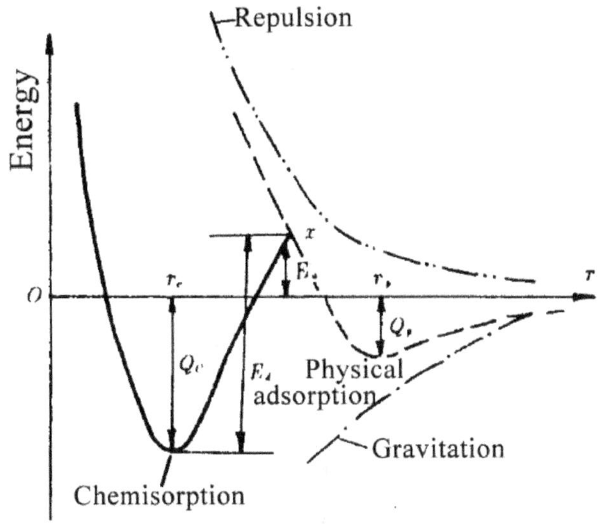

Figure 5.1: Energy curve of the adsorption process.

adsorption can be expressed by Q_p. Chemical adsorption occurs when the kinetic energy of the vapor atom is larger but is less than or equal to the activation energy E_a. The difference Q_c between E_d and E_a is called the heat of chemical adsorption. Because Q_0 is larger than Q_p, only gaseous atoms with larger kinetic energy can produce chemical adsorption on the substrate surface. When the kinetic energy of this vapor atom is greater than that of E_d, it will not be adsorbed on the substrate surface, and will be converted into gas phase by re-evaporation or desorption. So E_d is also called desorption energy.

The relationship between the average residence time τ_a of the adsorbed gas phase atoms on the substrate surface and the adsorption energy E_d is

$$\tau_a = \tau_0 \exp(E_d/kT), \tag{5.1}$$

where τ_0 is the vibration period of a single layer of atoms, the value is about 10^{-11} to 10^{-12} seconds, k is the Boltzmann constant, and T is the absolute temperature.

The relationship between adsorption energy E_d and average residence time τ_a at room temperature is shown in Table 5.1. As seen

Table 5.1: Relationship between adsorption energy E_d and average residence time τ_a.

E_d (kcal/mol)	τ_a (s)	E_d (kcal/mol)	τ_a (s)
2.5	6.6×10^{-12}	20	3.8×10
5	4.4×10^{-10}	25	1.7×10^{5}
10	1.6×10^{-6}	30	7.3×10^{8}
15	8.5×10^{-3}		

from Table 5.1, when E_d is greater than 20 kcal/mol, the τ_a value sharply increases.

5.1.2 *Surface diffusion process*

When the gas phase atoms incident on the substrate surface form adsorbed atoms on the surface, they lose kinetic energy along the normal direction of the surface, and only have kinetic energy moving parallel to the horizontal direction of the surface. By virtue of this kinetic energy, the adsorbed atoms can perform surface diffusion motion in different directions on the surface. During surface diffusion, a single adsorbed atom collides with each other to form an atomic pair before condensation can occur. Therefore, condensation refers to the process by which adsorbed atoms are combined into atoms later. Therefore, the surface diffusion motion of the adsorbed atoms is a necessary condition for the formation of condensation. The surface diffusion energy E_0 is much smaller than the adsorption energy E_d, which is about 1/6 to 1/2 of the adsorption energy E_d.

The residence time of the adsorbed atoms at one adsorption site is called the average surface diffusion time and is expressed by τ_D. The relationship between it and the surface diffusion energy E_D is

$$\tau_0 = \tau_0' \exp(E_0/kT), \tag{5.2}$$

where τ_0' is the period in which the surface atoms vibrate in the horizontal direction of the surface, approximately 10^{-13}–10^{-12} seconds, k is the Boltzmann constant, and T is the absolute temperature. It is generally considered that $\tau_0 = \tau_0'$.

The distance (the interval from the starting point to the end point) at which the adsorbed atoms move over the surface by the diffusion motion is called the average surface diffusion distance and is represented by $\overline{x}$.

$$\overline{x} = (D \cdot \tau_a)^{1/2}, \tag{5.3}$$

where D is the surface diffusion coefficient; if a_0 represents the interval between adjacent adsorption sites, then the surface diffusion coefficient is defined as $D = a_0^2/\tau_D$. Thus, the average surface diffusion distance can be expressed as

$$\overline{x} = a_0 \exp[E_d - E_D)/kT. \tag{5.4}$$

It can be seen from Equation (5.4) that the magnitude of E_d and E_D has a large influence on the condensation process. The larger the surface diffusion energy E_d, the more difficult the diffusion and the shorter the average diffusion distance $\overline{x}$. The larger the adsorption energy E_d, the longer the residence time τ_a of the adsorbed atoms on the surface, and the longer the average diffusion distance $\overline{x}$. This is very beneficial to the formation of the condensation process.

5.1.3 *Condensation process*

Hypothesize that the velocity of the evaporation source of the heat source follows the Boltzmann distribution, and the number of atoms in the velocity v to $v + vdv$ is

$$Nf(v)dv = C \exp\left(-\frac{mv^2}{2kT}\right) dv, \tag{5.5}$$

where N is the number of atoms per unit volume, $f(v)$ is the probability that the atom has a velocity v, m is the atomic mass, and C is a constant. If the velocity component of the vertical substrate surface is v_z, the number of atoms incident on the surface area ds of the substrate in the time dt is

$$dN = Nf(v)v_z ds dt. \tag{5.6}$$

The atomic rate deposited on the substrate surface is R_d, which can be expressed as

$$R_d = C \int v_z \succ 0 v_z \exp\left(-\frac{mv_2}{2kT}\right) dv. \tag{5.7}$$

When the atom collides with the surface elastically, the momentum transmitted by each atom is $2mv_z$, and the pressure is thus obtained as

$$P = C \int v_z \succ 0 2mv_z^2 \exp\left(-\frac{mv^2}{2kT}\right) dv. \tag{5.8}$$

In Equations (5.7) and (5.8), the integral to dv ranges from 0 to ∞; that is, atoms are incident from one side to the substrate surface. Thus, the ratio of deposition rate R_d to pressure P is

$$\frac{R_d}{P} = \frac{\int_0^\infty v_z \exp\left(-\frac{mv^2}{2kT}\right) dv}{\int_0^\infty 2mv_z^2 \exp\left(-\frac{mv^2}{2kT}\right) dv} = \left(\frac{1}{2\pi mkT}\right)^{\frac{1}{2}}. \tag{5.9}$$

The relationship between the deposition rate R_d and the pressure P is given by Equation (5.9).

The vaporized gas phase atoms are incident on the substrate surface, and the ratio of the number of gas phase atoms completely condensed on the substrate surface to the total number of gas phase atoms incident on the substrate surface is called the condensation coefficient, except for the atoms which are evaporated by the elastic reflection and adsorption. And the condensation coefficient is expressed by a_c.

When there are already condensed atoms on the substrate surface, the ratio of the number of recondensed gas phase atoms to the total number of gas phase atoms incident on the substrate surface is called the adhesion coefficient, and is represented by a_s

$$a_s = \frac{1}{J}\frac{dn}{dt}, \tag{5.10}$$

where J is the total number of gas atoms incident on the substrate surface, and n is the number of atoms present on the substrate surface during t time. When n approaches zero, $a_c = a_s$.

A physical quantity that characterizes the degree of energy exchanged between an incident gas phase atom (or molecule) and a surface of a substrate is called the thermal adaptation coefficient and is represented by a

$$a = \frac{T_i - T_T}{T_i - T_s}, \tag{5.11}$$

where T_i, T_τ and T_s represent the temperature of the incident gas phase atoms, the re-evaporating atoms and the substrate, respectively.

During the residence of the adsorbed atoms on the surface, if the energy exchange with the substrate reaches the thermal equilibrium ($T_\tau = T_s$), $a = 1$ indicates complete adaptation; if $T_s < T_\tau < T_i$, $a < 1$ indicates incomplete adaptation; if $T_i = T$, then there is no heat exchange between the incident gas phase atoms and the substrate, and the gas phase atoms are totally reflected back. $a = 0$ means completely unsuitable and belongs to the elastic reflection condition (no loss of energy).

When $a = 1$, all incident atoms are adsorbed on the substrate surface. When an atom incident on the substrate loses all its excess energy, its energy state is completely determined by the substrate temperature. If the energy of the incident atom is close to the adsorption energy of the substrate surface, the thermal adaptation coefficient is actually equal to 1. Generally, the adsorption energy is 1–4 eV, and the equivalent temperature of the corresponding incident atom is about 10^5 K, which is much higher than the temperature of all the evaporation sources. The atoms are physically adsorbed at the surface, and the residence time of the adsorbed atoms on the surface is

$$\tau_B = \frac{1}{v_s} \exp\left(\frac{E_d}{kT}\right), \tag{5.12}$$

where v_s is the surface vibration frequency of the adsorbed atoms; E_d is the adsorption energy of the atoms on a given substrate; T is the equivalent temperature of the atoms, and the value is usually between the evaporation source temperature and the substrate temperature.

When it has high adsorption energy, that is, $E_d \gg kT$, it is very large, so that the incident atom can quickly reach the temperature balance, the resident atoms on the surface are confined to a certain position, or will jump and migrate; if $E_d \approx kT$, the resident atoms cannot reach the equilibrium temperature quickly, thus maintaining a superheated state, with the result that the resident atoms diffuse on the surface.

During the condensation process, the adsorption energy E_d and the surface diffusion energy E_D are important parameters. The E_d and E_D values for some materials are given in Table 5.2. The relationship between the condensation coefficient a_c, the sticking coefficient a_s, the substrate temperature, evaporation time, and film thickness was obtained from experimental studies, as shown in Table 5.3 and Figure 5.2, respectively.

In order to ensure the formation of the condensed primary nucleus, the deposition rate must be large enough that the atoms will re-evaporate before encountering another particle. If the incident atomic rate is R_d, the atomic concentration on the substrate changes with time as

$$\frac{dn}{dt} = R_d - \frac{n}{\tau_s}. \tag{5.13}$$

Table 5.2: Adsorption energy (E_d) and surface diffusion energy (E_D) of some elements and substrate materials.

Condensation material	Substrate	E_d (eV)	E_D (eV)
Ag	NaCl		0.2
Al	NaCl	0.6	
Al	Mica	0.9	
Ba	W	3.8	0.65
Cd	Ag	1.6	
Cd	Glass	0.24	
Cu	Glass	0.14	
Cs	W	2.8	0.61
Hg	Ag	0.11	
Pt	NaCl		0.18
W	W	5.83	1.21

Table 5.3: Relationship between the condensation coefficient of gas phase atoms, substrate temperature, and film thickness.

Condensation material	Substrate	Substrate temperature (°C)	Film thickness (Å)	Condensation coefficient a_c
Cd	Cu	25	0.8	0.037
			4.9	0.26
			6.0	0.24
			42.2	0.26
Au	Glass, Cu, Al	25	The film thickness can just be observed	0.90–0.99
	Cu	350		0.84
	Glass	360		0.50
	Al	320		0.72
	Al	345		0.37
Ag	Ag(0)	20	The film thickness can just be observed	1.0
	Au(0.18)	20		0.99
	Pu(3.96)	20		0.86
	Ni(13.7)	20		0.64
	Glass	20		0.31

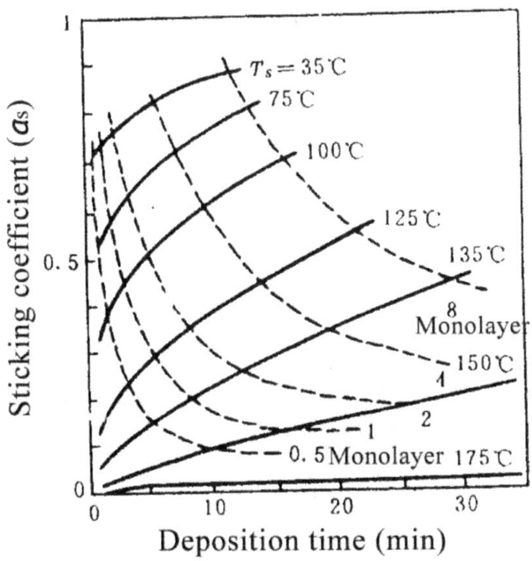

Figure 5.2: Sticking coefficient a_s at different substrate temperatures as a function of deposition time (dashed line is the equal average line of film thickness).

Assuming that at the beginning of deposition, $t = 0$, $n = 0$, integrating Equation (5.13) above, the atomic concentration on the substrate surface is

$$n = R_d\tau \left[1 - \exp\left(-\frac{t}{\tau_s}\right)\right], \tag{5.14}$$

where t is the adsorption time. If $t \to \infty$, then $n = R_d\tau_s$, which means that the atomic incident rate R_d is equal to the atomic re-evaporation rate R_e. In this case, the deposition rate is obtained by Equation (5.9) and Equation (5.12) as

$$R_d = nv_8 \exp\left(-\frac{E_d}{kT}\right) = P(2\pi mkT)^{-1/2}. \tag{5.15}$$

The ratio of deposition rate to re-evaporation rate, R_d/R_e, is called supersaturation. This is an important parameter for the condensation process. The deposition rate is determined by the evaporation rate of a given source (for example, when the evaporation rate of Ag is 1 Å/s, approximately one atom per second is dropped on one substrate atom, which corresponds to the pressure $P \approx 10^{-4}$ Pa). The rate of re-evaporation depends on the equilibrium vapor pressure at the substrate temperature (Ag is 300 K, pressure $P \approx 10^{-38}$ Pa). For the atoms deposited on the substrate, the condensation is determined by the adsorption energy E_a and the sublimation energy E_{8u} of the incident atoms. The following may occur:

(1) If $E_a \ll E_{8u}$, condensation can occur without supersaturation.
(2) If $E_a \cong E_{8z}$, condensation can occur with moderate supersaturation. The nucleation problem in this range can be discussed using thermodynamic theory.
(3) If $E_a \gg E_{8u}$, condensation can occur only with high supersaturation; usually only a small area of the substrate surface can be covered. The discussion of nucleation in this case requires atomic theory.

5.2 Formation and Growth of Thin Film Nuclei

After the gas phase atoms adsorbed on the substrate surface condense, the adsorbed atoms diffuse and migrate on the substrate surface to form a crystal nucleus. The crystal nucleus and other adsorbed gas atoms grow up gradually, and finally a thin film is formed. Therefore, the formation of the film starts from nucleation, and the film is formed by nuclei growth.

5.2.1 *Physical process of nucleation and growth*

The physical process of nucleation and growth is shown in Figure 5.3. It can be seen from Figure 5.3 that there are four stages of nuclei formation and growth:

(1) The gas phase atoms evaporated from the evaporation source are incident on the substrate surface, and a part thereof is elastically reflected back due to the large energy, and the other portion is adsorbed on the substrate surface. A small portion of the adsorbed gas phase atoms re-evaporate due to the greater energy and return to the gas phase.
(2) The adsorbed gas phase atoms diffuse and migrate on the substrate surface, collide with each other to form atomic pairs or small atomic groups and condense on the substrate surface.
(3) The atom pair or small atomic group and other adsorbed atoms continue to collide and combine. Once the number of atoms in

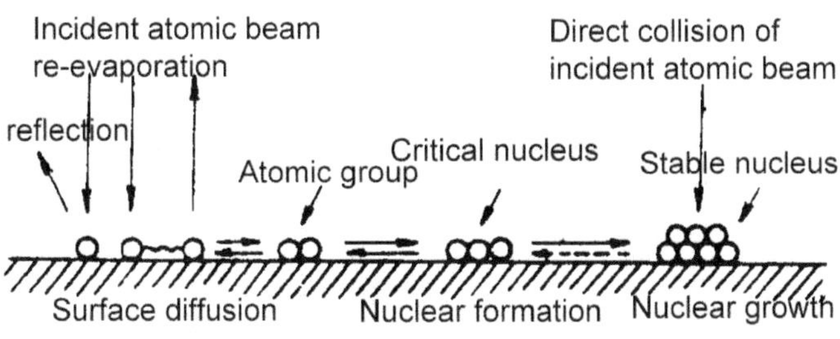

Figure 5.3: Physical process of nuclei formation and growth.

the atomic group exceeds a certain critical value, the atomic group further collides with other adsorbed atoms to form a stable atomic group toward the growth direction. An atomic group containing a critical number of atoms is called a critical nucleus, and a stable atomic group is called a stable nucleus.

(4) The stable nucleus recaptures other adsorbed atoms or combines with the incident gas phase atoms to grow it further into islands.

The nucleation process, if performed in a homogeneous phase, is referred to as uniform nucleation; if it is performed in a heterogeneous phase or a different phase, it is referred to as non-uniform nucleation. Non-uniform nucleation occurs when nucleation occurs at the interface of solid or impurities.

5.2.2 *Theory of crystal nucleation*

The main contents of the theoretical study of the film growth process include the conditions of nuclei formation and the growth rate. In the study of the nuclei formation process, the more mature theories are thermodynamic interface theory and atomic aggregation theory.

5.2.2.1 *Thermodynamic interface energy theory*

The basic idea of this theoretical model is based on the concept of thermodynamics, using macroscopic physical quantities to discuss nucleation problems, also known as capillary action theory), and the theory of nucleation forming a general gas on a solid surface into microdroplets is applied to the formation of nuclei in the film formation process. This theory uses macroscopic physical quantities such as vapor pressure, interfacial energy, and wetting angle, and is suitable for clusters with a large number of atoms. However, the meaning of some macroscopic physical quantities is unclear when the number of atoms contained in the atomic group is small.

(1) **Critical nucleus**

It is assumed that the nucleus formed on the substrate surface is a stud type shape, as shown in Figure 5.4. The radius of curvature of the nucleus is r, the wetting angle of the surface of the nucleus

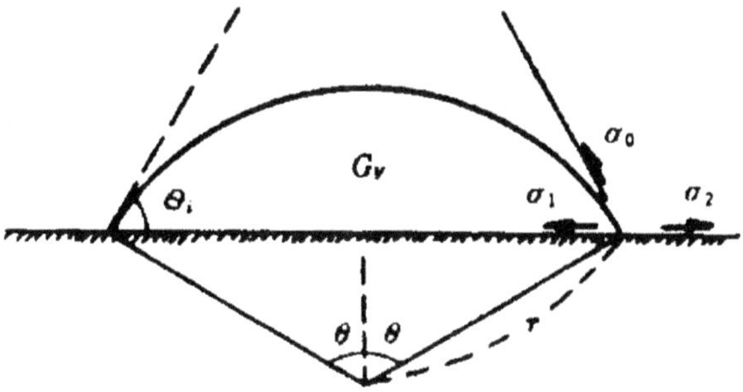

Figure 5.4: A spherical cap nucleus formed on the surface of the substrate.

and the substrate is θ, the free energy per unit area of the nucleus is ΔG_v, the free energy per unit area of the interface between the nucleus and the gas phase is σ_1, and the free energy per unit area of the substrate surface and the gas phase interface is σ_2.

The nuclei–gas phase interface area is $2(1 - cos\theta)$, and the interface area between the nucleus and the substrate surface is $\pi r^2 \sin^2\theta$, so the total free energy of the nuclei surface and interface changes to

$$\Delta G_s = 2\pi r^2(1 - \cos\theta)\sigma_0 + (\sigma_1 - \sigma_2)\pi r^2 \sin^2\theta. \tag{5.16}$$

Under thermal equilibrium,

$$\sigma_1 + \sigma_0 \cos\theta = \sigma_2. \tag{5.17}$$

Substituting Equation (5.17) into Equation (5.16) after conversion, the following results are obtained:

$$\Delta G_s = 4\pi r^2 \sigma_0 \cdot f(\theta), \tag{5.18}$$

where

$$f(\theta) = \frac{2 - 3\cos\theta + \cos^3\theta}{4},$$

and $f(\theta)$ is called the geometry factor.

In addition, due to the formation of the atomic group, the volume free energy ΔG_v will change, and the volume of the atomic group is

$4\pi r^3/3$, so the volume free energy of the stud type nucleus changes to

$$\Delta G_v = \Delta G_v \cdot \frac{4}{3}\pi r^3 \cdot f(\theta) \tag{5.19}$$

The total free energy of the system can be obtained by adding Equations (5.18) and (5.19) as follows:

$$\Delta G = 4\pi r \cdot f(\theta) \cdot (r^2 \cdot \sigma_0 + 1/3r^3 \cdot \Delta G_v). \tag{5.20}$$

The above formula is used to find the derivative of the radius of nucleus curvature r and make it equal to zero, then the critical nuclei radius r can be obtained as

$$r' = \frac{-2\sigma_0}{\Delta G_v} \quad (\Delta G_v < 0). \tag{5.21}$$

The change value of total free energy can be obtained by substituting r' into Equation (5.20).

$$\Delta G^* = \frac{16\pi \cdot \sigma_0^3 \cdot f(\theta)}{3(\Delta G_v)^2}. \tag{5.22}$$

From Equation (5.21), the critical nuclei radius r' is independent of the wetting angle θ because the effect of the wetting angle θ on the surface interface energy σ is the same as the effect on the volume free energy ΔG_v. However, the total free energy change value ΔG^* is related to the wetting angle θ.

When $\theta = 0°$, $\Delta G^* = 0$. This is the case of complete infiltration. When $\theta = 180°$, $f(0) = 1\Delta G^*$ is the largest, which is the case of no infiltration at all. It shows that the barrier to overcome in order to form stable nuclei is the highest at this time.

The functional relationship between ΔG and r is depicted as a curve as shown in Figure 5.5. As can be seen from Figure 5.5, when the atomic group radius is larger than r', the larger the r is, the smaller the ΔG is, indicating that the atomic group is more likely to grow. On the contrary, if r is smaller than r', the increase of r causes an increase in free energy, which is unstable, so the atomic group easily disappears. An atomic group having an r' size is called a critical nucleus, and a nucleus larger than a critical nucleus is stable and easily grows. ΔG^* can be regarded as the activation energy required

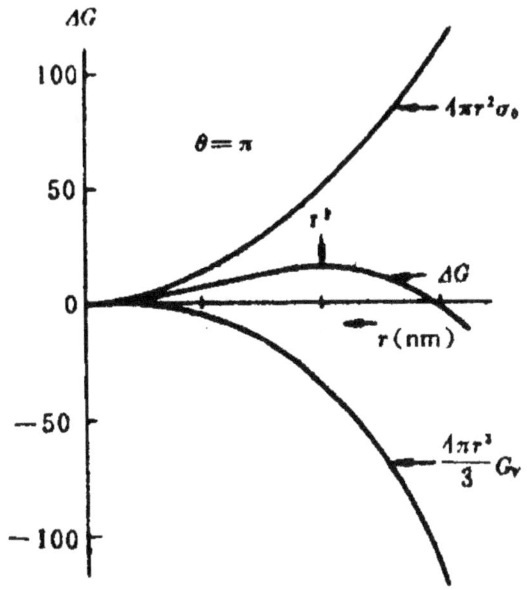

Figure 5.5: Total free energy change ΔG vs. nuclei radius r.

to grow a stable nucleus, and ΔG_v can be regarded as the energy required to generate saturated vapor during evaporation,

$$\Delta G_v = -\left(\frac{kT}{\Omega}\right)\ln\left(\frac{P}{P_e}\right), \tag{5.23}$$

where k is the Boltzmann constant, T is the absolute temperature, V is the gas phase atom volume, P is the supersaturated vapor pressure, P_e is the vapor pressure in equilibrium, and P/P_e is called supersaturation.

It can be seen that when the supersaturation P/P_e is large, the critical nuclei radius r' is small. Conversely, when the supersaturation is small, the critical nuclei radius r' is large. This is because the vapor phase atomic intensity J incident on the surface of the substrate is related to the supersaturated vapor pressure P, that is,

$$J = \frac{P}{(2\pi \cdot m \cdot kT)^{1/2}}, \tag{5.25}$$

where m is the gas phase atomic mass.

(2) **Nucleation rate**

The rate of nucleation refers to the rate at which a stable nucleus is formed or the rate at which a critical nucleus grows. Therefore, the nucleation rate is defined as the number of stable nuclei formed on the surface of unit substrate per unit time.

There are two ways to increase the critical nucleus: one is that the incident vaporized gas phase atoms are directly combined with the critical nuclei collision, and the other is that the adsorbed atoms are combined with the diffusion and collision on the surface of the substrate. If the number of critical nuclei on the surface of the substrate is small, the growth of the critical nuclei mainly depends on the surface diffusion of the adsorbed atoms combined with the migration collision. In this case, the nucleation rate is related to the number of critical nuclei per unit area, the capture range of each critical nuclei, and the total velocity of all adsorbed atoms toward critical nuclei motion.

Critical nuclei density refers to the number of critical nuclei per unit area. It is assumed that the adsorption sites with the same adsorption energy on the substrate surface are uniformly distributed. Aggregates of all sizes are in a quasi-equilibrium state in which a single atom is adsorbed. The number of atoms i^* in the critical nucleus is

$$i^* = \frac{4}{3}\pi \cdot r'^3 \cdot \frac{f(\theta)}{V}. \tag{5.26}$$

According to the Boltzmann equation, the critical nuclei density n_i^* can be calculated as

$$n_i^* = n_1 \cdot \exp(-\Delta G^*/kT), \tag{5.27}$$

where n_1 is the density of adsorbed single atoms, that is, the number of adsorbed single atoms per unit area.

At the beginning of the formation of the nucleus, $n_1 = J \cdot \tau_a, \tau_a$ is the average surface residence time of the adsorbed atoms, and J is the total number of vaporized gas atoms incident on the surface of the substrate.

The critical nuclei capture range is

$$A - 2\pi \cdot r' \cdot \sin\theta. \tag{5.28}$$

The diffusion rate of the adsorbed atoms on the surface of the substrate is

$$v = \frac{a_0}{\tau_0} = \frac{a_0}{\tau_0}\exp(-E_D/kT), \tag{5.29}$$

where a_0 is the distance between the adsorption points. From this, the total velocity v of all adsorption unit atoms moving toward the critical nucleus can be calculated as

$$v = J \cdot a_0 \exp[(E_d - E_D)/kT]. \tag{5.30}$$

In the above formula, E_d is the activation energy required for the atoms to detach from the substrate, and E_D is the activation energy required for the atoms to diffuse on the substrate. Multiplying Equations (5.27), (5.28), and (5.30) to obtain the nucleation rate I:

$$I = Z \cdot n_i^* \cdot A \cdot V = Z \cdot n_i 2\pi r' \cdot \sin\theta \cdot J \cdot a_0 \exp\left(\frac{E_d - E_D - \Delta G^*}{kT}\right), \tag{5.31}$$

where Z is the Zeldovich non-equilibrium correction factor, which is approximately 10^{-2}. The above formula gives the nucleation rate of a non-uniform nucleus. It was launched under balanced conditions. Although there is a non-equilibrium factor for correction, it is still processed according to the equilibrium state when n_1 is obtained, so there is still a certain gap between the final result and the actual situation.

5.2.2.2 *Atomic aggregation theory*

In the study of nuclei formation by the theory of atomic aggregation, the nucleus is regarded as a macromolecular aggregate, and the binding energy between the atoms of the aggregate or between the atoms of the aggregate and the atoms on the surface of the substrate is used instead of the thermodynamic free energy. In the theory of atomic aggregation, the relationship between the shape of critical and

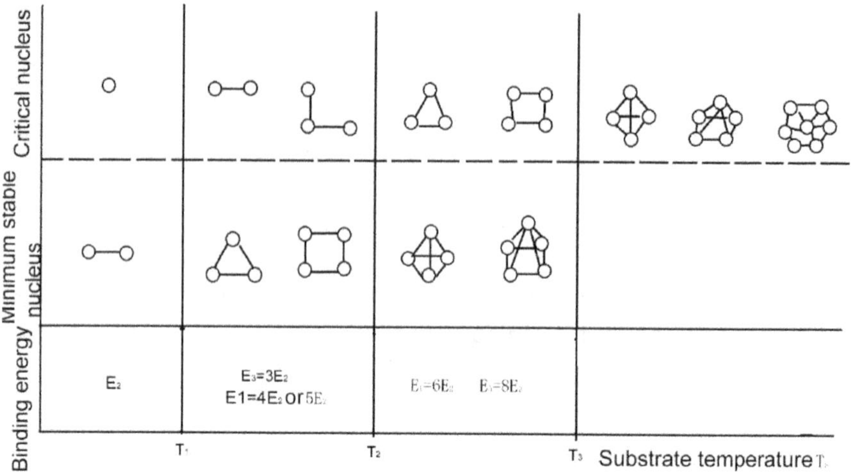

Figure 5.6: Shape of the critical and the minimum stable nucleus.

minimum stable nuclei and the binding energy is shown in Figure 5.6. As shown in Figure 5.6, the binding energy is not a continuous change, but a discontinuous change with the minimum unit of the binding energy of the atom pair.

(1) **Critical core**

When the critical nucleus size is small, the binding energy E_i will exhibit a discontinuity change and the geometry cannot be kept constant. Therefore, the mathematical analytical formula of the critical nucleus size cannot be obtained, but it is possible to analyze all possible shapes when it contains a certain number of atoms. The test difference method is then used to determine which atomic group is the critical nucleus. Below we assume that the deposition rate is constant, taking the face-centered cubic structure metal as an example to analyze the variation of the critical nucleus size with the substrate temperature.

(1) At lower substrate temperatures, the critical nucleus is a single atom adsorbed on the surface of the substrate. In this case, each adsorbed atom, once combined with other adsorbed atoms, forms a stable atom-to-shape stable nuclei. Since there is a collision

with another atom anywhere around the critical nuclei atom, the stable nuclei atom pair will not have a single orientation.

(2) After the temperature is greater than T_1, the critical nucleus is an atom pair. Because at this time each atom is unstable only by the constraint of a single bond, it must have a double bond to form a stable nucleus. In this case, the least stable nucleus is a three-atom atomic group. At this point the stable nucleus will be parallel to the substrate with the (111) plane.

(3) When the temperature rises above T_2, the critical nucleus is a tri-atomic or tetra-atomic group. Because at this time the double bond can no longer stabilize the atom in the nucleus, to form a stable nucleus, it must have at least three bonds per atom, thus its stable nucleus is a tetra-atomic group or a five atomic group.

(4) When the temperature is further increased to reach T_3, the critical nucleus is obviously a tetra-atomic group and a five-atom group, and some may be a seven-atom group.

The above situation is reflected in Figure 5.6. The temperatures T_1, T_2, and T_3 in Figure 5.6 are referred to as transition temperatures or critical temperatures. In the thermodynamic interface energy nucleation theory, the concept of a critical nucleus radius is described for the conditions of forming a nucleus. From this we can see the difference between the two theories in describing the critical nucleus.

Detailed theoretical calculations can be used to find T_1 and T_2 as follows:

$$\begin{aligned} T_1 &= \frac{-(E_d + E_2)}{k \cdot \ln(\tau_0 J / n_0)}, \\ T_2 &= \frac{-(E_d + \frac{1}{2}E_3)}{k \cdot \ln(\tau_0 J / n_0)}. \end{aligned} \tag{5.32}$$

(2) Nucleation rate

The nucleation rate is equal to the critical nucleus density multiplied by the capture range of each core and multiplied by the total velocity of the adsorbed atoms moving toward the critical nucleus.

The calculation of the critical nucleus density n_i is as follows: assume that there are n_0 positions on the surface of the substrate that can form aggregates, and several single atoms ($n_0 \geq n$) are adsorbed at any one position. These several atoms are respectively adsorbed by aggregates formed by several single atoms, adsorbed by aggregates composed of n_2 diatoms, adsorbed by aggregates composed of n_3 three atoms, etc. aggregated by n_i atoms body adsorption. So there is

$$\sum(n_i \times i) = n. \tag{5.33}$$

For n_i, if there are n_i and (n_0–n_i) aggregates at n_0 arbitrary adsorption positions, the attenuation of n_i is as follows:

$$n_0 C_{n_i} = \frac{n_0!}{n_i!(n_0 - n_i!)}. \tag{5.34}$$

If $n_0 \gg n_i \gg 1$, then the above equation is approximately equal to $n_0^{n_i}/n_i 1!$. If $n_0 \gg \sum n_i$, then the above formula holds for all i.

If the binding energy is E_i when a single atom is adsorbed, and the binding energy is $n_i E_i$ when i atoms are aggregated, the number of states in the aggregate is

$$\frac{n_0^{n_i}}{n_i!} \cdot \exp\left(\frac{n_i E_i}{kT}\right). \tag{5.35}$$

The number of states of all aggregates is

$$w = \prod\left(\frac{n_0^{n_i}}{n_i!}\right) \exp\left(\frac{n_i \cdot E_i}{kT}\right). \tag{5.36}$$

Assuming that n_0 is the maximum value of W, the actual state, the number of atoms in the total film system is n, the following results are obtained:

$$\sum i \cdot n_i = n. \tag{5.37}$$

Therefore, calculate the critical nucleus density n_i as the maximum value of W or $\ln W$ under the condition of Equation (5.37). To do this, suppose C is an unknown constant and

$$\ln W + n \cdot \ln C = L. \tag{5.38}$$

This becomes the maximum value of L. If the W and n values are substituted into Equation (5.38), then the differential is obtained as

$$\frac{\partial L}{\partial n_i} = \ln n_0 + \frac{E_i}{kT} - \ln n_i + i \ln C.$$

Let $\partial L/\partial n_i = 0$, we get

$$n_i = \left[n_0 \cdot \exp\left(\frac{E_i}{kT}\right)\right] \cdot C^i. \tag{5.39}$$

Assume $i = 1$ to get

$$C = \frac{n_1}{n_0} \exp\left(\frac{-E_1}{kT}\right). \tag{5.40}$$

Substituting Equation (5.40) into Equation (5.39) yields

$$n_i = n_0 \left(\frac{n_1}{n_0}\right)^i \cdot \exp\left(\frac{E_i - iE_1}{kT}\right). \tag{5.41}$$

Since E_1 is the potential energy in the single-atom adsorption state, if it is used as the energy reference (zero point), then the critical nucleus density n_i can be expressed as

$$n_i = n_0 \left(\frac{n_1}{n_0}\right)^i \exp\left(\frac{E_i}{kT}\right). \tag{5.42}$$

It corresponds to the thermodynamic interface energy theory obtained by the critical nucleus density Equation (5.27).

The total velocity of the adsorbed atoms moving toward the critical nucleus can still be obtained by Equation (5.30), and the critical nucleus capture range is A, then the nucleation rate

$$\begin{aligned} I &= n_i \cdot V \cdot A \\ &= n_0 \left(\frac{n_1}{n_0}\right)^i \exp\left(\frac{E_i}{kT}\right) \cdot J \cdot a_0 \exp\left(\frac{E_d - E_D}{kT}\right) \cdot A \\ &= A \cdot J \cdot n_0 \cdot a_0 \left(\frac{J \cdot \tau_a}{n_0}\right)^i \exp\left(\frac{E_i + E_d - E_D}{kT}\right) \\ &= A \cdot J \cdot n_0 \cdot a_0 \left(\frac{\tau_0 \cdot J}{n_0}\right)^i \exp\left[\frac{E_i + (i+1)E_d - E_D}{kT}\right]. \end{aligned} \tag{5.43}$$

There is no non-equilibrium correction factor Z in Equation (5.43) because the super-saturation is relatively small and the influence of non-equilibrium factors can be ignored.

As can be seen from the above discussion, the basic concepts underlying the two theories are the same. The form of the resulting nucleation rate formula is also substantially the same; the differences are the energy used by the two and the model used by the two.

The thermodynamic interface energy theory is suitable for describing large-scale critical nuclei. Therefore, this theory is suitable for materials with less free energy of aggregation or deposition with less super-saturation. On the contrary, the atomic aggregation theory is suitable for small-sized critical nuclei.

5.2.3 *Nucleation and growth on the surface of the substrate*

The above nucleation theory can be extended to the process of nucleation, growth and film formation of foreign atoms deposited on the surface of the substrate.

5.2.3.1 *Free energy function*

If the total number of atoms deposited per unit area of the substrate is N, the number of atoms per unit area is p, assuming that each atomic group contains n atoms, and the number of single atoms per unit area is N_D, then the atomic density on the surface of the substrate is

$$N = pn + N_D.$$

The binding energy of the system can be written as

$$U_b = p\lfloor a_1r^2\sigma w + \sigma_2r^2\sigma_{8e} + a_3r^3\Delta Gv\rfloor + \lfloor 1 - pa_2r^2\rfloor\sigma_{8v} - (V_c)^{2/3}\sigma_{8v}N_D + (V_c)^{2/3}\sigma_{cv}N_D, \quad (5.44)$$

where a_1r^2 is the surface area exposed to the vapor deposition phase; a_2r^2 is the contact area between the atomic group and the substrate; a_3r^3 is the volume of the atomic group; a_1, a_2, and a_3 are geometric constants describing the shape of the atomic group; σ_{cv} is the surface free energy of the atomic group; σ_{sc} is the interface free energy of the atomic group; σ_{8v} is the surface energy of the substrate; V_c is the average volume of the atom; ΔG_v is condensation free energy of the material deposited under the condition of super-saturation and can be written as

$$\Delta G_v = -\frac{kT}{V_c} \ln \left(\frac{R_d}{R_e} \right). \tag{5.45}$$

Free energy can be written as

$$F = E_0^8 - E_8^s pn + E_4^s pn^{2/3} - kT \ln \frac{N_c^{N_D}}{N_D} - kT \ln \frac{N_0}{p!} \tag{5.46}$$

$$E_0^s = \sigma_{8v} + (V_c)^{2/3} \sigma_{cv} N - (V_c)^{2/3} \sigma_{8v} N + (H_0 - E_5) N \tag{5.47}$$

$$E_8^s = (V_0)^{2/3} \sigma_{cv} - (V_c)^{2/3} \sigma_{8v} - V_c \Delta G v + (H_0 - E_5) + E_6 \tag{5.48}$$

$$E_4^8 = a_1 \left(\frac{V_c}{a_3} \right)^{2/3} \sigma_{cv} + a_2 \left(\frac{V_c}{a_3} \right)^{2/3} \sigma_{8c} - a_2 \left(\frac{V_c}{a_3} \right)^{2/3} \sigma_{8v}. \tag{5.49}$$

Here, H_0 is thermal energy, E_5 and E_6 are potential energies, and N_0 is the number of atoms in a single atomic layer per unit area.

5.2.3.2 *Nucleation critical conditions*

The critical conditions for nucleation can be obtained from $\partial F/\partial n = 0$ and $\partial F/\partial p = 0$. At the activation temperature T_A, the critical values of n, p and N_D are expressed as n_c, p_c and N_{Dc},

$$p_c = N_0 \exp \left(-\frac{E_4^3 n_c^{2/3}}{3kT_A} \right), \tag{5.50}$$

$$N_{Dc} = N - p_c n_c = N_0 \exp \left(-\frac{E_8^s - \frac{2}{3} E_4^s n_c^{-1/3}}{kT_A} \right), \tag{5.51}$$

$$\frac{N}{N_0} - n_c \exp \left(-\frac{E_4^8 n_c^{2/3}}{3kT_A} \right) = \exp \left(-\frac{E_8^s - \frac{2}{3} E_4^s n_c^{-1/3}}{3kT_A} \right). \tag{5.52}$$

Solving from the last equation gives the number of atoms n_c contained in the critical nucleus, which is a function of N/N_0, T_A, E_4^s, E_8^s, ΔG_v, and so on. The critical radius of the atomic group $r_0 = (V_c/a_3)^{1/3} n_c^{1/3}$. The physical condition of nucleation is

$$E_4^s > 0 N_{Dc} > N_0 \exp\left(-\frac{E_8^s}{kT_A}\right) \tag{5.53}$$

5.2.3.3 *Ball cap atomic group*

Assuming that the substrate is composed of elements A and B, for the ball cap atomic group, the relevant parameters can be written as

$$\begin{aligned} \sigma_{cs} &= \left(\frac{1}{V_t}\right)^{2/3} \cdot \frac{1}{2}(2\Phi_{\mathrm{AB}} - \Phi_{\mathrm{AC}} - \Phi_{\mathrm{DC}}), \\ \sigma_{sv} &= \left(\frac{1}{V_t}\right)^{2/3} \Phi_{\mathrm{AB}}, \\ \sigma_{cv} &= \left(\frac{1}{V_c}\right)^{2/3} \Phi_{\mathrm{cc}}. \end{aligned} \tag{5.54}$$

Here, V_t is the average volume of the base atoms, and Φ_{AB}, Φ_{AC}, Φ_{BC}, and Φ_{CC} are the binding energies of AB, AC, BC, and CC atom pairs, respectively.

$$a_1 = 2\pi(1 - \cos\theta), a_2 = \pi \sin^2\theta, \cdot a_3 = \pi\left[\frac{2}{3} - \frac{1}{3}\cos\theta(\sin^2\theta + 2)\right]. \tag{5.55}$$

The physical condition ($E_4^s > 0$) becomes

$$\frac{1}{2}\left[1 + \left(\frac{V_c}{V_t}\right)^{2/3}\left(\frac{\Phi_{\mathrm{AB}} + \Phi_{\mathrm{BC}}}{2\Phi_{\mathrm{CC}}}\right)\right] \cdot \left(\frac{V_c}{V_t}\right)^{\frac{2}{3}}\left(\frac{\Phi_{\mathrm{AC}} + \Phi_{\mathrm{BC}}}{2\Phi_{cc}}\right) < 1. \tag{5.56}$$

Here,

$$\left(\frac{V_c}{V_t}\right)^{2/3}\left(\frac{\Phi_{\mathrm{AC}} + \Phi_{\mathrm{BC}}}{2\Phi_{\mathrm{CC}}}\right) = \cos\theta \le 1. \tag{5.57}$$

5.2.3.4 *Atomic group growth*

The atomic group growth is related to surface diffusion, and the surface diffusion coefficient can be written as

$$D = D_0 \exp\left(-\frac{Ed}{kT}\right). \tag{5.58}$$

The number of atoms that diffuse into the atomic group with n atoms per unit time is

$$I = \frac{2}{3}\pi D \ln\left(\frac{\bar{n}}{n}\right) \cdot N_0 \exp\left(-\frac{E_8^s}{kT_A}\right) \exp\left(\frac{\frac{2}{3}E_4^3\bar{n}^{-1/3}}{kT_A}\right) - \exp\left(\frac{\frac{2}{3}E_4^8 n^{-1/3}}{kT_A}\right) \tag{5.59}$$

where $\bar{n}$ is an atomic group with $\bar{n}$ atoms, $\bar{n}$ can be solved by the following formula:

$$N_0 \exp\left(-\frac{E_8^s}{kT_A}\right) \cdot \exp\left(-\frac{\frac{2}{3}E_4^3\bar{n}^{-1/3}}{kT_A}\right) = N_D \tag{5.60}$$

When $n > \bar{n}$, I is positive and this atomic group will grow up. When $n < \bar{n}$, I is negative, and this atomic group will decrease in size and even disappear. Its growth rate can be written as

$$\frac{dr}{dt} = \left(\frac{V_c}{a_3}\right)^{1/3} \frac{2\pi D}{9n^{2/3}} \ln\frac{\bar{n}}{n} \cdot N_0 \exp\left(-\frac{E_8^8}{kT_A}\right) \times \left[\exp\left(\frac{\frac{2}{3}E_4^3\bar{n}^{-1/3}}{kT_A}\right) - \exp\left(\frac{\frac{2}{3}E_4^3 n^{-1/3}}{kT_A}\right)\right]. \tag{5.61}$$

If E_D, D and E_8^s are small, T_A and E_4^s are large, and the larger atomic group will grow up quickly and smaller ones will shrink rapidly.

5.2.3.5 *Epitaxial growth conditions*

Epitaxial growth can be discussed using nucleation theory. For heteroepitaxial growth, the following three conditions should be satisfied; that is, the first condition is $E_4^8 < 0$, which is obtained from Equation (5.49).

$$a_2 \left(\frac{V_c}{a_3}\right)^{2/3} \sigma_{sv} > a_1 \left(\frac{V_c}{a_2}\right)^{2/3} \sigma_{cv} + a_2 \left(\frac{V_c}{a_3}\right)^{2/3} \sigma_{8c}. \tag{5.62}$$

As discussed for the stud type atomic group, when

$$\frac{V_c^{2/3}}{V_t} \left(\frac{\Phi_{\mathrm{AC}} + \Phi_{\mathrm{BC}}}{2\Phi_{\mathrm{CC}}}\right) > 1,$$

the contact angle θ is 0; that is, the atoms on the substrate surface can only be arranged in a single atomic layer.

The second condition $E_8^S \leq 0$. If $E_8^S \leq 0$, then $N_{Dc} \geq N_0$ from Equation (5.53); the external growth formula is obtained as

$$kT \ln \frac{Rd}{Re} \leq (V_c)^{2/3}(\sigma_{sv} - \sigma_{sc}) + \beta kT - H_0, \tag{5.63}$$

where β is a constant (under a first approximation); R_e is an exponential function of the substrate temperature. For a given temperature, R_d should be below the upper limit given by Equation (5.63), and as the temperature increases, R_d/R_e decreases.

The third condition is surface diffusion. Because for a perfect epitaxial growth, the deposited atoms should migrate to the proper position, so there is a lower limit of the substrate temperature below which the foreign atoms cannot migrate to the position where the nucleation grows into crystals. The substrate temperature for epitaxial growth is determined by considering the value of the diffusion energy E_d and the binding energy of the deposited atoms to the substrate, but the substrate temperature should not be too high. Otherwise, more defects may occur in the film, and may even become amorphous.

When $E_4^s > 0$ and $E_8^s \leq \frac{2}{3} E_4^3 n_c^{-\frac{1}{3}}$, this means that $N_{Dc} \geq N_0$, Equation (5.63) becomes

$$kT \ln \frac{Rd}{Re} \leq (V_c)^{2/3}(\sigma_{sv} - \sigma_{cv}) + \beta kT - H_0 + \frac{2}{3} E_4^8 n_c^{-\frac{1}{3}}. \tag{5.64}$$

If this condition is satisfied, the epitaxial film will grow stably when the substrate temperature is within the epitaxial temperature range.

5.3 Film Formation Process and Growth Mode

The formation process of a thin film refers to the process after the formation of a stable nucleus; the growth mode of a thin film refers to the macroscopic form of thin film formation. The growth mode of a thin film can be roughly divided into three types: an island growth mode, a layer growth mode, and layer island combination growth mode.

5.3.1 *Three modes of film growth*

A schematic diagram of the three growth modes of the film is shown in Figure 5.7.

5.3.1.1 *Island growth mode*

In the island growth mode (Volmer–Weber mode), the formation of the film is thus the atoms reaching the substrate first coalesce

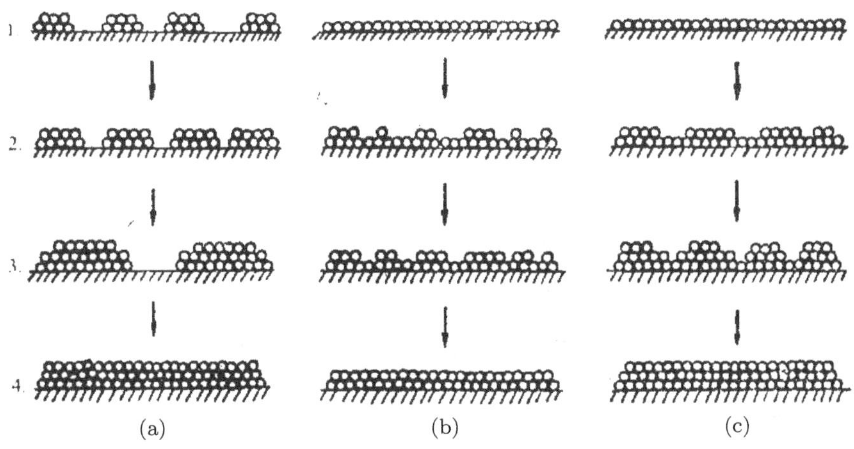

Figure 5.7: Schematic diagram of the film growth process. (a) Island growth mode; (b) layered growth mode; (c) layer-island combined growth model.

into a number of discrete small nuclei, and the subsequent flying atoms are continuously concentrated in the vicinity of the nucleus, so that the nucleus grows in three dimensions and eventually forms a continuous film. Therefore, the island growth mode is also called a nucleus growth mode. Most of the film formation process belongs to this type of growth. The process of this growth mode is shown in Figure 5.7(a).

The growth process of the island growth mode can be divided into the following stages.

(1) **Nucleation stage**

The atoms that reach the substrate, some of which exchange energy with the substrate atoms, still have considerable energy and can therefore return to the gas phase. The other part is adsorbed on the surface of the substrate. This adsorption is mainly physical adsorption, and the atoms will stay on the surface of the substrate for a certain period of time. Since the atom itself has a certain energy, and heat energy can also be obtained from the substrate, the atom may migrate or diffuse on the surface. In this process, the atoms may re-evaporate, may also chemically interact with the substrate to form chemisorption, and may also encounter other evaporating atoms to form atomic pairs or radicals. In the latter two cases, the possibility of re-evaporation and migration of the atom is extremely small, and it gradually becomes a stable condensed nucleus.

(2) **Small island stage**

When the condensed crystal nucleus reaches a certain concentration, no new nucleation is formed after evaporation continues. The newly evaporated adsorbed atoms will accumulate on the existing nuclei by surface migration, causing the nuclei to grow and form small islands. These small islands are usually three-dimensional structures, and most of them have the crystal structure of the substance, that is, microscopic grain.

(3) **Network stage**

As the island grows up, the adjacent islands will contact each other and combine with each other. The process of combining is similar to the case where two small droplets are combined into one large droplet, because the island releases a certain amount of energy when combined. These energies are sufficient to instantaneously melt the contact microcrystalline islands, and after bonding, the islands formed due to the temperature drop will recrystallize. With the continuous integration of the islands, some network-like films with a maze structure will be formed.

(4) **Continuous film**

As evaporation or sputtering continues, the adsorbed atoms will fill the gap between the islands, and it is also possible to generate new islands between the islands, filling the empty channels by the growth of the islands, and finally forming a continuous film.

Different materials experience these four stages in different situations. For example, aluminum film and silver film are island growth type, but aluminum film only presents island structure at the initial stage of growth, and then can form continuous film when the film is very thin, while silver film can form continuous film when the film is relatively thick. Figure 5.8 shows the formation process of silver film on MoS_2 substrate. In Figure 5.8, (a) is the nucleation stage, (b) is the island stage, (c) is the network stage, (d) is the ditch gradually filled to form a continuous film.

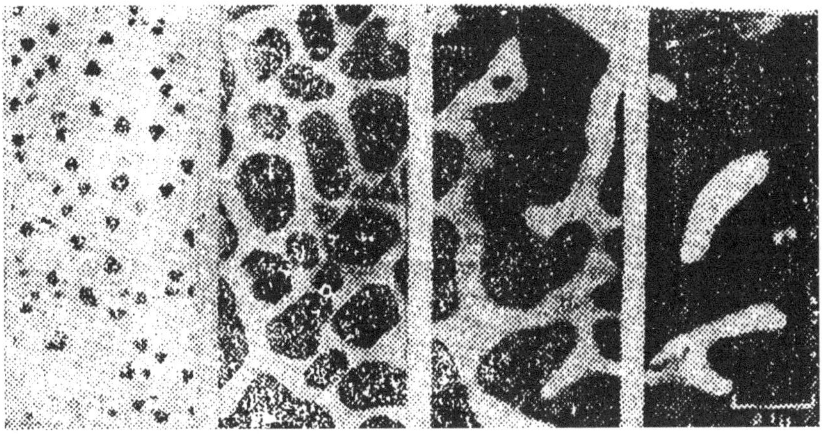

(a) (b) (c) (d)

Figure 5.8: Formation process of silver film on MoS_2 substrate.

5.3.1.2 *Layered growth mode*

When the wettability between the deposited substance and the substrate is good, the atoms of the deposited substance are more likely to be bonded to the substrate atoms. Therefore, the film adopts a two-dimensional expansion mode from the nucleation stage (Frank–van de Merwe mode). This growth mode is characterized by this: the evaporated atoms are first uniformly covered in the form of a single atomic layer on the surface of the substrate, and then the second layer, and the third layer are grown in three dimensions. Most of this growth occurs when the binding energy between the substrate atoms and the evaporating atoms is close to the binding energy between the evaporating atoms. For example, Pd is grown on an Au single crystal substrate, PbSe on a PdS single crystal substrate, and a Cu thin film on a Fe single crystal substrate. The most typical examples are homoepitaxial growth and molecular beam epitaxy.

The general process of layered growth is as follows: the atoms incident on the substrate surface form a two-dimensional nucleus after surface diffusion and collision with other atoms, and the adsorbed atoms around the two-dimensional nucleus capture grow into two-dimensional islands. The island concentration formed on the surface of such materials is generally saturation concentration; that is, the distance between the islands is roughly equal to the average diffusion distance of the adsorbed atoms. In the process of island growth, the radius of the island is smaller than the average diffusion distance. Therefore, the adsorption atoms arriving at the island are captured by the edge of the island after diffusion. The concentration of adsorbed atoms on the island surface is low, and it is not easy to grow in the three-dimensional direction. That is to say, only when the islands in the nth layer have grown large enough, or even the islands have combined with each other, and the nth layer is nearly fully formed, the two-dimensional crystal nucleus or two-dimensional islands in the nth + 1 layer can be formed, so the film is grown in the form of layers.

In layered growth, the crystal structure of the film near the substrate is usually similar to that of the substrate. The growth process of layered mode is shown in Figure 5.7(b).

5.3.1.3 *Layer-island combined growth model*

The layered-island growth mode (or Stranski–Krastanoy mode) is shown in Figure 5.7(c). After the first one or two atomic layer thickness layer growth, the growth mode is transformed into island mode. The physical mechanism leading to this mode transition is complex, and this growth mode is easy to appear only when the interaction between the substrate and film atoms is particularly strong. Firstly, 1–2 layers of single atom layers are grown on the surface of the substrate. This two-dimensional structure is strongly affected by the substrate lattice, and the lattice constant has a large distortion. Then, the incident atoms are adsorbed on the atomic layer, and small islands are formed in the manner of nucleus growth, and finally thin films are formed. When metal films are formed on the surface of semiconductors, it is usually the layer-island growth type, such as the evaporation of Gd on the surface of Ge, Bi, and Ag on the surface of Si.

5.3.2 *Film formation process*

The formation process of the film is discussed in detail in the island growth mode. After the stable nucleus is formed, the formation process of the island-shaped film is as shown in Figure 5.9. As can be seen from Figure 5.9, the formation process of the island-shaped film can be divided into four main stages:

5.3.2.1 *Nucleation stage*

In the photograph of the film formation process observed by transmission electron microscopy, the size of the smallest nucleus can be observed to be about 20–30 Å. In the process of further growing the nucleus into a small island, the growth rate parallel to the surface of the substrate is greater than the growth rate in the vertical direction. This is because the growth of the nucleus is mainly due to the diffusion–migration collision bonding of the adsorbed atoms on the surface of the substrate, rather than the collision of the incident vapor phase gas atoms. These nuclei, which continuously capture the growth of adsorbed atoms, gradually change from a ball-shaped,

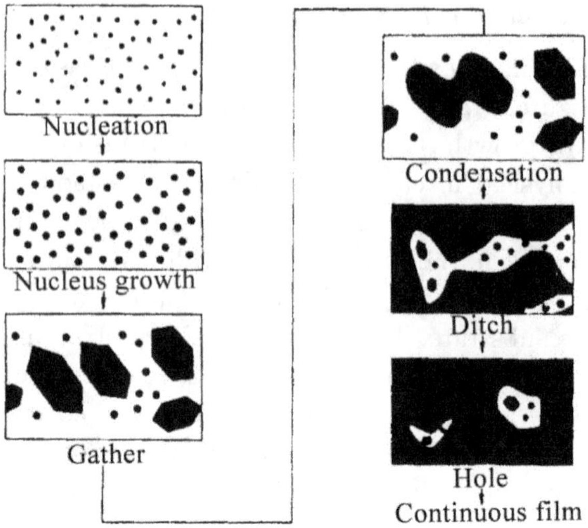

Figure 5.9: Formation process of island film.

circular shape to a polyhedral island. Thermodynamic macroscopic physical quantities such as surface free energy can be used for the formation of islands, and microscopic physical quantities such as binding energy can also be used for discrimination.

The following expressions have been obtained in the study of nucleus formation using the thermodynamic interface energy:

$$\sigma_2 = \sigma_1 + \sigma_0 \cdot \cos\theta, \tag{5.65}$$

thus we get

$$\cos\theta = \frac{\sigma_2 - \sigma_1}{\sigma_0}. \tag{5.66}$$

Because the angle of θ should satisfy $0 < \theta < \pi/2$, so $\cos\theta < 1$, the following formula must be true:

$$\sigma_2 - \sigma_1 < \sigma_0. \tag{5.67}$$

In the case where the substrate and the film cannot form an alloy, $\sigma_1 > 0$, because if $\sigma_2 < \sigma_0$, the above relationship is of course satisfied. If it is clearly known that the film and the substrate cannot form a compound, even if the size of σ_1 is unclear, it is expected that it will grow in a three-dimensional island manner. Therefore, Equation (5.67) is a condition for predicting the growth of three-dimensional islands using macroscopic physical quantities.

When the microscopic physical quantity is used to discriminate the growth of the three-dimensional island, it is considered that the lattice constant between the film and the substrate is different. The energy E_s causing lattice mismatch at the interface between the film and the substrate is negligible, and the adsorption energy of the adsorbed atoms on the surface of the substrate E_{ad} can be expressed by the following formula:

$$E_{\text{ad}} = (\sigma_2 + \sigma_0 - \sigma_1) \cdot S + E_s \cdot S, \tag{5.68}$$

where S is the projected area of the atom.

The relationship between the binding energy E_b between the adsorbed atoms and the surface free energy σ_0 of the nucleus has the following relationship:

$$E_b = \frac{2 \cdot \sigma_0 \cdot S}{Z_C}, \tag{5.69}$$

where Z_C is the number of empty bonds (dwelling bonds) on the surface of the nucleus, and substituting Equations (5.68) and (5.69) into (5.67):

$$E_{\text{ad}} < Z_C \cdot E_b + E_S \cdot S. \tag{5.70}$$

Since the E_S is small and negligible, the above formula becomes

$$E_{\text{ad}} < Z_C \cdot E_b. \tag{5.71}$$

Equation (5.71) shows that when the binding energy between the nucleus and the adsorbing atom is larger than the adsorption energy of the adsorbing atom and the substrate, a three-dimensional island can be formed, and thus the above formula becomes a condition for discriminating island growth by microscopic physical quantity.

5.3.2.2 *Island stage*

As the island continues to grow, the distance between the islands gradually decreases, and finally the adjacent islands can be combined

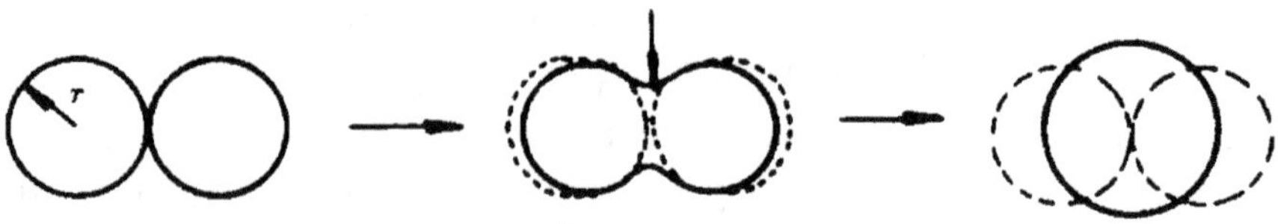

Figure 5.10: Changes in the island during the merger.

to form a large island, as shown in Figure 5.10. After the islands are combined and grown, the area occupied on the surface of the substrate is reduced, the surface energy is reduced, and the space on the surface of the substrate can be nucleated again. The combination of islands is similar to solid phase sintering. Suppose that the two islands are spherical with radius r, the radius of curvature of the joint is r', and the time after the island contact is t, then the relationship between them can be expressed by

$$\frac{r'^n}{r^m} = \frac{56 \cdot \sigma \cdot V^{4/3}}{kT} \cdot D \cdot n \cdot t, \tag{5.72}$$

where V is the atomic volume, n is the surface density of the adsorbed atom on the island, D is the diffusion coefficient of the adsorbed atom, σ is the surface free energy, both n and m are constants, k is the Boltzmann constant, and T is the absolute temperature.

For surface diffusion, $m = 3$, $n = 7$; for bulk diffusion $m = 2$, $n = 5$. The substrate temperature plays an important role in the merging of the islands. Although the initial phase of the small island merger was very fast, the new island continued to change its shape for a long time after the merger. Therefore, the area of the new island is constantly changing during the merger and after the merger. In the initial stage, the area of coverage on the surface of the substrate is reduced by the combination and then gradually increased. In the initial stage of merging, in order to reduce the surface free energy, the area of the new island is reduced and the height is increased. Depending on the surface of the substrate, the surface of the island, and the free energy of the interface, the island will have a shape of the lowest energy, which is a configuration with a certain height to radius ratio.

5.3.2.3 *Network stage*

During the growth of the island into a new island, the tendency of its shape to become circular is reduced. Only when the island is further merged, large deformations continue to occur; when the islands reach a critical state, they coalesce to form a network structure. In this structure, irregularly distributed trenches with a width of 50–200 Å are distributed. As the deposition continues, two or three nucleations occur in the trench. When the nucleus grows up to contact the edge of the trench, it merges into the film of the mesh structure. At the same time, in some places, the ditch is bridged and quickly filled in a liquid-like form. As a result, most of the ditches are quickly eliminated, and the film becomes a continuous structure with small holes from the ditch. Two or three additional nucleations occur at these small holes. Some of the nuclei merge directly with the film, and some of the nuclei grow to form secondary islands that are then incorporated into the film.

Because the combination of nucleus or island has a similar liquid character, this property allows the ditches and holes to disappear quickly. Finally, the high surface curvature area is eliminated, so that the total surface free energy of the film is minimized.

5.3.2.4 *Continuous film stage*

After the trenches and holes are eliminated, the gas phase atoms incident on the surface of the substrate are directly adsorbed on the film, and a film of a different structure is formed by the combination. Some films undergo a significant change in the orientation of the island during the consolidation phase of the island. For the formation of epitaxial films, the orientation of the islands is quite important. In the formation of a polycrystalline film, in addition to having a certain orientation to each other when the islands are combined in the epitaxial film, some recrystallization occurs during the combination. The grain in the film is larger than the distance between the initial nuclei. A considerable amount of recrystallization occurs even when the substrate is at room temperature. Each grain contains approximately 100 or more initial nucleus regions. It can be seen from this that the size of the grain size in the film depends on the recrystallization process when the nucleus or island is combined, and does not depend on the density of the initial nucleus.

5.3.3 *Growth characteristics of sputtered films and epitaxial films*

Films obtained by sputtering and epitaxy have special properties due to different film formation conditions.

5.3.3.1 *Sputtering film growth*

When the film is prepared by the sputtering method, the energy of the sputtered particles reaching the substrate is much larger than that of the evaporation method, and thus a series of effects are exerted on the formation of the film, except that the adhesion of the film to the substrate is increased. It also changes the structure of the film due to high-energy ions bombarding the surface of the film to increase its temperature, or increases the internal stress of the film. The energy of the sputtered particles is closely related to the sputtering voltage, the distance between the substrate and the target, the pressure of the vacuum chamber gas, and the like, and thus the morphology and structure of the film are related to the above factors. Figure 5.11 shows the relationship between working gas pressure, substrate temperature, and film structure. Sputtered films often exhibit a columnar structure, as if many small cylinders having a diameter of about a few hundred angstroms are tightly packed together. This columnar structure is believed to be caused by the limited mobility of atoms or molecules on the substrate. This columnar structure will easily change the properties of the film by leaving a lot of pores like capillary pores between the small cylinders.

(1) **Process of depositing particles**

Vacuum evaporation is a thermal process, a process in which a material changes from a solid phase to a liquid phase and then a gas phase, or a process of sublimation from a solid phase to a gas phase. The deposited particles (atoms) produced by this thermal

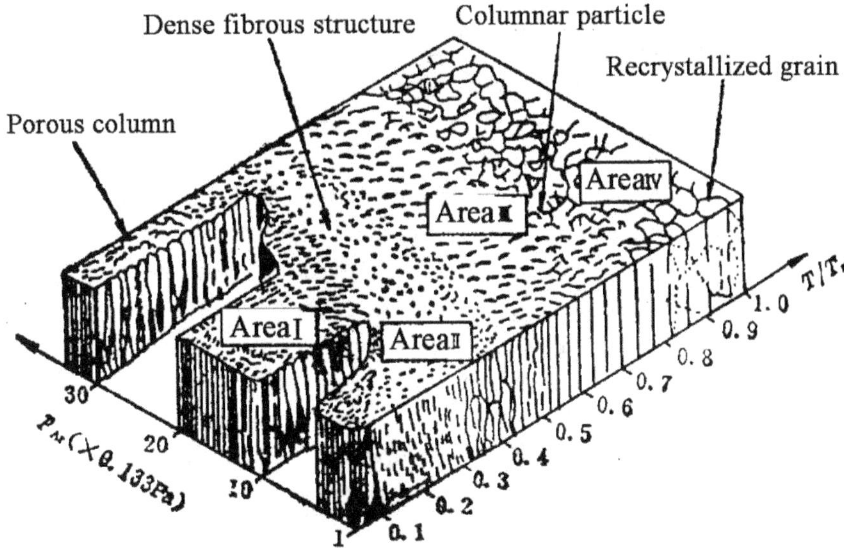

T: Substrate temperature; *T*m: Melting point of film material

Figure 5.11: Structure of a sputtered film.

process all have lower thermal motion energy. At normal evaporation temperatures, the energy is about 0.1–0.2 eV.

The sputtering process is a dynamic process based on momentum-driven ion bombardment. The incident ions with high energy collide with the target atoms, and through the energy transfer, the target atoms are splashed out of the surface of the target after obtaining a certain kinetic energy. Therefore, particles sputtered from the target have a higher function. The kinetic energy of the gas phase atom evaporating from the evaporation source is 1–2 orders of magnitude higher.

For point or small-area evaporation sources, the vapor phase atoms are directed toward the surface of the substrate and are oriented according to the law of cosine. For cathode sputtering, when the incident Ar ion energy is large and the target is composed of polycrystalline material, the sputtering target can be regarded as a point source, and the sputtered atoms fly to the surface of the substrate to conform to the cosine law distribution, or a symmetric distribution with the target surface normal to the axis. For a single crystal target, the surface bonding energy is different due to different atomic arrangement densities on different crystal faces, and the sputtering intensities of different crystal faces are also different. This phenomenon is called the preferential sputtering effect.

The gas phase atoms evaporating from the evaporation source are almost all uncharged neutral particles, or there are few charged particles (caused by thermal electron emission). However, the sputtering process is different. In addition to sputtering a neutral atom or a radical from the target, a plurality of particles such as a positive ion, a negative ion, a secondary electron, and a photon of the target can be sputtered.

When the alloy material is evaporated, fractional distillation occurs due to the difference in vapor pressure of each component in the alloy. The vapor pressure component evaporates at a high rate, causing a deviation of the film composition from the evaporation source material component. However, when sputtering alloy materials, although the sputtering rate of each component is different (the difference in sputtering rate of various metals is much smaller than the difference in their vapor pressure), the alloy film component formed at the initial stage of sputtering is slightly different from the target component. However, due to the low temperature of the target, after a short period of time, the components that are easily sputtered on the surface of the target are insufficient, so the components having a small sputtering rate are gradually increased in the film. A sputtered film consistent with the target composition is finally obtained.

(2) Migration process of deposited particles

The vacuum is higher in vacuum evaporation, generally 10^{-2}–10^{-4} Pa, and the mean free path of the gas molecules is larger than the distance between the evaporation source and the substrate. On evaporation of gas phase atoms during the flight to the substrate, there is little chance of evaporating gas phase atoms colliding with residual gas molecules. They will essentially retain the energy, energy distribution, and straight flight trajectory they had when leaving the evaporation source.

At the time of cathode sputtering, the degree of vacuum is low due to the filling of the working gas Ar gas, and the average free path of the gas molecules is smaller than the distance between the target and the substrate at about 10^0–10^{-2} Pa. When the sputtered atoms fly from the target surface to the substrate, they collide with each other and the Ar atoms and other residual gas molecules collide with each other, which not only reduces the initial energy of the sputtered particles, but also changes the direction in which the sputtered particles are separated from the target surface. The sputtered particles that reach the surface of the substrate can come from all directions of the entire hemispherical space directly in front of the substrate. Therefore, it is easier to prepare films with uniform thickness by sputtering than evaporation.

(3) **Film formation process**

After the deposited particles from the evaporation source or the sputtering target reach the surface of the substrate, the stable crystal nucleus is formed by adsorption, condensation, surface diffusion, migration, and collision. Then, the nucleus grows into small islands by adsorption, and the islands grow and then join together to form a continuous film. In such a film formation process, the main differences between the evaporation method and the sputtering method are:

For the vacuum evaporation method, the gas phase atoms incident on the substrate have no influence on the surface of the substrate, and the nucleation conditions are not changed. During the evaporation process, the substrate and the surface of the film are less bombarded by residual gas molecules or atoms, about 10^{13} times/$\text{cm}^2 \cdot \text{s}$. Therefore, it is less likely that the impurity gas is incorporated into the film. In addition, the vaporized gas phase is less likely to chemically react with the residual gas. The temperature changes of the substrate and the film were also not significant.

For the sputtering method, the ions incident on the surface of the substrate and the high-energy neutral particles have a great influence on the surface of the substrate, which makes the surface of the substrate rough, ion implantation, temporary charging of surface islands, and chemical reaction with residual gas molecules. Therefore, the nucleation conditions have changed significantly, the formation process of the nucleation center is accelerated, and the nucleation density is significantly improved. Working gas molecules, residual gas molecules, atoms and ions can produce bombardment on the surface of the substrate of the order of 10^{17} times/cm$^2\cdot$s. This is obviously much larger than what happens during the evaporation process. Therefore, there is a greater chance that the impurity gas or the external material is incorporated into the film, and a chemical reaction such as activation or ionization is likely to occur in the film. In addition, since the incident sputtered particles have a large kinetic energy, the temperature changes of the substrate and the film are also remarkable.

5.3.3.2 *Growth of epitaxial film*

One of the basic parameters of epitaxial growth is the substrate temperature. There is a critical epitaxial temperature between the substrate and the film material. Epitaxial growth above this temperature is good, and epitaxial growth below this temperature is imperfect. In addition, the growth rate R is related to the substrate temperature T, which is obtained according to experimental results

$$R = Ae^{-E_D/(kT)}, \tag{5.73}$$

where A is a constant and E_D is the surface diffusion activation energy. The above formula shows that the rate of epitaxial film growth is related to the diffusion capacity of the adsorbed atoms.

If the adsorbed atoms can enter a stable equilibrium position before they collide with another atom, the epitaxy can proceed smoothly. Obviously, the increase of the substrate temperature promotes epitaxial growth, which gives the surface atoms more opportunities to reach the equilibrium position, and the surface diffusion and bulk diffusion are enhanced, so the crystallization process becomes easier, thereby promoting the growth of the single-crystal film.

The crystal structure of the substrate has a very important influence on the structure and orientation of the epitaxial film. The structure of the two is the same when homoepitaxial, and the

Table 5.4: Epitaxially grown films on various substrates.

Film	Substrate	Lattice mismatch (%)	Epitaxial growth temperature (°C)
Cu	Split in the air NaCl(100)	–36	300
Ag	Split in vacuum NaCl(100)	–31	Room temperature
Au	Mica	–44	400
Ge	CaF_2	—	225
Si	Si(111)	0	450–550
GaAs	GaAs	0	700–850

structure of the two is closely related. The degree of mismatch is often used in heteroepitaxy to describe the relationship between substrate structure and film structure. The degree of mismatch is

$$m = \frac{b-a}{a} \times 100\%, \tag{5.74}$$

where a is the lattice constant of the sheet and b is the lattice constant of the film material. When homogenously epitaxial, $m = 0$. The m can be regarded as an indicator that the evaporation substance and the substrate substance have similarities in crystal structure. Epitaxial growth can also be achieved when m is relatively large.

The case of epitaxially growing a film on various substrates is shown in Table 5.4.

5.4 Computer Simulation of Film Formation

With the development of computer science, since the 1970s, many researchers in the world have used computer simulation methods to study the formation of thin films.

Two methods can be used to simulate the film formation process using a computer: the Monte Carlo method and the molecular dynamics method.

The Monte Carlo method is also called the stochastic simulation or statistical test. When dealing with problems in this way, we first need to build a stochastic model, then generate a series of random numbers to simulate the process, and finally do statistical processing. In the process of simulating the film formation, the gas phase atoms are incident on the substrate, adsorbed and desorbed; the condensation of the adsorbed atoms, surface diffusion, nucleation, formation of aggregates and formation of small islands are regarded as independent processes and treated as random phenomena.

The molecular dynamics method is an older method. The evolution of typical samples of the system in this method is performed on a microscale of time and distance.

In both methods, the spherically symmetric Lennard–Jones potential $V(r)$ is used to deal with the interaction between atoms and atoms

$$V(r) = 4\varepsilon\left[\left(\frac{\sigma}{r}\right)^{12} - \left(\frac{\sigma}{r}\right)^{6}\right], \tag{5.75}$$

where r is the distance between the atom and the atom, ε is the height of the Lennard–Jones potential, and σ has the same dimension as r. The potential energy $V(r)$ is truncated at $r = 2.5\sigma$, and the interatomic interaction time interval $\Delta t = 0.03\sigma/(m/\varepsilon)^{1/2}$, where m is the mass of the thin film atom.

Repulsive Moliere potential energy $\Phi(r)$ is used when dealing with ions and atoms, especially inert gas ions and atoms.

$$\Phi(r) = \frac{Z_1 \cdot Z_2 \cdot e^2}{r} = \left[0.35e^{\frac{-0.3r}{a}} + 0.55e^{\frac{-1.2r}{a}} + 0.1e^{\frac{-6.0r}{a}}\right], \tag{5.76}$$

where a is the Firsoov shielding length, $a = 0.4683[Z_1^{1/2} + Z_2^{1/2}]^{-2/3}$, Z_1 and Z_2 are the atomic numbers of the ions and thin film atoms, respectively, and r is the distance between atoms.

5.4.1 *Monte Carlo method computer simulation*

5.4.1.1 *Monte Carlo algorithms classification for simulating film growth*

Film growth is a stochastic process. Using Monte Carlo algorithm to simulate the growth of films is an effective way to study its growth mechanism. Currently, Monte Carlo mainly includes three algorithms for simulating film growth:

(1) Kinetic Monte Carlo (KMC) of the specified event

This method specifies several types of events that may occur, but does not allow other types of events to occur. For example, it is stipulated that the adsorbed atoms can be migrated in a certain crystal plane or along the edge, and other types of events such as migration across crystal planes or across boundaries are not allowed. The advantage of this method is that we can know exactly what has happened and its probability; its disadvantage is that it is possible to ignore certain important types of events and it can cause inaccurate simulation results.

(2) Key Count KMC

This method uses the activation energy to determine the probability of an event during atomic migration. The activation energy is determined by the number of nearest neighbors initial position of the migrating atom, or by the difference between the number of initial nearest neighbors positions of the migrating atom and the number of nearest neighbors target positions of the migrating atom. The advantage of this method is that it can accurately study the environmental changes around the atom when the established model is reasonable. The calculation of the activation energy is not simply determined by both the initial environment around the migrating atoms and the environment around the target position, because saddle points of energy changes may occur during the migration process. As this method does not actually take into account the intermediate state in the migration process, the calculation of the activation energy is not accurate.

(3) Exhaustive method KMC

In this method, every possible atomic migration should be enumerated, and a large table that includes various possible events and each event corresponding to the activation energy should be created. It is a more accurate method, but this method is very complicated in specific applications. For example, when both the initial position and the target position have 12 nearest neighbors, a total of 18 crystal positions should be involved in the migration process, and then there are 218 possible events and corresponding probabilities that need to be calculated (there is no consideration of the impact of the second nearest neighbor position). So it is very difficult to build the table, and this method is not very versatile.

For the above three algorithms, the applicability of the key count KMC is stronger and more common than that of the other two.

5.4.1.2 *Model of film growth simulated by the Monte Carlo method*

Generally, there are three types of growth of films, namely the Volmer–Weber (VW) type (i.e., nuclear growth type), the Frank–van der Merwe (FM) type (i.e., monolayer growth type), and the Stranski–Krastanov (SK) type. The VW type is one in which a continuous film is formed on the surface of the substrate by nucleating, growing and merging, and most of the deposited films belong to this type; the FM type is one in which the deposited atoms are uniformly covered on the surface of the substrate to successively form single atomic layers; SK type is carried out in the manner of nucleation and growth after initially depositing 1–2 monoatomic layers.

(1) **Establishing model**

Thin film deposition is a very complicated change process of the physicochemistry. The entire process of film growth involves not only the mechanism of grain growth, grain–grain interaction, boundary motion and other factors, but also the topological geometry mechanism of space filling. Therefore, it is impossible to be carry out a simulation of the process, but we can only establish a suitable ideal model based on the purpose of research for understanding the events in the microworld. The Monte Carlo simulation method is used to examine the random process of particles depositing on the surface of the substrate. The entire deposition process needs to be processed as follows:

(a) Although the kinetic energy and binding energy released by the deposited particles will cause a local temperature increase near the deposited atoms, the temperature rise of the entire substrate is extremely small, so film growth can be considered to be an isothermal growth process.

(b) When the energy of the incident particle on the surface of the substrate is sufficiently large, the particle not only can perform migration movement on the surface of the film layer, but also can move in the vertical direction. If there are vacancies in the lower layer, it can also fill theses vacancies by crossing the lower layer.

(c) In the actual process, the re-evaporation rate of single particles is high, but if a particle meets and combines with other particles, its re-evaporation probability will be greatly reduced, so it can be approximated that only a single particle can be re-evaporated to the outside of the substrate, and the possibility of multiple particles being re-evaporated and decomposed is very small.

(d) The movement of particles on the surface of the substrate can be divided into three cases:

    (1) **Adsorption:** The particles are adsorbed on the surface of the substrate;
    (2) **Diffusion:** Particles migrate on the surface of the substrate, and the direction of migration should be to reduce the overall energy of the system. Diffusion includes free diffusion of individual particles and diffusion along the cluster edges;
    (3) **Evaporation:** The particles detach from the surface of the substrate under specific conditions.

(2) **Model processing method**

(a) **Periodic boundary:** Monte Carlo mostly uses periodic boundary conditions to simulate the growth of thin films, i.e., the calculated square substrate has to be infinite and repetitive, so it should cover the entire plane. Since the exact same motion occurs in each square substrate, that is, if there is a particle coming out of a square to the right, the corresponding position of the square would be occupied by a particle at the left of the square. So macro-scale (micro-infinite) samples can be described using the calculation results in a limited range. Therefore, in the actual operation process, when a particle diffuses from the left, the corresponding position will be occupied by one particle from the right; when the particle goes out from the top, the corresponding position will be occupied by one particle from the bottom. Similarly, when there are particles in the relative positions of the left and right and the upper and lower boundaries, it is considered that there are also particles existing in the corresponding neighboring positions of these particles, which should be taken into account when statistics are clustered.

(b) **The interaction potential between particles:** Monte Carlo can be used to simulate film growth by either Lenard–Jones potential or Morse potential. Here, Morse potential would meet the following formula:

$$\Phi_{\mathrm{ij}} = D\left\{\exp\left[-2a\left(\frac{r_{ij}}{r_0}-1\right)\right] - 2\exp\left[-a\left(\frac{r_{ij}}{r_0}-1\right)\right]\right\}, \tag{5.77}$$

where r_{ij} is the distance between two particles numbered i and j, D and a are the intensity and range of the interaction, respectively, and r_0 is the position of the lowest point of potential energy, namely the equilibrium position. Among them, D can be expressed by parameters E_{ff}, E_{fs}, and E_{ss}. E_{ff} represents the binding energy between film-deposited particles; E_{fs} represents the binding energy between film-forming particles and substrate particles; E_{ss} represents the binding energy between substrate particles.

The film growth mode is mainly determined by energy parameters rather than deposition conditions. When $E_{ff} < E_{fs}$, it is FM type; when $E_{ff} > E_{fs}$, it is VW type; when E_{ff} is close to E_{fs}, the formed nucleus becomes very flat, and the difference between the two growth modes becomes less obvious. Only when the value of a was changed was a typical SK growth mode obtained.

(3) **Scope of action**

When a has different values, the range R of Morse potential is also different. Therefore, when considering the interaction between particles, we should not simply take into account its nearest neighboring or sub-nearest neighboring point, but also consider all the points in the entire range R. Here, the specific value of R can be determined by the specific conditions and accuracy requirements.

(4) **Substrate structure**

Generally, the substrate can be composed of a 100×100 two-dimensional grid lattice under periodic boundary conditions. According to the previous discussion, the coordinates of the incident particles are randomly selected lattice coordinates, so the length of the particle diffusion is the lattice constant. When the substrate is a specific material during the discussion, the substrate structure is not necessarily a grid lattice. At this time, a new substrate structure can be constructed by changing the coordinate system to make it more near practice.

Scientists have used the Montre Carlo method to carry out computer simulations of the film growth process, and have obtained some satisfactory results, such as the annealing process of the film, the growth process of Au (001) film, the growth of the polycrystalline materials in normal and abnormal particles, the morphology and structure of silver oxide thin film, infrared optical characteristics of amorphous carbon film, the nucleation process of the two-dimensional Ge island, the growth process of semiconductor quantum dots by strain control, the nucleation and growth pause process of semiconductor quantum dots, etc. Visualization of the growth process of the film has been achieved, and continuous and discontinuous morphologies in the film growth have been simulated.

5.4.2 *Computer simulation of molecular dynamics*

The initial purpose of molecular dynamics is to define the microscopic description of a physical system. The system can be a mono-body system or a multi-body system. Hamiltonian formulation or Lagrangian formulation can be used to describe the system, and the system formulation can be directly expressed by Newton's equation of motion. Its essence is to calculate the properties of the system by using the equation of motion. As a result, both the static and dynamic characteristics of the system can be obtained. The molecular dynamics method can calculate the phase space trajectory of a group of molecules, in which the motion of each molecule is described by the equation of motion, obeying the classical laws of motion, which can be solved by means of a computer.

For a microparticle in a system, its equation of motion is

$$m\frac{dr_i}{dt} = P_i, \tag{5.78}$$

$$\frac{dp_i}{dt} = \sum_{i<j} F(ij), \tag{5.79}$$

where m is the mass of the particle, P_i is the momentum of the ith particle, r_i is the position of the ith particle, F_{ij} is the action of the jth particle on the ith particle, and t is time. The above equations are differential equations, which cannot be solved by a computer. They must be converted into a numerical addition form by using a discretization method (Taylor transformation).

Therefore, Equations (5.78) and (5.79) can be processed to obtain their discrete form:

$$\frac{d^2r_i}{dt^2} = \frac{1}{h^2}[r_i(t+h) - 2r_i(t) + r_i(t-h)] = \frac{1}{m}F_i(t), \tag{5.80}$$

where h is the time interval, $F_i(t)$ is the force on the ith particle at the time t; Equation (5.80) shows that the position at time $t+h$ can be derived from the force of the particle and the position of $t-h$ and t. Equation (5.80) can be transformed into (5.81) and (5.82), where v_i^n is the speed of the ith particle and the nth step

$$r_i^{n+1} = 2r_i^n - r_i^{n-1} + \frac{1}{m}h^2F_i^n, \tag{5.81}$$

$$v_i^n = (r_i^{n+1} - r_i^{n-1})/2h. \tag{5.82}$$

As mentioned above, the simulation method of molecular dynamics can be used to measure the measurable variables which can be expressed as a function of the position and momentum of the particles in the system. Finally, the particle motion described by Newton's motion equation in the system will be transformed into discrete numerical solutions which consist of a series of time intervals of h. Using Equations (5.81) and (5.82), the position and momentum of the N particles in the system can be calculated after time t, and the average values for the observables can also be calculated by specifying a certain initial condition, and selecting a certain potential energy function, such as potential, universal function, combined potential energy, and so on.

Molecular dynamics simulation methods are often used to study the properties of large block materials, but the method cannot usually be performed in systems of almost infinity. In order to enable the simulation in a finite volume extending to a real large system, periodic boundary conditions will be usually applied to construct a quasi-infinite volume, which can represent the macrosystem more accurately. The periodic boundary condition is expressed as

$$A(x) = A(x + nl), \tag{5.83}$$

where A is a random observable, n is any integer, and l is the length of a primitive cell. The boundary condition is to make every primitive cell in the elementary molecular dynamics repeat infinite times.

The simulation conditions for molecular dynamics also need to know the interaction potential between atoms. The choice of the interaction potential directly affects the accuracy of simulation results. Hypothesize that the deposited gas phase spherical atoms or molecules reach the surface of the substrate randomly, and then they either adhere to the position where they can reach the surface of the substrate (their mobility is zero) or move to the minimum energy position supported by three atoms (corresponding to very limited mobility). For research on two-dimensional growth, the deposited atoms are not spherical atoms supported by three points, but circles supported by two points. For the assumption of zero mobility, a loosely aggregated chain-like structure film can be simulated, and the chain-like branching and merging are random. For the assumption of the finite mobility, a dendritic structure can be simulated as growing outward from the substrate with a diameter of several molecular scales. This structure is similar to the actual columnar structure. Molecular dynamics simulation not only can simulate the growth process of thin films, but also can simulate the film deposited process with doping or ion-assisted deposition.

Molecular dynamics has been used to study many problems in growth processes of thin film, i.e., the deposition and infiltration processes of cluster particles can be studied on the thin film; dislocations and stacking faults that can be caused by lattice mismatch in epitaxial thin films; the energy of deposition particles and the thickness of critical layer effect on film growth mode; energy particle effect on film texture in particle-beam-assisted deposition, and films can be grown from liquid ionic oxides.

Molecular dynamics simulations were used to calculate the growth process of hydrogen-containing diamond-like carbon (DLC) films using CH_3 as radical groups of the deposition source. The structural characteristics of hydrogen-containing DLC films were investigated by statistical analysis, density analysis and hybridization analysis of deposited atomic number. Hypothesize there are 250 CH_3 radical groups to collide with the surface of diamond (100) in order to deposit a DLC film, where the calculation step is set to 0.5 fs; a canonical ensemble is used; the temperature is controlled at 293 K; a Berendsen thermal bath is used; the time interval for each CH_3 group to collide with the diamond substrate is taken as 3 ps. The third-order precise prediction-correction method is used to solve the motion equation, and the periodic boundary conditions are applied in the $\pm x$ and $\pm y$ directions. The diamond substrate has a total of 1024 atoms and its size is $14.24\,\text{Å} \times 14.24\,\text{Å} \times 27.59\,\text{Å}$. Meanwhile, it is composed of carbon atoms with 32 layers, and atoms of the bottom two layers are fixed to simulate a thick substrate. Atoms of the top three layers are not limited by temperature, and are only affected by the potential function. Atoms of the middle 27 layers are coupled to the thermal bath, so that the substrate temperature fluctuates within the controlled range.

Figure 5.12 shows that the structure of a hydrogen-containing DLC film grows under different incident energy conditions. In Figure 5.12(a), it can be clearly observed that a layer of hydrogen DLC film is formed on the substrate surface because of only chemical adsorption at 1 eV incident energy; because the incident energy is

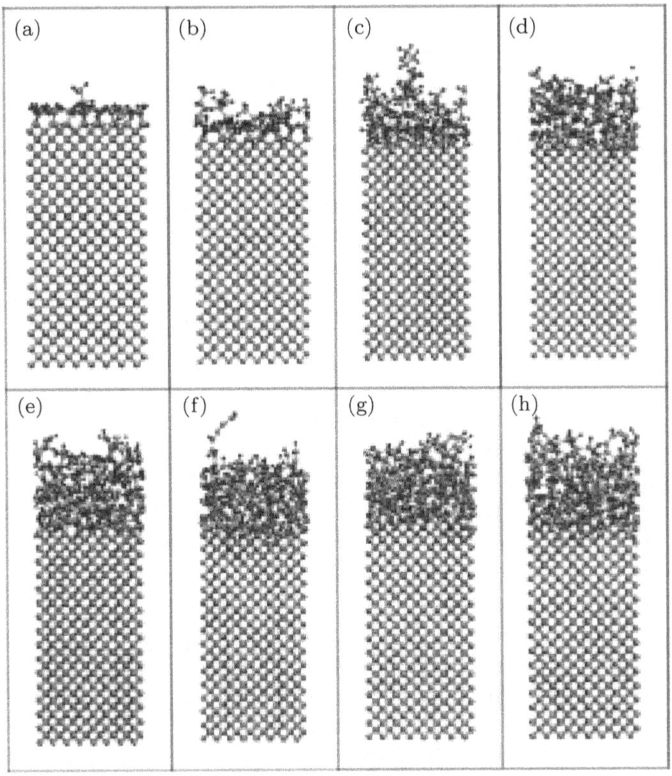

Figure 5.12: Structure diagrams of the grown hydrogen-containing DLC films using CH_3 as the deposition source radical group at different incident energies. (a)–(h) Structure diagrams of the film at incident energies of 1, 10, 20, 35, 50, 65, 80, and 100 eV, respectively, where gray atoms represent carbon atoms, small black atoms represent hydrogen atoms; the substrate is a diamond (100) plane.

too low, both the injection and the low injection behavior cannot happen.

Figure 5.13 reflects the changes in the structure of the hydrogen-containing DLC film. At 50 eV, the growth in the density of the film tends to be flat. At this time, the hydrogen content reaches 56.1%, which indicates that optimal deposited parameter is near the 50 eV incident energy (35–65 eV); the total number of deposited atoms reach the maximum value; the average density value of the film is close to the saturation value, and the hydrogen content is also

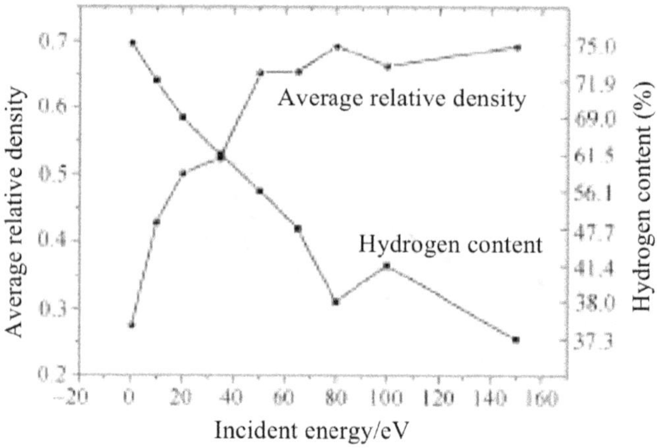

Figure 5.13: Relationship between the average relative density and hydrogen content in the hydrogen-containing DLC film as incident energy changes.

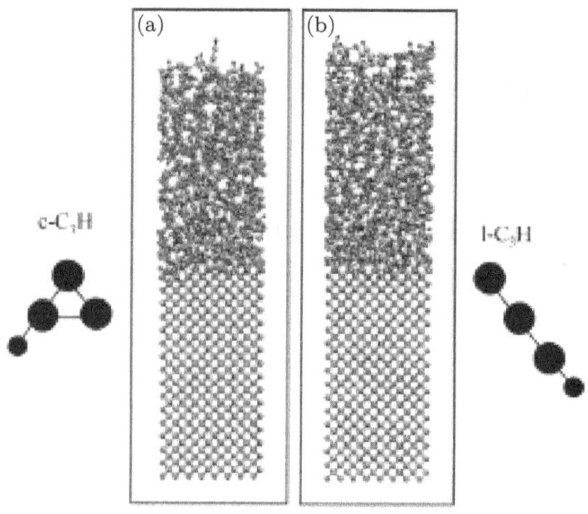

Figure 5.14: Final illustrative structural diagrams of a hydrogen-containing DLC film grown with c-C_3H (a) and l-C_3H (b) as the source group.

high (47.7–61.5%), which is the necessary condition for growing the structure of a hydrogen-containing DLC film.

Figure 5.14 shows the final illustrative structural diagrams of a hydrogen-containing DLC film grown at 50 eV incident energy using a circular C_3H (c-C_3H) and a linear C_3H (l-C_3H) as radical groups of the deposition source. Since carbon atoms are more reactive than hydrogen atoms, the hydrogen-containing DLC film becomes more dense and uniform as the carbon content in the source radical group increases. However, since a large amount of hydrogen is existing in the hydrogen-containing DLC film, the higher hydrogen content makes the ultra-low friction characteristics become more obvious.

5.5 Factors Affecting Film Nucleation and Growth

The parameters affecting film growth include vaporization rate, atomic kinetic energy, adhesion coefficient, surface migration rate, nucleation density, condensation rate, bonding rate, impurity and defect concentration, and charge strength. In this section, we will discuss some of the main influencing factors that affect the nucleus formation, growth, particle bonding, continuous film formation, defect morphology, film density, and final structure.

5.5.1 *Deposition rate*

The deposition rate is related to the degree of super-saturation achieved during nucleation and is related to the nucleation orientation. With a higher deposition rate at the beginning of the deposition of the film, nucleation super-saturation and higher nucleation density are easily formed. The slow deposition before the particle bonding facilitates the alignment of the crystallographic direction of the particles. Usually the low deposition rate ranges from 0.02 to 10 Å/s. For the heating evaporation source, the molar weight W of evaporation per unit time per unit area is M, the evaporation pressure is P, and the temperature is T, then there is

$$M = KP\sqrt{\frac{W}{T}}, \tag{5.84}$$

where K is a constant.

For glow discharge sputter deposition, when the theoretical deposition rate is calculated, the sputtering yield parameter is used, and since the backscattering coefficient of the emitting atom cannot be accurately determined, the result has a large error. However, under low pressure conditions, at least the yield data of pure metals is accurate, and the theoretically calculated sputtering rate is very close to the experimentally measured results.

For film workers, more attention is paid to growth rates than to evaporation rates. Because this parameter will affect the nucleation and growth mechanisms, the relationship between growth rate and evaporation rate is characterized by an adhesion coefficient which is a function of the substrate composition, structure, temperature, and film thickness. Due to the different vaporization rate, the structure of the film is different, which in turn changes the optical and electrical properties of the film.

5.5.2 *Kinetic energy of deposited atoms*

During thermal evaporation, the energy of evaporated particles (atoms, ions, or radicals) follows the Maxwell distribution, and the energy peak varies with the equivalent temperature of the evaporation source. When the equivalent temperature of the evaporation source is 1500 K, the typical peak value is 0.2 eV. At the time of sputtering deposition, the atomic energy distribution peaks at about 4 eV. Incident atoms with large energies cause changes in nucleation and growth characteristics. At the same time, since these energies are similar to the lattice binding energy of the substrate used for preparing the film, the incident atoms or ions disturb the surface bonds of the substrate, promote the chemical reaction between the film and the substrate, and significantly affect the structure and characteristics of the film.

If the deposited atoms have higher kinetic energy, they will increase the aggregation of the atoms on the surface of the substrate and reduce the directional growth. Since the sputtering deposition method has higher kinetic energy, the epitaxial growth of Ag and Au and the reactive sputtering of oxygen in Al and Ta have a substrate temperature much lower than that at the time of evaporation deposition of the same material. This is due to the higher kinetic energy of the sputtered atoms, which is equivalent to a higher substrate temperature.

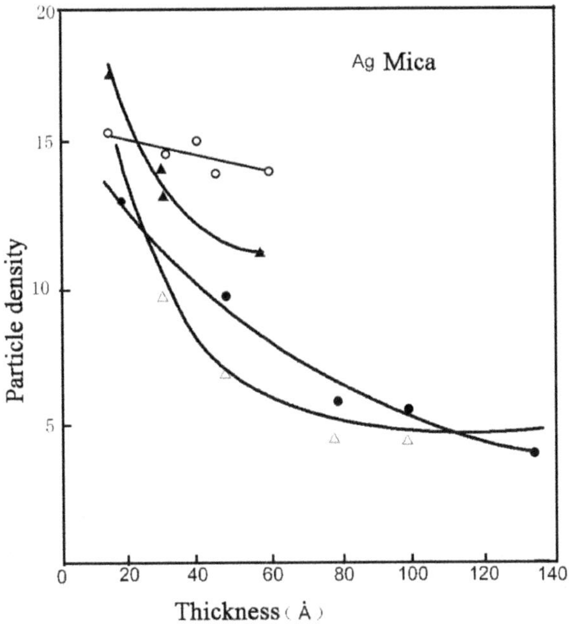

Figure 5.15: Relationship between particle density and thickness of Ag on a mica substrate at different substrate temperatures and deposition methods.

As the kinetic energy of the particles increases, the diffusion rate of the atoms on the surface of the substrate increases, thereby increasing the condensation rate and the growth rate of the particles, and also accelerating the bonding of the particles, so that the density of the particles is rapidly reduced. Figure 5.15 shows the deposition of an Ag film on a mica substrate with the density of the particles as a function of thickness at different substrate temperatures and different deposition methods. At the same substrate temperature, the density of the sputtered film particles decreases more rapidly with increasing deposition. When the vapor deposition thickness is further increased, the density of the sputtered particles approaches a constant value. At a sufficiently low substrate temperature (such as 20 K), the influence of the thermal diffusion of atoms on the substrate can be neglected, and the infiltration, compounding and alloy formation between the film and the substrate can be studied to further understand the influence of the kinetic energy of the incident atom. In the experiment on the Au film, it was found that the particle size of Au decreased as the pressure of the sputtering gas Ar increased; the same phenomenon occurred in the evaporation of the Au film in the Ar gas, indicating the influence of the kinetic energy of the particles. Because the number of collisions is increased, the incident atoms reduce the energy, which facilitates the rapid thermal equilibrium between the vapor atoms and the substrate, resulting in a fine-grained structure.

5.5.3 *Oblique incidence deposition*

When the angle of incidence of atoms deposited on the surface of the substrate is different, it causes a change in mobility, affecting nucleation and subsequent growth. The incident atom deviates from the vertical direction and will affect the direction and rate of diffusion of foreign atoms on the surface.

Most experiments have confirmed that the crystal phase structure and grain orientation undergo significant changes in the later stages of film growth. The electron microscopic topography of the vaporized Au and Ag films showed an increase in the rate of condensation as the angle of incidence increased. For atoms incident at an angle of 80° from the direction of the vertical substrate, it was found that the formation of the particles and the initial growth distribution were uniform. As the particle size increases, it becomes apparent from the shadow region, and columnar growth occurs in the incident direction of the atom. The critical thickness of the electrical continuous film increases rapidly with the atomic incident angle, as shown in Figure 5.16. Figure 5.16 shows the relationship between the normalized critical thickness and the incident angle. The critical thickness of normal incidence is represented by $t_c(N)$, and the critical thickness of oblique incidence is represented by $t_c(0)$. It can be seen that when the incident angle of the atom exceeds a certain value, the critical thickness thereof increases rapidly and tends to produce columnar growth. The film has been observed to have non-uniform mechanical stress and magnetic properties, as well as significantly varying optical reflectance and the like.

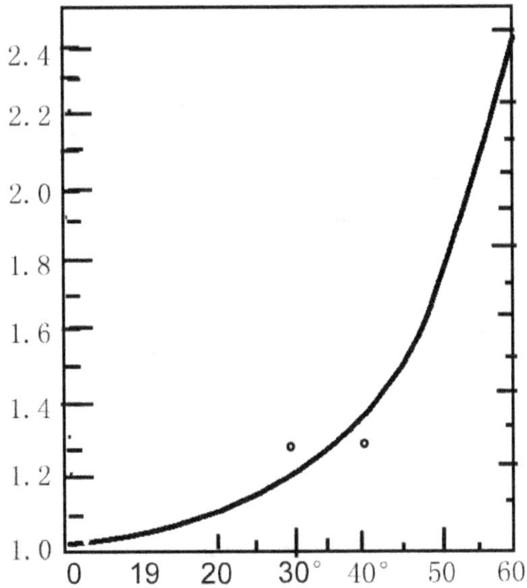

Figure 5.16: Relationship between the incident angle of Ag film and the critical origin angle.

The anisotropy and columnar growth of the film are controlled by the incident direction of the deposited atoms and the surface migration rate of the foreign atoms. The higher the surface migration rate, the smaller the anisotropic growth. Au and Ag are vaporized on a substrate of elevated temperature, and have small anisotropic growth despite large angle incidence. When low migration rate Al is vaporized on glass and NaCl substrates at room temperature, it has significant non-uniformity and columnar growth.

The incident angle effect is not significant in sputter deposition because the larger area of the target is often used as the sputter source material, and the atomic collision at high pressure changes the small emission cone angle. When having a small source area and low pressure sputtering, as with the vapor deposited film, anisotropic structural and physical properties will result.

5.5.4 *Applied electric field*

It is found that bonding with lower average thickness occurs when a transverse DC field of 100 V/cm is applied during the deposition

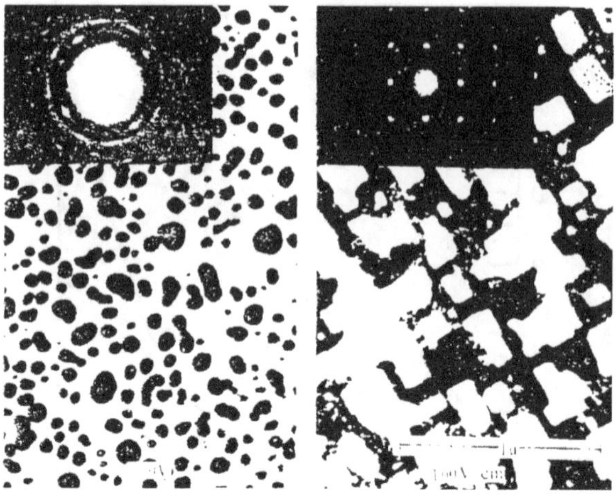

(a) (b)

Figure 5.17: TEM image and diffraction pattern of Ag film (50 Å) on NaCl substrate. (a) No electric field; (b) electric field.

of metal films with higher atomic migration rate. The morphology and electron diffraction of Ag evaporated on NaCl with and without transverse electric field are shown in Figure 5.17. The substrate temperature is 200°C, and the applied electric field strength is 100 V/cm (DC). The film with DC electric field has relatively average separated particles, which increases their surface area, promotes bonding, and forms a grid structure with an approximate square parallel to the NaCl surface. The diffraction pattern shows that it is a good single crystal structure (see Figure 5.17(b)). The critical thickness of the film is decreased by adding the electric field, and the decrease is more significant at high substrate temperature. Figure 5.17(a) shows that the covering area of particles on the substrate is small when the film thickness is the same, and the electron diffraction pattern shows that it is a polycrystalline structure.

Increasing the electric field can enlarge the area of particles, but there is a limit. If it is exceeded, an electric arc will be generated and the film will be damaged.

The effect of electric field is mainly electrostatic characteristics (ignoring the current), and only during the initial particle bonding (discontinuous film). After the bonding, the effect of electric field is weakened. It is generally believed that the effect of electric field is not due to Joule heat, because Joule heat will lead to the aggregation of particles and increase the discontinuity of the film. If a large current is allowed to pass through the film in the network forming stage, Joule heat will significantly affect the growth behavior of the film.

The resistance of the continuous film prepared by adding electric field is usually lower than that of the film prepared with no electric field. When the applied electric field increases, the resistance of the film decreases and approaches a constant value (it is very close to the block value). In the case of high substrate temperature, the decrease of resistivity is quite obvious. The reason for the decrease in resistivity is believed to be due to the fact that there are fewer "frozen" defects in the structure when the islands are joined in the film.

The electric field is applied to charge the particles, and the charge comes from the ionized vapor or the interface potential of the substrate. The charge of spherical particles with radius r is q, which increases the free energy, and the energy is the sum of surface energy ($4\pi r^2\sigma$) and electrostatic energy ($q^2/(8\pi\varepsilon_0 r)$. Due to the increase of surface area, the total energy is increased correspondingly; that is, the ball will become a flat sphere, and the exact form depends on the balance of various free energy. Further increasing the charge, the particles may be broken. This process can be described by the charged mercury drop. When $r = 100\,\text{Å}$, $q = 10^3$ unit charge, the electrostatic energy is almost the same as the surface energy, which makes the particles quite flat, thus increasing the surface area. If q is proportional to the number of atoms in the particle, the spherical particle cluster will become flatter with the increase of the particles.

Now let's calculate the interaction between two charged particles. If the radii of the two particles are r_1 and r_2, respectively, the charges are q_1 and q_2, respectively, and the separation distance is d; considering the effect of the mirror image force, the net force given is

$$F_0 = \frac{q_1 q_2}{4\pi\varepsilon_0 d^2} - \frac{q_1^2 r_2}{4\pi\varepsilon_0 (r_1^2 - d^2)^2} - \frac{q_2^2 r_1^2 d}{4\pi\varepsilon_0 (r_2^2 - d^2)^2} \cdots . \tag{5.85}$$

Suppose two particles with positive and negative charges, one charge being much larger than the other, such as $r_1 = 100$ Å, $q_1 =$ 100 unit charge; $r_2 = 20$ Å, $q_2 = 1$ unit charge; $d = 200$ Å, then the attraction between the two charged particles is $F_0 \approx 10^{-4}$ dyn. The shear stress on the surface of the ball is 10^{10} dyn/cm^2 = 10^4 atm/cm^2, which is large enough to move the small ball towards the big ball and make the two join. It must be noted that for the same charged particles, the repulsion force of the same magnitude will hinder the particles from joining. If the substrate is irradiated by an electron beam before or during the deposition of the film, particle bonding will be promoted in the initial stage and further growth of the film will be affected.

External electric fields often affect the orientation of the film. Some researchers have found that an applied electric field (or magnetic field) can monitor the dominant orientation of the film. In the sputtering-deposited Ni–Co–Fe film, the effect of the applied magnetic field on the orientation of the film was observed. Under the action of the electric field, the dendrite growth behavior of the vaporized Ag–Fe film was observed. Its growth direction is from cathode to anode. The applied electric field perpendicular to the substrate increases the growth rate of the epitaxial Ge, Si, and GaAs films.

5.5.5 *Substrate temperature*

The substrate temperature will affect various parameters of the film growth, such as adhesion coefficient, surface diffusion rate, nucleation conditions, critical thickness (the film reaching this thickness becomes continuous), and the crystallization and orientation of the film on the amorphous substrate. The epitaxial growth quality of the single crystal film depends on the crystal substrate, the difference in thermal expansion coefficient between the substrate and the film material, the generation of residual stress, and the like. Due to the presence of residual stress, an increase in defects in the film is caused, such as a magnetostrictive effect, a film anisotropy, a change in the transition temperature of the superconducting film, and a change in the position of the edge of the semiconductor film.

A very low substrate temperature, such as liquid nitrogen temperature, will cause certain metal films (such as vaporized Ag film) to become amorphous films; when the temperature rises, the film can be converted into a polycrystalline film. In the medium temperature region, a polycrystalline film is often formed; in the higher temperature region, a single crystal film may appear; and at a higher temperature, a film having more defects will be formed.

The influence of high substrate temperature mainly includes: (1) the condensation coefficient of foreign atoms changes, most materials reduce the condensation coefficient; (2) the high migration rate makes the two-dimensional gas transmission replace the jump between local positions, and the atomic rotation is enhanced; (3) there is significant diffusion between islands; (4) the foreign atoms enter the lattice of the substrate and produce the interface diffusion, which makes the interface between the island and the substrate blurred; (5) the critical size of nucleation increases; that is, there will be more atoms in the critical nucleus. On the contrary, with the decrease of substrate temperature, the stable critical nucleus size decreases. At a very low substrate temperature, the relaxation time of the adsorbed atoms on the substrate may become so large that the equilibrium between the vapor and the concentration of the adsorbed atoms cannot be established.

5.5.6 *Substrate structure*

The composition and structural form of the substrate (amorphous, polycrystalline, and single crystal) significantly affect the structural type of the film. Polycrystalline films typically obtained on amorphous substrates have a weaker grain orientation. The orientation can be strengthened by raising the substrate temperature or oblique incidence. A single-crystal thin film can be epitaxially grown on a single-crystal substrate at a sufficiently high temperature. The use of surface cleaning or vacuum cleavage of the substrate for a material to obtain an atomic clean surface can significantly alter some of the film-forming parameters such as adhesion coefficient, grain orientation type, epitaxial growth conditions, and the like.

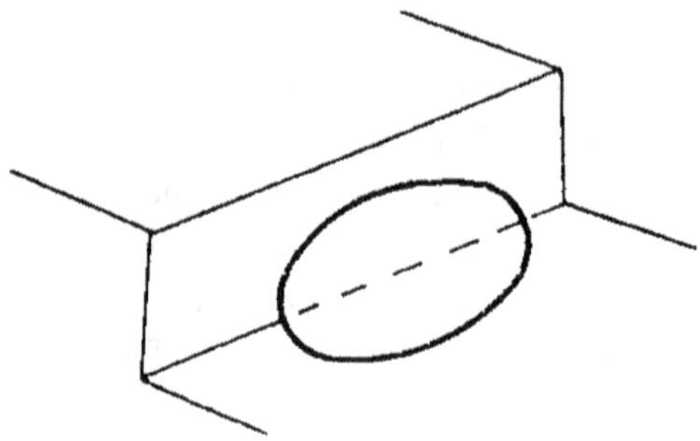

Figure 5.18: Nucleation at the step.

The effect of imperfect structures on the nucleation process is extremely significant compared to a uniformly smooth substrate surface. Imperfections in the surface of the substrate mainly refer to macroscopic cracks, protrusions, grooves and microscopic steps, single atom defects, and the like. As an example, consider nucleation on a 90° step, as shown in Figure 5.18, which increases the probability of nucleation and growth rate. If the substrate adsorbs impurities, it will generally reduce the binding energy of the adsorbed extraneous atoms (which may of course increase the binding energy) and the diffusion activation energy, and correspondingly also affect the growth rate of nucleation.

5.5.7 *Environmental pollution*

Contamination is mainly derived from evaporation and sputtering sources, or in residual gases, and its effects are strongly dependent on the interaction of the contaminant with the material. Usually oxygen and moisture are the main pollutants. Oxygen not only directly contaminates the film, but also forms volatile oxides in the evaporation source, which can significantly change the W, Mo, and Ta films.

The source of the cathode used in sputtering must be pure and dense, as impurities in the solids and the oxygen and moisture in the pores will be the main source of contamination.

5.5.8 *Chemical decomposition*

During the vaporization process, chemical decomposition will cause a change in composition, which is mainly produced in the vapor deposition alloy, oxide, and semiconductor compound film. These changes may be due to the different vapor pressures of the alloy, or the nature of the different components of the compound. If the vaporization rate of each component in the compound is proportional, the effect of decomposition in the vapor phase is not critical. Because compounds or alloys tend to re-grow on the substrate; if the vaporization rate is not proportional, the difference is large, and a layered film structure will be produced due to the selection of the volatile component. Some films can be remedied by post-steam annealing. This is especially important for materials containing low-valent oxide and semiconductor compounds with low decomposition energy and high-valent Group V or Group VI components. Decomposition of the compound on the substrate also occurs during and after the evaporation. This may be due to a large difference in adhesion coefficient, which may result in the coexistence of two phases in the film, resulting in non-uniformity and the like.

5.6 Exercises and Questions

(1) The solid–liquid interface rate of an atomic model during solidification should meet the following equation:

$$v = a_0^2 \upsilon \exp\left(-\frac{E_D}{RT}\frac{2\Delta G}{a_0 RT}\right).$$

Calculate the v value of Si: hypothesize that the diffusion rate of the atom in the liquid state is $10^{-4}\text{cm}^2/\text{sec}$, $\nu = 10^{13}\text{s}^{-1}$, $a_0 = 2.7$ Å, and the difference between the free energies of the Si atom in the liquid and solid states is equal to the latent heat of fusion. Try to find the interface rate.

(2) Analysis of the possibility of surface alloying by the laser process on the following 2000-Å thin film composite substrates: Zn–Mo, Mo–W, Fe–Ni, and Ta–Cd.

(3) A Kr ion laser pulse with 50 ns, 37.5 mW/μm^2 is injected into TeGe, where TeGe's $R = 0.65$, $\rho = 6.2$ g/cm^3, $k = 0.015$ cal/cm $\cdot$ s $\cdot$ °C, $c = 0.08$ cal/g $\cdot$ °C, and $2\sqrt{K_d \tau_P} > \alpha^{-1}$.

(a) The curve of the surface temperature in the heating and cooling process changing with time should be drawn;
(b) How many times can the maximum surface temperature be achieved?
(c) How much is the maximum quenching rate of the surface?

(4) An amorphous Si film is annealed on a flat single crystal Si wafer. Draw sketches of the following two cases and distinguish the grain structure of the product:

(a) Random nucleation growth;
(b) Solid phase epitaxial growth.

(5) The corrosion resistance of the distributed Cr on the steel surface is a thickness-dependent constant. This effect can be obtained by the implanted successively overlapping ions. Considering the following three injections, the energy of Cr ions is 50, 100 and 200 keV, respectively. If the peak concentration of Cr is maintained at 20% (i.e., the composition of stainless steel), how much should be the proportion of ingredients? Draw the cross-section graphic of the overlapping distribution of the products.

(6) If the diffusion time of an atom is t, the result of the annealed thin film atom should meet the following diffusion equation:

$$C(z,t) = \frac{\phi}{\sqrt{2\pi(\Delta R_p^2 + 2Dt)}} \exp\left[-\frac{1}{2}\left(\frac{z - R_p}{(\Delta R_p^2 + 2Dt)^{1/2}}\right)^2\right].$$

Explain the physical meaning of the term $2Dt$ and the term $(\Delta R_p^2 + 2Dt)$.

Chapter 6

Structure and Defects of Thin Films

Due to the particularity of the film manufacturing method, the structure and defects of the film have their special features compared with block materials. For example, the crystal structure of the microcrystal in the film is not necessarily the same as that of the block material. It can be disordered, or it may be in an abnormal structure due to the influence of the substrate. Film defects are usually more than block materials, which will have a certain impact on the performance of the film.

6.1 The Structure of Thin Films

6.1.1 *The tissue structure of thin films*

The structure of thin films is the crystalline form of the thin films, and can be divided into four types: amorphous structure, polycrystalline structure, fibrous structure, and single crystal structure.

6.1.1.1 *The amorphous structure*

In terms of the arrangement of the atoms, it is a nearly ordered structure. That is, the atoms are arranged in order at a distance of 2–3 atoms, and beyond that, they are arranged in disorder. This structure shows no crystalline properties and is sometimes called amorphous or glassy. For example, each atom of amorphous silicon is still tetragonal, covalently bonded and forms a tetrahedron with its nearest neighbor, which is still regular. Another characteristic of the amorphous state is that its free energy is higher than that of the crystalline state of the same material, it is in a metastable state, and it has a tendency to change to the equilibrium state (stable state). However, a certain barrier should be overcome when transitioning from the metastable state to low free energy. Therefore, the amorphous state and its structure are relatively stable. This relative stability is directly related to the service life and application of amorphous materials. The most typical amorphous material is glass. In addition, many thin films prepared by evaporation, sputtering, CVD, and other methods at lower substrate temperature are all amorphous.

Compared with bulk materials, it is easier to obtain amorphous structure when preparing thin films. This is because the preparation method of thin film can easily cause the external conditions of amorphous structure, that is, a higher supercooling degree and a lower atomic diffusion ability. These two conditions are also the conditions to improve the supercooling degree of the phase transition process and inhibit atomic diffusion, thus forming amorphous structure. The above conditions can be achieved by lowering the substrate temperature, introducing reactive gas, and doping. For example, for sulfide and halide films, amorphous films can be formed when the substrate temperature is lower than 77 K. Some oxide films ($Ti0_2$, ZrO_2, Al_2O_3, etc.) tend to form amorphous films at room temperature. An example of the introduction of reactive gas is the evaporation of aluminum, iridium, steel, tin, and other thin films at 10^{-2}–10^{-3} Pa oxygen partial pressure. In doped films of 83% ZrO_2–17% SiO_2 and 67% ZrO_2–33% MgO, amorphous films can also be formed due to different atomic sizes of the two types of deposition.

In addition to the preparation conditions, the ability of a material to form amorphous states depends largely on its chemical composition. In general, metallic elements do not easily form amorphous structures. This is because there is no directional bonding between metal atoms, so we need a larger degree of supercooling to inhibit the formation of an ordered arrangement of metal atom. The tendency of alloys or compounds to form amorphous structures is obviously higher than that of single components, because the structures of compounds are generally more complex, and there are certain differences in crystal structure, lattice constants, chemical properties, etc. And the interaction between different components greatly inhibits the diffusion ability of atoms. In a single component, Si, Ge, C, S and other nonmetallic elements have a greater tendency to form amorphous structures. This is because such elements tend to form covalent bonds, as long as the neighboring atomic coordination meets the requirements and the energy difference between amorphous and crystalline materials is small. For example, in the presence of hydrogen, Si atoms will form a large number of H bonds, so Si films deposited at 800 K may still have amorphous structures.

The amorphous films are stable at ambient temperature. They contain both irregular network structures (the glassy state) and structures stacked with secrets. The former mainly appear in oxide film, element semiconductor film and sulfide thin, and the latter mainly appear in alloy film. It can be considered that the irregular network structure is composed of two interconnected classified structures. When studied by diffraction, the structure shows a wide scattering peak in the X-ray diffraction pattern and a wide dispersion ring in the electron diffraction pattern.

The growth of amorphous thin film materials can also adopt a columnar growth mode. For example, Si, SiO_2 and other materials can form an amorphous columnar structure at low temperature. The development of columnar morphology of Ge, Si and other thin films can be divided into three kinds of columnar structures: nano, micro and macro. Figure 6.1 is a schematic diagram and a structure diagram of the columnar morphology of each layer in the amorphous Ge film. 30% Au–70% Co alloy film is a typical example of amorphous film. In the equilibrium state, the solid alloy structure of this component should be a mixture of Au and Co solid solutions. In order to inhibit the formation of a crystal core, the substrate temperature was reduced to 80 K and then evaporated. After deposition, the film was heated to different temperatures and its microstructure was observed.

Figure 6.2 (a) shows the morphology and selected area electron diffraction of as deposited alloy. At this time, the morphology of the alloy has no characteristics, and the corresponding diffraction pattern is a halo ring. This shows that the structure of the film is

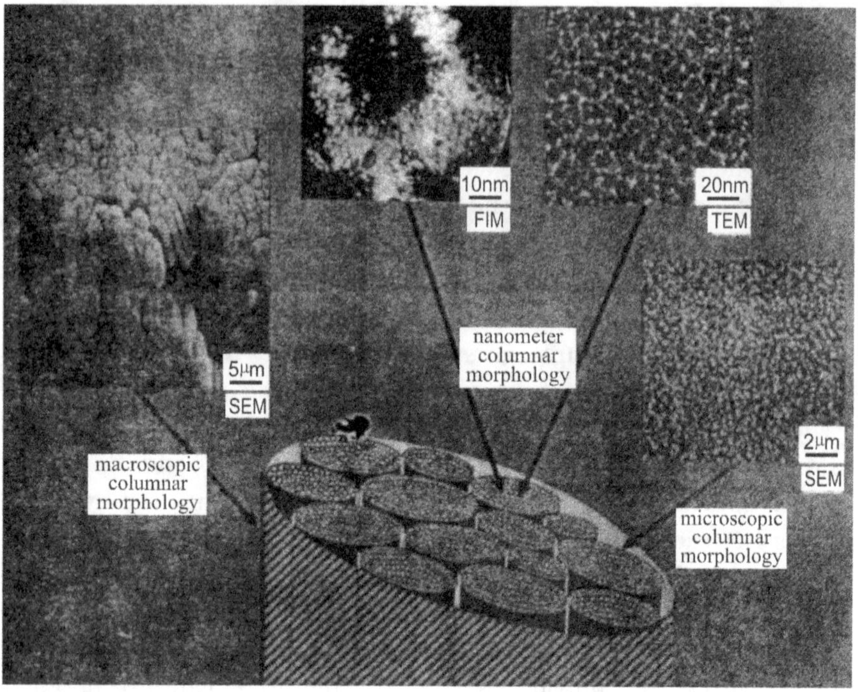

Figure 6.1: Columnar micrograph and schematic diagram of various layers in amorphous Ge films prepared by sputtering.

(a) (b) (c)

Figure 6.2: Microstructure and electron diffraction pattern of Au-Co alloy as deposited at after treatment: (a) 300 K; (b) 470 K and (c) 650 K.

amorphous, and the ordered range of the structure does not exceed a few atoms spacing. However, it is unclear whether the film structure is completely amorphous or microcrystalline.

The effect of annealing at different temperatures on the morphology of the films is shown in Figures 6.2(b) and 6.2(c). After

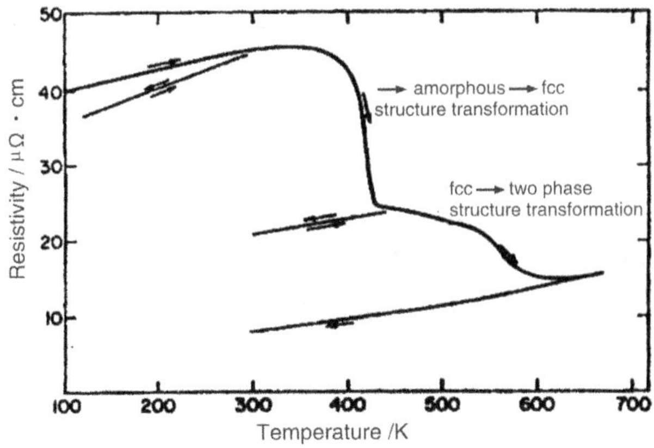

Figure 6.3: Temperature dependence of resistivity of 38 mol% Au–62 mol% Co alloy film.

470 K treatment, the structure of the film is transformed into a microcrystalline state with a face centered cubic structure. The analysis of the diffraction ring shows that it is a metastable structure not found in the phase diagram. After 650 K treatment, the film structure changed into a stable Co, Au two-phase structure.

Corresponding to the above structural change is the change of resistivity ρ with temperature of the film, as shown in Figure 6.3. From the resistance temperature curve in Figure 6.3, it can be seen that in the process of temperature increase, when the temperature is 420 K and 550 K respectively, there is an irreversible change of resistivity. These two change points correspond to the change of film structure. It is proved that the alloy film is amorphous. The resistivity of amorphous films is higher because of the strong scattering of electrons by the disordered state of atom arrangement.

6.1.1.2 *The polycrystalline structure*

Polycrystalline films are composed of several grains of different sizes. The islands formed during film formation have the characteristics of crystals (atoms are arranged in a regular way). A film formed by the coalescence of small islands is a polycrystalline film. The thin films made by vacuum evaporation or cathode sputtering are grown through island structure, which inevitably produces many intercrystalline boundaries and forms a polycrystalline structure.

In polycrystalline films, metastable phase structures are often found which are not in bulk materials. For example, the metastable quadric phase often exists in ZrO_2 films. When the tetragonal phase is transformed to the monoclinic phase, there is a 4% volume expansion. The incorporation of Y_2O_3 into ZrO_2 can prevent the formation of metastable phase and the formation of stable cubic phase. The formation of metastable phase may be caused by deposition conditions, substrate, impurities, electric and magnetic fields, etc. However, the metastable phase can be transformed into a stable normal structure by annealing.

The interface between different grains in the polycrystalline film is called the grain boundary (or intergrain boundary). The atomic arrangement state of the grain boundary is actually the intermediate structure from the atomic arrangement state in one grain to the atomic arrangement state in the other grain. Therefore, the grain boundary is an incomplete structure of "plane shape". Thus, a series of characteristics different from those inside the grain are shown.

(1) Due to the large lattice distortion in the grain boundary, the average energy of the atoms on the grain boundary is higher than the average energy of the atoms inside the grain, and their difference is called the grain boundary energy. High grain boundary energy indicates that it has a tendency to transform to the lower energy state spontaneously. Both grain growth and grain boundary flattening can reduce the grain boundary area and thus the grain boundary energy. So, as long as the atom has enough kinetic energy, the grain will grow and the grain boundary will be straight when it migrates.
(2) Due to the irregular arrangement of atoms in the grain boundary, there are more vacancies. When there are trace impurities in the grain, the free energy of the system increases less than the free energy of the grain when it fills the vacancy in the grain boundary. Therefore, trace impurity atoms are often concentrated at

grain boundaries, and it is much easier for impurity atoms to spread along grain boundaries than through grains.

The atom diffusion at the grain boundary in polycrystalline thin films is an important issue in thin film physics, including the establishment of analytical mathematical models and experiment.

6.1.1.3 *The fiber structure*

Fiber structure films are films with a preferred grain orientation. According to the orientation direction, they can be divided into single fiber structure and double fiber structure. The former has the preferred orientation of each grain in only one direction, while the latter has the preferred orientation in two directions. Sometimes the former is called the one-dimensional film and the latter the two-dimensional film. One-dimensional films may have the characteristics of two-dimensional homogeneity and one-dimensional anisotropy. The structure of a two-dimensional oriented film is similar to that of a single crystal, and It has the properties of a single crystal. The characteristics of several piezoelectric films with a preferentially oriented fiber structure are shown in Table 6.1.

Most polycrystalline films tend to show preferential orientation on amorphous substrates. Because the surface of [111] has the

Table 6.1: Characteristics of piezoelectric films with fiber structure.

Material	Substrate	Crystalline morphology	[0001] preferred orientation
ZnO	Glass	Hexagonal	Perpendicular to the film surface
	Glass	Hexagonal	Inclined to the film surface
	Glass	Hexagonal	Parallel to the film surface
CdS	Mica, rock salt	Hexagonal	[111] Perpendicular to the film surface
	Mica, rock salt	Hexagonal	Perpendicular to the film surface
	Glass	Hexagonal	Perpendicular to the film surface
	Glass	Hexagonal	Perpendicular to the film surface
ZnS	Glass	Hexagonal	The angle is 40° with the film surface
	Glass	Hexagonal	[111] Perpendicular to the film surface
AlN	Glass	Hexagonal	Perpendicular to the film surface
	Glass	Hexagonal	Parallel to the film surface

lowest surface free energy in the face-centered cubic structure, the preferred orientation of the polycrystalline film of this structure on the amorphous substrate is [111]. Therefore, it can be expected that the hexagonal densely packed polycrystalline film on the amorphous substrate shows the preferred orientation [0001], and the polycrystalline film of the body-centered cubic structure shows the preferred orientation [110].

A ZnO piezoelectric film on a glass substrate is a typical example of fiber structure films. The excellent piezoelectric properties of the films are related to the same orientation of the tiny grains. The *c*-axis of each microcrystal belonging to the hexagonal system in the film is perpendicular to the surface of the film or substrate.

The preferential orientation of grains in films can occur at various stages of film growth: initial nucleation stage, island coalescence stage, and final stage. If the adsorbed atoms have a higher diffusion rate on the substrate surface, the preferential orientation of the grains may occur in the initial stage of film formation. The arrangement of atoms in the initial layer depends on the substrate surface, substrate temperature, crystal structure, atomic radius, and melting point of the thin film material. If the surface diffusion rate of the adsorbed atom is small, the initial membrane layer will not generate preferential orientation. When the film layer is thicker, it forms a strong orientation towards the direction of the evaporation source. The inclination of the grain toward evaporation source depends on the substrate temperature, the incident angle of gas phase atom, and the deposition rate.

6.1.1.4 *The single crystal structure*

Films with a single crystal structure are prepared by the epitaxial process. The first basic condition for epitaxial growth is that the adsorbed atom must have a high surface diffusion rate, so substrate temperature and deposition rate are very important. At a certain evaporation rate, the lowest epitaxial growth temperature (epitaxial growth temperature) exists between most substrates and thin films. The second basic condition is the crystal compatibility between

Table 6.2: The growth of epitaxial films on various substrates.

Thin film	Substrate	m (%)	Epitaxial temperature (°C)	Thin film	Substrate	m (%)	Epitaxial temperature (°C)
Au	NaCl(100) A	-28	400	GaAs	GaAs	0	700–850
	NaCl(100) V	−28	200	Cu	NaCl(100)A	−36	300
	mica	−44	400		NaCl(100)V	−36	100
	MoS_2	+9	180		Ag(100) F		0
	Ag(100) F		−100	Ag	LiF		340
	MgO		−190		NaCl(100)A	−31	150
Ge	CaF		225		NaCl(100) V	−31	Indoor temperature
Si	Si(100)	0	Indoor temperature		KCl		130
		0	380		KI		80
	Si(11)	0	450		Ag(100 F)		-196
		0	550				

the substrate and the film material. If the lattice constant of the substrate is a, the lattice constant of the film is b, and the lattice mismatch number $m = (b - a)/a$, then the smaller the value of m, the easier it is to achieve epitaxial growth in general. The third condition requires the substrate surface to be clean, smooth, and chemically stable. Table 6.2 lists the specific physical parameters of epitaxial growth of single crystal films on various substrates, where A represents cleavage in air, V represents cleavage in vacuum, and F represents thin films.

In most cases, the crystal structure of the grain in the film is the same as that of the block crystal. Only the grain orientation and grain size are different from those of block crystals. In addition to the crystal type, the lattice constants of grains in the film are often different from those of block crystals. There are two reasons for this phenomenon: one is that the lattice constant of the film material itself does not match the lattice constant of the substrate material; second, the film has a greater internal stress and surface tension. Because of the mismatched lattice constants, the crystal lattice at the interface of the thin film and the substrate is distorted to form a new crystal lattice to match the substrate. If the binding energy between the film and the substrate is large, and the percentage difference between the lattice constant $(\alpha_f - \alpha_s)/\alpha_f$ is approximately equal to 2% (α_f and α_s represent the lattice constants of the film material and the substrate respectively, and $(\alpha_f - \alpha_s)/\alpha_f$ is called lattice mismatch), the thickness of the lattice distortion layer at the interface between the film and the substrate is several angstroms. When the mismatch is about 40%, it can reach several hundred angstroms. When the mismatch is more than 12%, the lattice distortion is completely mismatched.

The change of membrane lattice constant due to surface tension can be expressed by the following theoretical calculation. Suppose there is a semispherical grain on the substrate surface with a radius of r and the surface free energy per unit area is σ_0. Due to surface tension, the pressure on this grain is $f = 2\pi r \cdot \sigma$, and the area under this force is $S = \pi r^2$, so the pressure intensity is

$$P = 2\pi r \cdot \sigma / \pi r^2 = 2\sigma / r, \tag{6.1}$$

by Hooke's law

$$\frac{\Delta V}{V} = 3\frac{\Delta a}{a} = -\frac{1}{E_v} \cdot P, \tag{6.2}$$

the change ratio of lattice constant can be obtained as

$$\frac{\Delta a}{a} = -\frac{2\sigma}{3 \cdot E_v \cdot r}, \tag{6.3}$$

where E_v is the elastic coefficient of the film and a is the lattice constant.

It can be seen from the above equation that the change ratio of lattice constant (strain) is inversely proportional to the grain radius; that is, the smaller the grain, the greater the change of lattice constant. This clearly indicates that the lattice constant of the grain in the film is different from that of the grain in the block material.

6.1.2 *The surface structure*

From thermodynamics, it is known that in order to achieve the lowest total energy, the thin film should have the smallest surface area;

that is, it should be in the ideal plane state. In fact, this kind of film is not available. In the process of film deposition, the incident gas phase atoms on the substrate surface are irregular, so the film surface has a certain roughness. Assume that the incident gas phase atoms deposit on the substrate and remain in place. The thickness of the resulting film is uneven throughout the substrate surface. If the average thickness of the film is d, it is distributed according to the Poisson's ratio without regular changes, and the average deviation value $\Delta d = \sqrt{d}$ of the film thickness can be obtained. The surface area of the film increases with the square root of its thickness. But in fact, once the incident gas phase atoms are deposited on the substrate, the energy released for re-evaporation is absorbed onto the substrate surface. And then by the spreading of energy across the surface, taking up some of the vacancies on the surface, there is adsorption of atoms on the surface of diffusion, so that the trough on the surface of the film is filled, the peak is cut, resulting in that the film surface area shrinks, and the surface energy is gradually reduced. In addition, some low-energy crystal faces (low-exponential crystal faces) can be developed due to the diffusion of adsorbed atoms on the surface. However, under the action of surface atomic diffusion, the fastest growing crystal faces can consume these slower growing crystal faces, causing the film to show a further increase in roughness. Such a structure often occurs at a higher substrate temperature.

When the substrate temperature is low, the energy of adsorbed atoms is small and the surface area is large. The surface area of the film increases linearly with the thickness of the film. It indicates that the membrane is porous. This micropore has a large internal surface area and can extend to the lowest layer. This situation is consistent with the cylindrical structure in the microstructure.

If the vacuum degree is low in the deposition of the film, due to the high pressure of the residual gas, the incident gas phase atoms collide with the residual gas molecules, then it condenses into smoke in the gas phase and then reaches the surface of the substrate and deposits into the film. So this membrane is also porous. This porous structure is easy to appear in the case of low substrate temperature.

6.1.3 *The factors affecting the membrane structure*

The structure of thin films is affected by evaporation conditions, which mainly include evaporation rate, substrate temperature, atomic incidence direction, substrate surface state, and vacuum degree.

For different types of metals, the degree of influence of evaporation rate is different, because the mobility of vacuum deposited metal atoms on the substrate is related to the nature of the metal and the surface condition. Even for the same metal, the effect of evaporation rate on film structure is not completely consistent under different process conditions. In general, evaporation rate will affect grain size, grain distribution uniformity, and defects in the film.

Under the condition of low evaporation rate, the metal atom migrates on the substrate for a long time, and it is easy to reach the adsorption point, or it is captured by small islands at other adsorption points and forms coarse grains, which makes the structure of the film rough and the film is not dense. At the same time, after the evaporating atoms reach the substrate, the subsequent atoms have not arrived in time, so the exposure time is relatively long, and it is easy to get contaminated by residual gas molecules or impurities introduced in the evaporation process, as well as various defects. Therefore, higher evaporation rate is better. High evaporation rate can make the film grain have a small, dense structure, but because of the condensation of the core at the same time, the cores are in the higher state of energy, so the film has a larger internal stress and more defects.

The low evaporation rate makes the membrane structure loose, the electron's ability to cross its potential barrier to generate conductance is weak, coupled with oxidation and adsorption, so the resistance value is high, and the resistance temperature coefficient is small, even negative. With the increase of evaporation rate, the value of resistance increases from large to small, while the temperature coefficient increases from small to large, from negative to positive. This is due to the low evaporation rate of the film having oxidation and semiconductor characteristics, so the temperature coefficient appears negative; The high evaporation rate film tends to metal characteristics, so the temperature coefficient is positive. Therefore, in general, a higher evaporation rate is also desired. But for a specific material, the best evaporation rate should be selected from specific experiments.

The substrate temperature also has a great influence on the structure of the film. Because of the high temperature of the substrate, increasing the kinetic energy of adsorbed atoms and the probability of crossing the surface barrier, the film is easy to crystallize and the defects of the film are fewer, as well as the internal stress of the film is lower. The substrate temperature is low, it is easy to form amorphous structure of the film. The choice of substrate temperature depends on the specific situation. Generally speaking, if the evaporation of the film layer is relatively thin, when the substrate temperature is relatively low, the metal atoms in the evaporation chamber quickly lose kinetic energy, and on condensation on the surface of the substrate, at this time the film layer is relatively uniform and dense; when the substrate temperature is too high, there will be large grains, making the film surface rough. If we evaporate the thicker film, the general requirement of the substrate temperature is appropriately higher, which can reduce the internal stress of the film.

The film structure is also related to the incident direction of the evaporated atom. The size of the incident angle of the atom has a great influence on the structure of the film. With the increase of crystal particles, the incident evaporated atoms gradually grow along the incident direction of the atoms, which will produce the so-called self-influence effect, resulting in uneven surface of the film, more defects, and anisotropy. This tendency to grow along the incident direction of the atom is more serious when the evaporation angle is larger.

The structure and properties of evaporative deposited films can be improved by heat treatment. For the single-metal film, the lattice arrangement is more orderly after heat treatment. For alloy materials, after heat treatment, the alloy components will be diffused with each other to obtain the required solid solution. Heat treatment can partially eliminate lattice defects, improve the thermal stability of the film, remove internal stress, enhance the adhesion between the film and the substrate, and also eliminate the adsorption of gas molecules in the film layer, forming a layer of oxidation protection layer on the surface of the film, so as to protect the film layer from erosion and pollution.

The surface state of the substrate also has a great influence on the structural quality of the film. If the substrate is smooth and clean, the obtained film layer structure will be dense and easy to crystallize, otherwise it will be the opposite and has poor adhesion.

The vacuum degree also directly affects the structure and performance of the film. The vacuum degree is low, the material is seriously polluted by residual gas molecules, and the performance of the film is poor. Even in the case of high vacuum, it is inevitable to adsorb gas molecules in the film, but increasing the temperature facilitates the desorption of the gas.

6.2 Defects of Film

There are more defects in thin films than in block materials, and these defects directly affect the properties of thin films. There are four kinds of defects in thin films: point defect, dislocation, grain boundary and stacking fault.

6.2.1 *Point defect*

In the process of rapid cooling, there will be many spot defects in block metal materials, and many point defects will also be generated in the process of evaporation, condensation and other drastic changes in the process of film formation. For example, because of the rapid condensation of the metal film, the deposited atomic layer is covered by a new layer before it can reach thermal equilibrium with the substrate, resulting in many atomic vacancies in the film. Although this kind of point defect can't be directly observed by an electron microscope, it has an effect on the properties of the film.

There are many kinds of point defects, and the main consideration in thin films is the atomic vacancy. Taking the one-valence metal with a face-centered cubic structure as an example, the formation energy of atomic vacancy in Cu, Ag, and Au is about 1 eV, while the energy required for the interstitial atom of Cu to enter the center of the unit cell is about 2.5–3.3 eV. It can be seen that the energy of forming an atomic vacancy is lower than that of forming other defects, so the possibility of producing atomic vacancy is the greatest.

The vacancy effect in the crystal is shown in the macroscopic table of the volume and density of the present crystal. One vacancy reduces the volume of the crystal by about half the volume of the atom. There is usually the concentration of vacancies higher than the equilibrium concentration in the evaporation film, so the density of the film material is less than the density of the block material. As time goes on, the vacancy concentration decreases because of the diffusion, so the thickness of the film also decreases.

When the concentration of point defects in the film is higher than the equilibrium concentration, it will gradually disappear through diffusion and so on, so that the resistivity is irreversibly reduced, and the degree of reduction is more significant with the increase of heat treatment temperature. Point defects in thin films are often studied by measuring resistivity over time. From the change of resistivity with time, the expression for defect concentration distribution in the initial stage of the film is obtained.

At a certain time t, the formation energy of the defect is between E and $E+dE$, the defect concentration is $N(E,t)$, and the resistivity p_1 related to N will disappear with the passage of time. The resistivity that does not change with time is ρ_0, so the electricity of the film after evaporation is

$$\rho_t = \rho_i + \rho_0. \tag{6.4}$$

By measuring ρ_i and ρ_0, we can determine ρ_t.

Assume that the resistivity caused by a defect with the formation energy of E is $r(E)$, then the resistivity caused by the defect concentration $N(E,t)$ that varies with time is

$$\rho_i(t) = \int_0^{\infty} r(E)N(E,t)dE \tag{6.5}$$

At T, the change of $N(E,t)$ with time is expressed as the change in ρ_i with time, so

$$\frac{dN(E,t)}{dt} = -CN(E,t)e^{-E/(kT)} \tag{6.6}$$

where C is a constant related to lattice vibration frequency.

In the experiment, if the T is unchanged, defects will take a long time to disappear, so ρ_0 cannot be determined quickly. Therefore, the method of heating is often adopted, $t = \alpha \cdot (T - T_0)$ is used to heat the film at a certain heating rate. α is the time required for every 1°C rise in temperature, t is the heating time, and T_0 is the temperature when $t = 0$. Substitute this relation into Equation (4.42), of Chapter 4, integrate and calculate it, and finally get

$$r(E)N(E,T_0) = -\frac{1}{kU}\frac{d\rho_i}{dT} \tag{6.7}$$

$U = u\frac{u+2}{u+1}$, $u = \frac{E}{kT}$. Once T, U and $d\rho_i/dT$ have been determined, you just need to know $r(E)$, and the distribution of defect concentration can be calculated. For example, in metals such as Ag, the resistivity caused by a 1% atomic vacancy is known to be about 1–2 $\times$ $10^{-9}\Omega \cdot$cm. According to this, the relation curve between N and E can be made. The peak value of this curve appears near 1.2 eV, indicating that the defect is mainly atomic vacancy.

6.2.2 *Dislocation*

Dislocation is one of the most frequently encountered defects in thin films. It is a kind of "linear" incomplete structure in the lattice with a density of about 10^{12}–10^{13}/cm^2. In massive high-quality crystals, the dislocation density is about 10^4–10^6/cm^2.

Most of the dislocations in the film extend from the surface of the film to the surface of the substrate and produce distortion around the dislocations. There are two main causes of dislocation in thin films: dislocation caused by the substrate and dislocation caused by the aggregation of small islands.

If there is a lattice mismatch dislocation between the film and the substrate, there will be a bit of a fault when growing into a single crystal layer. If there is an error on the substrate, the film formed on the substrate will cause dislocation defects due to the dislocation of the substrate.

However, in general, the dislocation density of the substrate is very small; for example, there are only about $10^5/\text{cm}^2$ dislocations on the cleavage surface of NaCl.

The main cause of dislocation in thin film is the growth and aggregation of small islands. In most small islands the direction of crystallization is arbitrary. In particular, when two small islands with slightly different crystal directions aggregate and grow together, they will form a small inclination grain boundary in the form of dislocation. The relationship between dislocation density and film thickness in the Au film is shown in Figure 6.4 when the substrate temperature is 300°C on MoS_2 substrate. As can be seen from Figure 6.4, many dislocations occur during island coalescence and subsequent growth. In addition, when the islands have just coalesced, there is considerable stress in the membrane. Sometimes the stress is concentrated in the place where the gap is formed during the island coalescence and makes the dislocations. This phenomenon can be used to study the properties of dislocation by measuring the internal friction.

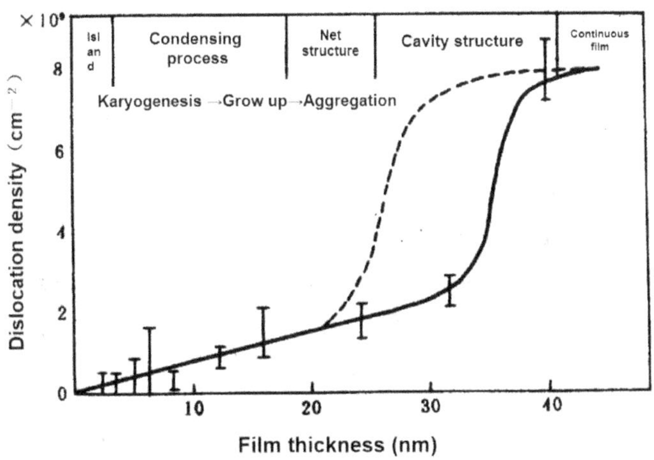

Figure 6.4: Relation between dislocation density and film thickness in Au films.

The internal friction of the film is that if the film is allowed to vibrate at random, its amplitude will be attenuated at a certain ratio, which is mainly due to the vibration energy converted into heat energy through the dislocation movement in the film, and this energy conversion process is called internal friction of the film.

For the block face-centered cubic metal, when it is strongly processed and the change of internal friction with the temperature is measured by reducing the temperature, there will be a so-called Bordoni peak, whose existence indicates the movement process of dislocation. When the temperature goes up to a higher temperature this peak disappears. But the dislocation in the film is more stable than in the block material. Figure 6.5 shows the relationship between internal friction and temperature in polycrystalline Ag films. When the polycrystalline Ag film evaporated at normal temperature was reduced to measure internal friction at low temperature, it, like the processed block material, showed the peak corresponding to the Bordoni peak at 50 K. However, when the film was heated at 300 K for three hours and measured again, it was found that the peak did not change. However, if the same condition is applied to the block

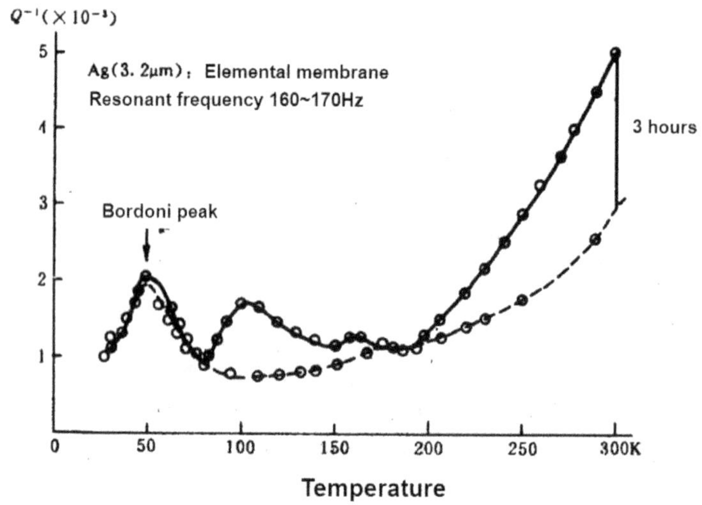

Figure 6.5: Relation between internal friction and temperature in polycrystalline Ag films.

material, this peak basically disappears. It can be concluded that the dislocation in the film is quite stable compared with that in the block material. A large number of experimental data also show that most of the dislocations in the film appear in the direction perpendicular to the surface of the film and are locked as stable dislocations.

6.2.3 *Grain boundaries*

Because there are many small grains in the film, the area between the grains of the film is larger than that of the block material. When the surface diffusivity of the adsorbed atom is very small, there is no significant difference between the grain size and the critical size. In general, however, the surface diffusivity of adsorbed atoms is relatively large, so the grain size is much larger than the critical size when the small islands are large enough to contact and coalesce. The relationship between grain size and deposition process conditions in thin films is shown in Figure 6.6.

It can be seen from Figure 6.6 that a common feature is that grain size changes with the change of deposition process parameters, and finally tends to a saturated grain size. When the film thickness is taken as the deposition variable, the grain size reaches saturation, which means that after the film thickness reaches a certain value, a new grain is produced on the original grain. There may be two reasons for the formation of a new grain: first, there is a pollution layer on the old grain surface, so that the new grain does not grow continuously with the original grain; second, the upper surface of the original grain has become a nearly perfect closed stacking surface, and it is difficult for the newly incident gas phase atoms to enter it again, so they can only be rearranged to form the new grain.

The particle size can also be changed by annealing at a temperature higher than the deposition temperature. The higher the annealing temperature, the larger the grain size. This effect is more obvious when the thick film is heat treated. It should be pointed out that the effect of grain increase is not significant when the film is deposited at annealing temperature. The reason is that the condensation process of adsorption atom with surface diffusion in

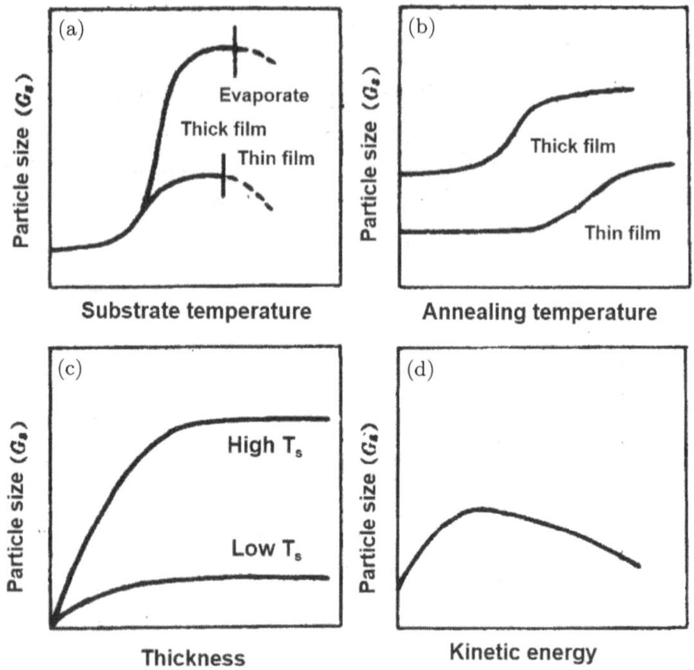

Figure 6.6: Relationship between grain size and process conditions.

deposition and the thermal diffusion process of condensation atom with high activation energy in annealing cancel each other.

However, the relationship between grain size and deposition rate is not as obvious as the above factors. When the deposition rate is high, although the gas phase atoms incident on the surface have greater surface diffusion motion ability, they are buried by the later incident gas phase atoms before they can spread far. When this happens, the deposition rate must exceed a critical value. When below this critical rate, the grain size is only limited by the substrate temperature. When above this critical rate, the grain size decreases with the increase of deposition rate.

6.2.4 *Stacking fault*

Another important defect in vacuum evaporation coating is stacking fault. It's caused by a misarrangement of atoms. In the complete face-centered cubic crystal, it should be piled up in the order of ABC, repeatedly and repeatedly in every three layers, that is, ABC ABC .... If the A layer is missing in the arrangement of atoms (such as the fourth layer A), the stacking relationship becomes ABC BC ABC .... Thus, stacking fault is generated. An example of this defect is still the observation of Au films deposited on MoS_2 substrates. When the coalescence and coalescence of two small islands in the thin film were studied by an electron microscope, a strong diffraction contrast could be observed at the boundary of the crystal lattice diffraction fringe of each island as soon as the islands were coalesced. This reflects the existence of stacking defect. Delamination defects occur at the boundary of a small space. However, when the coalesced islands grow up, the diffraction contrast reflecting delamination defects will completely disappear. However, this does not mean that delamination defects will completely disappear when the film forms a continuous structure.

When the substrate temperature is 80 K, we evaporate an Ag film of thickness 450 nm; using the X-ray diffraction method for observational studies, we can know that there are about 6% stacking fault in [111] and [200] from the displacement of the diffraction peak. If from the whole of the film, the face-centered cubic metal layer stacking defects are different, and most are in the form of a zonal distribution, its density is 10^9–$10^{10}/cm^2$.

The stacking fault in Si and Ge is very different from that of face-centered cubic metal in appearance. Because they are closely related to epitaxial growth processes, we won't discuss them in detail here.

In addition, there are other types of defects in thin films in addition to the four typical defects described above, such as a dislocation ring, a twin crystal, and so on. A dislocation ring is a face-centered cubic metal film with a large defect energy in the lamination. A blade-shaped dislocation ring is defined as follows: the vacancies in the film agglomerate in to a single layer along a certain crystal plane, and when the number of vacancies increases, the upper and lower parts of the layer are destroyed, so a blade-shaped dislocation ring is formed outside the contour of the layer. The blade-like dislocation ring is equal to an empty absorption port, which can clear the empty port. There are dislocation rings in a face-centered cubic metal film, and a large number of [111] twin crystals are also found.

6.3 Analyzing Method of Thin Film Structure and Composition

6.3.1 *Overview*

Different research methods are often needed for different research objects, and so is the study of the structure or composition analysis of thin film materials. For the early optical coating film materials, the research mainly involves the thickness, uniformity, and optical properties of the film. With the development of electronic technology, research on thin films has been extended to various structural characteristics, composition distribution, interface properties, and optical properties. At present, with the diversified use of thin film materials, the research methods and objects have become more and more extensive. In particular, a series of new methods for analyzing the structure and composition of thin films have been developed on the basis of the utilization of various micro-physical phenomena, providing a realistic route for the further study of thin film materials. Table 6.3 lists some common thin-film research methods and the physical processes and energy ranges involved, according to the type of particles used.

The methods commonly used to detect the structure and composition of thin films include X-ray diffraction, electron diffraction, scanning electron display, micromirror, Auger electron spectroscopy, X-ray photoelectron spectroscopy, and secondary ion mass spectrometry.

6.3.2 *X-ray diffraction*

The theoretical basis of this method is ray crystallography. Elements, compounds, and mixtures can be verified and quantitatively analyzed by this method. It is used in phase analysis and phase boundary determination and related research; crystal structure analysis; single crystal orientation; lattice parameter measurement; measurement of grain size and lattice deformation; and the study of crystal defects.

Table 6.3: Physical analysis methods used in thin film technology.

Incident particle	Particle energy/keV	Produced particle	Analysis method	Abbreviation	Analysis purposes
Electron	0.02–0.2	Electron	Low Energy Electron Diffraction	LEED	Surface structure
	5–100	Electron	Reflective High Energy Electron Diffraction	RHEED	Surface structure
	0.3–30	Electron	Scanning Electron Microscopy	SEM	Surface morphology
	1–30	X-ray	x-ray Electron Energy	EDX	Surface composition
	0.5–10	Electron	Auger Electron Spectroscopy	AES	Surface composition
	100–400	Electron	Transmission Electron Microscopy	TEM	Structural analysis
	100–400	Electron	Scanning Transmission Electron Microscopy	STEM	Structure and composition
	100–400	Electron	Electron Energy Loss Spectrum	EELS	Regional composition
Ion	0.5–2.0	Ion	Ion Scattering Spectrum	ISS	Surface composition
	1–15	Ion	Secondary Ion Mass Spectrometry	SIMS	Microelement depth
	>1	X-ray	Ion-induced X-ray Emission Spectrum	PIXE	Microelement
	5–20	Electron	Scanning Ion Microscopy	SIM	Surface structure
	>1000	Ion	Rutherford Backscattering	RBS	Component depth profile
Photon	>1	X-ray	X-ray Fluorescence Analysis	XRF	Composition
	>1	X-ray	X-ray Diffraction	XRD	Crystal structure
	>1	Electron	X-ray Photoelectron Spectroscopy	XPS	Surface composition
	Laser	Ion	Laser Probe	—	Selection composition
	Laser	Photon	Laser Induced Light Emission Probe	LEM	Microelement analysis

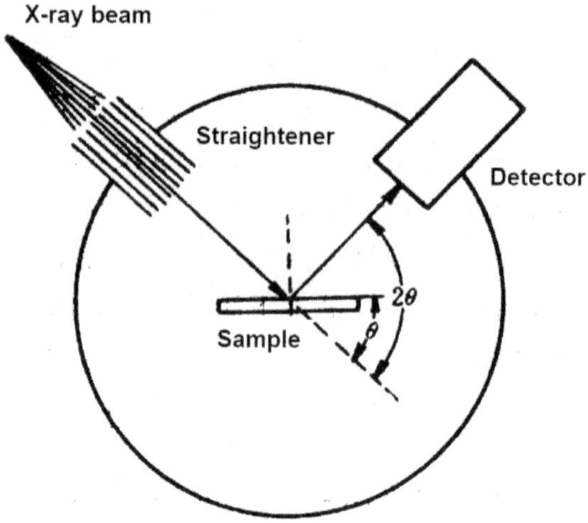

Figure 6.7: Schematic diagram of a X-ray diffractometer with the rotating crystal method.

Depending on the sample, the X-ray diffraction method can be divided into Laue method (suitable for single crystal), rotating crystal method (powder material and thin film) and powder method (powder material). The rotating crystal method is suitable for the analysis of thin film samples. Its working principle is shown in Figure 6.7.

The ray beam in Figure 6.7 is a characteristic ray with a certain wavelength emitted from a special X-ray tube. The commonly used characteristic rays are K_α ray of Al (8.34 Å), K_α ray of Cu (1.542 Å), K_α ray of Cr (2.29 Å) and K_α ray of Fe (1.94 Å). After the X-ray beam is incident on the surface of the analysis sample, reflection occurs and the detector collects reflection X-ray information. When the wavelength of the incident X-ray λ, the angle θ between the sample and the X-ray beam, and the spacing of the sample crystal plane d satisfy the Bragg equation, the detector can detect the strongest information. Since λ is known, θ can be measured, and the above equation can be used to calculate the lattice surface spacing d, which can be used in various analysis and research methods.

$$2d \cdot \sin\theta = n \cdot \lambda. \tag{6.8}$$

Because the characteristic X-ray cannot be focused by electromagnetic lens, its beam spot size is large. In addition, when the ray analyzes the sample, it is weakly scattered by electrons from the atomic shell layer, so it has a strong penetration ability. This method is suitable for analyzing films with larger grain sizes and thicker film layers.

6.3.3 *Electron diffraction*

Electron diffraction has more advantages than ray diffraction in studying and analyzing thin films. Firstly, the structure of a thin film is observed and analyzed by electron diffraction in a transmission electron microscope. Secondly, the wavelength of the electron has the following relationship with the accelerating voltage V:

$$\lambda = \frac{12.27}{\sqrt{V(1 + 0.978 \times 10^{-6} \cdot \mathrm{V})}}. \tag{6.9}$$

When the acceleration voltage is greater than 1000 V, the wavelength of the electron beam is less than 0.37 Å, which is much smaller than the wavelength of the characteristic X-ray. Instead of using Bragg's equation, we use the following equation:

$$L \cdot \lambda = R \cdot d, \tag{6.10}$$

where L is the distance from the sample to the base plate of the placement photo and R is the radius from the center point of the diffraction pattern to the diffraction point, both of which can be measured. Since the electron wavelength λ is known, the spacing between the lattice faces can be calculated by $d = L \cdot \lambda / R$. Since the product of L and λ is determined by the instrument itself, $(L \cdot \lambda)$ is usually referred to as the instrument constant. Thirdly, the beam of electrons incident into the film is scattered by electrons around the atom, making it less permeable, so the method is suitable for the film and surface. Fourth, the beam can be focused by electromagnetic lens to make the beam on the sample smaller in diameter, up to several hundred Å. Therefore, this method is suitable for the analysis of films with smaller grain size and thinner film thickness. Low energy electron diffraction (LEED) and high energy electron diffraction (HEED) are commonly used.

6.3.4 *Scanning electron microscopy analysis*

A scanning electron microscope (SEM) is a large-scale electron optical instrument for the analysis of film morphology and composition. The working principle is shown in Figure 6.8. The part from the electron gun to the sample is called the electron optical system or the lens barrel; the vacuum system diagram connected with it is not drawn; the detection and display system is composed of the operator, amplifier, and oscilloscope; the fourth part is the power supply and operation control system. From Figure 6.8(b), it can be seen that the interaction between the incident electron beam and the sample surface can produce many kinds of information. Among them, backscattered electrons, secondary electrons, and characteristic X-rays are often used for thin film analysis. The first two kinds of information can be used to observe the surface morphology, especially the secondary electrons, which are from the sample itself and have a small kinetic energy, can best reflect the surface morphology information of the sample, and are generally used to observe the

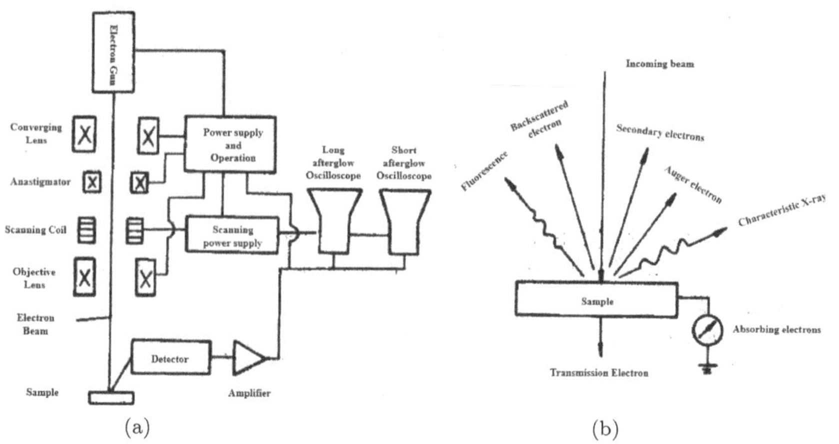

Figure 6.8: Working principle of a scanning electron microscope: (a) instrument structure; (b) information generated by the interaction of the incident electron beam with the sample.

sample morphology. Characteristic X-rays can be used to analyze the chemical components of the samples.

The main advantages of SEM are: high resolution, up to 50 Å; large depth of field, up to 7.5 μm, so it has a strong sense of three dimensions when observing the surface; the magnification can be adjusted continuously from tens to tens of thousands of times; in combination with the X-ray chromatograph or the X-ray energy spectrometer, it can analyze the distribution of point, line, and surface components while analyzing the morphology; it can do a dynamic analysis of fracture, heating, cooling, tension, voltage and so on; a sample frame with simple sample preparation and partial freedom can analyze different parts of the sample.

In the scanning electron microscope, the characteristic X-ray chromatograph or X-ray energy spectrometer emitted by the sample can be used for chemical composition analysis, and its working principle is shown in Figure 6.9.

The theoretical basis of chemical composition analysis by characteristic X-ray is Moseley's law, which reflects the direct relationship between characteristic X-ray wavelength and atomic number Z

$$\sqrt{\frac{c}{\lambda}} = K(Z - \sigma), \tag{6.11}$$

where C is the speed of light and K and Σ are constants. It can be seen from the above formula that the atomic number Z can be

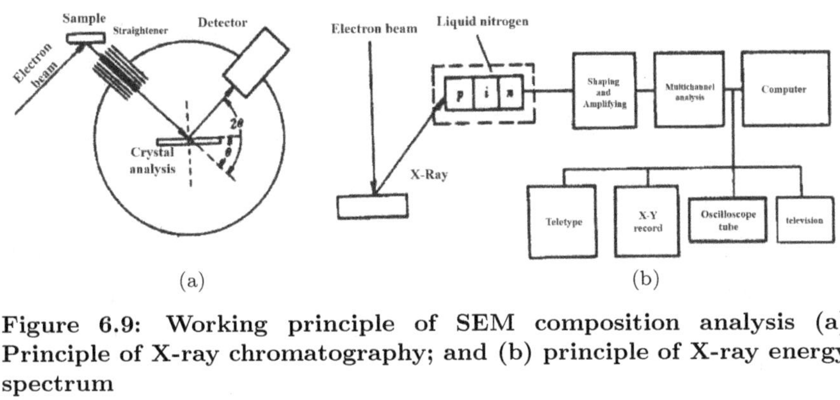

(a) (b)

Figure 6.9: Working principle of SEM composition analysis (a) Principle of X-ray chromatography; and (b) principle of X-ray energy spectrum

obtained by measuring the wavelength of the characteristic X-ray, so as to determine its chemical composition.

When the characteristic X-ray wavelength is measured by X-ray chromatography, the Bragg equation $2d \sin\theta = n \cdot \lambda$ is still used. As can be seen in Figure 6.9(a), the d value of the spectroscopic crystal is known, and the angle can be measured through the dial, so the characteristic X-ray wavelength λ can be obtained. The detector converts the characteristic X-ray information of different wavelengths received into electric pulse and transmits it to the display recording system, so as to obtain the sample component information.

Because the characteristic X-ray has not only the characteristics of light wave but also the characteristics of photon energy, it is called the X-ray energy spectrum analysis technology to use the compliance relationship between photon energy and wavelength $E = h \cdot c/\lambda$, based on the characteristic X-ray photon energy, and then determine the atomic number to obtain the sample component information by using the Moseley law. The pin diode in Figure 6.9(b) is a counter whose output pulse is highly proportional to the incident X-ray photon energy. When these pulses of different heights are analyzed and identified by the multi-channel pulse height analyzer, the characteristic X-ray energy of each element in the representative sample is expanded in turn, so as to obtain the content of each component. Because the excitation probability of the same element in the sample is equal, the number of photons output by the element with higher content is also higher. Record the number of pulses of the same height to obtain the content of this element in the sample. Compared with X-ray chromatography, it has the advantages of fast analysis speed, high sensitivity, good stability, full spectrum analysis at one time, and so on.

6.3.5 *Auger electron spectroscopy*

The Auger electron was discovered by French scientist Auger in 1926. Auger electron spectroscopy (AES) is one of the advanced analytical techniques for the analysis of thin film samples. Its characteristics are: suitable for the analysis of light and ultra-light elements with atomic number less than 13; small beam spot of the incident electron beam, suitable for the analysis of micro area components; with the help of a scanning electron microscope, Auger electron spectroscopy analysis can be carried out at the same time as analyzing the sample morphology; according to the chemical displacement effect, the chemical state of elements can be identified; with the help of a sputtering ion gun, the sample can be sputtered and peeled at one side. The information on the distribution of elements along the depth can be obtained by analyzing the elements from one side; because the kinetic energy of Auger electrons is small, all the information obtained comes from the near surface region, which is only a few nanometers away from the surface.

Figure 6.10 is a schematic diagram illustrating the generation process and analysis principle of Auger electrons with Si as an example. When the atom is in the ground state, the electrons outside the nucleus are arranged on the energy levels K, L, and M with a certain energy. Under the action of the incident electron, an electron in the K level is excited and ionized, leaving a vacancy, while the

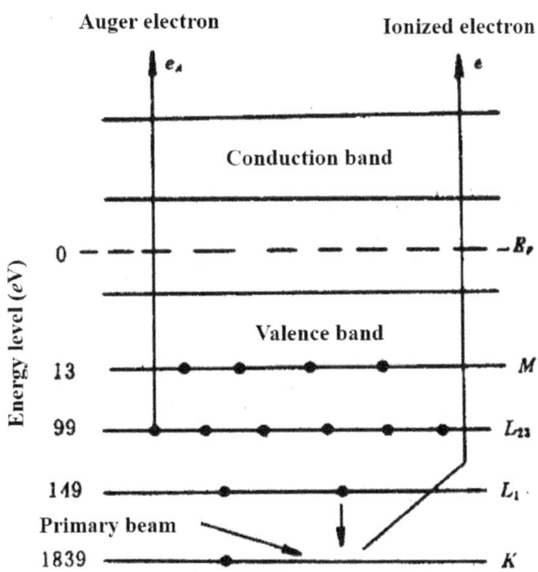

Figure 6.10: Generation process of Auger electrons in Si.

atom is in an unstable excited state. To reach a stable state, assume that an electron in the L_1 layer transitions to the K layer to fill the vacancy, then it will release extra energy $(E_K - E_{L_1})$. This part of the energy can be consumed in two ways: one is to emit characteristic X-ray photons; the other is to excite and ionize the electron on the L_{23} layer, which is the Auger electron. If the kinetic energy of the Auger electron is E, the following relationship can be obtained:

$$E_K - E_{L_1} = E_{L_{23}} + E. \tag{6.12}$$

Auger electron kinetic energy

$$E = E_K - E_{L_1}\text{-}W_{L_{23}}. \tag{6.13}$$

If E_{L_1} is considered approximately equal to $E_{L_{23}}$ and expressed in terms of E_L, the kinetic energy of the Auger electron can be further expressed as

$$E = E_K - 2E_L. \tag{6.14}$$

For each element's atom, energies E_{L_1}, $E_{L_{23}}$ have their own characteristic values. Therefore, as long as the Auger electron kinetic energy E is measured, it can be used for element identification. Therefore, Equation (6.14) becomes the basic principle formula of Auger electron spectroscopy.

If the Si atom is not in the zero valence state Si^0, but in the oxidation state Si^{+4}, such as SiO_2, then the energy levels of K, L, M will change, which also affects the change in the energy E of the Auger electron. If the differentiation of E, K, L_1, L_{23} from Si to Si^{+4} is ΔX, ΔY, ΔZ and ΔE respectively, the kinetic energy of L_{23} Auger electron in Si^{+4} is

$$\begin{aligned} E + \Delta E &= (E_K + \Delta X) - (E_{L_1} + \Delta Y) - (E_{L_{23}} + \Delta Z) \\ &= E + (\Delta X - \Delta Y - \Delta Z). \end{aligned} \tag{6.15}$$

So,

$$\Delta E = \Delta X - \Delta Y - \Delta Z. \tag{6.16}$$

The variation ΔE of Auger electron kinetic energy E comes from the change of Si atom from the zero-valence atom state to

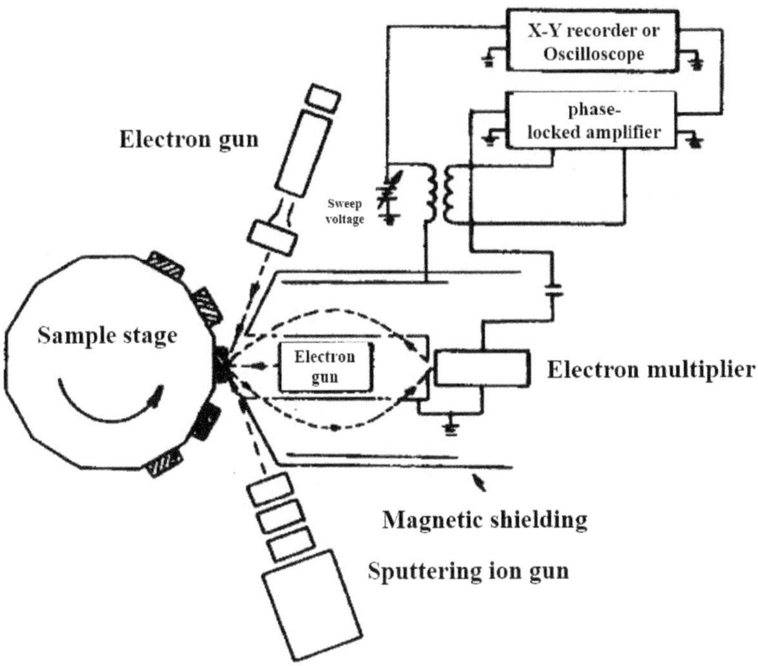

Figure 6.11: Structure principle of an Auger electron spectrometer.

the +4 valence ion state. This is caused by the change in chemical environment of the Si atom, so ΔE is called the energy chemical displacement, or chemical displacement for short. The chemical shift effect of Auger electron spectroscopy can not only identify the component elements in the sample, but also determine its chemical state. This is the unique advantage of Auger electron spectroscopy.

Figure 6.11 is a structure diagram of an Auger electron spectrometer, which is composed of electron gun, sputtering ion gun, electron energy analyzer, and vacuum system.

6.3.6 *X-ray photoelectron spectroscopy*

X-ray photoelectron spectroscopy (XPS) uses characteristic X-rays as a beam to excite and ionize the inner shell electrons after interacting with the atoms on the sample surface. The electron ionized by the incident characteristic X-ray is called a photoelectron. The kinetic energy of the photoelectron is measured to identify the elements and their chemical states in the sample. Therefore, this method is called X-ray photoelectron spectroscopy. This method was first used for chemical analysis. With the development of surface science, it is not only used in chemical analysis, but also widely used in surface science and material science. It is called X-ray photoelectron spectroscopy.

XPS analysis technology is characterized by the following: no damage to the sample; no sample consumption; identification of elements and their chemical state; shallow analysis depth, about 25–100 Å below the surface; analysis of electron density around atoms in molecules; wide analysis of elements, from He to U.

The working principle of X-ray photoelectron spectroscopy is shown in Figure 6.12. The incident X-ray photon energy $h\nu$ is known, the commonly used Mg target K_α X-ray photon energy $h\nu = 1254.6$ eV, and Al target K_α X-ray photon energy $h\nu = 1486$ eV. When the X-ray is incident on the inner shell of the free atom, an electron is excited and ionized into a photoelectron.

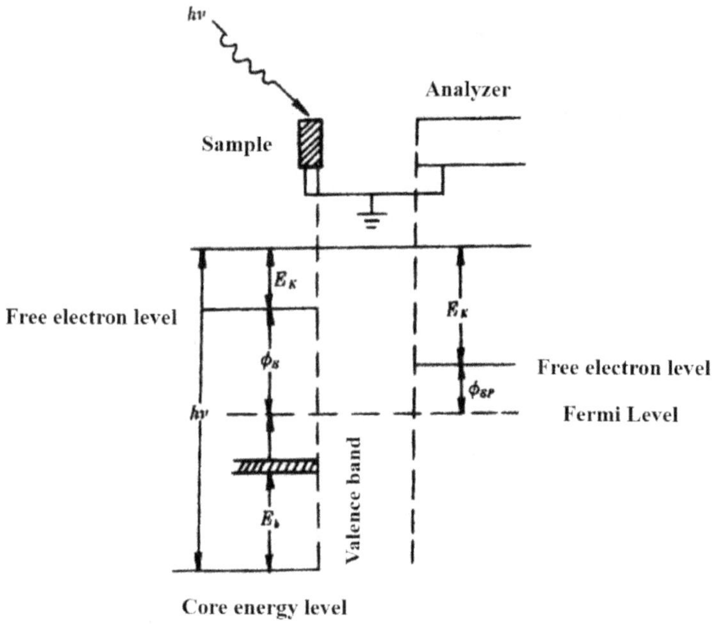

Figure 6.12: Operation principle of X-ray photoelectron spectroscopy.

If the binding energy of electrons is E_b and the kinetic energy after ionization is E_K, the famous Einstein photoemission law can be obtained according to the law of conservation of energy:

$$h\nu = E_b + E_K. \tag{6.17}$$

Since $h\nu$ is known, the photoelectron kinetic energy E_K can be measured by an electron energy analyzer, and then the photoelectron binding energy E_K can be obtained. For different atoms or different shells of the same atom, E_b has different numbers of eigenvalues. Element identification can be carried out by using the E_b value.

In addition to the binding energy E_b, the work function φ_s and atomic recoil energy E_r should also be considered for the photoelectron produced by solid samples such as thin films. If the kinetic energy of the photoelectron is expressed byE'_k, then

$$hv = E_b + \varphi_s + E_r + E'_k. \tag{6.18}$$

Because the E_r value is small and ignored, the above formula can be changed to

$$E'_k = hv - E_b - \varphi_s. \tag{6.19}$$

When the photoelectron enters the spectrometer, because the two Fermi levels are the same, there are

$$E'_k + \varphi_s = E_k + \varphi_{sp}. \tag{6.20}$$

So

$$E'_k = E_k + \varphi_{sp} - \varphi_s, \tag{6.21}$$

where φ_{sp} is the work function of the spectrometer material, and E_K is the kinetic energy of photoelectron in the spectrometer. Substituting the E'_k into Equation (6.19) we can get

$$E_b = hv - \varphi_{sp} - E_k. \tag{6.22}$$

Equation (6.22) is the basic theoretical formula for the measurement of solid samples. As introduced in AES, when the chemical environment of atoms changes, the binding energy E_b also changes. This change in binding energy E_b is also known as chemical shift.

Because the production of photoelectron in X-ray photoelectron spectroscopy only involves one electron and the production of the Auger electron involves three electrons, the chemical shift effect in X-ray photoelectron spectroscopy is more significant than that in Auger electron spectroscopy, so X-ray photoelectron spectroscopy is often used to study the chemical state of atoms in samples.

The X-ray photoelectron spectroscopy instrument is basically the same as the Auger electron spectroscopy instrument. As long as the electron gun is changed to the characteristic X-ray generator and the phase-locked amplifier is removed, most of it is an X-ray photoelectron spectroscopy instrument. At present, in addition to the single function X-ray photoelectron spectrometer, there is generally a multi-functional surface analysis instrument which combines the X-ray photoelectron spectrometer with the Auger electron spectrometer in the world. By using two functions of a device to analyze a sample at the same time, we can get various information.

6.3.7 *Secondary ion mass spectrometry*

Secondary ion mass spectrometry (SIMS) is a surface ion profiling technique. The incoming beam interacting with the surface is a kind of ion beam with high energy (primary beam). After energy and momentum exchange with the solid surface, it can emit neutral particles, positive and negative secondary ions (from the sample), electrons, photons, and elastic scattering incident ions from the surface. Secondary ion mass spectrometry is to use a mass spectrometer to identify surface elements by analyzing the quality of positive and negative secondary ions emitted from the sample surface. Because the secondary ions emitted are from the atoms below the surface, this analysis method is an advanced method to analyze the film and surface. Its information depth can be varied in one or several atomic layers, which mainly depends on the ion collision characteristics, incident ion beam energy, atomic number of incident ion and atom, atomic weight and cohesive energy of the sample, etc.

The advantages of secondary ion mass spectrometry are that it can detect all elements including hydrogen; it can analyze isotopes; it has a high detection sensitivity and can detect trace level impurities;

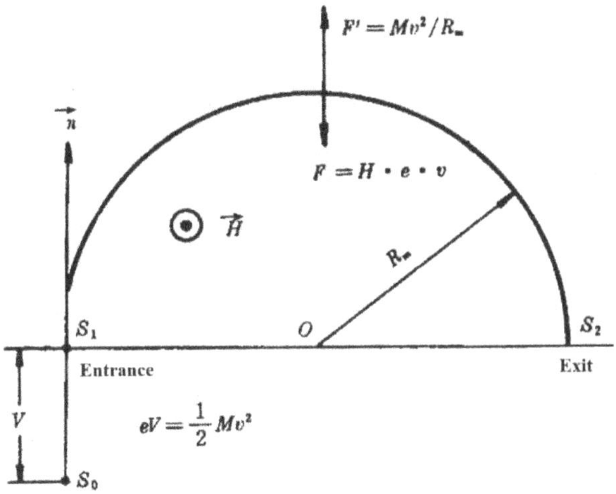

Figure 6.13: Schematic diagram of SIMS.

it is destructive to the sample but requires fewer samples; it is simple to prepare the sample and does not need pretreating the sample; and it can measure the composition distribution in the depth direction.

The working principle of SIMS is shown in Figure 6.13. Between two parallel planar electromagnets is the working area for mass analysis of secondary ions. It is assumed that the secondary positive ion with mass M and charge e is driven by accelerating voltage V from the sample S_0 to the parallel electromagnet entrance S_1. If the initial energy of this positive ion is ignored, the kinetic energy of its arrival at S_1 is

$$eV = \frac{1}{2}Mv^2, \tag{6.23}$$

where v is the velocity of the positive ion. If there is no external magnetic field ($\boldsymbol{H} = 0$), it enters S_1 and moves in a straight line in the vertical direction. When a magnetic field is produced in an electromagnet ($\boldsymbol{H} \neq 0$), it moves in a curve under the action of Lorentz force. When the centrifugal force F' is equal to the centripetal force F', the positive ions move in a circle. Centrifugal force $F' = M\nu^2/R_m$, centripetal force $F = H.e.v$, and R_m is the radius of the positive ion's circular motion track. Because $F' = F$:

$$\frac{Mv^2}{R_m} = H \cdot e \cdot v. \tag{6.24}$$

The mass to charge ratio of positive ions is

$$\frac{M}{e} = \frac{R_m^2 \cdot H^2}{2V}. \tag{6.25}$$

After unit conversion, there is

$$\frac{M}{e} = 4.82 \times 10^{-5} \frac{R_m^2 \cdot H^2}{V}, \tag{6.26}$$

where

$$R_m = \frac{144}{H}\sqrt{\frac{M \cdot V}{e}}, \tag{6.27}$$

where the unit of V is volt, the unit of H is Auster, the unit of R_m is centimeter, the unit of e is electronic charge, and the unit of M is atomic mass unit (amu). The relation is 1amu $= 1.6599 \times 10^{-24}\,g$.

Equations (6.26) and (6.27) are the basic formulas for the composition analysis of elements by the mass of elements in secondary ion mass spectrometry. It can be seen from the formula that when the accelerating voltage V and the external magnetic field H are determined, the orbital radius R_m of the positive ion motion is directly proportional to the square root of the positive ion mass $\sqrt{M/e}$ load ratio. If you adjust V and H make one $\sqrt{M/e}$ meet a certain R_m equal to OS_1 and OS_2, all positive ions entering from S_1 can come out from S_2 at the outlet. No ions smaller than or larger than $\sqrt{M/e}$ can come out from S_2. In this way, it can play the role of quality filtration, which is the basis of sample composition identification by using the quality difference of ions.

The structure principle of the secondary ion mass spectrometer is shown in Figure 6.14. The primary ion generated from the ion gun, after mass focusing and electromagnetic focusing, is incident to the sample system, which is called the primary ion optical system; the secondary ion is emitted from the sample, after mass spectrometry

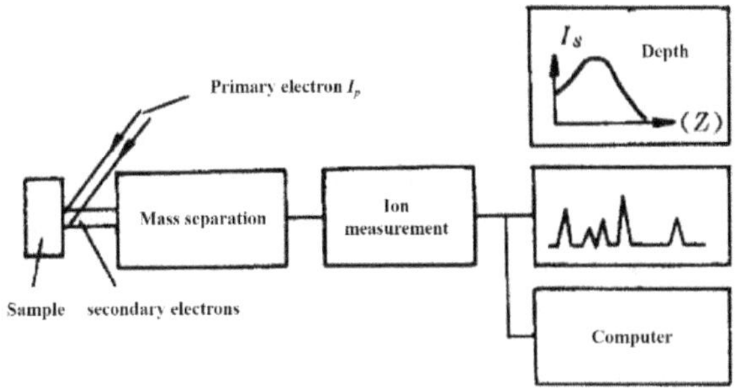

Figure 6.14: Structure principle of SIMS.

analysis and measurement, which is called the analysis and measurement system; the computer processing, recorder recording and spectrum display are called the operation display system; in addition, the vacuum system and power system and so on.

6.3.8 *Rutherford backscattering*

6.3.8.1 *Basic principle*

Rutherford backscattering (RBS) is a simple, quantitative, reliable and non-destructive method for the analysis of solid surface layers and thin films. The principle of Rutherford backscattering analysis is simple: a beam of MeV ions (usually ^{4}He ions) is incident on the target sample. An elastic collision occurs with the target atom (nucleus) (Figure 6.15(a)); some ions are backscattered. By measuring the energy of these backscattered ions with a semiconductor detector, the mass of the target atom and the depth of the colliding target atom in the sample can be determined. The concentration of the target atom can be determined from the scattering ion count.

In the elastic collision (Figure 6.15(b)), when the energy E_0 of the incident ion and its mass M_1 are fixed and the scattering angle is determined, the scattered ion energy E_1 is only related to the target atomic mass M_2 ($M_1 < M_2$):

$$E_1 = \left(\frac{M_1 \cos\vartheta + \sqrt{M_2^2 - M_1^2 \sin^2\vartheta}}{M_1 + M_2} \right)^2 E_0 = KE_0, \qquad (6.28)$$

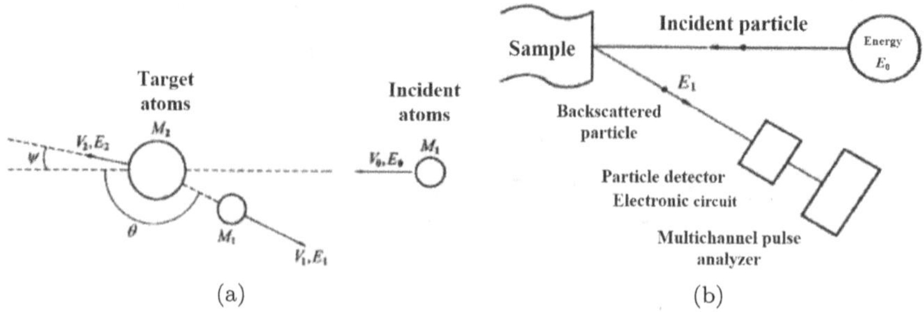

Figure 6.15: Schematic diagram of ion backscattering. (a) Elastic collision of ions with target atoms and (b) backscatter system.

where K is called the backscattering motion factor, which is only related to M_1, M_2, and θ. The heavier the target atom is, the bigger the quantity of scattering ion energy is. The larger the energy difference of scattered ions is, the easier it is to distinguish target impurity atoms with different masses.

6.3.8.2 *Analysis method*

(1) **Backscattering energy loss factor**

As the ions pass through the sample, they lose energy by inelastic collisions (ionization and excitation) with the target atom electrons. If the included angle between the incident beam and the scattered beam and the normal line of the sample surface is θ_1 and θ_2, and the energy of the incident ion on the target sample surface is E_0, before the collision at a certain depth t, the ion energy of the sample is E:

$$E = E_0 - \int_0^{t/\cos\theta_1} \left(\frac{dE}{dX}\right)_1 dx, \tag{6.29}$$

where $(dE/dX)_1$ is the stopping power of incident ions in the target material, which is a function of ion energy. At the depth of t, the energy of an ion with energy E becomes KE after backscattering with the target atom. These backscattered ions also suffer energy loss as they pass through the sample. The energy E_1 when penetrating the sample surface in the theta direction is

$$E_1 = KE - \int_0^{t/\cos\theta_2} \left(\frac{dE}{dX}\right)_2 dx, \tag{6.30}$$

where $(dE/dX)_2$ is the stopping power of the outgoing ion in the target material. Therefore, the difference in the energy KE_0 of the ion whose backscattering is on the surface of the samples and the energy E_1 of the ion whose backscattering is in the depth t is $\Delta E = KE_0 - E_1$

$$\Delta E = K\int_0^{t/\cos\theta_1} \left(\frac{dE}{dX}\right)_1 dx - \int_0^{t/\cos\theta_2} \left(\frac{dE}{dX}\right)_2 dx. \tag{6.31}$$

ΔE is related to the depth. Therefore, the depth distribution of atoms in the sample and the thickness of the film can be obtained by ion backscattering analysis. $(dE/dX)_1$ and $(dE/dX)_2$ can be considered constants for analysis of samples with depths (or film thickness) less than a few hundred nanometers. Replace $(dE/dX)_1$ with $(dE/dX)_{E_0}$ whose energy is E_0, and replace $(dE/dX)_2$ with $(dE/dX)_{KE_0}$ whose energy is KE_0, which means calculating by surface energy approximation. Hence ΔE can be written as

$$\Delta E = \left[\frac{K}{\cos\vartheta_1}\left(\frac{dE}{dX}\right)_{E_0} - \frac{1}{\cos\vartheta_2}\left(\frac{dE}{dX}\right)_{KE_0}\right]\cdot t = [S_0]\cdot t, \tag{6.32}$$

where $[S_0]$ is referred to as the backscattering energy loss factor. Thus, there is a linear relationship between the energy of the RBS spectrum and the depth of the film. $(dE/dX)_1$ and $(dE/dX)_2$ can also be used as the average energy of the incident and outgoing distances, when $E_{\text{in}} = 1/2(E_0 + E)$ and $E_{\text{out}} = 1/2(E_1 + KE)$. The value of (dE/dX) is calculated as an approximation. On further approximation, E_{in} and E_{out} can be written as $E_{\text{in}} = E_0 - 1/4(\Delta E)$ and $E_{\text{out}} = E_1 + 1/4(\Delta E)$. The average energy method is more accurate than the surface approximation method. In many backscattering experiments, incident ions are incident perpendicular to the surface of the sample: $\theta_1 = \theta$, and $\theta_2 = \pi - \theta$. The backscattering yield and the energy spectrum are assumed to be in perpendicular incidence.

(2) **Backscattering cross-section**

The probability of backscattering of ions can be calculated by the Rutherford cross-section formula

$$\sigma = \frac{Z_1^2 Z_2^2 e^4}{4\pi^2 \sin^4 \vartheta} \times \frac{\left(\cos\vartheta + \sqrt{1 - \left(\frac{M_1}{M_2}\sin\vartheta\right)^2}\right)^2}{\sqrt{1 - \left(\frac{M_1}{M_2}\sin\vartheta\right)^2}}, \tag{6.33}$$

where, Z_1 and Z_2 are the atomic numbers of incident ions and target atoms; e electron charge; σ differential scattering cross-section in cm^2. The bigger the Z_2, the bigger the σ. Therefore, ion backscattering is sensitive to the analysis of heavy impurities in a light matrix.

(3) **Backscattering yield and energy spectrum**

If the beam integrator to determine the number of incidentious as Q, the solid angle of the detector to the sample is Ω, and the sample atomic number per unit area is Nt, in the direction of scattering angle θ, then the scattering particle count (yield) A recorded by the detector is

$$A = Q \cdot \sigma \cdot N \cdot t \cdot \Omega. \tag{6.34}$$

If a thin sample contains the same number of O, Si, Cu, Ag, and Au atoms per unit area, the backscattering energy spectrum (scattering ion energy on the horizontal axis and yield count on the vertical axis) is divided into five small peaks. According to the K factor formula and α formula, the energy backscattered with the heavy element Au is the largest and the count is the largest; the ion energy after scattering with O is the least and the count is the least. Due to the inherent energy resolution of the detector system, the ion energy spectrum scattered from the elements on the thin sample shows small gaussian peaks.

For samples with a certain thickness, the backscattering energy spectrum is rectangular. The high energy side of the spectrum (the upper half of the front edge) is the ion energy KE_0 corresponding to the scattering from the sample surface, and the low energy side (the upper half of the back edge) is the energy E_1 reaching the detector after scattering from the back surface of the sample (or a certain depth). According to Equations (6.31) and (6.32), sample thickness can be obtained by the energy width ΔE. According to Equations (6.32) and (6.34), the yield (spectral height) $H(E_1)$ corresponding to the scattering from a thin layer at the sample depth t in the energy spectrum can be written as:

$$H(E_1) = Q\sigma(E)\Omega N \times \frac{\delta(KE)}{[S(E)]}; \quad \delta(KE) = \frac{S(KE)}{S(E_1)}\delta E_1, \tag{6.35}$$

where δE_1 is the energy corresponding to each line in the energy spectrum, which is called channel width and usually several keV. S (KE) and S (E_1) are, respectively, the stopping powers of the ions of energy KE and E_1 in the sample. $[S(E)]$ is the backscattering energy loss factor when the ion energy is E. $H(E_1)$ is related to N. Therefore, the information of element depth distribution can be got by Rutherford backscattering analysis. Similarly, due to the influence of the intrinsic energy resolution of the detector, the intrinsic energy dispersion of the incident beam and the energy divergence effect of the outgoing beam in the sample (statistical fluctuation of energy loss), the front and rear spectral lines of the RBS energy spectrum of the thick sample are not very steep. For infinitely thick samples, the low-energy side of the RBS spectrum continued to extend to the lowest number of channels. As the depth increased, the energy of incident particles gradually decreased, the scattering cross section increased, and the spectral height gradually increased. The shape and depth distribution of the backscatter spectrum can be simulated and analyzed by a computer program.

6.3.8.3 *Laboratory equipment and samples of RBS*

The experimental equipment for Rutherford backscattering analysis consists of a small particle accelerator, a vacuum target chamber, a particle detector, and an electronic circuit. A single or serial electrostatic accelerator provides a beam of ^{4}He ions or protons with MeV energy. The beam spot diameter is less than 1 mm, and the flow intensity is about several to dozens of nA. The vacuum degree of the target chamber is 10^{-4}–10^{-5} Pa. Samples, sample translation and rotation devices, and gold–silicon surface barrier semiconductor detectors are installed in the target chamber. The detected backscattered particle signal is sent to the microcomputer multichannel pulse analyzer through the preamplifier and the main amplifier to record the particle energy spectrum. At the same time, the beam integrator was used to record the beam incident to the sample (number of incident ions).

The samples analyzed by RBS are generally thin films or solid samples of semiconductors and metals with smooth surfaces. For insulating materials such as ceramics, the surface should be evaporated to several to dozens of nanometers conductive layer (such as Al film) to prevent the accumulation of surface charge. The organic membrane material is prone to become damaged by beam bombardment, so the experiment should be conducted with a very small beam strength. Because the analysis bundle spot is small, the sample size only needs to be 5 mm×10 mm or so. The thickness of the sample (including the substrate) is generally between 0.5 and 2 mm. The conventional backscatter analysis can be used to analyze samples at depths ranging from a few hundred nanometers to 1 micron. Several samples can be installed at the same time; the sample racks need to be moved so that the samples can be analyzed one by one and the vacuum of the target chamber is not damaged. Figure 6.16 shows the RBS spectrum of a 90nm PtSi thin film sample on a silicon matrix. From the signal spectrum width of the Pt ΔE_{Pt} and $[S]_{\mathrm{Pt}}^{\mathrm{PtSi}}$ factor, according to Equation (6.32) the thickness of Pt film can be obtained. The accuracy of thickness measurement is ±5%. The spectral heights of Pt and Si increase gradually with the decrease of energy, mainly because of the decrease of energy and the increase of scattering cross-section ($\sigma \propto 1/E^2$) with the increase of depth, and are also influenced by the $[S]$ factor with the change of energy. In addition, the RBS spectrum can also be used for component analysis, impurity analysis, surface concentration analysis, particle injection analysis, etc., for which the reader may refer to other specialized books.

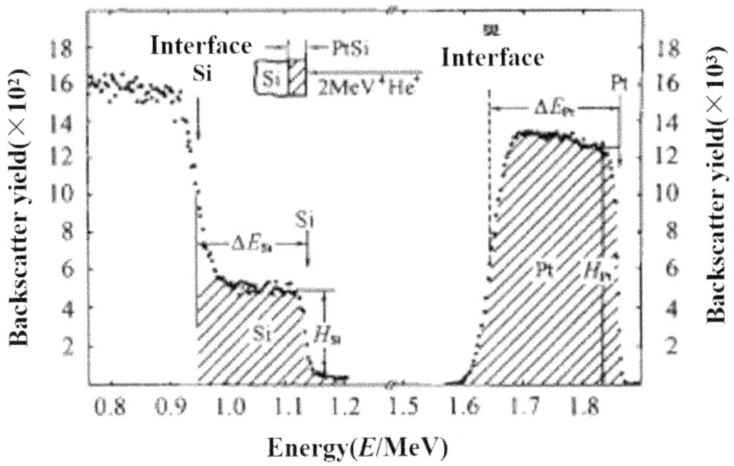

Figure 6.16: RBS spectrum of PtSi film on a Si matrix.

6.3.9 *Atomic force microscope*

An Atomic force microscope (AFM) is one of the most important detection instruments in the field of surface morphology of modern thin films. Developed primarily from the scanning tunneling microscope (STM) developed by *g.* Binnig *et al.*, who won the 1986 Nobel Prize in Physics, the AFM is better at analyzing the surface topography of thin films. The AFM is based on the electron tunnel emission theory proposed in the late 1920s, and does not have any electronic optical system, so you can obtain very high magnification and no aberrations. A microscopic image can be obtained by measuring the tunnel current of a tiny needle about 10 angstroms from the surface of the sample. The instrument can measure different substances under a variety of conditions, measure the position of atoms on a single surface, and has a three-dimensional resolution. This instrument can also be used to measure the localized and non-localized electron energy on the surface, as well as the energy profile of the work function. AFM mainly includes probe, sample, scanner, scan, feedback control circuit, deflection measurement, and computer data processing.

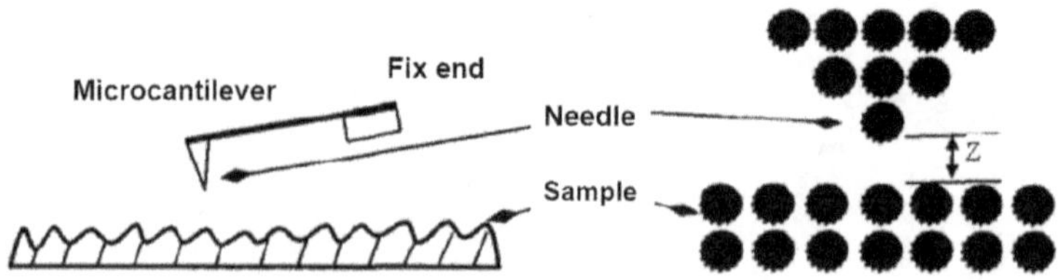

Figure 6.17: Atomic force microscope imaging principle.

6.3.9.1 *Fundamentals of AFM*

Atomic force microscopy uses the interaction between the force-sensitive probe tip and sample atoms to achieve surface imaging. As shown in Figure 6.17, the end of an elastic microcantilever that is extremely sensitive to force is fixed. When the probe tip of the other end approaches the sample surface in the Z direction, a certain force, namely atomic force, will be generated between the atoms of the probe tip and the atoms of the sample, whose size is about 10^{-12}–10^{-9} N. During the scanning, the atomic force is controlled to be constant, so that the microcantilever with the tip corresponds to the equipotential of the interatomic force between the tip and the sample surface and fluctuates in the direction perpendicular to the surface of the sample. Then, by measuring the deformation of the microcantilever corresponding to the changes in the position of the scanning points, the undulating height of the sample surface is measured. The surface topography of the sample can be obtained by plotting the horizontal position of the tip corresponding to the local fluctuation height of the sample.

6.3.9.2 *Imaging patterns in AFM*

Atomic force microscope imaging mode mainly includes the contact mode, non-contact mode, and tapping mode.

The non-contact tip always vibrates above the surface of the sample without touching the sample. This mode is more difficult to operate and the resolution is not high, so it is less used.

The contact mode is a commonly used scanning mode in which the tip is always in contact with the sample to produce stable, high-resolution images. However, for some biomolecules, samples

Figure 6.18: Contact mode image (silicon sample).

with low elastic modulus and easy to move and deform, the contact mode is obviously deficient. These samples can be displaced or damaged by probe movement. The two pictures in Figure 6.18, respectively, show the images after two scans on an epitaxial film with contact mode. It is not difficult to see from Figure 6.18 that the resolution of the image is not high due to shear force, and the surface of the sample has been obviously damaged after the first scan. Of course, for most samples, it is not difficult to get a clear and stable picture without damaging the sample.

In atomic force microscopy, the tapping mode is the most common mode. The tapping mode is actually by the forced vibration of the piezoelectric crystal probe near the resonance frequency of the cantilever; when there is no contact with the sample, the pinpoint piezoelectric drive will make the vibration amplitude of the cantilever beam larger, then due to the large amplitude vibration the tip will come close to or strike the sample surface; when the tip and the sample come into contact the energy of the beam is reduced, the amplitude decreases, and so it is lifted. During the scanning process, the tip vibrating vertically is in intermittent contact with the sample surface at a frequency of 51,500,000 times per second. In tapping mode, due to coming into contact with the surface of the tip the sample time is very short, the interaction force between them is too

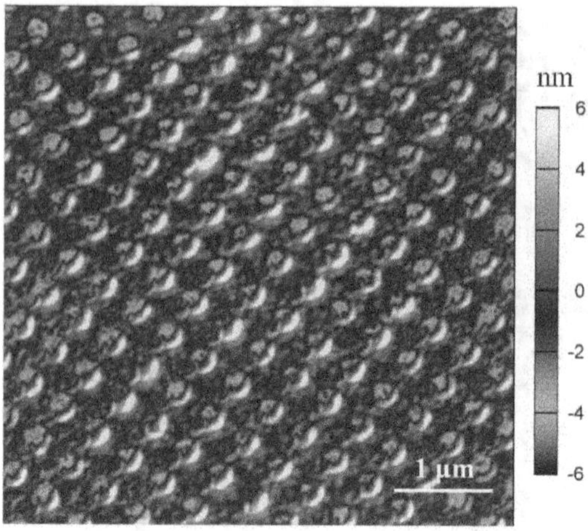

Figure 6.19: Tapping mode image (silicon sample).

negligible to remember, so almost no shear force will damage the sample surface and make the image resolution. In Figure 6.19 the two images, respectively, are after the first scan and the second scan epitaxial film results. Compared with the contact pattern images, the resolution of the images obtained by tap mode is improved, and does not destroy the sample after the first scan. This greatly increases the scope of the tap mode.

6.3.9.3 *Piezoelectric response force microscope*

Due to surface heterogeneity and mechanical constraints, ferroelectric materials exhibit electrical domain structures. Electric domains are regions with consistent spontaneous polarization orientations in ferroelectric crystals. They are the result of minimizing the free energy of the system. They are the basis for understanding many properties of ferroelectric materials. The domains of ferroelectric thin films are mainly characterized by piezoresponse force microscopy (PFM). Piezoresponse mode is a type of contact mode imaging technology. The principle of imaging ferroelectric domains is based on local piezoelectric vibration caused by the inverse piezoelectric effect under the effect of applied AC voltage. The sample is connected to the base through conductive silver glue, and then grounded. The needle tip and the sample must remain in contact so that a continuous current loop can be formed between the needle and the sample after a voltage is applied to the needle tip. When a voltage is applied between the probe and the bottom electrode of the sample, the probe actually acts as a movable upper electrode, and the microcantilever signal that generates resonance with the sample can be detected by phase-locked technology. In piezoelectric response imaging mode, the imaging voltage frequency is much lower than the microcantilever resonance frequency to avoid mechanical resonance of the microcantilever. The applied voltage at the frequency ω will cause the sample to generate the same frequency vibration (inverse piezoelectric effect) and 2ω vibration (electrostrictive effect). The ferroelectric domain structure can be observed by controlling the first harmonic signal (piezoelectric response signal). The phase of the piezoelectric response depends on the sign of the piezoelectric coefficient (depending on the direction of polarization) and reverses as the piezoelectric coefficient reverses, This means that the regions with opposite polar orientations vibrate against each other under the AC electric field and appear as regions with different contrast on the piezoelectric response image. The reverse piezoelectric effect causes the ferroelectric sample to deform such as expansion, contraction, and shear under the action of the AC voltage $U = U_0 \sin(\omega t)$ between the needle tip and the sample (see Figure 6.20), These deformations will act on the micro cantilever in the form of friction, causing the micro cantilever to generate lateral vibration. The amplitude and phase determine the direction of polarization. The laser reflects the

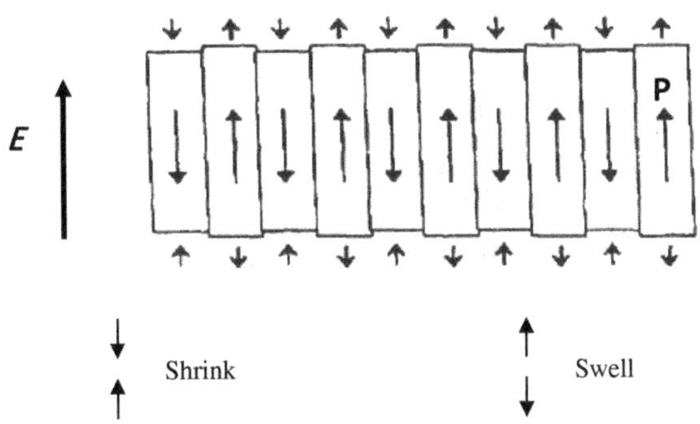

Figure 6.20: Schematic diagram of the shrinkage and expansion of a ferroelectric film under an applied electric field.

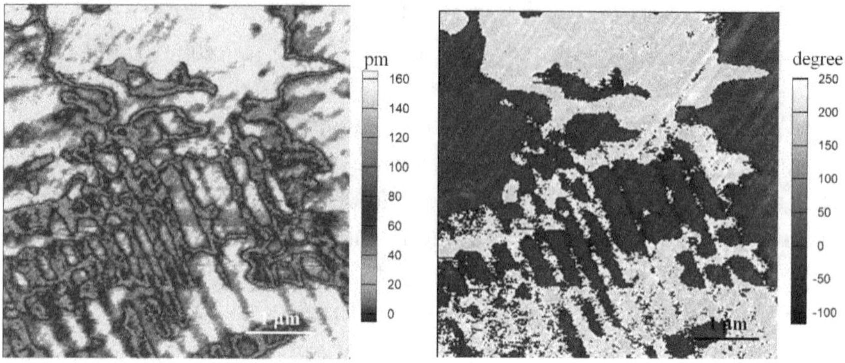

Figure 6.21: The domain structure of a ferroelectric sample, (a) amplitude image; (b) phase image.

vibration signal of the micro cantilever to the photodetector. Then it is input into a lock-in amplifier, and the nanodomain structure of the ferroelectric thin film can be obtained. The domain structure of a ferroelectric sample are shown as Fig. 6.21.

In addition to the analysis methods introduced above, related technologies suitable for thin-film analysis include low-energy ion scattering spectroscopy, and low-energy electron energy loss spectroscopy, and due to space limitations, they are not described here.

6.4 Exercises and Questions

(1) What are the main structures of films? How to distinguish between amorphous and crystalline films?
(2) What are the defects of the thin films? How to determine the different defects of the film?
(3) What properties of the film can be analyzed using XRD, and how accurate is the analysis?
(4) Which properties of the film can be analyzed by SEM, and what principles should be noted?
(5) What properties of the film can be analyzed by AFM, and how can it be related to the analysis results of SEM and TEM to obtain more information about the film?
(6) What is the principle of RBS? How to get the elemental analysis results of the film?

Chapter 7

Physical Properties of Thin Films

Different films (such as metal films, dielectric films, semiconductor films, etc.) have various properties. It is very important to understand the mechanical properties, electrical properties, optical properties, and magnetic properties of thin films for their application.

7.1 Mechanical Properties of Thin Films

The mechanical properties of thin films mainly include adhesion, stress, elasticity, and mechanical strength. The adhesion force between the film and substrate will directly affect the performance of the film. In some cases, other mechanical properties of the film should also be considered, such as for the superhard film, which is mainly used to enhance the hardness and friction resistance of the matrix, the compressive strength should be considered.

7.1.1 *Adhesion force of films*

The adhesion property of a film on a substrate determines the possibility and reliability of the film's application. Adhesion is defined as the state in which the atoms of the film and the substrate remain in contact and are adsorbed by each other. The force required for the quasi-static peeling off of unit area of film from the substrate is the adhesion force, and the required energy is the adhesion energy.

7.1.1.1 *Adhesion phenomenon and mechanism*

The adhesion of the film on the substrate is a result of van der Waals force, diffusion adhesion, mechanical locking, electrostatic attraction,

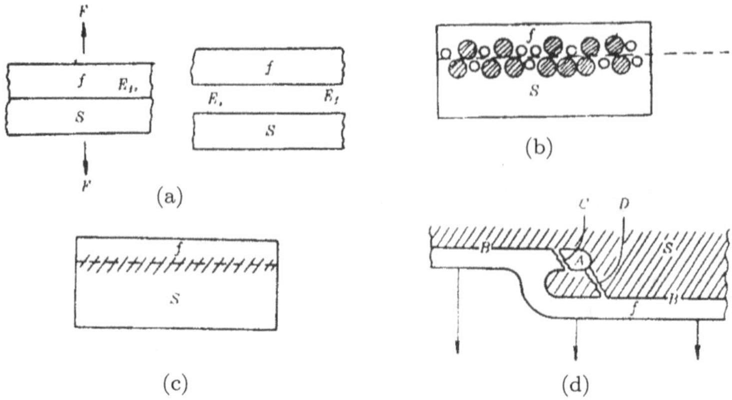

Figure 7.1: Four types of attachment: (a) simple adhesion; (b) diffusion adhesion; (c) adhesion through intermediate layer; (d) macroscopic effect adhesion.

and other factors. For example, the chemical bond force becomes the main adhesion force when the film material forms a compound with the substrate. The adhesion of the film can be divided into four types: simple adhesion, diffusion adhesion, adhesion through the middle layer, and macro effect adhesion, as shown in Figure 7.1.

Simple adhesion is the existence of a clear interface between the film and the substrate. The adhesion is formed by the mutual attraction of two contact surfaces. When two dissimilar or incompatible surfaces come into contact with each other, such an adhesion is prone to form.

Diffusion adhesion is a gradual interface formed by diffusion or dissolution between the film and the substrate. When the atoms reach the substrate, they have large kinetic energy (e.g., sputtering particles), and when they are deposited on the substrate, they can have deep longitudinal diffusion and form diffusion adhesion.

The adhesion through the intermediate layer forms a compound intermediate layer between the film and the substrate, and then the film forms a firm adhesion between the intermediate layer and the substrate. Because there is such an intermediate layer between the film and the substrate, there is no simple interface between them.

Through the macroscopic effect, there are mechanical locking and double electric layer attraction. Mechanical locking is a kind of macroscopic mechanical action. When the surface of the substrate is rough and there are all kinds of micropores or microcracks, in the process of film formation, the gas atoms incident on the substrate surface will enter into all kinds of defects, micropores or cracks on the rough surface to form mechanical locking. If the distribution of all kinds of micro defects on the substrate surface is uniform and appropriate, the adhesion performance of the film can be improved by mechanical locking.

The main mechanism of adhesion is adsorption. According to the different adsorption energy, it can be divided into physical adsorption and chemical adsorption.

Physical adsorption includes van der Waals force adsorption and electrostatic adsorption. Van der Waals force is a kind of short-range force. When the distance between the adsorbed atoms increases slightly, it tends to zero rapidly. Electrostatic force is a long-range force; even if there is a small displacement between the film and the substrate, its attraction will not change greatly. Therefore, although the value of electrostatic force is small, its contribution to adhesion is large. The adsorption energy of physical adsorption is in the range of 0.001–0.1 eV.

Chemical adsorption is a kind of adsorption produced by the chemical bond between the film and the substrate. There are three kinds of bonds: covalent bond, ionic bond, and metal bond. Chemical bond attraction is a kind of short-range force, but it is much larger than van der Waals force. The adsorption energy of chemisorption is in the range of 0.1–0.5 eV.

The phenomenon of adhesion appears on the surface of two kinds of materials. It cannot be separated from the specific surface free energy of the substrate and the specific surface free energy of the film. It is related to the free energy of the interface between the film σ_f, and the substrate σ_{sf}. The adhesion energy (E) per unit area can be expressed as follows:

$$E = \sigma_s + \sigma_f - \sigma_{sf}. \tag{7.1}$$

The free energy of the interface, σ_{sf} is related to the type of atoms, the distance between atoms and the bonding characteristics of the two materials.

7.1.1.2 *Methods of increasing the adhesion*

To increase the adhesion of the film, the following methods are adopted.

(1) **Cleaning the substrate**

The surface state of the substrate has a great influence on the adhesion. If there is a layer of pollution on the surface, the film will not come into contact the substrate directly, the van der Waals force will be greatly weakened, the diffusion adhesion will not be possible, and the adhesion performance will be very poor. The solution is to clean the substrate thoroughly, remove the inorganic and organic impurities attached to the substrate, and pretreat the substrate surface by ion bombardment.

(2) **Increase the substrate temperature**

When the film is deposited, increasing the substrate temperature is conducive to the mutual diffusion of atoms between the film and substrate, accelerating chemical reaction, forming diffusion adhesion and chemical bond adhesion, and increasing the adhesion. However, if the substrate temperature is too high, the grain size of the film will be coarse and the thermal stress in the film will be increased, which will affect other properties of the film. Therefore, a comprehensive consideration should be taken when increasing the substrate temperature.

(3) **Introduction of intermediate transition layer**

The adhesion between one material and some substances is large, while the adhesion with other substances may be small. For example, when silicon dioxide is deposited on the glass, it has great adhesion, but the adhesion on KDP crystal is not good; when gold is deposited on the glass substrate, the adhesion is very poor, but the adhesion on platinum, nickel, titanium, chromium, and other metal substrates is very good. Therefore, in order to improve the adhesion of the film, an additional material can be added between the film and the substrate to form a middle transition layer. If the gold film does not adhere well to the glass, a very thin layer of chromium or nickel–chromium can be evaporated on the glass first. Chromium can capture oxygen from the oxide substrate to form oxide, which has strong adhesion, and then evaporate the gold film. Between the gold film and chromium film, a metal bond would be formed, and also a great adhesion.

(4) **Increase adhesion by sputtering**

After selecting the film material and substrate, the adhesion of the films deposited by sputtering is better than that by evaporation. This is due to the high energy of sputtered particles, which can not only remove the gas adsorbed on the surface and increase the surface activity, but also facilitate the diffusion of film atoms to the substrate, so the adhesion of the film is significantly improved.

7.1.1.3 *Test methods of adhesion*

Adhesion testing methods can be broadly classified into the adhesive method and the direct method.

(1) **Adhesive method**

The adhesive method is suitable for a film having a relatively small adhesion; that is, the adhesion between the film and the substrate must be less than the adhesion between the film and the adhesive. The volume shrinkage of the selected adhesive after curing should be small, and the epoxy resin adhesive can be generally used. The adhesive method includes pulling method, peeling method, etc., and the pulling method and the peeling method are mainly introduced here.

(a) *Pulling method*

The principle is to bond a columnar rod on the film and apply a force perpendicular to the film surface on the rod. The experimental setup is shown in Figure 7.2. If the minimum pulling force of the film is F,

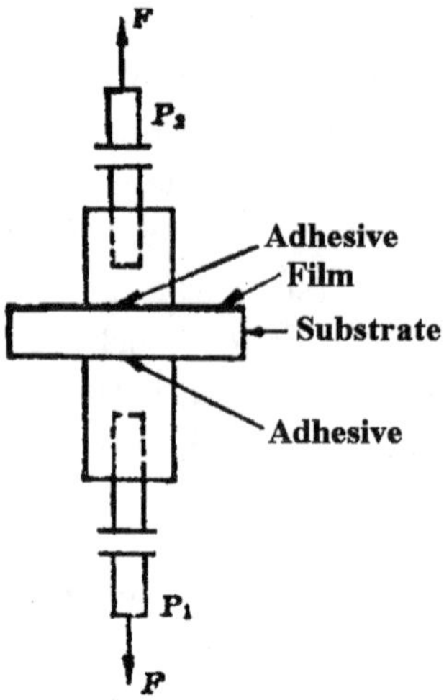

Figure 7.2: Pulling test device.

and the bottom area of the bond is A, then the adhesion per unit area is

$$f = \frac{F}{A}. \tag{7.2}$$

When measuring by the pulling method, the direction of the pulling force must be consistent with the normal direction of the film surface; that is, the force applied to P_1 and P_2 in Figure 7.2 is ensured to pass through the center point of the plane of the substrate, otherwise the torque will be generated and measurement errors will occur. In addition, the bonding thickness is required to be uniform and does not affect the film properties.

(b) *Peeling method*

This method is to adhere a certain width of adhesive tape on the surface of the film, and then apply a pulling force to the adhesive tape at a certain angle. After the adhesive tape is pulled down, the adhesion can be judged according to the condition of the film being peeled off, as shown in Figure 7.3. If the adhesion energy per unit area of the film is γ, then the total adhesion energy of the film with

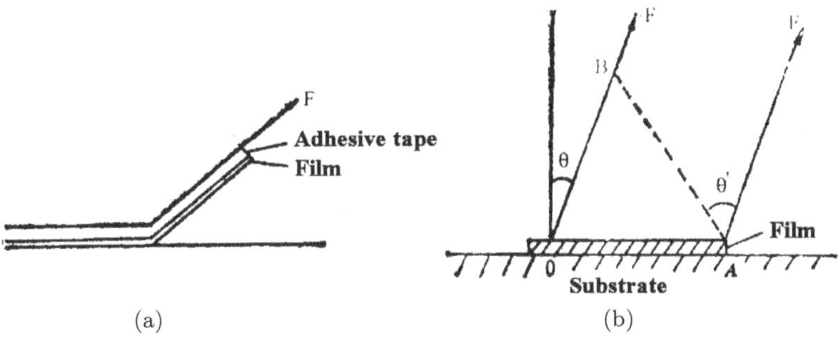

Figure 7.3: Schematic diagram of the peeling method. (a) The film is pulled up and (b) the film is attached to the substrate.

length a and width b is

$$E = ad\gamma. \tag{7.3}$$

The work performed by the force F used to peel the film is

$$W_p = (F\cos\theta')2a\cos\theta' = Fa(1 - \sin\theta). \tag{7.4}$$

If it is considered to be static stripping and we ignore the elastic energy generated when the film is bent, the work done by F is approximately equal to the total adhesion energy of the film; that is, $W_p = E$. Then

$$ab\gamma = Fa(1 - \sin\theta). \tag{7.5}$$

Therefore,

$$F = \frac{b\gamma}{1 - \sin\theta}. \tag{7.6}$$

Where F varies with the angle θ and does not truly reflect the adhesion properties of the film. When the applied peeling force is perpendicular to the film, that is, $\theta = 0$, then Equation (7.6) can be simplified to

$$F = b\gamma. \tag{7.7}$$

According to the measured F, the adhesion energy $r = F/b$ can be calculated. If the adhesion f per unit length needs to be calculated directly, according to the definition and peeling by the above method

($\theta = 0$), $f = r$ can be obtained. It can be seen that the magnitude of the adhesion and the adhesion energy r are the same.

(2) Direct method

If the adhesion of the film is strong, even if the bonding surface is peeled off, the film remains attached. Then it's necessary to apply a force directly to the film to peel off the substrate. This method of measuring adhesion is called the direct method. The direct method includes scratch method, rubbing method, centrifugal method, and so on. The commonly used scratch method is highlighted here.

The principle of direct measurement of adhesion by the scratch method is shown in Figure 7.4. The tip of a hard needle is placed perpendicular to the surface of the film; the radius of the tip of the steel needle is known (typically 0.5 mm). Then gradually increase the vertical load on the steel needle until the film is scratched. The load that the film has just been scratched is generally referred to as the critical load and is used as a measure of film adhesion. The scratches were observed with an optical microscope to determine the critical load, and the value was generally from several grams to several hundred grams.

Under the vertical load, the surface of the substrate under the tip of the steel needle is severely deformed. The film extends due to deformation of the substrate so that a shear force is generated between the film and the substrate, wherein the shear force at the indentation L is the largest. When the vertical load reaches the critical value W, the shearing force at the indentation L is increased

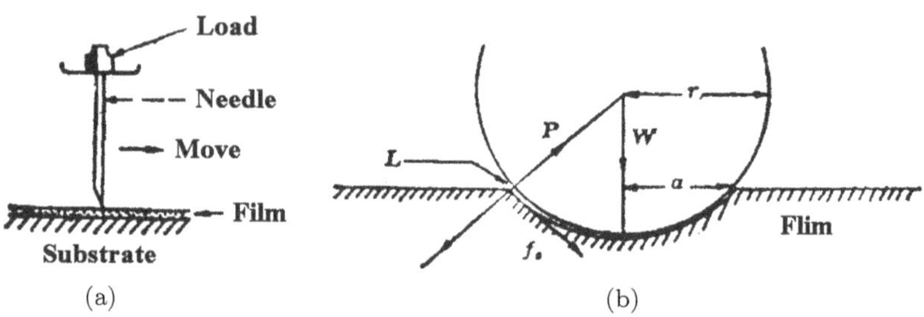

Figure 7.4: Principle of the scratch method (a) Schematic diagram of the measuring device; (b) thin film pressing by a spherical needle tip.

enough to break the adhesion of the film to the substrate, and the critical shear force f_s required to peel the film per unit area from the substrate is equal to the adhesive force; f_s can be calculated according to the following formula:

$$f_s = P\sqrt{\frac{W}{\pi r^2 P - W}}, \tag{7.8}$$

where r is the curvature radius of the tip and P is the reaction force of the substrate to the needle at point L. It can be seen from the above formula that the greater the vertical load applied to peel the film, the greater the adhesion of the film. When the P value is unknown, P can be obtained by $d = 2a = \sqrt{W/(\pi P)}$ according to the measured indentation width d. However, the scratch method is also affected by the hardness of the film, so it can only be a qualitative method.

Figure 7.5 shows the results of the adhesion test on various metal films on a glass substrate, in which the horizontal axis represents the formation energy of the oxide of the film material. As shown in Figure 7.5, the film which is prone to form an oxide has a greater adhesion.

Figure 7.6 shows the measurement results of the adhesion of various metal films on different substrates. When a film such as Ag, Al, Cd, Cr, Cu, Mo, W and Zn is appropriately combined with a substrate such as NaCl, KCl, KBr, TiO_2, MgF_2, Bi_2O_3, Al_2O_3, SnO_2, Al, and Cu, the main role in adhesion is by van der Waals

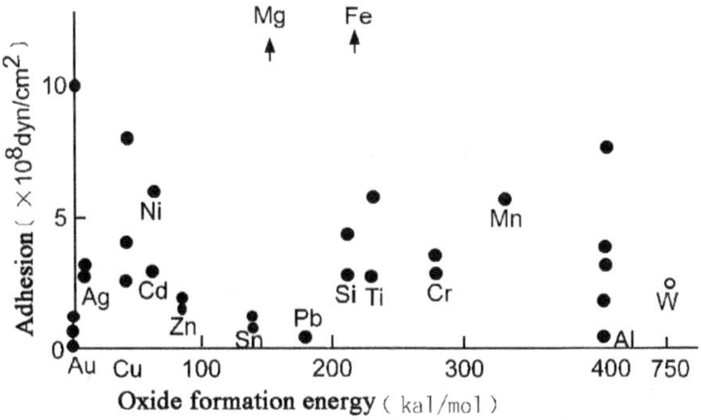

Figure 7.5: Adhesion of various metal films on glass substrates.

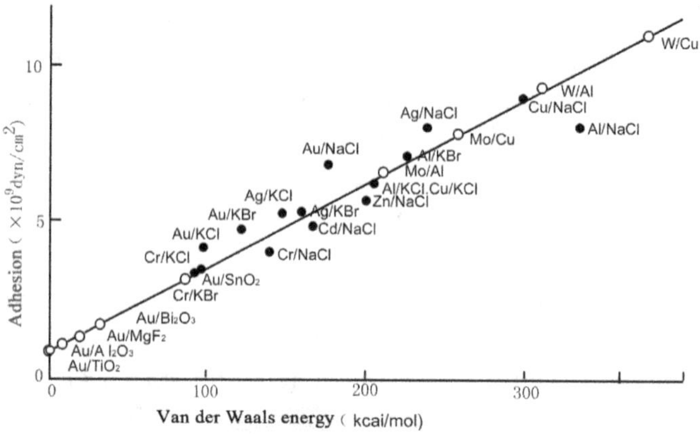

Figure 7.6: Adhesion relationship between various metal films and different substrates.

force. In Figure 7.6, black dots indicate that adhesion and van der Waals energy have been obtained; white dots indicate that although the adhesion is obtained, the van der Waals energy is still unknown.

7.1.2 *Internal stress of thin films*

7.1.2.1 *Definition and classification of internal stress*

On any section inside the film, the force exerted by one side of the unit section on the other side is called the internal stress of the film. The internal stress is generally expressed by σ and its unit is N/m^2. Films prepared by vacuum evaporation, sputtering, vapor growth, etc. all have internal stresses, and internal stresses can reach $10^9\,N/m^2$ at the maximum. Excessive internal stress can cause cracking or wrinkling of the film, which can invalidate the film components.

Internal stress can be divided into two major categories: tensile stress and compressive stress. When one side of the section receives a force from the stretching direction of the other side, it is called a tensile stress; and when a force of the pressing direction is received, it is called a compressive stress. Excessive tensile stress causes the film to crack; excessive compressive stress causes the film to wrinkle or fall off. It is customary to take a positive sign for the tensile stress and a negative sign for the compressive stress.

From its origin, internal stress can be divided into thermal stress and intrinsic stress. In the process of preparing the film, the film and the substrate are at a relatively high temperature, and when the film is prepared, the film and the substrate are restored to a normal temperature state. Since the thermal expansion coefficients of the film and the substrate are different, stress is generated inside the film. The stress caused by the thermal effect is called thermal stress. The thermal stress σ_T can be expressed by the following formula

$$\sigma_T = E_f(a_f - a_s)\Delta T, \tag{7.9}$$

where a_s is the thermal expansion coefficient of the substrate, a_f is the thermal expansion coefficient of the film, E_f is the elastic modulus of the film material, and ΔT is the difference between the measured temperature and the film deposition temperature.

The internal stress caused by defects or the like during the formation of the film is called the intrinsic stress. The subtraction of thermal stress from the total internal stress is the value of intrinsic stress, as shown in Figure 7.7. The intrinsic stress is related to the film thickness and the typical changes are shown in Figure 7.8. When the film thickness is small, the islands constituting the film are not connected to each other. Even if they are connected, they have a network structure, and the internal stress at this time is small. As the

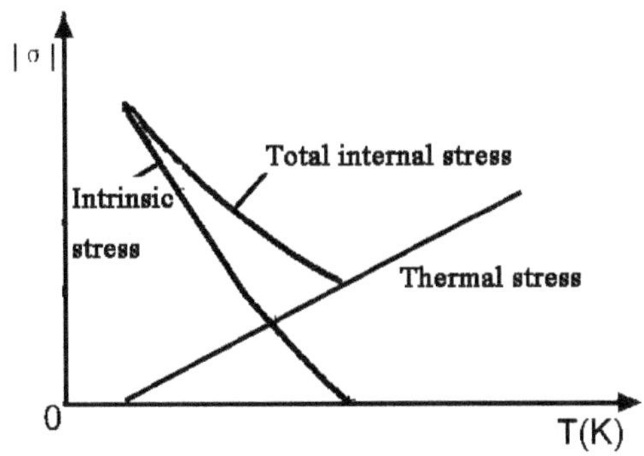

Figure 7.7: The intrinsic stress of the silver film on the mica sheet varies with thickness.

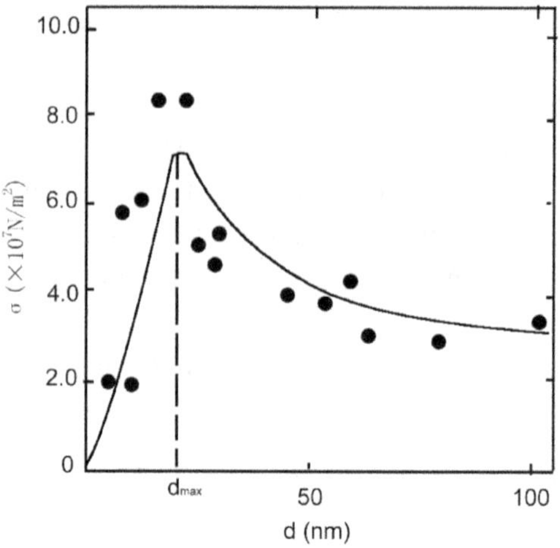

Figure 7.8: Relationship between the internal stress and the substrate temperature.

film thickness increases, the islands are connected to each other. Due to the difference in lattice arrangement between the islands and the presence of small holes, the internal stress rapidly increases and a maximum occurs. When the film thickness is further increased and a continuous film is formed, there are no more small holes in the film. The stress decreases and approaches a stable value.

7.1.2.2 *Origin of internal stress*

It is generally believed that the internal stress of a film is a result of the combination of thermal effects, phase transitions, interfaces, and impurities.

(1) **Thermal effect**

Thermal effects are the main source of internal stress. Therefore, when selecting the substrate, the material whose thermal expansion coefficient is close to that of the film should be selected as much as possible, or select the material that can make the film under a compressive stress state. The substrate temperature has a great influence on the internal stress of the film, which directly affects the migration ability of the adsorbed atoms on the surface of the substrate, thereby affecting the structure of the film, the size of the crystal grains, and the number and distribution of defects, and these are related to the internal stress.

(2) **Phase transition**

The formation process of the film is actually a phase transition process, that is, from the gas phase to the liquid phase and then the solid phase. This phase transition causes a change in volume, resulting in the internal stress. In the formation process of the film, firstly, a short-range ordered liquid-like solid phase is agglomerated on the substrate, and there is no intrinsic stress in the film. After the liquid-like solid phase changes to a stable crystalline phase, intrinsic stress is generated due to the difference in the density of the two phases. For example, when Ga changes from a liquid phase to a solid phase, the volume will expand, and thus internal stress will be generated in the film. The film formed by Sb at normal temperature is generally amorphous. When the film thickness exceeds a certain critical value, crystallization occurs, the volume shrinks during crystallization, and tensile stress is generated in the film.

(3) **Interface stress**

When the lattice structure of the film material is different from the lattice structure of the substrate material, the structure of the first few layers of the film will be affected by the substrate, forming a lattice size that is close to or similar to the substrate, and then gradually transits to the lattice size of the film material itself. Structural distortion in the transition layer will cause internal stress in the film. This internal stress due to the mismatch of the lattice on the interface is called interfacial stress. In order to reduce the interfacial stress, the lattice structure of the substrate surface should be matched to that of the film as much as possible.

(4) **Impurity effect**

When the film is deposited, the ambient atmosphere greatly affects the internal stress. Residual gases or other impurities in the vacuum chamber enter the membrane structure to create compressive stress.

When the prepared films are placed in the atmosphere, they will be affected by the ambient atmosphere. For example, when the surface of the copper film encounters air, it will form a very thin oxide layer on the surface, which will also increase the compressive stress of the film. The residual gas entering the film may also run out again, leaving vacancies or micropores in the film, causing tensile stress. Compressive stress also occurs when impurities enter the film due to grain boundary diffusion. In addition to the above reasons, the combination of islands, grain, and the diffusion of micro-holes and defects will cause a change in surface tension and internal stress during the growth of thin films.

7.1.2.3 *Measurement method of internal stress*

The measurement method of internal stress can be broadly classified into two types: the mechanical method and the diffraction method. The mechanical method measures the bending degree of the substrate under stress, and the diffraction method measures the distortion of the lattice constant of the film. There are many different methods under these two categories. Here are some commonly used methods.

(1) **Cantilever beam method**

When the film is deposited on the substrate, the substrate will bend under the stress of the film. When the internal stress of the film is tensile stress, the surface of the substrate should be concave and bends toward the direction of evaporation. When the internal stress of the film is compressive, the surface of the substrate on which the film is vaporized becomes convex and bends toward the direction of evaporation. According to this principle, one end of the elongated substrate is fixed and the other end is suspended to form a so-called cantilever beam, as shown in Figure 7.9.

When the substrate is evaporated onto the film, the substrate is deformed under the influence of the film stress, and the displacement will occur at the end of the suspension. The displacement is measured and the internal stress is calculated by applying a given formula. The internal stress in the film can be calculated by measuring the displacement and applying the given formula. In order to facilitate

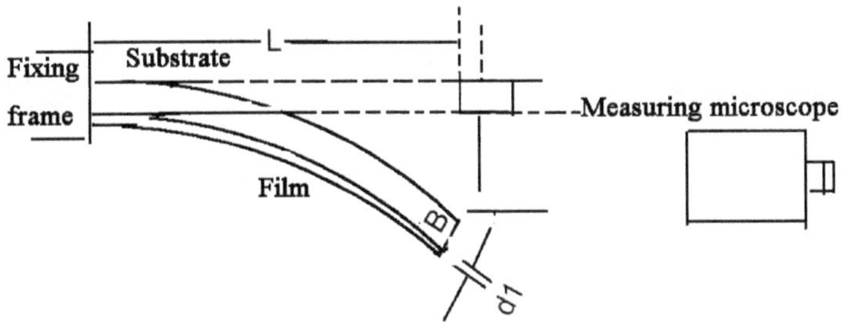

Figure 7.9: Schematic diagram of the cantilever method.

measurement and achieve higher sensitivity, the substrate is required to have good elasticity, uniform thickness, and a small ratio of thickness to length. The commonly used substrates are Yuntian and glass, silicon, copper, aluminium, and nickel. The sensitivity of this method depends on the minimum displacement at one end of the substrate that can be detected.

(2) **Bending plate method**

The basic principle of the curved disc method is the same as that of the cantilever beam method, and the internal stress is also measured by the deformation of the substrate. The difference is that the substrate is circular. The radii of curvature R_1 and R_2 of the substrate are measured before and after deposition. Then the internal stress in the film is calculated according to the following formula:

$$\sigma = \frac{ED^2}{6}\left(\frac{1}{R_2} - \frac{1}{R_1}\right), \tag{7.10}$$

where E and D are the elastic modulus and thickness of the substrate, respectively.

In order to ensure a high sensitivity of the test system, it is necessary to select a suitable substrate and detection method. The substrates currently used are glass, quartz, single crystal silicon, etc., and the surface of the substrate needs to be optically polished.

Before and after deposition, the substrate is placed on the optical plane of the quartz plate, and its radius of curvature is measured. The most commonly used method is the Newton loop method, and

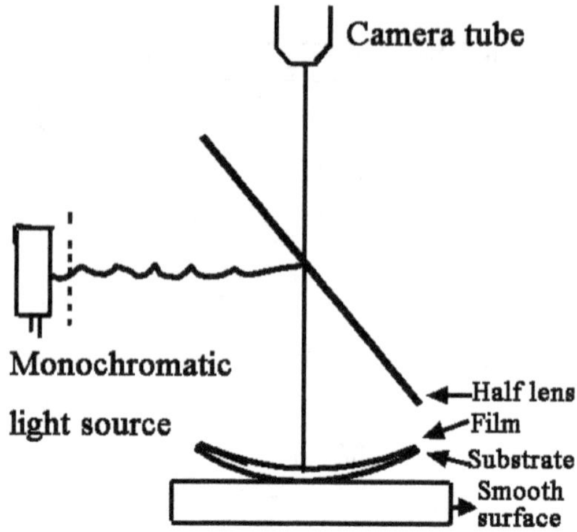

Figure 7.10: Schematic diagram of the Newton ring method.

Figure 7.10 is the schematic diagram. After placing the sample on the optical plane, it is illuminated with light perpendicular to the surface. Then Newton rings are generated between the surface of the sample and the optical plane due to the interference of light. The curvature radius of the substrate can be measured by measuring the interval between Newton rings, and the internal stress can be calculated.

(3) **X-ray diffraction method**

The interplanar spacing of the film structure can be measured by X-ray diffraction. The strain of the film can be determined by comparing the measured value with the standard interplanar spacing of the material, and the internal stress of the film can be obtained. In order to keep the diffraction peaks from dispersing, the thickness of the film must be at least tens of nanometers. Figure 7.11 shows the X-ray diffraction pattern of the gold film. It can be seen from Figure 7.11 that the peak position (Bragg angle) of the film before heat treatment (dashed line) is larger than the angle value given by the standard card. According to the measured angle, the interplanar spacing d can be calculated by the Bragg formula

$$2d \sin \theta = \lambda. \tag{7.11}$$

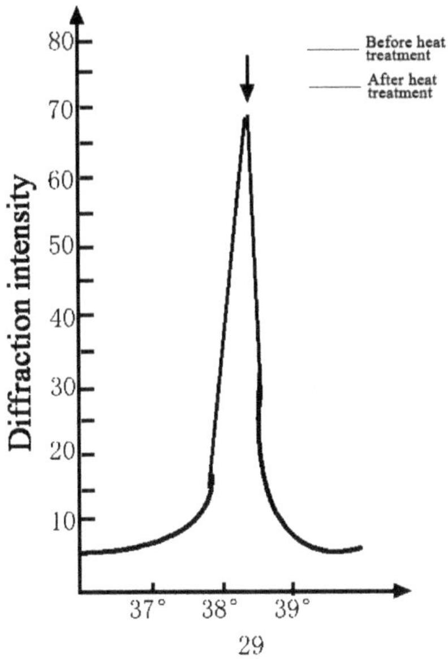

Figure 7.11: X-ray diffraction pattern of the Au film.

Let the interplanar spacing given by the standard card be d, which can be used to calculate the strain in the direction perpendicular to the film surface:

$$\varepsilon = \frac{d - d_0}{d_0}. \tag{7.12}$$

The internal stress of the film can be obtained from the relationship between stress and strain.

$$\sigma = \frac{E_f}{2\upsilon_f} \cdot \frac{d - d_0}{d_0}, \tag{7.13}$$

where E_f and ν_f are the elastic modulus and Poisson's ratio of the film material.

Figure 7.11 also shows that the diffraction peak and the diffraction integral intensity increase after the film is annealed, and the Bragg angle moves to the value of the bulk material, which means that the internal stress of the film becomes smaller.

The internal stress of the crystal grains in the film was measured by X-ray diffraction, which does not include the internal stress of the amorphous region and the crystallite region. Therefore, the value of the internal stress measured by this method is smaller than the value measured by the cantilever beam method.

7.1.3 *Hardness of thin films*

The hardness of a substance is defined as the ability of one substance to resist friction or scratch resistance relative to another substance. Some films are very hard, and they can be deposited on the surface of the workpiece to protect them. Some films can significantly improve the friction resistance of the workpiece, so sometimes such films are also referred to as superhard films. For example, titanium nitride films are evaporated on the surface of gears or knives to extend their life. The diamond-like film with extremely high hardness, excellent corrosion resistance, and excellent light transmission properties can be used for surface protection of optical devices and solar cells.

When performing the hardness test, we use a hard material (such as diamond) as the indenter, and the sample is pressed at a certain speed and weight, and then the hardness is judged according to the size of the indentation left on the sample. For different test methods, especially when the shape of the indenter is different, the results are also different. The most commonly used is the Vickers hardness, followed by the Knoop hardness and the Brinell hardness. Here we mainly introduce the Vickers hardness test method.

The Vickers hardness tester calculates the hardness value using the length value (the unit is generally micrometer) of the diagonal of the obtained indentation after the diamond is pressed into the sample. The shape of the diamond indenter is a square pyramid with an angle of 136° and its indentation is a pressed square pyramid. If the average length of the diagonal line is $d(\mu\text{m})$, the load applied to the indenter is $P(g)$, so the Vickers hardness is

$$HV = \frac{1854P}{d^2} \quad (\text{kg/mm}^2). \tag{7.14}$$

The hardness test of the film is generally carried out using a microhardness tester, and the length of the diagonal of the indentation is measured with a microscope. In addition, attention should be paid to the selection of the load size, the speed of loading, and the measurement of the diagonal.

When the indenter is loaded to a certain load, the indentation will appear on the surface of the specimen. Due to the deformation of the indentation, plastic deformation and elastic deformation will occur in the surrounding area. The influence range of indentation deformation varies with the material, and generally reaches a hemispherical area with a depth of more than 10 times the indentation depth. For the same material, the greater the load on the indenter, the deeper the indentation. For the film, the deformation of the too deep indentation may reach the substrate, and the hardness measured is inaccurate. Therefore, in order to accurately measure the hardness of the film, the thickness of the substrate should be at least 10 times the depth of the indentation; that is, the load weight should be as small as possible.

When performing a hardness test on a film sample, the mass of the applied load is often less than 1g in many cases, and it is not easy to load such a small load onto the test surface at the slowest speed. In order to apply the load vertically to the test surface, the hardness tester also has a load mechanism to load the weight, and the load mechanism itself has a weight. If the weight is m and the moving speed is v, the momentum mv will bring errors in the measurement. The higher the speed, the deeper the indentation. The research studies show that if the mass of the movable part of the load mechanism is 1g and the weight of the load mechanism is 10 mg, the loading speed on the test surface should be less than 10 μm/s in order to make the load error below 1%. Therefore, the loading speed must be carefully controlled during the test.

When measuring with a Vickers hardness tester, the diagonal d should be accurately measured. Since the Vickers hardness is inversely proportional to d^2, an error in d will strongly affect the hardness value. The measured diagonal length is often only a few microns or even shorter, so even if the measurement accuracy is 0.1 μm, the measurement error is quite large. Especially for some superhard films such as diamond films, the diagonal length is very small, and the measurement error is very large with an ordinary optical microscope. At this time, a scanning electron microscope can be used for measurement.

7.2 Electrical Properties of Thin Films

The electrical properties of thin films, such as resistivity, temperature coefficient of resistivity, dielectric constant, etc., and their relationship with film thickness, applied electric field, and ambient temperature directly determine the performance of thin films in various practical applications. This section mainly introduces the electrical properties of metal films, dielectric films, semi-conducting films, and superconductor films.

7.2.1 *Electrical properties of metal thin films*

7.2.1.1 *Conductive properties of block metal materials*

For a wire with length L and cross-sectional area S, its resistance R can be expressed as

$$R = \rho \frac{L}{S}, \tag{7.15}$$

where ρ is the resistivity, the coefficient of proportionality, which is only related to the nature of the material and is independent of the geometry of the conductor. The reciprocal of ρ is called conductivity and is expressed as

$$\sigma = \frac{1}{\rho}. \tag{7.16}$$

In the absence of phase transition, the resistivity of a metal increases with the increase of temperature T as follows:

$$\rho_T = \rho_0(1 + aT), \tag{7.17}$$

where ρ_T is the resistivity at temperature T, ρ_0 is the resistivity at temperature $0°$, and α is the temperature coefficient of resistance.

The average resistance temperature coefficient from 0°C to temperature T is

$$a_T = \frac{\rho_T - \rho_0}{\rho_0 \cdot T}. \tag{7.18}$$

The true resistance temperature coefficient for temperature T is

$$a_T = \frac{d\rho}{dT} \cdot \frac{1}{\rho_T}. \tag{7.19}$$

According to the theory of quantum mechanics, the expression for metal resistivity ρ is

$$\rho = \frac{2m}{n \cdot e^2 \cdot \tau}, \tag{7.20}$$

where m is electronic quality, e is the electron charge, n is effective electron concentration involved in the conducting process, and τ is the time interval between two times of electron wave scattering, also using scattering probability $P = 1/\tau$ (the number of scattering per unit time) to represent the electron wave scattering.

There are many scattering mechanisms in metal crystals, such as phonon scattering, ionized impurity scattering, neutral impurity scattering, dislocation scattering, carrier scattering, and grain boundary scattering. If the above scattering probability is expressed as P_1, P_2, P_3, P_4, P_5, and P_6, respectively, there will be

$$P = P_1 + P_2 + P_3 + P_4 + P_5 + P_6. \tag{7.21}$$

Among them, only phonon scattering is related to temperature T, and the contribution of phonon scattering to resistivity at different temperatures is as follows:

At high temperature,

$$\rho_T = \frac{2m}{n \cdot e^2}\rho_1 = \frac{C \cdot T}{4\theta_D}. \tag{7.22}$$

At low temperature,

$$\rho_T = \frac{124.4CT^5}{\theta_D^5}, \tag{7.23}$$

Table 7.1: Physical parameters for the electrical conductivity of block metals.

Metal	θ_D (K)	Resistivity $\rho(\mu\Omega\cdot\text{cm})$ $T = 0°\text{C}$	Resistance temperature coefficient $\alpha(10^{-3}/\text{K})$ T = Room temperature	Electron concentration $n(10^{22}/\text{cm}^{23})$	Fermi level E_F (eV)
Ag	220	1.49	4.30	5.9	5.5
Al	400	2.50	4.60	18.5	11.8
Au	185	2.06	4.02	5.9	5.5
Bi	80	100	4.45	2.75×10^{-5}	0.0276
Cr	630	13.2	3.01	—	6.3
Cu	310	1.55	4.33	8.5	7.0
Fe	355	8.6	6.51	—	4.4
K	99	6.1	6.73	1.3	1.9
Na	160	4.28	5.46	2.6	2.5
Ni	320	6.14	6.92	—	4.7
Pt	225	9.81	3.96	—	
Ti	278	42	5.46	6	5.6
W	315	4.89	5.10	—	

where C is a constant, and θ_D is the Debye temperature. If the contribution of other scattering mechanisms to resistivity ρ is summed up as P_i, then resistivity can be expressed as

$$\rho = \rho_T + \rho_i. \tag{7.24}$$

As temperature T approaches 0 K, ρ_T approaches zero and resistivity ρ approaches ρ_i. Equation (7.24) is called Matthiessen's law. Table 7.1 shows some representative physical parameters of bulk metal conductivity.

7.2.1.2 *Conductive properties and shape effects of continuous metal films*

The various micro-defects in a continuous metal film make it quite different from a block metal in conducting properties, thus forming the following characteristics:

(1) Film resistivity ρ_F is closely related to film thickness d, and the resistivity gradually decreases with the increase of film thickness and tends to a stable value. Film resistivity ρ_F is always greater than block metal resistivity ρ_B.
(2) The temperature coefficient α_F, Hall coefficient R_H, and thermoelectric potential of the film resistivity ρ_F are all related to the film thickness.
(3) The resistivity of the film is ρ_F, which gradually decreases and tends to stabilize with the increase of the grain size of the film.

The phenomenon that the conductive property of a continuous metal film is related to the thickness of the film is called the shape effect, as shown in Figure 7.12. Suppose that the thickness of the film d is along the z axis, numerically close to the average free path (λ) of the conducting electron. The length is along the x-axis, and the width is along the y-axis. Compared with the thickness, the length and width of the film are approximately infinite. Appling an electric field E in a certain direction, the electrons in the film will move in the x direction. If this electron is at an angle α or β to the x-axis, it has a velocity component v_z in the z-axis. It is therefore possible to change the direction of motion by colliding with the surface of the thin film in the z direction less than the average free path of the

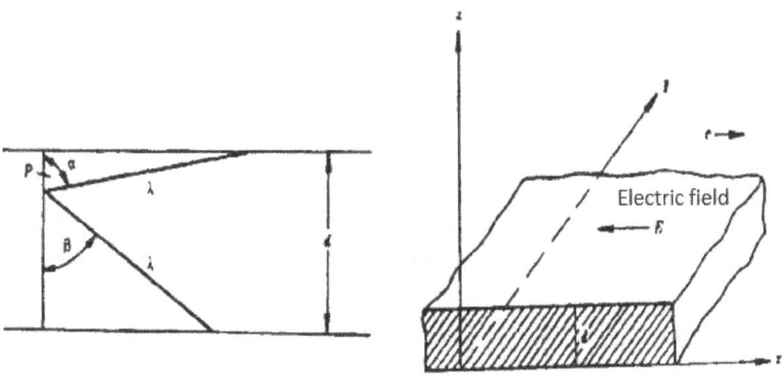

(a) (b)

Figure 7.12: Sectional view of a continuous film. (a) Cross-section and (b) three-dimensional view.

electron, and affecting the velocity along the x-axis. The greater the effect, the smaller the thickness of the film.

There are two kinds of reflections that can occur when a conducting electron collides with the surface of a thin film: elastic reflection and inelastic reflection. The ratio of the total number of electrons reflected in an elastic reflection (or specular reflection) to the total number of reflected electrons is called the specular reflection coefficient and is expressed as P, while the ratio of the number of electrons reflected in an inelastic reflection (or diffuse reflection) to the total number of reflected electrons is expressed as $(1 - P)$.

According to this physical model, the Boltzmann equation of the conductive electron distribution function $f(\boldsymbol{r}, \boldsymbol{v}, \boldsymbol{t})$ is solved to obtain the functional relationship between continuous metal film resistivity ρ_F and film thickness d:

$$\text{When } d \gg \lambda, \quad \rho_F = \rho_B \left[1 + \frac{3\lambda}{3d}(1 - P)\right]. \tag{7.25}$$

$$\text{When } d \ll \lambda, \quad \rho_F = \rho_B \cdot \frac{3}{4} \cdot \frac{\lambda}{d(1 + 2P)} \cdot \frac{1}{1n(\lambda/d)}, \tag{7.26}$$

where ρ_B is the block metallic resistivity and P is the elastic scattering coefficient.

The following conclusions can be obtained from the above two equations:

Film resistivity ρ_F is greater than block metal resistivity ρ_B; film resistivity ρ_F is related to film thickness d, and its relation curve is shown in Figure 7.13.

7.2.1.3 *The conductive properties of discontinuous metal films*

A film goes through several stages before it becomes a continuous film. The film is usually discontinuous if its thickness is less than 5 nm.

Experimental results show that the main characteristics of discontinuous thin film conductivity are the following:

(1) The resistivity (or block resistance) of discontinuous films is much higher than that of continuous films and block materials.

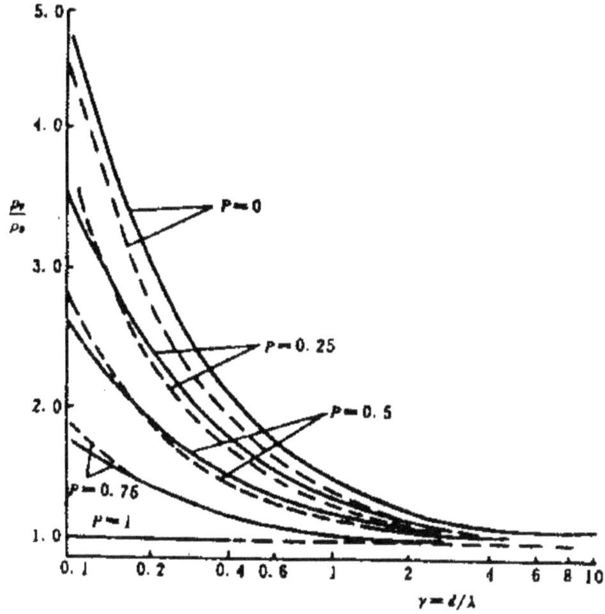

Figure 7.13: Theoretical curve of the shape effect of a continuous metal film.

(2) The relation between the resistance and temperature of discontinuous films is completely different from that of continuous films. It has the temperature characteristic of a semiconductor (or dielectric), the temperature coefficient α_F of resistance is negative.
(3) The electrical conductivity of the discontinuous film is related to the applied electric field, which presents ohmic conductivity at low electric field and non-ohmic conductivity at high electric field, and is related to the activation temperature. Electron emission and light emission are observed under high electric field.
(4) Activation energy of discontinuous films is related to film thickness.

The relationship between the conductivity and temperature of discontinuous metal films is exponential. When the variation of the island interval with temperature is ignored, the temperature coefficient of resistivity is

$$a = \frac{1}{\rho}\frac{d\rho}{dT} = -\frac{B}{Tk^2}, \tag{7.27}$$

where T is temperature, k is the Boltzmann constant, and B is the function of the work function. At this point, the temperature coefficient of the resistance of the discontinuous metal film is negative. The conduction of discontinuous films is related to electron emission; when the temperature rises, the number of emitted electrons increases, so the conductivity increases and the resistivity decreases, and the temperature coefficient of resistance is negative. However, when considering the influence of the temperature variation of island spacing (especially the large thermal expansion coefficient of the substrate) on the resistance temperature coefficient, the resistance temperature coefficient of the discontinuous film is not always negative, but may change from negative to positive. When the linear expansion coefficient is $\beta_c \gg \beta_f$, the general form of the resistance temperature coefficient can be obtained as

$$a = -\frac{d\ln\sigma}{dT} = A\frac{\Delta b}{\Delta T} - \frac{B}{kT^2}, \tag{7.28}$$

where A is the function of the work function, Δb is the change in temperature with the distance between the islands. It can be positive, negative, or zero.

7.2.2 *Electrical properties of dielectric films*

The electrical properties of dielectric thin films mainly include conductance properties, dielectric properties, piezoelectric properties, pyroelectric properties, and ferroelectric properties.

7.2.2.1 *Conductance properties of dielectric films*

The conductance of dielectric films includes dc conductance and ac conductance. When measuring the conductance of a dielectric thin film, a metal electrode must be made on its upper and lower surfaces to form a metal–media–metal MIM structure. When an ac voltage is applied to the structure, the measured impedance is the sum of the interface impedance and the dielectric thin film impedance in series. When applying a dc voltage, the measured resistance is the sum of two interface resistors and the dielectric thin film resistors in series. Therefore, the impedance or resistance measured only when the dielectric film of the electrode is in ohmic contact is the impedance or resistance of the dielectric film. According to the properties of conducting carriers, the conductance of dielectric films can be divided into ionic conductance and electronic conductance. According to the source of conducting carriers, it can be divided into the intrinsic conductance from the dielectric film itself and the non-intrinsic conductance from impurities and defects.

In dielectric films, ionic conductance and electronic conductance usually coexist. It is generally believed that the ionic conductance conforms to the following formula:

$$\frac{\sigma}{D} = \frac{N \cdot Z^2 \cdot e^2}{kT}, \tag{7.29}$$

where σ is the conductivity, D is the diffusion constant, Z is the ionic valence, e is the electron charge, and N is the ion concentration with a charge of Z_e.

The dc conductance in the dielectric film is much larger than that of bulk dielectric material because the former contains more defects and impurities than the latter. Under the strong electric field, the conductance in the dielectric film includes electronic conductance and ionic conductance. Electronic conductance mainly comes from electrons in the conduction band, including conduction electrons in the conduction band, conductance caused by the tunneling effect, electron conductance of impurity level, and space charge at the interface between the medium thin film and the metal electrode. The ionic conductance has the ion defect caused by the foreign impurity ion and the deviation in stoichiometric ratio. Under a weak electric field, its conductance comes from impurity level electronic conductance and ionic conductance. Because there are almost no free electrons in the conducting band of the dielectric film, the electron conductance of impurity level occupies the main position.

When the electric field intensity $E < 10^5\,\mathrm{V/cm}$, the ionic conductance conforms to Ohm's law. If the energy needs to be activated when the ions move, the movement distance is L; when $(e \cdot E \cdot L) \ll kT$(weak current field), the ion current density J_1 is

$$J_1 = \frac{C}{kT}\exp\left(\frac{-\phi}{kT}\right), \tag{7.30}$$

where C is a constant.

When $(e \cdot E \cdot L) = kT$, the external electric field increases the excitation and energy, and the ion current density is related to the electric field and temperature. At this point, the conductance is non-ohmic

$$J_1 = A \cdot \exp\left[-\left(\frac{\phi}{kT} - \frac{e \cdot E \cdot L}{2kT}\right)\right]. \tag{7.31}$$

When $(e \cdot E \cdot L) \gg kT$, Under the action of an external electric field, such as defect ions, high energy is obtained, which results in avalanche collision ionization and induced electron current. The current density formula is more complicated.

Under the general electric field, the conductivity of the dielectric film increases with rising temperature. Figure 7.14 shows the relationship between $\ln \sigma$ and $1/T$ of a typical dielectric thin film.

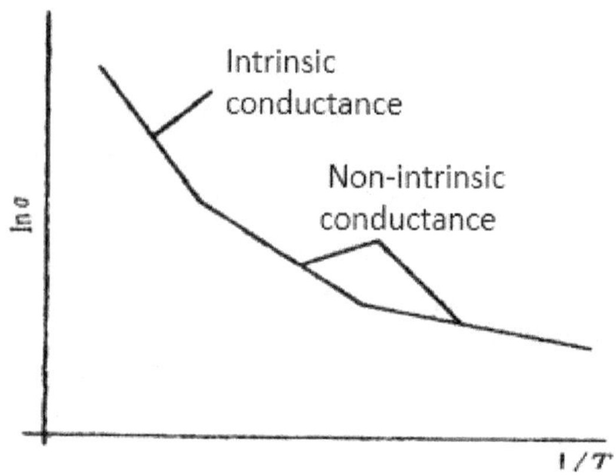

Figure 7.14: The relation between σ and T of thin films under a weak electric field.

It indicates different activation energies in different temperature ranges. When the slope of the curve of the same material is equal at high temperature, its conductance is called intrinsic conductance. In the case of medium and low temperature, the activation energy of different temperature ranges is not equal, which reflects different conductive mechanisms. Such a conductance is called non-intrinsic conductance.

7.2.2.2 *Dielectric properties of dielectric films*

Although the vacancy, dislocation, crack, oxide, stoichiometric deviation, impurities, and other defects of the dielectric film are more than those of the corresponding block material, the dielectric mechanism, loss mechanism and breakdown mechanism of the dielectric film are the same as those of the block dielectric material.

(1) **Polarization of dielectric films**

When an electric field is added perpendicular to the dielectric film, an induced bound charge appears on the surface of the dielectric film, which is called polarization of the dielectric film. Its essence is that the nuclei, electron clouds, and ions in the dielectric film move in a directional manner under the action of the electric field, and some molecules with dipole moments are arranged in a directional manner under the action of the electric field. The polarization intensity P is often used to represent the bound charge. P is the electric moment per unit volume, and the size of P of different charge carriers is proportional to the applied electric field. The carriers involved in polarization in dielectric materials are mainly electrons, ions, and dipoles.

The polarizability α_e and polarization intensity p_e of the electron are as follows:

$$a_e = 4\pi\varepsilon_0 r^3; \quad Pe = N\alpha_e E, \tag{7.32}$$

where ε_0 is the vacuum dielectric constant, r is the atomic radius, N is the number of atoms per unit volume, and E is the strength of the applied electric field.

The polarization rate a_i and the polarization intensity P_i of the ions are

$$a_i = q^2/(2f), \tag{7.33}$$

$$P_i = Na_iE, \tag{7.34}$$

where q is the ionic charge, f is the interionic force constant, and N is the concentration of the positive and negative ion pairs. This polarization is also called atomic polarization.

Dipole switching polarizability α_d and polarization intensity p_d are

$$a_d = \mu_0^2/(3kT), \tag{7.35}$$

$$P_d = Na_dE, \tag{7.36}$$

where μ_0 is the intrinsic dipole moment of the molecule, k is the Boltzmann constant, T is the temperature, and N is the dipole concentration.

Polarization intensity of dielectric polarization is

$$P_f = \frac{\sigma_1\varepsilon_2 - \sigma_2\varepsilon_1}{\sigma_1 + \sigma_2}E, \tag{7.37}$$

where σ_1, σ_2, ε_1, ε_2 are the conductivity and dielectric constant of the two dielectric films, respectively. When $\sigma_1\varepsilon_2 - \sigma_2\varepsilon_1 = 0$, $P_f = 0$.

The relation between the relative permittivity ε and the polarizability α and the polarizability P of the dielectric film is

$$P = \varepsilon_0(\varepsilon - 1)E, \tag{7.38}$$

$$\frac{\varepsilon + 1}{\varepsilon + 2} = \frac{Na}{3\varepsilon_0}. \tag{7.39}$$

The above equation is called the Clausius–Mossodei (CM) equation. In addition, due to the different structure, the dielectric constant of the dielectric film is not necessarily equal to the dielectric constant of the corresponding bulk material.

(2) **Dielectric constant of dielectric films**

According to the different polarization properties, dielectric films can be divided into two types according to the size of dielectric constant

ε_r: non-polar dielectric films and polar dielectric films. The dielectric constant of non-polar films is about 2–45, among which the lower ones are mostly organic polymer films and the higher ones are mostly inorganic oxide films. The permittivity of polar dielectric films is about 3–1000 or greater, and low permittivity is mostly found in organic polymer films, while high permittivity is mostly existing in inorganic ferroelectric films.

Generally speaking, the dielectric constant of a dielectric film is composed of two parts: the intrinsic dielectric constant and the non-intrinsic dielectric constant. The intrinsic permittivity is derived from the electron state, intrinsic dipole moment and lattice structure of the atoms in the film. The non-eigen permittivity is derived from the inhomogeneity, impurity, vacancy, interstitial ion, stress, segregation on the grain boundary layer, oxide, etc. The intrinsic permittivity is determined by various polarization mechanisms inside the film. Among them, the main ones are electron polarization, ion lattice vibration polarization and ion deformation polarization. In addition, there are dipole polarization, thermionic polarization and electron relaxation polarization.

Since the strength of polarization is related to the total charge in the dielectric film and the strength of the interaction between charges, the intermediate electric constant of the dielectric film is related to the atomic number, which is shown in Figure 7.15. The average molecular atomic number in Figure 7.15 is the sum of the atomic numbers of the atoms in each oxide molecule divided by the number of atoms in it. The dielectric constants ε_r of some common dielectric thin films are given in Table 7.2.

(3) ac losses of dielectric films

Under an alternating electric field, a dielectric thin film will inevitably produce energy loss due to the reasons of conductance and polarization. The size of the loss value is related to the crystal structure of the dielectric film itself and various defects. Therefore, the loss value of the dielectric film is an important parameter for characterizing the quality and performance of the dielectric film, and

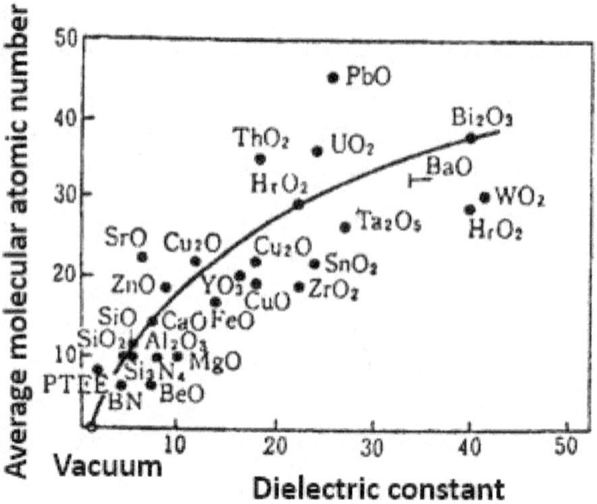

Figure 7.15: Relationship between dielectric constants and molecular mean atomic numbers of thin films.

Table 7.2: Relationship between average atomic number and preparation of common dielectric thin films.

Film	Preparation methods	ε_r	Film	Preparation methods	ε_r
BN	Reaction sputtering	4.0	Ta_2O_5	Anodic oxidation	45.0
A1N	Reaction evaporation	8.5	Ta_2O_5	Anodic oxidation Ta/N	27.0
Si_3N_4	RF sputtering	6.8	Bi_2O_3	RF sputtering	25.0
Al_2O_3	Anodic oxidation	6.0	Bi_2O_3	Reaction sputtering	38.0
SiO_2	Vapor phase growth	4.0	$MnTiO_3$	Reaction sputtering	$20 \sim 120$
TiO_2	Vacuum evaporation	100.0	$BaTiO_3$	Vacuum evaporation	$130 \sim 820$
ZrO_2	Reaction sputtering	20.0	$PbTiO_3$	Reaction sputtering	$200 \sim 400$

Table 7.3: Preparation methods and dielectric losses of various dielectric thin films.

Film	Preparation methods	tgδ (%)	Film	Preparation methods	tgδ (%)
Si_3N_4	RF sputtering	0.1	ZrO_2	Anodic oxidation	1.0
Al_2O_3	Anodic oxidation	0.5	ZrO_2	Reaction sputtering	0.8
Al_2O_3	RF sputtering	0.4	Ta_2O_5	Anodic oxidation	1.0
SiO	Vacuum evaporation	0.1	Ta_2O_5	Reaction sputtering	0.9
SiO_2	Reaction sputtering	0.01	$MnTiO_3$	Reaction sputtering	3–17.0
SiO_2	RF sputtering	0.1	$BaTiO_3$	Vacuum evaporation	15.0
TiO_2	Anodic oxidation	3.0	$PbTiO_3$	Reaction sputtering	0.6
TiO_2	Reaction sputtering	6.5			

is expressed by tgδ (%). Table 7.3 lists the losses tgδ (%) at 1 kHz of various dielectric thin films.

The dielectric constant and loss tangent of dielectric films with various polarization and conductivity are related to the frequency:

$$\varepsilon = \varepsilon_\infty + \frac{\varepsilon_s - \varepsilon_\infty}{1 + (\omega\tau)^2}, \tag{7.40}$$

$$tg\delta = \frac{\gamma + \varepsilon_0(\varepsilon_s - \varepsilon_\infty)\frac{\omega^2\tau}{1+(\omega\tau)^2}}{\omega\varepsilon_0\varepsilon}, \tag{7.41}$$

where ε_∞ is the dielectric constant of the dielectric film at very high frequencies, ε_s is the dielectric constant, and ε_0 is the vacuum dielectric constant, τ is the relaxation time of dielectric film, and ω is angular frequency.

The loss of dielectric films consists of three parts. The first part is the conductivity loss, which always exists under the action of dc or ac electric fields. dc conductance loss is significant at low frequency and does not vary with frequency. The second kind of loss is the relaxation loss, which is closely related to the frequency of the alternating electric field. At high frequency dissipation, the peak frequency is above 1 MHz. The peak frequency of the low frequency relaxation loss is below 100 Hz. The third kind of loss is non-relaxation loss. For the vast majority of dielectric membranes, non-relaxation losses can be observed at room temperature and even when the electrical conductivity is completely ohmic at a low electric field. The characteristic of this loss is that tgδ is almost independent of frequency. The frequency characteristics of non-dissipation of various dielectric films are shown in Figure 7.16. This non-relaxation loss is caused by the inhomogeneity of the dielectric film, which is caused by the inhomogeneity of various microscopic defects and impurities in the dielectric film, resulting in the different microscopic environments of electrons, ions, and atoms.

The relationship between the above three losses and frequency is shown in Figure 7.17. The relationship between the dc conductance and temperature of the dielectric film is exponential. Therefore, the dielectric thin film loss is dominated by the conductivity loss at high temperature.

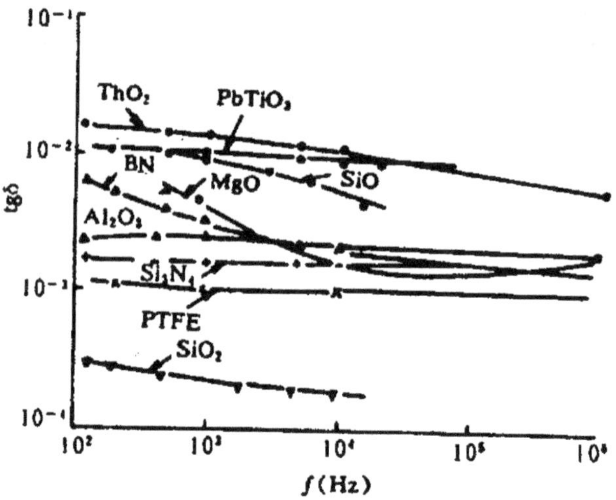

Figure 7.16: Non-dissipation frequency characteristics of dielectric films.

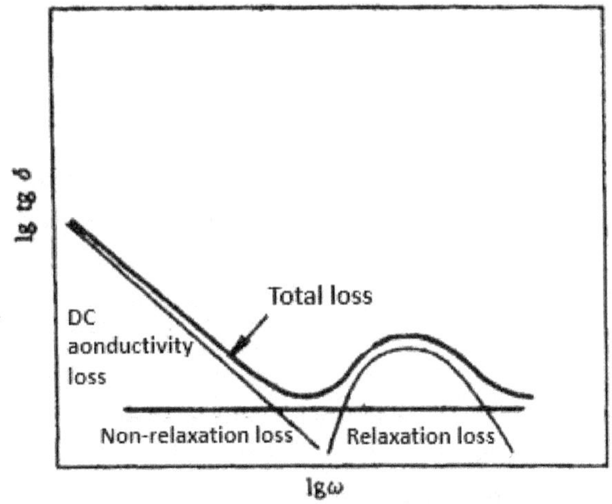

Figure 7.17: The relation between dielectric thin film loss and frequency.

7.2.2.3 *Piezoelectric properties of thin films*

The phenomenon that an electric charge on the crystal surface is caused by stress is called the positive piezoelectric effect. In contrast, when a crystal is subjected to an electric field, strain occurs in one of its directions in a linear relationship with the strength of the field, a phenomenon known as the inverse piezoelectric effect. These two phenomena are commonly referred to as the piezoelectric effect. This property of a crystal is called its piezoelectric property. In addition to some piezoelectric single crystals (quartz, $LiNbO_3$, $LiTaO_3$, CdS) and piezoelectric ceramics ($BaTiO_3$, $PbTiO_3$, $Pb(Zr, Ti)O_3$) and other block materials, some dielectric thin films also have piezoelectric properties, for example, CdS, ZnS, AlN, ZnO, $LiNbO_3$, PZT, and other thin films. This type of film is mainly used to make surface acoustic wave devices.

(1) **Structure of piezoelectric films**

Whether a crystalline material has piezoelectric properties depends on the symmetry of the crystal structure. Structurally, it must have positively charged and negatively charged ions or ionic clusters, so piezoelectric crystals are ionic crystals. There must be no center of symmetry in the crystal structure. If there is a symmetric center, the positive and negative charge centers of the crystal always coincide with each other without the effect of stress. In order to present good piezoelectric properties, piezoelectric films must have the above characteristics in the crystal structure. Piezoelectric films deposited in vacuum are polycrystalline structures. There are monocrystals of varying degrees among the micro-grains. Therefore, in order to make the piezoelectric film have excellent piezoelectric properties, the structure of the first has to ensure that the microcrystals basically have the same orientation. Secondly, the atomic structure of microcrystals should be perovskite with a cubic crystal structure, sphalerite with a cubic crystal structure, and wurtzite with a hexagonal crystal structure. The atomic surface index and microcrystalline structure of wurtzite are shown in Figure 7.18. Typical structures of piezoelectric thin films are listed in Table 7.4.

(2) **Piezoelectric property parameter**

The following physical parameters are often used to characterize the piezoelectric properties of thin films.

(a) *Electromechanical coupling coefficient k*

The electromechanical coupling coefficient k reflects the coupling relationship between mechanical energy and electrical energy of piezoelectric films. Mechanical stress applied to piezoelectric materials deforms them. The positive piezoelectric effect converts a portion

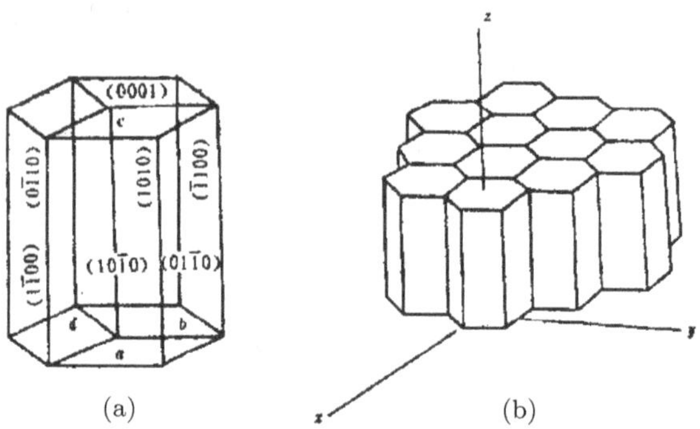

Figure 7.18: (a) Atomic number and (b) crystal structure of hexagonal wurtzite.

Table 7.4: The crystal structure of typical piezoelectric films.

Dielectric film	Substrate	Crystal structure	[0001] Preferred orientation
ZnO	Glass	Hexagonal	Perpendicular
			Inclined
			Horizontal
CdS	Mica, rock salt	Hexagonal	—
	Mica, rock salt	Hexagonal	Perpendicular
	Glass	Hexagonal	Perpendicular
	Glass	Hexagonal	40° angle
ZnS	Glass	Hexagonal	Perpendicular [111] direction
		Hexagonal	Perpendicular
AlN	Glass	Hexagonal	Perpendicular
		Hexagonal	Horizontal

of the input mechanical energy into electrical energy, which is then returned to the electrical load of the piezoelectric material. The remaining mechanical energy deforms the piezoelectric material and is stored in the piezoelectric material as elastic energy. Conversely, applying an external electric field to a piezoelectric material will polarize it. A portion of the input electrical energy is converted into mechanical energy by the reverse piezoelectric effect to deform the piezoelectric material, while the rest is stored in the electrical material in the form of polarization energy. The electromechanical coupling coefficient is used to describe the above energy conversion process, which is defined as

$$k^2 = \frac{\text{electrical energy converted to mechanical energy}}{\text{input electrical energy}} \tag{7.42}$$

or

$$k^2 = \frac{\text{mechanical energy converted to electrical energy}}{\text{input mechanical energy}} \tag{7.43}$$

(b) *Piezoelectric coefficient d*

When stress σ and electric field E are used as variables, the piezoelectric coefficient d is used to describe the polarization intensity under unit stress or the strain under unit electric field. The unit of the piezoelectric coefficient d is coulomb/newton or meter/volt.

The positive piezoelectric effect can be described by the following formula:

$$P = d\sigma, \tag{7.44}$$

where σ is the stress, P is the polarization intensity, and d is the piezoelectric constant.

The negative piezoelectric effect can be described by the following formula:

$$S = dE, \tag{7.45}$$

where S is elastic strain and E is electric field strength.

(c) *Mechanical quality factor*

Mechanical quality factor is used to describe the mechanical energy loss of piezoelectric materials at resonance, expressed by Q_m. It is defined as

$$Q_m = 2\pi \frac{\text{maximum elastic energy stored by a pizeoelectric vibrator } w_m}{\text{loss of mechanical energy in each cycle}}. \tag{7.46}$$

(d) *Electrical quality factor*

Piezoelectric materials lose some energy through conductance and polarization relaxation under the action of an alternating electric field. The electrical quality factor describes this energy loss and expresses it as Q_e. Electrical quality factor is defined as the ratio of the reactive current I_e to the active current I_R through a piezoelectric material or the reciprocal of the tangent of the dielectric loss angle:

$$Q_e = \frac{I_e}{I_R} = \frac{1}{tg\delta}. \tag{7.47}$$

7.2.2.4 *Pyroelectric properties of thin films*

The phenomenon of spontaneous polarization intensity change due to temperature change in crystals with spontaneous polarization is called the pyroelectric effect. This property of a crystalline material is called pyroelectric.

The pyroelectric effect is the relative displacement of positive and negative charges in some piezoelectric crystals to form electric polarization caused by the change of temperature, which results in different sign bound charges on the surfaces of both ends of the crystal. On the contrary, when the external electric field is applied to the pyroelectric transistor, the change in electric field will cause a change in crystal temperature, which is called the electrothermal effect (also called the electrocaloric effect).

In addition to the absence of a symmetrical center in the structure, pyroelectric crystals also require spontaneous polarization. Therefore, it is only when there is a single polarization axis in the crystal that is different from the other polarization axes, that the pyroelectric property appears.

(1) **Basic parameters of the pyroelectric effect**

When thermal radiation is irradiated to the pyroelectric crystal (see Figure 7.19), the change in spontaneous polarization intensity P_s of the crystal with temperature T, dP_s/dT, will cause a change in the surface charge density of the crystal. The voltage signal V_p is then

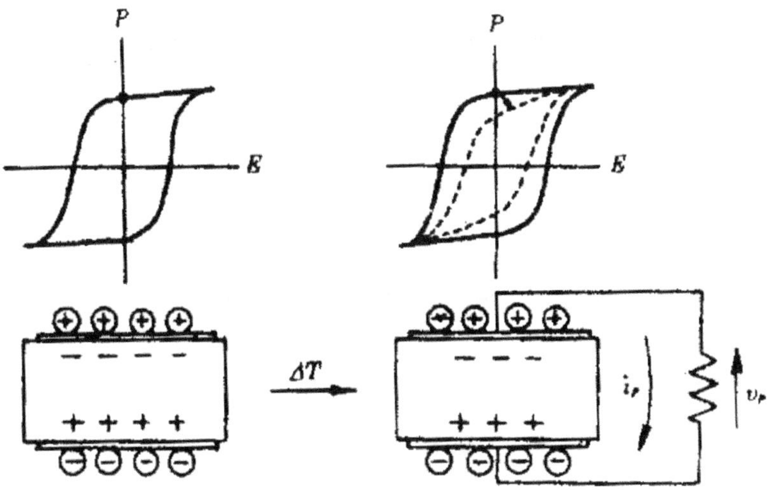

Figure 7.19: Principle of the pyroelectric effect test.

generated over the load resistor R_s

$$V_p = S \cdot R_s \cdot \frac{dp_s}{dT} \cdot \frac{dT}{dt}, \tag{7.48}$$

where S is the effective area of the electrode, R_s is the load resistance, dP_s/dT is the pyroelectric coefficient ($\mathrm{C/cm^2 \cdot K}$), and dT/dt is the temperature change rate. The heat balance current generated in the circuit is

$$i_p = \frac{V_p}{R_s} = S \cdot \frac{dP_s}{dT} \cdot \frac{W}{L \cdot \rho \cdot j \cdot C_p}, \tag{7.49}$$

where L is the crystal thickness, W is the thermal radiation power, ρ is the crystal density, J is the thermal equivalent, and C_P is the specific heat of the crystal.

The rate of change in spontaneous polarization intensity P_s with temperature T, dP_s/dT, is called the pyroelectric coefficient and is denoted by p.

$$p = \frac{dP_s}{dT}. \tag{7.50}$$

(2) **Pyroelectric thin films**

The bulk pyroelectric materials include single crystal materials ($LiTaO_3$, $LiNbO_3$, etc.), ceramic materials ($PbTiO_3$, PZT, PLZT, etc.) and organic polymer materials (polyvinylidene fluoride PVDF). Among them, $PbTiO_3$-based materials have successfully prepared various $PbTiO_3$ pyroelectric thin films due to their advantages of high pyroelectric coefficient, low dielectric constant, high Curie temperature, and low manufacturing cost (see Table 7.5). Generally speaking, pyroelectric thin films should have the following properties: small heat capacity; large pyroelectric coefficient; that is, the surface charge variable caused by temperature should be large; small dielectric constant to improve the output voltage value; small thermal diffusion coefficient, especially in the production of pyroelectric camera device.

Table 7.5: Physical properties of pyroelectric materials.

Materials	T_c(°C)	ε_r	Properties $p \times 10^{-8}$ (C/cm$^2 \cdot$ K)	Thermal diffusion coefficient 1×10^{-13} (cm^2/s)	T(°C)
Triglycine sulfate (TGS)	49	50	3.5	2.6	25
Triglycine sulfate doped with alanine (LATGS)	49.5	35	7.0	—	25
Deuterated triglyceride sulfate (DTGS)	62	20	2.7	~2.6	25
Triglyceride of fluobyllium acid (TGFB)	73.8	15	2.1	2.0	30
Deuterated triglyceride of fluobyllium acid (DTGFB)	74.5	12	2.5	2.9–1.7	25
$Sr_{0.48}Ba_{0.52}NB_2O_5$	105	380	6.5	—	25
$Pb_5Ge_3O_{11}$	178	40	1.1	8	25
$PbTiO_3$	470	200	6	9.9	25
$PbTi_{0.07}Zr_{0.93}O_3$	~240	300	2.6	—	25
$LiTaO_3$	618	43	1.7	13	25
Poly-difluoroethylene (PVF_2)	120	11	0.3	0.53	25
$PbSr_{0.5}Ta_{0.5}O_3$	−10	1800	24	—	25

7.2.2.5 *Ferroelectric properties of thin films*

(1) **Characteristics of ferroelectrics**

Some crystals have spontaneous polarization within a certain temperature range, and the direction of spontaneous polarization can change with the direction of the applied electric field. Ferroelectric crystals are called ferroelectric bodies. Of course, ferroelectrics do not always contain iron.

Ferroelectrics have three important characteristics: one is that it has a hysteresis loop, which is an important basis for judging the crystal as ferroelectrics. The second is that there is a critical temperature, that is, the ferroelectric Curie temperature T_c, which is the transition temperature between the paraelectric phase and the ferroelectric phase of the crystal. Thirdly, ferroelectric crystals have critical properties, such as dielectric properties, elastic properties, thermal properties, and optical properties in the vicinity of the critical temperature abnormal phenomenon.

Ferroelectrics themselves are a subfamily of dielectric crystals, piezoelectric crystals, and pyroelectric crystals, so ferroelectrics must have dielectric, piezoelectric, and pyroelectric properties, as well as electro-optical properties for translucency ferroelectrics.

(2) **Properties of ferroelectric thin films**

There are many kinds of ferroelectric crystals and ceramics, and their typical performance values are shown in Table 7.6. However, it is easier to make ferroelectric thin films and the dielectric materials studied more are $BaTiO_3$ and $PbTiO_3$.

Compared with the block material preparation process, the ferroelectric film is prepared at relatively low temperature, so the following problems need to be solved to reproduce the ferroelectric properties of the block materials.

(a) *Chemical component deviation*

If the binary or multicomponent materials are thin-film, due to the different sputtering rate of various atoms and the influence of local

Table 7.6: Properties of typical ferroelectric materials.

Material	T_c (°C)	P_s (μc/cm^2) [T(°C)]	Material	T_c (°C)	P_s (μc/cm^2) [T(°C)]
KH_2PO_4	−150	4.8(−177)	$PbTiO_3$	490	>50(23)
Glycosulphate sulphate	49	2.8(20)	$KNbO_3$	435	30(250)
SbSI	≈20	25(0)	$LiNbO_3$	1210	71(23)
$BaTiO_3$	135	26(23)	$LiTaO_3$	665	23(450)

electric field on the substrate, the molecular ion deposition states of different charge symbols are different. Too high a base temperature can cause low-melting point elements to evaporate. The interaction and diffusion between the substrate and the film also have an effect on the components of the interface layer. The deviation of components has a great influence on the properties of ferroelectric films. In $BaTiO_3$ and $PbTiO_3$ films, when the ratio of Ti and Ba or Pb is not equal to 1, the dielectric constant drops sharply. The electron exchange causes the insulation resistance to decrease when the Ti^{+3} ions are formed during lack of oxygen and when they transition to the Ti^{+4} ions. In order to prevent component deviation, the low-melting point component should be excessive in the sputtering target material. For example, when preparing PLZT films, PbO should be allowed to overdo 3–8% of Pb losses to compensate for the loss of Pb.

(b) *Substrate temperature*

The high temperature of the substrate causes the low-melting point material to evaporate or cause hypoxia. Low temperature also affects the crystallization of the film. For example, when the substrate temperature is below 300°C, the $PbTiO_3$ film is amorphous, the dielectric constant is small, and there is no ferroelectricity. Stable polymorphisms are often formed when the temperature increases. For example, $BaTiO_3$ of cubic and quad crystal structures, or the coking turquoise structure slabs appear in PLZT films. At a temperature above 500°C, the $BaTiO_3$ film forms a calcium–titanium structure with good ferroelectric property. PLZT can form a calcium–titanium ore structure in the range of 500–700°C. When the temperature is too high, the film forms a microcrystalline state.

Applying a certain bias voltage to the substrate or selecting an appropriate surface can make the iron-electric film opt-out orientation even for a single-crystal extension film. If the lattice constant is close to the platinum sheet substrate of PLZT, sputtering deposition of the PLZT film at a lower temperature can obtain better crystallization performance.

(c) *Uniformity*

The iron-electric film that has just sputtered its deposit is structurally uneven. In particular, the interface layer in contact with the substrate forms a transition zone with a high defect density due to diffusion. The disordered structure and unevenness of amorphous states can form micro-regions with different components. Non-chemical metering areas are often formed in the surface layer of the film due to lack of oxygen. This area is a semiconductor, and its application is oxygen deficiency. Microcracks in film sepsis is another form of uniformness; the reason is closely related to the large thermal stress of the film.

(d) *Dimensional effect*

The thickness of the iron-electric film d and the size of grain size D have a direct impact on the performance of the ferroelectric film. Experiments show that both d and D have a critical value, about 20 nm. When less than this value, the film does not exhibit ferroelectricity, spontaneous polarization intensity $P_s = 0$; even if d is slightly greater than the critical value, the film iron electricality is also weak. When $d < 50$–$100\,\text{nm}$, both P_s and ε_r are smaller. When the thickness of the $BaTiO_3$ film increases from $0.1\,\mu\text{m}$ to $0.8\,\mu\text{m}$, the dielectric constant increases from 200 to 1350. Studies on $BaTiO_3$ ferroelectric films prepared using the sol–gel process show that the quality and orientation of the film are related to the type of substrate and the temperature of heat treatment. $BaTiO_3$ film prepared on a quartz or MgO (100) substrate treated by heat treatment above 900°C to leads an extended iron electro-electric film with perovskite structure and good c-axis orientation, but only a polycrystalline film on microcrystalline glass and silicon substrates. For a PZT ferroelectric film deposited on the (100) orientation of a polished silicon substrate by radio frequency sputtering method, the dielectric constant is 100–200, the critical temperature is about 250°C, and its coercive field $E_c = 43\,\text{kV/cm}$, which was measured from the P–E loops, and the spontaneous polarization $P_s = 30\,\mu\text{C/cm}^2$, and the residual polarization $P_r = 11\,\mu\text{c/cm}^2$ at room temperature. The experimental results also show that the film is amorphous when the base plate temperature is less than 350°C, and when the temperature is higher than 450°C, the film has a perovskite structure. When using silicon wafers as a substrate, we can obtain a better quality PZT ferroelectric film when the substrate temperature is in the range of 450–600°C.

7.2.2.6 *Breakdown of dielectric films*

When the electric field strength applied to the dielectric film reaches a certain value, it immediately loses its insulation, a phenomenon known as breakdown. Although the operating voltage is not high in many cases, it is naturally high in strength due to the small thickness of the medium film. For example, in semiconductor MOS devices, a gate medium film thickness of about 1000 Å, plus a gate bias of 10 V, ignoring the surface space charge, in the media film produces up to 10^6 V/cm of electric field strength. The dielectric film has little insulation resistance when it breaks through. If the electric field is still continuously added to the dielectric film, there is a large current by burning it; this breakdown is called hard breakdown. Some dielectric films are not burned when they are pierced but maintain a long-term low-resistance state for a long time, which is called soft breakdown. Some commonly used media films have a strong breakdown field as shown in Table 7.7. As can be seen from Table 7.7, for the same dielectric film, due to different manufacturing methods, its breakdown field strength has a large difference. The reason for this difference is that the defects of pinholes, microcracks, filaments and miscellaneous substances produced by different preparation methods in the dielectric film are different.

If classified by the breakdown mechanism, it can be divided into two categories: when the external electric field exceeds the anti-electric strength of the dielectric film itself, the breakdown is called the intrinsic breakdown, and the breakdown caused by the film defect is called the extrinsic breakdown.

The intrinsic breakdown is a breakdown produced by the combined action of electric breakdown and heat breakdown. Electric breakdown is an electronic avalanche puncture formed by the production of electron multipliers under the action of a critical electric field

Table 7.7: Strong breakdown field of media film (10^6 V/cm).

Film	Manufacturing methods	Strong breakdown field	Film	Manufacturing methods	Strong breakdown field
Si_3N_4	RF sputter	1	ZrO_2	Anode oxidation	4
Al_2O_3	Anode oxidation	6–8	ZrO_2	Reaction sputtering	4
Al_2O_3	RF sputter	1	Ta_2O_5	Anode oxidation	6
SiO	Vacuum evaporation	2	Ta_2O_5	Reaction sputtering	1
Si_2O	Reaction sputtering	1.5	Bi_2O_3	Reaction sputtering	0.7
Si_2O	Thermal growth	10	$BaTiO_3$	Inner light evaporation	0.35
TiO_2	Anode oxidation	1	$PbTiO_3$	Reaction sputtering	2.3

E_c in a dielectric film, which usually occurs in a very short period of time. The energy that electrons derive from the electric field is mainly used in the collision ionization process. When an electrical breakdown occurs, the current avalanche increases, producing a large amount of joule heat, and the dielectric film temperature rises rapidly and turns into a hot puncture. The dielectric film conductivity increases exponentially with the increase of temperature, then the current increases, the joule heat increases, and the dielectric film temperature increases further. As the temperature of the dielectric film increases rapidly, resulting in local heat decomposition, volatilization or melting then further contribute to the production of heat breakdown.

7.2.3 *Electrical properties of semiconductor films*

Semiconductors are a class of substances with resistivity between those of conductors and insulators, and their resistivity is generally 10^{-3}–$10^9 \Omega \cdot$cm. Semiconductors are very sensitive to impurities and defects. Since impurities and defects introduce energy levels into the band, they will have a greater impact on the physical and chemical properties of semiconductors. Therefore, the electrical properties of semiconductors will change greatly under the action of light, heat, electricity, magnetism, and other fields. This section focuses on the electrical properties of semiconductor films.

7.2.3.1 *Semiconductor conductivity and photoelectric conductivity*

Conductivity is the inverse of resistivity, in Siemens/m, or S/m. The migration rate reflects the conductivity of carriers in semiconductors, which represents the average drift velocity of electrons under a strong unit field. The relationship between the conductivity and migration of semiconductors is

$$\sigma = nq\mu_n + pq\mu_p. \tag{7.51}$$

Here, n, p represent the concentration of electrons and holes, respectively, and the migration rates of electrons and holes are represented by μ_n and μ_p, and q is the battery.

As can be seen from Equation (7.51), the greater the migration rate of the carrier, the greater the conductivity of the material. In different semiconductors, the migration rates of electrons and cavities are not the same. For the same material, the migration rate of the carrier is also affected by the type and concentration of doped substances; if doping is different, the migration rate is also different in size. The migration rate of the carrier also varies with the temperature. Since the concentration of carriers in doped semiconductors remains basically unchanged in the device's operating temperature range, the conductivity varies with temperature mainly from the migration rate.

Photoelectric conducting refers to the phenomenon that an increase of the conductivity of semiconductors is caused by light; that is, light makes the semiconductor form unbalanced additional carriers, and this increase of the concentration of carriers will inevitably increase the conductivity of the sample.

Conductivity of semiconductor samples (also known as dark conductivity) when there is no light is

$$\sigma_0 = q(n_0\mu_n + P_0\mu_p), \tag{7.52}$$

where n_0, P_0 is the equilibrium carrier concentration.

With light, the conductivity of the specimen is

$$\sigma = q(n\mu_n + p\mu_p), \tag{7.53}$$

where $n = n_0 + \Delta n$, $p = P_0 + \Delta P$, and Δn and Δp are the additional carrier concentrations of light injection, respectively. So, the additional photoelectric guide (referred to as photoelectric guide) is

$$\Delta\sigma = q(\Delta n\mu_n + \Delta p\mu_p). \tag{7.54}$$

The relative value of the original photoelectric guide can be derived from Equations (7.52) and (7.54)

$$\frac{\Delta\sigma}{\sigma_0} = \frac{(1+a)\Delta n}{an_0 + p_0}, \tag{7.55}$$

where $a = \mu_n/\mu_p$. As can be seen from Equation (7.55), to make a highly sensitive photoresistor, n_0 and p_0 should be made smaller. As a result, photoresists are generally made of high-resistance materials or used at low temperatures.

7.2.3.2 *Photovoltaic effect*

When a non-uniform semiconductor (such as a PN junction) is irradiated with the appropriate wavelength of light, an electric potential, called photovoltage, is generated inside the semiconductor. If the PN junction is shorted, a current, called photoelectric current, appears in the outer circuit. This effect is called the photovoltaic effect. The photovoltaic effect should be the theoretical basis of photocells.

There are three types of currents when the photocell operates: photoelectric current I_L, a PN junction positive current I_F acting under the effect of photovoltage V, which flows through the outside circuit of the current I_O, I_L and I_F flowing through the inside of the PN junction, but in the opposite direction. According to the PN junction rectifier equation, under the action of positive bias V, the forward current passing by the junction is

$$I_F = I_S(e^{\frac{qv}{kT}} - 1), \tag{7.56}$$

where V is the photovoltage, I_S is the reverse saturation current, k is the Boltzmann constant, and T is the absolute temperature.

When a certain intensity of light irradiates the photocell, due to the existence of absorption, the intensity of light drops with the depth of light penetration according to the exponential law, so the production rate of photogenerated carriers also decreases with the depth of illumination; that is, the yield G is a function of depth x. For simplicity, $\bar{G}$ indicates the average yield of non-equilibrium carriers in the diffusion length $(L_P + L_n)$ on both sides of the junction, and the holes in the diffusion length L_P and electrons within the L_n can be diffused to the PN junction and into the other side. In this way, the optical current I_L is

$$I_L = q\bar{G}A(L_P + Ln). \tag{7.57}$$

A is the PN junction area, E is the electronic power. Optical current I_L flows from area N to P, opposite to I_F. If the photocell is connected to the load resistor, the current through the load should be

$$I = I_L - I_F = I_L - I_S(e^{qv/kt} - 1). \tag{7.58}$$

This is the relationship between current and voltage on the load resistor, that is, the Volte characteristics of the photocell; the curve is shown in Figure 7.20.

The curves (1) and (2) in Figure 7.20 are voltage–current characteristics of the photocell with and without light, respectively.

Available from previous discussions:

$$V = \frac{kT}{q} \ln\left(\frac{I_L}{I_S} + 1\right). \tag{7.59}$$

In the case of the PN junction opening ($R = \infty$) the voltage at both ends is the open-circuit voltage V_{oc}. At this point, the current that

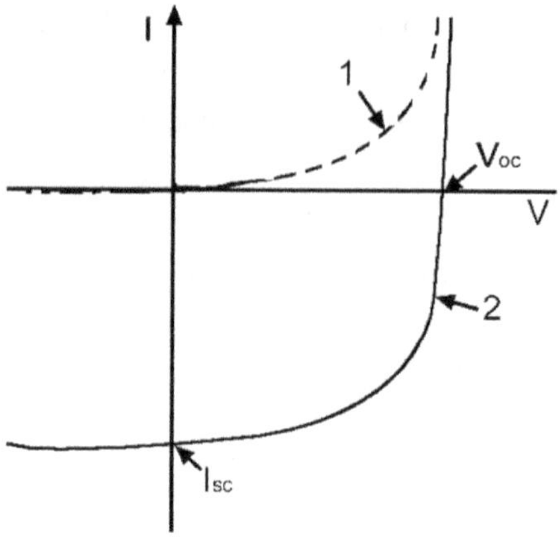

Figure 7.20: Voltage–current characteristic of the photocell.

flows through R is $I = 0$, i.e. $I_L = I_F$. Substitute $I = 0$ in Equation (7.59) to get the open-circuit voltage

$$V_{\text{oc}} = \frac{kT}{q} \ln\left(\frac{I_L}{I_S} + 1\right). \tag{7.60}$$

If the PN junction is shorted ($V = 0$), so $I_F = 0$, the resulting current is the short-circuit current I_{sc}. The short-circuit current should be equal to the photoelectric current, i.e.,

$$I_{\text{sc}} = I_L. \tag{7.61}$$

V_{oc} and I_{sc} are two important parameters of photocells, and their values can be obtained by the intercept of curve (2) on the V and I axis in Figure 7.21. Obviously, both increase with the increase in light intensity; the I_{sc} increases linearly with the intensity of light, and the V_{oc} increases in pairs, as shown in Figure 7.21. V_{oc} does not increase indefinitely with the intensity of light. When the photovoltage V_{oc} increases to the PN junction force disappear, the maximum photovoltage V_{max} is obtained, Therefore, V_{max} is equivalent to the PN junction height V_D, which is related to the degree of material doping. In practice, V_{max} is equivalent to the band width E_g.

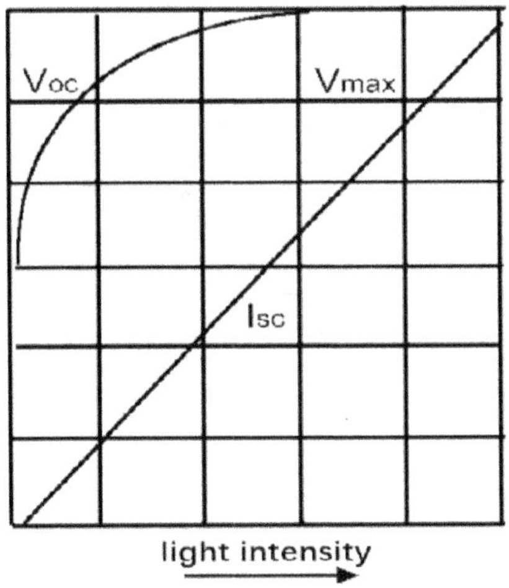

Figure 7.21: V_{oc} and I_{sc} vary with light intensity.

7.2.3.3 *Measurement of electrical properties of semiconductor films*

Common measurement methods are resistance measurement, capacitance measurement, voltage measurement, and other analytical methods.

(1) **Resistance measurement**

In the measurement of the electrical properties of semiconductor thin films, the simplest and most commonly used method is the four-terminal resistance test method. For example, the main method of detecting the resistance of silicon monocrystalline is the four-probe method; the principle is shown in Figure 7.22 using four electrodes including a pair of current ends and a pair of voltage ends. This method is not affected by contact resistors as long as a measuring instrument with high input impedance is used in the voltage measurement. Since the thickness of the film is relative to the probe spacing S, it can be considered as an infinite thin layer. When the current flow between the outside ends is I and the potential difference between the inner ends is V, the thin-film resistance (also known as block resistance) R_s can be calculated according to the

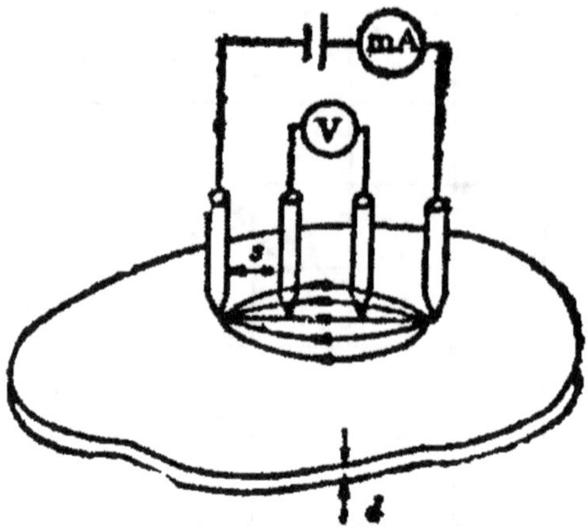

Figure 7.22: Schematic diagram of four-probe method to measure thin-film resistance.

following formula:

$$R_s = \frac{\pi}{\ln 2}\frac{V}{I}. \tag{7.62}$$

$\frac{\pi}{\ln 2}$ is due to the current expansion, R_s is expressed in ohm (Ω).

If the electrical properties of the film are uniform in the direction of film thickness, when the film thickness is d, the resistivity of the film is

$$\rho = R_s d = \frac{\pi}{\ln 2}\frac{V}{I}d. \tag{7.63}$$

If the electrical properties of the film are not uniform in the direction of film thickness, the resistivity can also be calculated by Equation (7.63); the result is an average resistivity.

(2) **Capacitor–voltage method**

A widely used method for measuring the distribution of impurity concentrations in semiconductor films is to measure the electrostatic capacity of the surface depletion layer. Since the exhaust layer that extends inward from the semiconductor surface potential does not conduct electricity in itself, a capacitor is formed between the inside

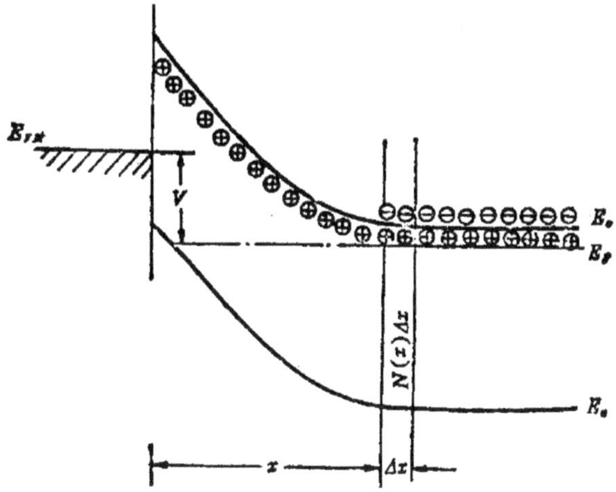

Figure 7.23: Measuring the distribution of impurity concentrations in n-type semiconductors.

of the semiconductor (i.e., the inside of the film) and the surface of the film. Its electrical capacity is inversely proportional to the width of the exhaust layer, which in turn reflects the concentration of impurities in the film.

In order to control the surface potential, special electrodes need to be made on the outer surface of the capacitor, and then a reverse voltage is added between the metals and semiconductors. As shown in Figure 7.23, the exhaust layer width is x, and when the bias increases to ΔV, the exhaust layer width increases to Δx, then the space charge change of the depleted layer is

$$\Delta Q = qN(x)\Delta x, \tag{7.64}$$

where $N(x)$ is the concentration distribution of impurities. Therefore, the power field increment in the depletion layer

$$\Delta E = \frac{\Delta Q}{\varepsilon}. \tag{7.65}$$

Among them, ε is the dielectric constant of the semiconductor film. Thus, the relationship between the change in voltage and the change in the width of the exhaust layer is

$$\Delta V = x \cdot \Delta E = qN(x) \cdot \frac{x}{\varepsilon} \Delta x. \tag{7.66}$$

As a result, the capacity of the exhausting layer is

$$C = \frac{\Delta Q}{\Delta V} = \frac{\varepsilon}{x}. \tag{7.67}$$

The depth of the depletion layer can be derived from the above two:

$$x = \frac{\varepsilon}{C}. \tag{7.68}$$

Thus, as long as we measure the electrical capacity changes with the voltage we can obtain the impurity concentration in the depth direction of the distribution, Figure 7.24 is a capacitor–voltage measurement of a GaAs film to obtain the impurity concentration results. The curve (1) is the distribution of impurity concentration in the n-type GaAs film with extended growth on a high-concentration n-type substrate; The curve (2) is the distribution of impurity concentration in the n-type GaAs film produced by the same method on the high resistivity GaAS substrate.

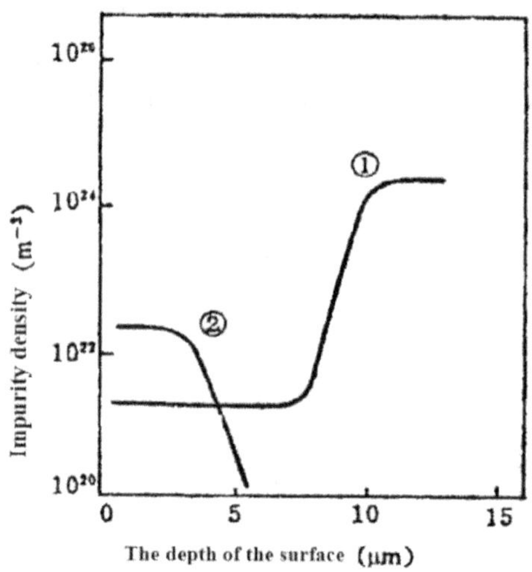

Figure 7.24: ***C*–*V* measurement of impurities in n-type GaAs films.**

7.2.3.4 *Superconductivity of films*

(1) **Superconductivity**

The resistance of a conductor is caused by the collision of electrons and lattices. When the temperature is decreased to reduce the lattice thermal vibration, the resistance decreases. However, even if the temperature is absolutely zero (0 K), the conductor will have residual resistance due to the lattice defect and other reasons. Many metals or alloys have been found for which in the liquid helium temperature range, there will be the zero resistance phenomenon; this is the phenomenon of superconductivity. The temperature at which it is transformed to the superconducting state from the usual conductive state is the critical temperature or transition temperature of the material, expressed as T_c. Each superconductor has its specific critical temperature, for example for mercury, T_c is 4.153 K.

Superconductivity was first discovered in mercury by K. Onnes in 1911, and is also known as the complete conductivity of materials. After the discovery of the superconducting phenomenon, many scientific and technological workers have done a lot of work and achieved gratifying results in order to find the superconducting mechanism, guide new superconducting materials, and study the practical application of superconductors. In 1972, a critical temperature of 23.2 K of sanmolys was discovered. After April 1986, scientists from all over the world announced the discovery of superconductors of more than 30 K, such as barium lanthanum copper ceramics, strontium lanthanum copper ceramics, calcium lanthanum copper ceramics, barium scandium copper and yttrium barium copper oxygen ceramics, etc. After just over a year, by the end of 1987, the conversion temperature of oxide superconducting materials had increased to 100 K, as shown in Table 7.8. Currently, the highest superconducting critical transition temperature is 250 K for lanthanum hydride at high pressure.

The superconducting state of a material can be destroyed by a magnetic field, which has a bounding value and is called a critical magnetic field, which is temperature-related and can be indicated by $H_c(T)$. At critical temperature T_c, $H_c(T) = 0$. The relationship

Table 7.8: T_c and H_c values of certain superconducting substances.

Elements	T_c (K)	H_c (Oe)	Alloys or Compounds	T_c (K)	H_c (Oc)**
Al	1.19	98.8	V_3Ga	14.8	25×10^4
In	3.41	285	V_3Si	16.9	24×10^4
La(β)	5.9	1000	Nb_3Sn	18.3	28×10^4
Nb	9.2	2,000,* 3,000**	Nb_3Ga	20.2	34×10^4
Pb	7.18	800	Nb_3Ge	22.5	38×10^4
Re	1.70	200	$PbMo_3S_B$	14.4	60×10^4
Sn	3.72	308	NbN	15.7	15×10^4
Ta	4.48	825	$Yba_2Co_3O_7$	93	—
Tc	8.22	—	Bis_rC_BCuO	107	—
Th	1.37	161	TlBaCaCuO	120	—
Tl	2.39	170			
V	5.13	1290,* 7000**			

Note: $^*H_{c1}$, $^{**}H_{c2}$.

between H_c and temperature of superconducting elements is

$$H_c = H_0\left[1 - \left(\frac{T}{T_c}\right)^2\right]. \tag{7.69}$$

H_0 in Equation (7.69) is the critical value of the magnetic field when $T = 0\,\text{K}$.

The magnetic properties of superconductors are shown in Figure 7.25, and when the temperature $T > T_c$, the superconducting material is in a normal lyonate state, and the flux passes through the material as shown in Figure 7.25(a). In the case when $T < T_c$, it becomes superconducting, as shown in Figure 7.25(b). Complete exclusion from materials is known as the Messner effect. The complete absence of flux in the material is regarded as an antimagnetic field in the material that completely cancels out of the external magnetic field, which can be called complete diamagnetism.

Of course, flux is not completely excluded from superconductors. The research studies show that the track can enter a certain depth of the superconductor's surface, that is, the penetration depth of the λ_s. The superconducting London theory gives λ_s the expression

$$\lambda_s(T) = \lambda_s(0)[1 - (T/T_c)^4]^{-1/2}. \tag{7.70}$$

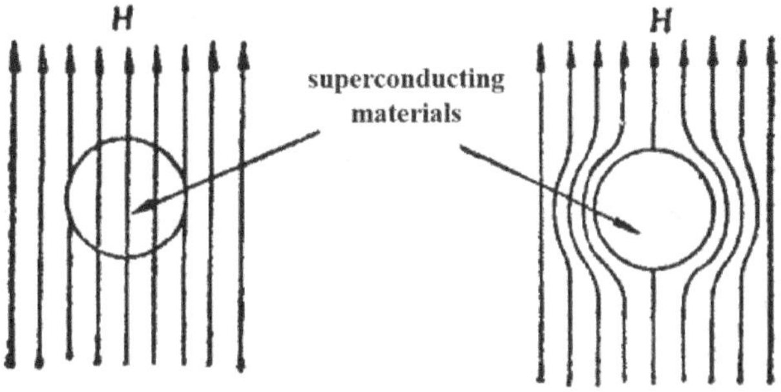

(a) (b)

(a) $T > T_c$; (b) $T < T_c$

Figure 7.25: The Meissner effect of materials in magnetic fields.

$\lambda_s(0)$ is the penetration depth when $T = 0\,\text{K}$. Generally, λ_s is 50–100 nm.

In 1957, Bardeen, Cooper, and Schrieffer gave theoretical explanations of the superconducting properties of the material (the BCS theory). BCS theory points out that in the superconducting state, the conductive electrons of the metal are in a loose connection and form an electron pair (also known as the Cooper pair), which has spin and momentum moments in the opposite direction, and the electron migration is carried out in pairs, which are formed by the gravitational pull of these spin-opposite electrons. The wave number vector K state is on both up and down energy levels, or both occupied by electrons or both. The superconductive state is the macro-physical phenomenon of these conductive electrons by coupling with the interaction of lattice sound.

Due to the weak electron coupling, the energy difference between normal and superconducting states is small, and the superconducting state is 2Δ lower than the normal energy level; thus, the band width is

$$2\Delta = 3.5\text{k}T_c. \tag{7.71}$$

This prediction has been confirmed by many superconductors.

As the temperature rises, the frequency of the amplitude of the lattice layer is increased, hindering the movement of the phonon between the Cooper electron pairs. As the attraction between electrons decreases, so does 2Δ. When $T = T_c$, $\Delta = 0$.

(2) **Superconductivity in the film**

Since the surface energy of the film state cannot be ignored, the critical magnetic field H_{cf} of the superconducting film is different from the H_{co} of the block superconductor. While the film thickness is d, and the flux into the film penetration depth is λ_L, the following relationship is established:

$$\frac{H_{\text{cf}}}{H_{\text{co}}} = \left[1 - \frac{2\lambda_L}{d} \cdot \tan\left(\frac{d}{2\lambda_L}\right)\right]^{-\frac{1}{2}}. \tag{7.72}$$

The relationship between the flux density B and the incoming depth x of the incoming film is

$$B(x) = B(0)e^{-x/\lambda_L}. \tag{7.73}$$

The flux penetration depth λ_L is approximately 10^{-6} cm. Therefore, the superconductor could not exhibit the Messner effect in the film state, when the thickness d of the film is smaller than λ_L, the flux almost uniformly through the film.

The critical temperature of some superconducting materials (such as In, Al, and other soft superconducting materials) increases with the decrease of its thin film thickness d, and this relationship of thin films is shown in Figure 7.26. The critical temperature T_{cf} of the superconductor film has been found to be much higher than the critical temperature T_c of the block-shaped superconductor of the same material. For example, for the metal Al, $T_c = 1.2\,\text{K}$, and for the film, $T_{\text{cf}} = 4.2\,\text{K}$. For the metal Bi, it does not appear superconductive under a block, but in the thin film state $T_{\text{cf}} = 15\,\text{K}$.

Table 7.9 lists the superconducting characteristics of certain metal films; the deposition method V in Table 7.9 refers to the evaporation method (under the cold substrate), CS refers to the cathode sputtering method, EB and refers to the electron beam evaporation method.

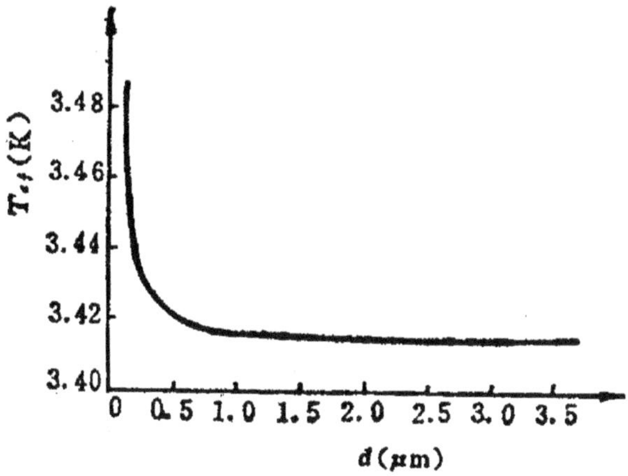

Figure 7.26: Relationship between film thickness and T_{cf}.

Table 7.9: Superconductivity properties of certain metal films.

Material	T_{cJ}/T_c	$D(\mu m)$	T (K)	Deposition method
Al	2.6	100	370–400	V
Mo	5	10–400	375	CS, EB
W	400	10–400	315	CS, EB
Sn	1.26	50	110–210	V
In	1.22		100–150	V
Be	800		925	V
Bi	600		111	V

It was also found that the relationship between film thickness and critical magnetic field strength H_{cf} is as follows:

When $d \ll \lambda_L$ (film is very thin):

$$H_{cf} = 2\sqrt{6}H_c\frac{\lambda_L}{d}. \tag{7.74}$$

When $d \gg \lambda_L$ (i.e., the film is thicker):

$$H_{cf} = H_c\left(1 + \frac{\lambda_L}{d}\right). \tag{7.75}$$

The above results are obtained in the case of a magnetic field in a direction parallel to the surface of the film; in both cases, H_{cf} all decrease with the increase of membrane thickness d. When the direction of the magnetic field is perpendicular to the surface of the film:

$$H_{\text{cf}} = \sqrt{2}xH_c, \tag{7.76}$$

where

$$x = \frac{\sqrt{2}qH_C\lambda_L^2}{hc},$$

called the Ginzburg–Landan parameter, q is the electronic power, h is the Planck constant, and c is the speed of light.

7.3 Optical Properties of Thin Films

Optical properties are one of the earliest properties used in thin films. In 1817, Fraunhofer made an optical antireflection film by acid corrosion. The optical properties of the film can be used to increase or decrease the reflection of light, can absorb or pass through the light, but also can be used for color synthesis and reduction. Common optical thin film systems are often composed of multiple layers of thin films, into which light beams will reflect at each interface, resulting in interference or absorption. Modern color photography, color television, solar cells, and integrated optics are inseparable from the film technology.

7.3.1 *Basic theory of thin film optics*

A light wave is an electromagnetic wave, and its propagation in three-dimensional space can be described by the energy vector ε and the magnetic vector H. For monochromatic plane waves with a wavelength of λ, if cosine (α, β, γ) is propagated in a given direction, the wave equation of the energy vector is

$$\varepsilon = \varepsilon_0 \exp\left\{i\left[\omega t - \frac{2\pi n}{\lambda}(ax + \beta y + \gamma z)\right]\right\}, \tag{7.77}$$

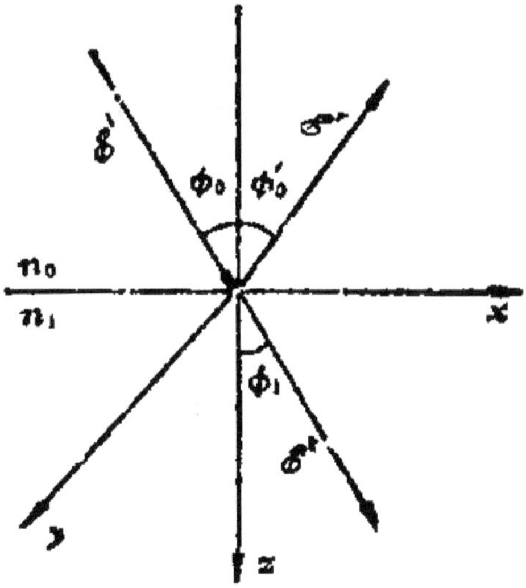

Figure 7.27: Reflection and refraction of light at the interface.

where ω is the angular frequency of light vibration, and n is the refractive index of the medium.

Suppose a beam of monochromatic plane light propagating along the xz plane is incident from the medium with a refractive index of n_0 to the medium with a refractive index of n_1 (as shown in Figure 7.27), whose boundary is the xy plane, and the incidence angle is φ_0, then the beam will reflect and refract on the interface. The incidence, reflection, and refraction of light are all in the same plane and follow the law of reflection and refraction. The law of reflection says that when light is reflected from one medium to another at the interface, the angle of reflection φ_0 is equal to the incident angle φ_0. The expression for the law of refraction is

$$n_1 \sin \phi_1 = n_0 \sin \phi_0, \tag{7.78}$$

where φ_1 is the angle of refraction; the law of refraction applies equally to transparent or absorbent media. If we know the refractive index n_0 and n_1 of the substances on both sides of the interface and the incident angle φ_0, we can obtain the refraction angle φ_1 from the law of refraction.

7.3.1.1 *Reflectivity and transmittance*

The optical properties of thin films are often expressed by their reflectivity and transmittance. First, the case of a light wave incident at the interface of two media is discussed, where the reflectivity R and transmittance T are

$$R = \frac{E_r}{E_i}; \quad T = \frac{E_t}{E_i}, \tag{7.79}$$

where E_i, E_r, E_t represent the energy of the incident light, reflected light, and transmitted light, respectively.

Based on the electromagnetic field theory, the reflectivity and transmittance of polarized light perpendicular to the incident plane can be calculated as follows:

$$R_s = \frac{\sin^2(\phi_0 - \phi_1)}{\sin^2(\phi_0 + \phi_1)}, \tag{7.80}$$

$$T_p = \frac{n_1 \cos \phi_1}{n_0 \cos \phi_0} \cdot \frac{4 \sin^2 \phi_1 \cos^2 \phi_0}{\sin^2(\phi_0 + \phi_1)}. \tag{7.81}$$

The reflectivity and transmittance of polarized light parallel to the incident plane are

$$R_p = \frac{\tan^2(\phi_0 - \phi_1)}{\tan^2(\phi_0 - \phi_1)}, \tag{7.82}$$

$$T_s = \frac{n_1 \cos \phi_1}{n_0 \cos \phi_0} \cdot \frac{4 \sin^2 \phi_1 \cos^2 \phi_0}{\sin^2(\phi_0 + \phi_1) \cos^2(\phi_0 - \phi_{1)})}. \tag{7.83}$$

As can be seen from the above

$$R_s + T_s = 1, \quad R_p + T_p = 1. \tag{7.84}$$

It shows that energy is conserved when light is reflected and refracted.

If the incident light is natural light, the light vector contains two components perpendicular to the incident plane and parallel to the incident plane, and the energy of both components is equal to half of the energy of the incident light, i.e.,

$$E_{is} = E_{ip} = \frac{1}{2}E_i. \tag{7.85}$$

The reflectivity of natural light

$$R_n = \frac{E_r}{E_i} = \frac{E_{rs} + E_{rp}}{E_i} = \frac{E_{rs}}{2E_{is}} + \frac{E_{rp}}{2E_{ip}} = \frac{1}{2}(R_s + R_p). \tag{7.86}$$

By substituting Equations (7.80) and (7.82) into the above equation, the relation between the change of reflectance of natural light and the incidence angle can be obtained as

$$R_n = \frac{1}{2}\left[\frac{\sin^2(\phi_0 - \phi_1)}{\sin^2(\phi_0 + \phi_1)} + \frac{\tan^2(\phi_0 - \phi_1)}{\tan^2(\phi_0 + \phi_1)}\right]. \tag{7.87}$$

As can be seen from the above equation, the reflectivity of natural light in the region of $\phi_0 < 45°$ is almost unchanged, and $R_n \approx 0.043$; that is, about 4% of the light energy is reflected. That is, in the case of only one reflective surface, the loss of energy is small, but in a complex optical system, the loss of light energy is considerable due to the number of reflective surfaces.

7.3.1.2 *Polarization resulting from reflection and refraction*

When the natural light is incident on the interface of two different medium, if the incident angle is to satisfy $\phi_0 + \phi_1 = \frac{\pi}{2}$, depending on Equation (7.34), then $R_p = 0$, namely no vibration component is parallel to the incident plane, the reflected light is polarized light, the electric vector vibration is only perpendicular to the reflecting surface, which is the equivalence of Brewster's law. The incident angle is called polarizing angle or the Brewster angle, as ϕ_B. Substituting the relation $\phi_\mathrm{B} + \phi_1 = \frac{\pi}{2}$ into the law of refraction, we can get

$$tg\phi_B = n_1/n_0. \tag{7.88}$$

The above formula is often referred to as Joust's formula. When the refractive index of the medium is known, the starting angle can be easily calculated by this formula. This formula also provides a simple algorithm for measuring the refractive index.

7.3.1.3 *Total reflection*

If a light wave moves from a dense medium to a thin medium (i.e., $n_1 < n_0$), according to the law of refraction

$$\frac{\sin\phi_0}{\sin\phi_1} = \frac{n_1}{n_0} < 1,$$

when $\sin\phi_1 = 1$, i.e., $\phi_1 = \frac{\pi}{2}$, all the light is reflected back to the first medium; this phenomenon is called total reflection. The incident angle ϕ_c that satisfies $\sin\phi_0 = \frac{n_1}{n_0}$ is called the critical angle, and the corresponding refraction angle $\phi_1 = 90°$.

Total reflection can make all light reflect back to the first medium, so many optical instruments use total reflection to change the direction of light. Total reflection is also used to transmit light energy in the fields of fiber-optic communications and integrated optics.

7.3.2 *Measurement of optical properties of thin films*

To determine whether the film performance meets the design requirements, the optical properties of the film need to be measured. Therefore, the measurement of the optical properties of thin films is a very important technology.

7.3.2.1 *Measurement of transmittance*

Transmittance is generally measured by a spectrophotometer, which mainly includes light source, spectrophotometer system, detection system, and so on. Figure 7.28 shows a simple transmittance measurement system. When measuring, first let the light from the blank space of the sample frame through the measurement of its light intensity be I_0. Then, the sample is inserted into the optical path to let the light pass through the sample, and its light intensity is measured as I_1. The ratio of the two measurement results is the transmittance T of the sample, i.e.,

$$T = \frac{I_1}{I_2} \times 100\%. \tag{7.89}$$

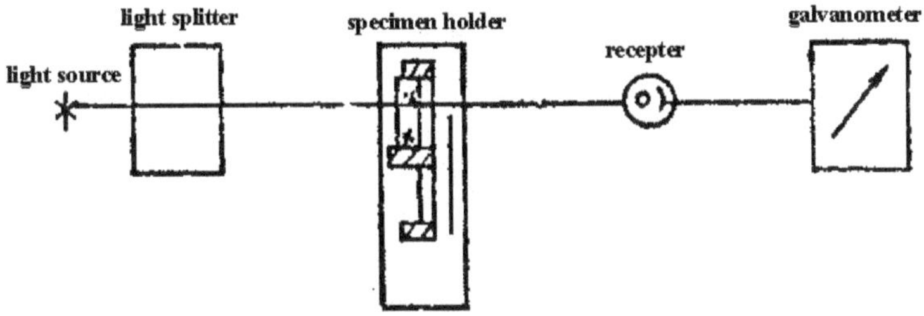

Figure 7.28: Schematic diagram of a transmission measurement device.

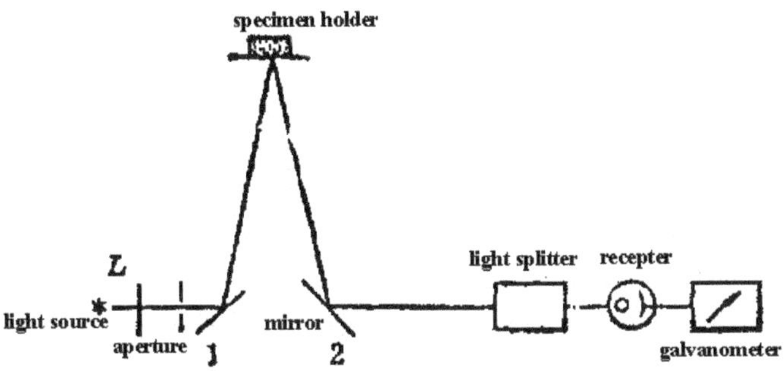

Figure 7.29: Schematic diagram of a low reflectivity measuring device.

7.3.2.2 *Measurement of reflectivity*

Reflectivity is another important parameter of optical thin films. For films with low reflectivity, the method of measurement can be compared with the standard sample. Figure 7.29 is a schematic diagram of the measuring device. The incident light is incident on the sample at a small angle through reflector 1, and the light reflected from the sample is passed into the spectrometer and detector through reflector 2. During the measurement, the standard sample is first put on the measuring rack to measure its reflector I_0, and then the sample to be tested is replaced to measure its reflected light intensity I_1. Then, the reflectance of the sample to be measured is $R = \frac{I_1}{I_0} R_0$, where R_0 is the reflectance of the standard sample, and the measurement accuracy is mainly affected by the measurement accuracy of the standard sample.

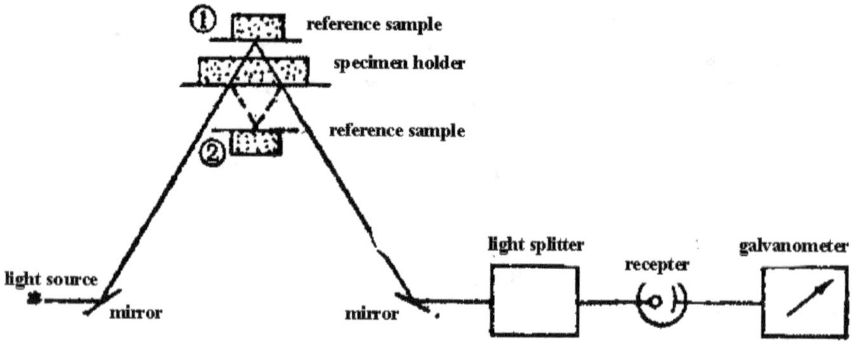

Figure 7.30: Schematic diagram of a high reflectivity measuring device.

For films with high reflectivity, the absolute method of measuring reflectance R, as shown in Figure 7.30, is generally used. When measuring, do not put the sample to be measured first, but put the comparison sample in the position ①. This time the light reflects from the sample and then enters the detection system, and the reflected light intensity is measured as I_1. Then, put the sample to be tested on the sample rack, and compare the position of the sample, so the incident light is reflected twice by the sample to be tested, and then enters the detection system, and the reflected light intensity is measured as I_2. The light intensity ratio was measured to be

$$\frac{I_2}{I_1} = \frac{I_0 R_1 R^2}{I_0 R_1} = R^2, \tag{7.90}$$

where I_0 is the incident light intensity, and R_1 is the reflectivity of the sample. From this, the reflectivity of the sample to be measured can be calculated as

$$R = \sqrt{I_2/I_1}, \tag{7.91}$$

From Equation (7.91), it can be seen that the measurement result is only related to the accuracy of the light intensity measurement, but has nothing to do with the reflectance R_1 of the comparison sample, so it is called the absolute measurement.

7.3.2.3 *Measurement of refractive index of transparent films*

When the beam is vertically incident to the surface of the monolayer, the reflectivity is

$$R = \frac{n_0 n_2 - n_1^2}{n_0 n_2 + n_1^2}. \tag{7.92}$$

It can be deduced that

$$n_1 = \sqrt{\left(\frac{1-R}{1+R}\right) n_0 n_2}. \tag{7.93}$$

Therefore, when the refractive indices n_0 and n_2 of the medium on both sides of the film layer are known (normally, n_0 is the refractive index of air and n_2 is the refractive index of the substrate), the refractive index n_1 of the film can be obtained as long as the reflectance R of the vertical incidence is accurately measured. Although the accuracy of this method is not high, it is more practical, so it is a commonly used method.

Another more accurate method is based on Brewster's law. When the linearly polarized light vibrating in the incident plane is incident at Brewster angle, its reflected light intensity is zero, and tangent $\tan \varphi_B = n_1/n_0$, where φ_B is the incident angle and n_1 is the refractive index of the medium. Thus, when the linearly polarized light vibrating in the incident plane is incident at Brewster angle to the monolayer optical film, the upper surface of the film makes no contribution to the reflected light intensity, and only the lower surface of the film, namely the substrate, has the reflected light intensity equal to that of the beam directly incident to the substrate. This law can be used to calculate the refractive index of the membrane. The method is as follows: the film is only partially coated on the substrate, with linearly polarized light incident, and the reflected light intensity is measured while the incident angle is changed until the reflected light intensity at the interface of coating and uncoating is equal (as shown in Figure 7.31). The angle of incidence is the Brewster angle (φ_B) at which light enters the film from the air, and from this the refractive index of the film can be calculated.

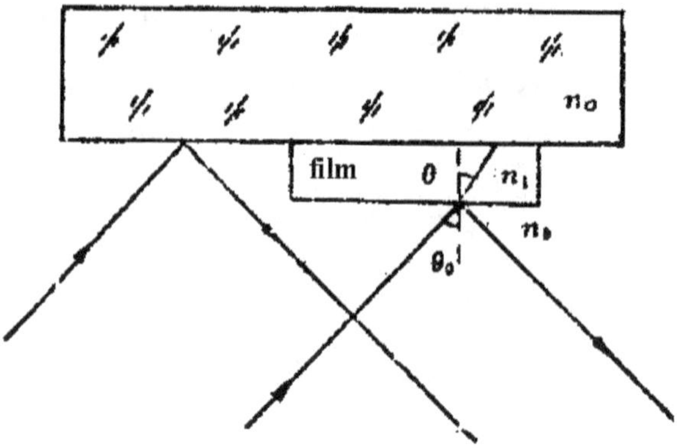

Figure 7.31: Refractive index measurement.

In recent years, due to the popularization of the computer and the use of a laser light source, the technique of measuring the refractive index and thickness of thin films by ellipsometry is receiving more and more attention. Its basic principle is to synthesize the light wave produced by the laser into the elliptically polarized light, and pass it at a certain angle of incidence to the sample under test; after the film reflects, the state of polarization of the reflected light changes, and the incident light amplitude values of p and s components and their phase difference, as well as the reflection amplitude values of p and s components and their phase difference between certain changes have to be measured through a set of more complex equations; with the aid of a computer, we can indirectly measure the refractive index of dielectric films and the film thickness.

7.3.3 *Thin-film waveguide and light coupling*

It is of great significance to modern optical communication technology that light propagates inside the film and can enter or leave the film in a controlled way.

7.3.3.1 *Thin-film waveguides*

A thin-film waveguide is a thin film evaporated on the substrate. Its refractive index n_1 is higher than the air refractive index n_0 and

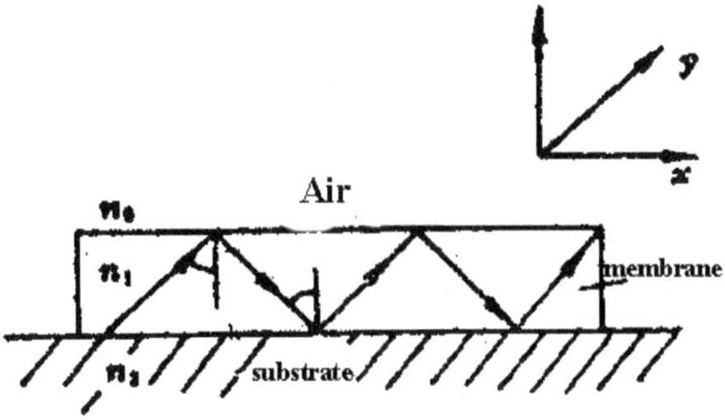

Figure 7.32: Schematic diagram of a thin-film waveguide.

the substrate refractive index n_2. Light can be transmitted in the membrane by total reflection, and can travel a long distance with minimum loss. As shown in Figure 7.32, there is a ray propagating upward in the film, because the refractive index n_1 of the film is higher than the refractive index n_0 of the air. When the incidence angle is greater than the critical angle $\phi_{c0} = \sin^{-1} \frac{n_0}{n_1}$, total reflection of the upper surface will occur. The reflected light travels all the way down to the surface. Similarly, when the incident angle is larger than the critical angle $\phi_{c1} = \sin^{-1} \frac{n_2}{n_1}$, total reflection will occur on the lower surface of the film. By doing this over and over again, the light in the film will travel along a "*Z*" path, also known as a "*Z*" wave.

7.3.3.2 *Optical coupling*

Because the film is very thin (about 1 nm), it is difficult to project the light from the outside directly into the edge of the film. In the same way, it is very difficult to propagate light in thin films accurately without losing ground, which is a very important coupling problem in integrated optics. The commonly used coupling method is prism coupling, whose principle is shown in Figure 7.33. The prism is placed on top of the film, and a small air gap is maintained between the bottom of the prism and the upper surface of the film, with a thickness of $\frac{\lambda}{8} \sim \frac{\lambda}{4}$. When the laser is incident on the bottom surface of the prism, total reflection occurs and standing waves are formed in the prism. The field is sinusoidal in the normal direction of the

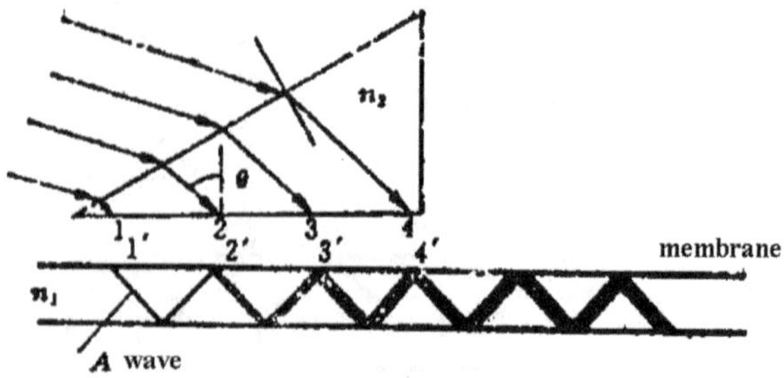

Figure 7.33: Prism coupling.

bottom, but the end of the decay is exponentially attenuated and can extend below the bottom of the prism, transferring the energy of the laser beam from the prism into the thin film.

Conversely, the energy in the film can also be transferred to the prism. For example, as shown in Figure 7.33, four rays from the laser beam entering the prism, respectively, hit four points, 1, 2, 3, 4, on the bottom surface of the prism. When the first ray of light reaches point 1, it whips up the A wave in the membrane at the point (1′) opposite to point 1, which propagates at a certain speed. When the second ray reaches 2, it also provokes A waves in the membrane at the point 2′. It takes some time for the A wave excited by the first ray to travel from point 1′ to point 2′, and the second ray arrives at point 2 later than the first ray arrives at point 1. If these two periods of time are exactly equal, then as the A wave travels along the membrane, it will become stronger and stronger because of the constant addition of in-phase waves. For the prism coupler to work effectively, this synchronization condition must be satisfied by adjusting the direction of the laser beam incident on the prism. In this way, the light waves in the prism are coupled to the membrane. Therefore, prism coupling is very efficient, and for uniform air gaps, 81% of the energy of the laser beam can be coupled to the film (theoretically up to 100%). For the prism coupling to have good performance, the right angle of the prism must be precisely polished. Even small irregularities in a rectangular prism can affect the performance of the coupler. In addition, there are grating coupling and wedge film coupling, but their coupling efficiency is lower than prism coupling.

7.4 Magnetic Properties of Films

Magnetic films refer to the thickness of micron scale magnetic film materials. The research and development of magnetic thin films is relatively early; especially those magnetic thin film materials that are used as magnetic recording and information storage materials are very important in the field of electronic information technology. The magnetic properties of thin films mainly include saturation magnetization, coercive force, magnetic anisotropy, domain structure, magnetic strain, magnetic resonance, and magnetic resistance effect.

7.4.1 *Magnetism of films*

The basic performance parameters of magnetic materials are their saturation magnetization M_s and Curie temperature T_C; when the temperature of a material is higher than its Curie temperature, it will lose spontaneous magnetization. The results show that as long as the membrane structure is continuous, the M_s and T_c are the same as those of block materials. Only those films having island discontinuous structure can cause M_s to deviate from the value of block materials. When the thickness of the film is less than 2.7 nm, its M_s value deviates from that of the bulk material. Figure 7.34 shows the relationship between the saturation magnetization and film thickness of permalloy at room temperature.

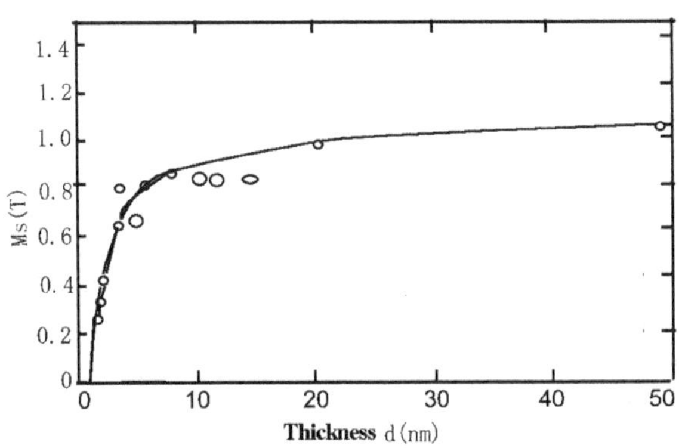

Figure 7.34: Relationship between the M_s and film thickness d of permalloy.

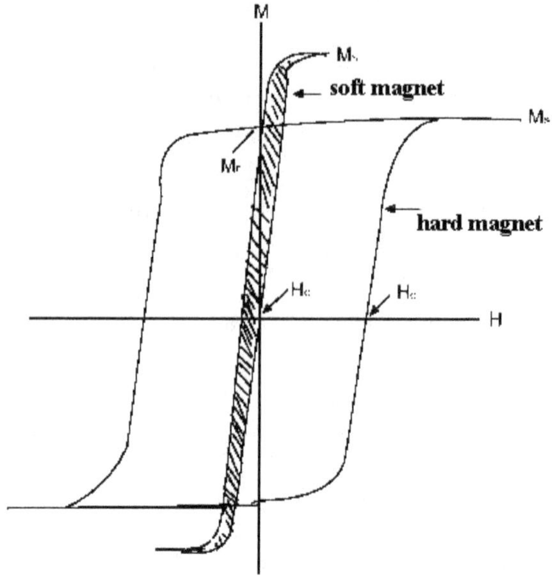

Figure 7.35: Soft and hard magnetic materials.

Elements (e.g., Fe, Ni, Co, etc.), alloys (e.g., Fe–Ni, Co–Ni, etc.), oxide insulators (e.g., Ni–Zn ferrite, Sr–Fe Ferrite, etc.) and ionic compounds (e.g., $CrBr_3$, EnS, EnI_3, etc.) are magnetic. In addition, amorphous crystals can also be magnetic, such as $Fe_{80}B_{20}$ prepared by dissolution quenching method and Co-Gd film deposited by evaporation. There is a hysteresis loop relationship between magnetization M and the external magnetic field H, as shown in Figure 7.35. According to the size of the coercivity field H_c, magnetic materials can be divided into soft magnetic materials (H_c is very small) and hard magnetic materials (H_c is very large). The properties of soft and hard magnetic materials are shown in Table 7.10.

7.4.2 *Magnetic anisotropy*

Magnetic anisotropy refers to the change in magnetization direction of the magnetization inside a magnetic body. This is called magnetic anisotropy. The cause of magnetic anisotropy may be the crystal structure or the shape effect of magnetism itself. Because the film is very small in the thickness direction, this shape effect causes uniaxial

Table 7.10: Magnetic properties of soft and hard magnetic materials.

	Material	$4\pi M_s$ (kG)	H_c (Oe)	H_k (Oe)	Application
Soft magnetic materials	Permalloy	10	0.5	5	Computer memory/reluctance sensor-recording head
	CoZr (amorphous)	14	<0.5	2–5	—
	$Fe_{80}B_{20}$ (amorphous)	15	0.04	7	—
	$Fe_{72}Sl_{28}$ (amorphous)	12	0.2	4	—
	$Fe_{72}Sl_{18}C_{10}$ (amorphous)	16	0.2	7	—
	$Y_3Fe_3O_{12}$	1	~0	1000	Magnetic bubble memory device
Hard magnetic materials	Co–Re	6–9	700		Parallel magnetic recording media
	Co–Pi	10–18	1100–1800		Same as above
	Co–Ni	10–15	1000–13000		Same as above
	Co–Ni–W	5.5	650		Same as above
	Fe_3O_4	5	300		Same as above
	$\gamma Fe_2O_3(Co)$	3	700		Same as above
	$\gamma Fe_2O_3(Co_3)$	3	2100		Same as above
	Cr–Co	4–7	H_c (perpendicular)		Vertical magnetic recording medium
	TbFe		2000–10000		Magneto-optical recording medium
	GdCo	~10	1000–2000		Same as above
	GdTbFe		1000–3000		Same as above

Note: $1\,\text{Oe} = 80\,\text{A/m}$; $1\text{G} = 10^{-4}$ T; H_k (anisotropic field) $= 2k_u/M_s$.

magnetic anisotropy in the film. Uniaxial magnetic anisotropy is typical in permalloy films. It only becomes easy to magnetize in a certain direction within the film. The magnetic film deposited in the magnetic field shows obvious uniaxial magnetic anisotropy, the easy axis of magnetization is parallel to the magnetic field, and the anisotropy energy per unit volume can be simply expressed as

$$E = K_u \sin^2 \theta, \tag{7.94}$$

where K_u is the magnetic anisotropy constant and is the included angle between the direction of magnetization and the axis of easy magnetization. Due to different preparation processes, the K_u value can vary from zero to $4 \times 10^{-2}\,\text{J/cm}^3$. In addition, different preparation techniques, different process parameters, internal stress, inverse magneto-strictive effect, and grain and grain boundary shape effect also lead to the magnetic anisotropy of the film.

7.4.3 *Magnetic domains of films*

In block ferromagnetic materials, there are many magnetic domains in which the magnetization has a certain direction. The magnetic domains are separated by the domain wall, the magnetization direction in the domain wall can be changed gradually, from the magnetization direction of one domain to the magnetization direction of another domain. In general, the domain walls in the film are parallel to the axis of easy magnetization, and the plane of the wall is perpendicular to the surface of the film. The direction of magnetization in the domain walls can take many forms as they shift from one domain to another. In block materials and very thick films, the magnetization in the domain walls will rotate around an axis parallel to the axis of magnetization because this transition has the lowest energy. Such domain walls are called Bloch walls. For reasons of symmetry, there are two kinds of Bloch walls: one in which the magnetization points up in the middle of the wall, and the other in which the magnetization points down in the middle of the wall, but they all rotate in a plane parallel to the domain wall. Because the free magnetic pole appears at the intersection of

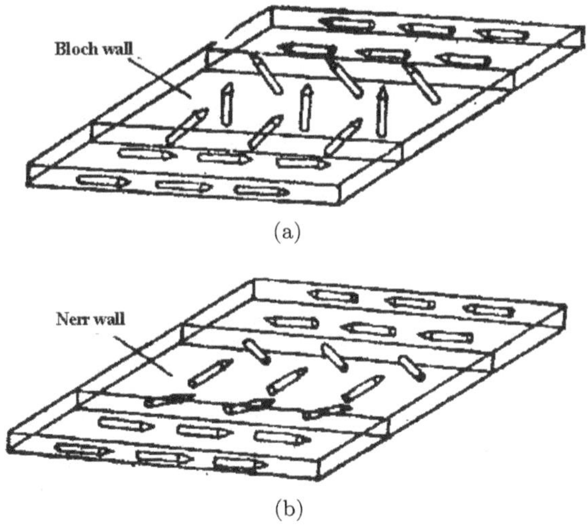

Figure 7.36: Bloch domain walls and Neel domain walls.

the domain wall and the surface of the thin film, stray fields with opposite magnetization direction are generated in the wall, and the magnetostatic energy related to these stray fields increases with the decrease of the thickness of the thin film. In very thin films, there is another domain wall with much lower magnetostatic energy, in which the magnetization rotates parallel to the plane of the membrane, called the Neel wall. Figure 7.36 is a schematic diagram of the Bloch and Nerr walls.

7.4.4 *Magnetoresistance effect*

In 1856, the famous British physicist W. Thomson discovered that magnetic fields can change the resistance of metals. This change in metal resistance caused by a magnetic field is known as the magnetoresistance (MR) effect. The formula to represent the size of the magnetoresistance effect is as follows:

$$\eta = \frac{R_H - R_o}{R_o} = \frac{\Delta R}{R_o} = \frac{\rho_H - \rho_o}{\rho_o} = \frac{\Delta \rho}{\rho_o}. \tag{7.95}$$

Among them, the η is the magnetic resistance coefficient, and $R_H(\rho_H)$ is the resistance (resistivity) when the magnetic field is H.

$R_0(\rho_0)$, is the resistance (resistivity) when the magnetic field is zero; for most of the metal material, η is very small, generally less than 3%.

In 1988, Baibich *et al.* found that $\eta_{\text{total}} > 50\%$ of y was obtained in the multilayer (Fe/Cr) (N is the number of cycles) formed by alternating deposition of Fe and Cr. Because the η of the Fe layer in the multilayer film layer is far less than the η_{total} of the measured values, this phenomenon is called the giant magnetoresistance (GMR) effect.

In 1993, Helmolt *et al.* found that $\eta = \Delta R/R$ can reach 10^3–10^6 for the class of perovskite structures such as rare earth manganese oxide; this is called the colossal magneto-resistance (CMR) effect.

In 1995, the tunneling magneto-resistance (TMR) effect was found in magnetic multilayer film; IBM and Fujitsu Company have developed TMR materials with $\eta = 22\%$ and 24%. At present, the GMR and CMR effects are found in multilayer films, spin valves, particle films, non-continuous multilayer films, oxide films, and other materials.

In 1994, IBM Company announced the development of the GMR head, which increased the storage density of computer hard disks 10-fold! The theoretical research and application development of GMR, CMR and TMR effects will be an important task for relevant scientists in the early 21st century.

7.4.5 *Influence of preparation conditions of films on magnetic properties*

The effect of preparation conditions on the magnetic properties of thin films, especially on anisotropy, is obvious. The magnetic field of a magnetic film with anisotropy is called the anisotropic field and is represented by H_K. Films with predetermined anisotropic fields can be obtained by different preparation conditions.

7.4.5.1 *Magnetic field controlling*

The magnetic field is alternately applied in two directions perpendicular to each other during the evaporation of the film, so that the easy magnetization axis of the film takes these two directions in turn during the deposition. Since there is a strong ferromagnetic exchange coupling between the spins of the magnetic film, and the magnetization vector cannot keep up with such a rapid change in the magnetic field, the magnetization has only an average anisotropy.

7.4.5.2 *Magnetic field annealing*

The uniaxial anisotropy of the prepared magnetic film can also be changed by magnetic field annealing. For example, the evaporated film at the substrate temperature of 270°C is placed in a magnetic field perpendicular to or parallel to the direction of the easy magnetization axis. The film is annealed at temperature T_1 for one hour. After cooling, H_K is measured by the B–H loop. When the film is annealed in a magnetic field parallel to the easy magnetization axis, no change can be detected in H_K as long as T_1 is lower than the preparation temperature T_0. When T_1 is higher than T_0, H_K decreases slightly. However, when the magnetic field is annealed perpendicular to the axis of easy magnetization, H_K will decrease even if the temperature is not high. The higher the temperature, the more it decreases. If the memory required by the magnetic film is less than that obtained by the ordinary evaporation process, controlled magnetic field annealing can be used. Recrystallization occurs when the annealing temperature is high and exceeds the preparation temperature, and the changes of H_K are irreversible.

7.4.5.3 *Substrate temperature*

Theoretical studies have shown that H_K is proportional to $T_c - T_0$, where T_c is the Curie point of the material and T_0 is the annealing temperature. This relationship is also observed for permalloy films if the substrate temperature during film preparation is used instead of annealing temperature. For the evaporated permalloy films with substrate temperature higher than 330°C, it is obvious that the method of controlling substrate temperature is more effective than the method of controlling annealing temperature to control the prepared H_K value. The fact that H_K decreases with substrate temperature can be used to control H_K, but other parameters of the film are also related to substrate temperature, so this method is limited.

7.4.5.4 *Anisotropy of oblique incidence*

Uniaxial magnetic anisotropy occurs even in the absence of a magnetic field when the substrate is at an angle to the incident beam. For permalloy, the axis of easy magnetization is perpendicular to the stream of vapor. The larger the angle between the steam beam and the normal line of the substrate, the lower the temperature of the substrate, and the larger the anisotropic field. Figure 7.37 shows the relationship between the anisotropic field of permalloy and the incident angle when the substrate temperature is 200°C. When the evaporation angle is slightly greater than 70°C, H_K changes the symbol and the easy magnetization axis is located in the evaporation surface. The results of electron microscopy show that the grain chain configuration perpendicular to the incident plane is generated at a small angle, while the chain parallel to the incident plane is observed at a large angle. Because magnetization parallel to the chain segment has little magnetization energy, these chains lead to easy magnetization axes perpendicular to and parallel to

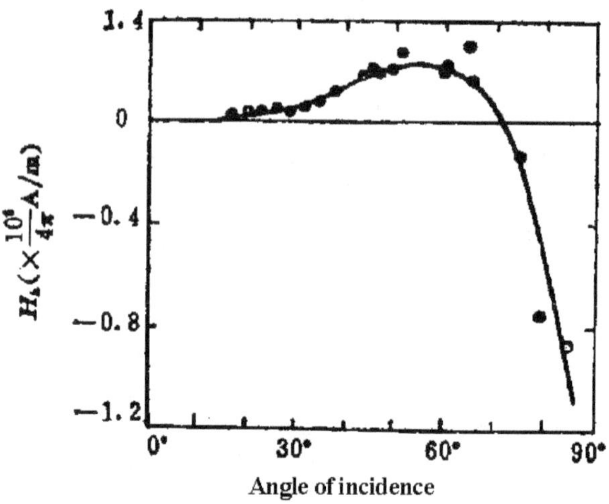

Figure 7.37: Relationship between H_K and incident angle.

the incident surface, respectively. Oblique incident anisotropy has not been used yet, because of the obvious thickness gradient of the film, so it is generally considered that oblique incident anisotropy is a disadvantage, which is sometimes overcome by increasing the distance between the evaporation source and the substrate.

7.5 Thermal Properties of Thin Films

Functional thin films exhibit thermal properties that are markedly different from those of bulk materials in many respects. There are two main reasons for this phenomenon: one is the extremely large surface volume ratio caused by the thin film thickness; second, the process of film formation is the condensation of individual particles, which makes the film have a special microscopic physical structure. Because of quantum effects, the surface and interface effect is huge, conductivity and thermal conductivity of thin films are below the corresponding values of the macro situation; one of the reasons for this phenomenon is when the mean free path of electrons near the surface is on the same order of magnitude as the minimum size of a given sample, the carrier transport process is dependent on the sample size effect.

Therefore, an accurate quantitative analysis of thin film thermal parameters is of great significance in the research and development of devices. At present, the measurement and characterization of thermal parameters of thin films mainly focus on thermal conductivity, thermal diffusion, specific heat capacity and thermal expansion coefficient.

7.5.1 *Measuring method for the thermal conductivity of film*

At present, a variety of thin-film thermal conductivity testing technologies have been developed: static method, 3ω method, transient reflection measurement, scanning thermal microscopy technology, photothermal deflection, infrared imaging technology and holographic interferometry, and photoacoustic photothermal technology. This section briefly introduces several common methods for measuring the thermal conductivity of thin films.

7.5.1.1 *Static method*

The static method is a steady-state method for solving thermal conductivity using one-dimensional Fourier heat transfer equations. Its relation is as follows:

$$Q = -\kappa \nabla T, \tag{7.96}$$

where Q is the heat flux vector, which represents the heat transmitted per unit time along the direction of temperature reduction on the unit isothermal surface area. Thermal conductivity κ reflects the thermal conductivity of a substance in the SI system $\mathrm{W/m \cdot K}$.

As the size of the film heat transfer L and the heat transfer time t can be compared to the phonon mean free path of L (10^{-8} – 10^{-7} m orders of magnitude) and mean free time τ (10^{-12} s orders of magnitude), it will not meet the Fourier heat transfer law. Therefore, to measure the thermal conductivity of the film in this way, the thickness of the measured film is required to be $L \gg 1$, that is, the thickness of the film should be at least in the order of micron.

The most common structure for testing the thermal conductivity of a thin film is shown in Figure 7.38. The bottom metal bridge A, the sample layer under test, and the top metal bridge B were successively grown on an insulating substrate (such as SiO_2) by the semiconductor process. Metal bridges are usually made of metal Al or Pt. Metal bridge A is connected with a large current and acts as both the heater and the temperature-measuring element; metal bridge B uses a small current as a temperature measuring element. Add constant current I to the 1 and 2 electrodes of the top metal bridge A, and

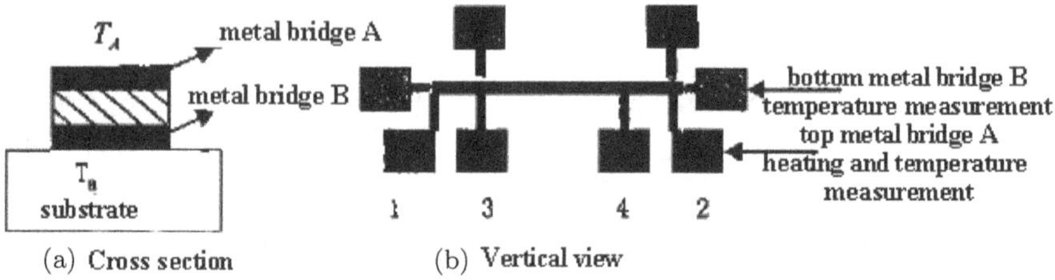

Figure 7.38: Schematic diagram of the longitudinal thermal conductivity test of thin films.

joule heat is generated on the metal strip. The temperature rises, and the heat flow passes through the sample medium layer to the bottom metal bridge B, and finally flows into the substrate. The heat loss at the boundary is small compared with the heat conduction, satisfying the one-dimensional heat transfer model. When the heat transfer of the system reaches a stable state, the temperatures of the top metal bridge and the bottom metal bridge are, respectively, T_A and T_B. In order to ensure the constant physical property measurement of thermal conductivity, the temperature rise of the top metal bridge should be within the range of 10 K.

The expression for thermal conductivity can be obtained from Equation (7.96):

$$\kappa = \frac{d \cdot I \cdot U}{l \cdot W(T_A - T_B)}, \tag{7.97}$$

where U is the voltage value on levels 3 and 4, d is the thickness of the dielectric layer, and l and W are the length and width of the dielectric layer, respectively. Temperatures T_A and T_B are obtained by the voltage–current four-terminal method.

The thermal conductivity of SiO_2 films measured by this method is about 20% less than that of the bulk material. The thermal resistance between SiO_2 film and metal is very small, and its influence can be ignored. This method has the characteristics of high precision and easy data analysis and processing, the error is about 5%, and it is suitable for measuring the thermal conductivity of insulating materials and poor conductor materials.

7.5.1.2 3ω *Methods*

The 3ω method was first used to measure the thermal conductivity of isotropic low thermal conductivity insulator materials, and later it was also used to measure the thermal conductivity of thin films deposited on good thermal conductivity substrates.

The measurement structure of the 3ω method is shown in Figure 7.39. A layer of insulating film to be tested (such as SiO_2) with a thickness of d is grown on the substrate of a good hot conductor (such as Si), on which a metal bridge of the shape shown

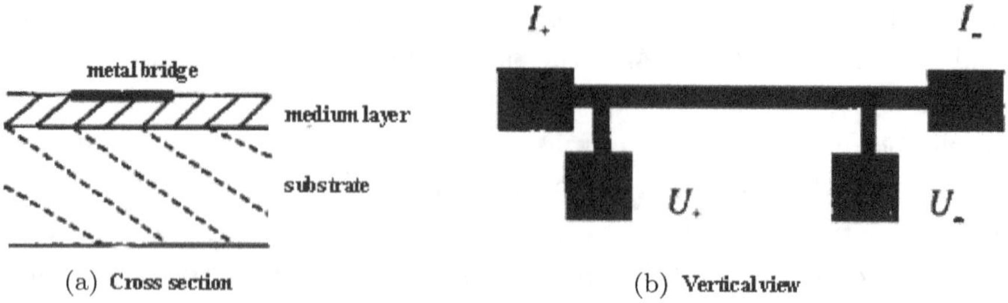

Figure 7.39: 3ω structure diagram of the test method.

in Figure 7.39(b) is made. Its width is b and meets the requirement of $b \gg d$, and its length is l. The metal bridge serves as both a heat source and a temperature-measuring device.

There is an alternating current on the $I+$ and $I-$ electrodes $I = I_0 \cos \omega t$, and the joule heat power generated by this current on the metal bridge is

$$P(t) = \frac{1}{2} I_0^2 R(1 + \cos 2\omega t). \tag{7.98}$$

And then you have the heat wave with a frequency of 3ω spreading down, and the wavelength is

$$q^{-1} = \sqrt{D/\omega} \tag{7.99}$$

where D is the thermal diffusivity of the substrate. When measuring thermal conductivity with the 3ω method, the value of wavelength q^{-1} is generally taken to be in the range of 10^{-3}–10^{-5} m. Only when the thickness of the film is $d \ll q^{-1}$ can the film be ignored and the temperature wave can be considered to have completely diffused into the substrate.

The 3ω method is useful when the film thickness is less than $10\,\mu$m. The advantages of the 3ω method are that it is not only insensitive to radiation loss but also it can effectively reduce the errors caused by black-body radiation; the measurement time is short; it is applicable to a wide temperature range, and can be measured at room temperature or higher temperature.

7.5.1.3 *Transient reflection measurement method*

The transient reflection measurement method is to obtain a layer of SiO_2 film of 10–200 nm by oxidation on the Si substrate which can be regarded as semi-infinite. On the SiO_2 layer is 20 nm chromium and 2 nm copper (or aluminum). The purpose of the chromium layer is to increase the bonding between copper and SiO_2, thus reducing the thermal resistance at the interface between them.

By heating the vertical surface Au with $\tau = 6\,\text{ns}$, $E = 50\,\mu\text{J}$ of Nd:YAG laser pulse, the surface shallow layer under the instantaneous temperature, the temperature of the surface reaches a maximum value. After this moment, the surface temperature attenuates with time, and its attenuating period is influenced by the thermal resistance of the composite film under the surface and attenuates exponentially. Surface temperature was measured by the reflection of a 1-mW He–Ne laser.

The SiO_2 layer should meet the following conditions:

(1) The thickness of the SiO_2 layer is only 10–200 nm, and the time of heat flow in SiO_2 film $t \approx d_2/D_{\text{SiO2}}$ (D_{SiO2} is the thermal diffusivity of SiO_2 layer) is very small, and its heat capacity can be ignored. Therefore, it can be treated as the non-thickness thermal resistance R_{th} between the Si substrate and the metal interface.
(2) Si substrate is a semi-infinite plane.
(3) The laser per unit of energy is absorbed by a very thin layer of metal Cu in a very short time to generate heat flow, the absorption depth is small compared with the thickness of the metal.

The change relation of surface temperature with time can be expressed as

$$T(t) = T_0 + \Delta T \exp\left(\frac{-tG}{hC_V}\right). \tag{7.100}$$

Among them, T_0 is the temperature before heating, ΔT is the maximum increase in temperature after the sample is heated, G is the thermal conductivity per unit area, h is the thickness of the metal layer, and C_V is heat capacity for the metal body. From the decay period of temperature, the decay time constant of temperature, hC_V/G, can be obtained, so the thermal conductivity G per unit area can be calculated, and the thermal conductivity of the SiO_2 layer can be obtained. The instantaneous reflection method is adopted to measure the thermal conductivity, which overcomes the error caused by the wire taking away the heat. Its biggest advantage is that there is no need to accurately measure the temperature value, but to only measure the change in temperature relative to time, namely the decay period.

7.5.1.4 *Micro bridge method*

An insulating material (such as SiN_x) is made into a microbridge structure as shown in Figure 7.40 by a semiconductor process. A thin layer of strip metal (Al/Pt) is evaporated in the middle of the suspension beam as a heating electrode. Make a thermistor or thermocouple at a distance from the heating electrode of x as a temperature-measuring device. In order to reduce convective heat dissipation, the measurement should be made in a vacuum environment. Since the heat loss from thermal radiation is less than 0.3%, the radiant heat loss is negligible. Let the ambient temperature at this point be T_0. On the heater on a direct current (dc) to produce a constant heat flux q, and a current whose size should meet: when it comes to a steady state heat transfer, the temperature of the

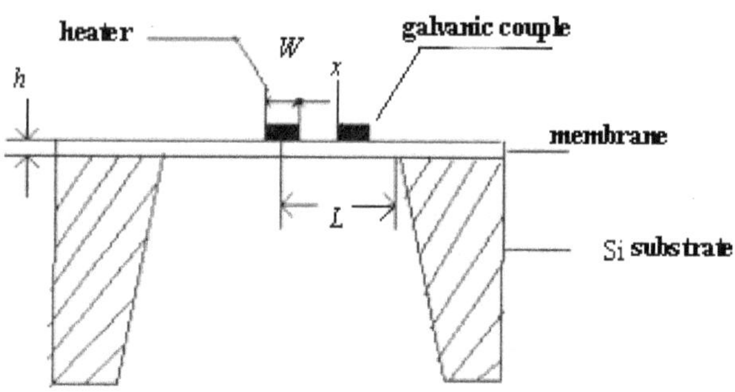

Figure 7.40: Microbridge test structure.

temperature measurement in ΔT tends to less than 10 K, which guarantees thermal conductivity measurement of physical properties, heat flow from the middle of the beam spreading on both sides. If the distance between the heater and the heat sink is large enough and the heat sink has good thermal conductivity, the temperature at the junction of the heat sink and the cantilever beam can be considered to be T_0. When the boundary heat dissipation is not considered, the heat conduction satisfies the one-dimensional heat transfer model when the stable state is reached. It can be deduced from Equation (7.101) that

$$\kappa = \frac{q}{2} \cdot \frac{(L - x)}{\Delta T}. \tag{7.101}$$

Among them, the $\Delta T = T_s - T_0$, T_s for the temperature-measuring device to measure temperature. Get the parameters L, x, q, T_s and T_0 and you can find the thermal conductivity.

On the basis of the above method, a double thermocouple method was developed to measure the thermal conductivity of thin films. Two temperature thermocouples were made on the SiN_x film x_1 and x_2 away from the heater, as shown in Figure 7.40.

$$\kappa = \frac{q}{2} \cdot \frac{(x_1 - x_2)}{T_2 - T_1}, \tag{7.102}$$

where T_1 and T_2 are the temperatures measured by the two thermocouples.

The two-temperature measurement method has the following advantages compared with single-temperature measurement: (1) It saves the distance between the measurement heater and the heat sink L, and reduces the measurement difficulty and measurement error. (2) The single temperature measurement thinks that the temperature at the junction of the heat sink and the cantilever beam is T_0. In fact, no matter how high the thermal conductivity of the heat sink is, it is impossible that there is no temperature gradient at the junction, so error must be introduced. Using a double thermocouple to measure the temperature can eliminate the systematic error caused by the measurement method itself.

7.5.1.5 *Scanning thermal microscopy*

A scanning thermal microscope (STHM), developed based on an atomic force microscope, can be used as a thin film surface thermal measuring device, and its resolution can be in the nanometer order of magnitude. The performance of the scanning heat microscope depends mainly on the different thermal probes. Thermal probes can be divided into thermocouple type and thermal resistance type according to different temperature sensing principles. Each can be classified by linestyle and thin film. Thermal resistance probes have been commercialized and thermocouple probes are still being studied.

When the scanning thermal probe microscope is in operation, the thermal probe and the sample surface can be measured in the contact or non-contact type of measurement. No matter which way it works, its resolution is directly determined by the size of the probe tip. At present, its size has been in the range of 10 nm–1 μm.

The thermal resistance probe has two working modes: passive mode and active mode. Passive mode means not in the test circuit with temperature feedback; the probe does not need a self-heating process, only a small current to induce temperature changes. When there is a temperature distribution on the surface of the sample, if the sample temperature is higher than the probe temperature, then the sample will transfer heat to the probe. As the temperature of the probe increases, the voltage–current method is used to measure the change in the probe thermistor resistance value to reflect the surface temperature distribution, which is also called the temperature model. At present, the scanning range of the scanning heat microscope reaches 200 μm × 200 μm, which can measure the thermal conductivity of the thin film. In addition, the thermal conductivity of the film can be also measured by laser scanning thermal microscopy.

7.5.2 *Measuring method of thermal diffusivity of thin films*

The thermal diffusion rate (also known as the temperature conductivity coefficient) reflects the relationship between the thermal conductivity of substances in the unsteady heat conduction process and the heat storage capacity of substances along the way. It is defined as

$$\alpha = \frac{\kappa}{\rho \cdot C_V}, \tag{7.103}$$

where α is the thermal diffusivity of the material; κ is the thermal conductivity of the material; ρ is the density of the material; C_V is the specific heat capacity of the material.

The greater value of α means that once a part of a substance gains heat, the heat can spread quickly throughout the substance, meaning that the temperature inside the substance tends to converge. Many methods have been developed to test the thermal diffusion coefficient of thin films. So far, many methods have been developed to test the thermal diffusivity of films. For example, O. Paul *et al.* proposed a method to test the thermal diffusivity of films based on microbridge technology. Morikawa *et al.* used Fourier transform thermal analysis to extract the thermal diffusivity and thermal capacity of the film based on the relationship between the square root of the frequency and the phase delay of the temperature wave. Irace *et al.* tested the thermal diffusion coefficient of the film by analyzing the transient characteristics of the double-ended solid-supported beam in a vacuum environment. In addition, there are phase sensitive technology to measure thin film thermal diffusivity, periodic heating for measuring thermal diffusivity of thin films, photoacoustic method for amplitude and phase of technology, the exchange quantity heat meter for measuring thermal diffusivity of thin films, one-point two-line method for gathering thermal wave technology, and ac heating technology, widely used in thin film materials such as thermal characteristics of the measurement. Most of these test methods require a vacuum test environment, the test structure uses a CMOS process and requires post-processing silicon corrosion, and the test process is complex or requires a special test device.

7.5.2.1 *Micro bridge method*

The film for measurement, the heating electrode and the temperature measuring electrode were prepared on the microbridge, as shown in Figure 7.40. Then at the temperature measuring electrode x

$$\begin{cases} \dfrac{\partial T}{\partial t} = \alpha \dfrac{\partial T^2}{\partial x^2} \\ T(x,0) = T_0 \\ T(L,t) = T_0 \\ -\kappa \dfrac{\partial T}{\partial x}\bigg|_{x=0} = 2A\cos^2\dfrac{\omega t}{2} = A(1+\cos\omega t), \end{cases} \tag{7.104}$$

$$T(x,t) - T_0 = \frac{A}{\kappa}\sqrt{\frac{\alpha}{\omega}}\exp\left(-\sqrt{\frac{\omega}{2\alpha}}x\right)$$

$$\cos\left(\omega t - \sqrt{\frac{\omega}{2\alpha}}x - \frac{\pi}{4}\right) - \frac{A}{\kappa}(x-L), \tag{7.105}$$

where L is the length of the bridge, T_0 is the temperature of the substrate, $A = V_0^2/(8RhL)$, V_0 is the amplitude of the voltage applied to the heating strip, R is the resistance of the heating strip, h is the thickness of the film, α is the thermal diffusivity of the film, and the heating frequency is $\omega/2$. It can be seen from the above equation that the phase change of the heat wave between the heating strip and the sensing strip is

$$\Delta\Phi = \sqrt{\frac{\omega}{2\alpha}}x + \frac{\pi}{4}. \tag{7.106}$$

The phase change measured experimentally can be used to calculate the thermal diffusivity.

7.5.2.2 *AC calorimetry*

AC calorimetry is one of the more mature methods developed in recent years for measuring the thermal diffusivity of thin films. Its basic principle is that a certain frequency of periodic heat flow is applied to one end of a long strip of thin film sample, which will form a temperature wave propagating along the length of the same frequency on the surface of the sample. During the propagation, the amplitude of the temperature wave will attenuate, and the thermal diffusivity of the sample can be determined by measuring the attenuation of the amplitude of the temperature wave at two

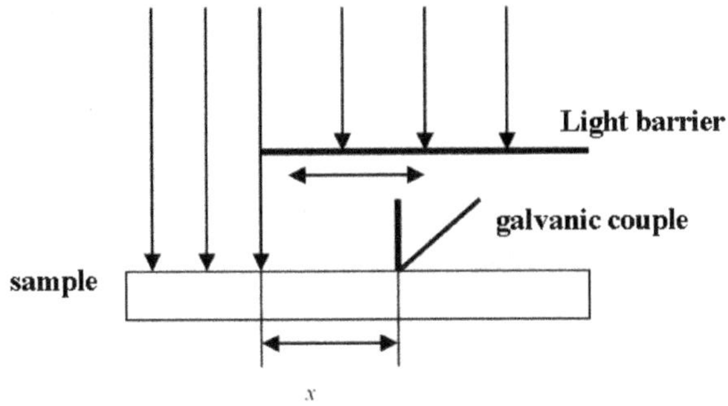

Figure 7.41: Schematic diagram of AC calorimetry.

distance points in the propagation direction. The physical model is shown in Figure 7.41.

The laser beam, as a heat source, is modulated and irradiated on the thin film sample. The area of the spot is larger than the surface area of the sample by adjusting the optical path, and the temperature wave with the same modulation frequency appears on the sample. The thermocouple pasted on the surface of the sample with silver paste is used to detect the temperature wave. When the thickness of the film is small, there is no temperature gradient in the thickness direction.

A one-dimensional heat conduction model can be adopted, and the temperature response on the thermocouple can be expressed as follows:

$$T(x,t) = \frac{Q\exp(i2\pi ft)}{4\pi fcd}\exp\left(-\sqrt{\frac{\omega}{2\alpha}}x - i\left(\sqrt{\frac{\omega}{2\alpha}}x + \frac{\pi}{2}\right)\right), \quad (7.107)$$

where f is the modulation frequency of the laser; $\omega = 2\pi f$; Q is the heat absorbed by the film; x is the distance between the thermocouple and the baffle edge.

According to the above formula, the measurement expression of thermal diffusivity can be obtained

$$\alpha = \frac{\pi f}{\left(\frac{d\ln|T|}{dx}\right)^2}. \quad (7.108)$$

According to Equation (7.108), the thermal diffusivity of the thin film can be obtained by measuring the size of the temperature response signal at two points on the thin film and determining the distance between the two points.

In the actual measurement, due to the limitations of the thin film manufacturing technology, the thin film samples are mostly attached to the substrate. Considering the influence of bonding layer or transition layer and other factors, the measured object is actually a multi-layer film composite structure. At this point, when the sum of the thickness of each layer and the product of its thermal diffusion length is much less than 1, the one-dimensional thermal conduction model is still applicable. At the same time, the temperature response at multiple points on the surface of the sample must be sampled. However, as the experimental conditions cannot completely meet the simplified model, the environmental heat dissipation, the air layer thermal conductivity, and the sample end effect will have a certain impact on the test results.

7.5.3 *Measuring method of the heat capacity of thin films*

At present, there are mainly the thermal pulse method, diffusion method, alternating current method, time delay method, and so on, among which the alternating current method and thermal time delay method are most commonly used.

7.5.3.1 *AC method*

The AC measurement's structure is similar to the single-temperature test's structure, as shown in Figure 7.41. First, the sample is made on a sample table with a total heat capacity of C and a thermal conductivity κ between the sample and the substrate. An ac current is added to the heating resistance to generate an ac heat flow, which then generates an ac temperature. The ac temperature amplitude contains information on the heat capacity C_V, and the ac temperature information is obtained through another temperature-measuring resistance. The heat capacity of the sample can be obtained by means of difference. Because of the phase-locked amplification technique, the measurement accuracy and resolution are very high, but the external thermal time constant of the structure and environment is much higher than the internal thermal time constant of the structure.

7.5.3.2 *Thermal time delay method*

The thermal time delay method is another commonly used measurement method. A schematic diagram of the test principle is shown in Figure 7.42. Thermal time constant meets: $\tau = C_V/\kappa$, predominate the total heat capacity C_V for the sample and the sample machine T and κ the product of the predominate. Thermal conductivity κ is generally determined by means of thermal stability. Thermal time constants τ can be obtained by ac or dc methods; the principle of this method is simple, but due to measuring the transient response, to achieve an accurate measurement is more troublesome.

7.5.4 *Measuring method of thermal expansion coefficient of films*

The thermal expansion coefficient of thin films is very important for the design of thin film devices, especially MEMS devices. On the one hand, the thermal expansion of the thin film material has a great influence on the device performance. On the other hand, thermal expansion is the power source of a micro-thermal actuator. Generally speaking, the coefficient of thermal expansion of bulk materials is not exactly the same as that of thin film materials. Therefore, the

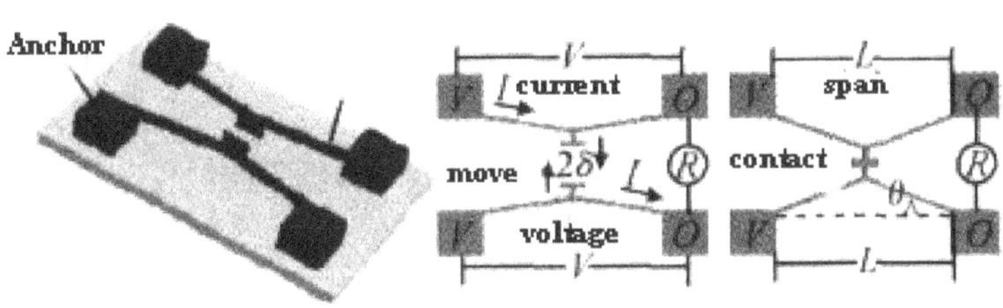

(a) (b) (c)

Figure 7.42: Top view of the test structure with double bending beams.

accurate measurement of the thermal expansion coefficient of thin films is of great significance for the practical application of thin film devices.

There are many methods to measure the thermal expansion coefficient of thin films, for example, X-ray diffraction, T-type Vernier caliper method, thermal actuator method, double-layer film method, thermal stress method, etc.

In 2005, a method was developed to test the film thermal expansion coefficient of a double curved beam structure; this structure uses the surface micromachining technology, convenient testing, good accuracy, and output to electrical signal. This structure does not need to predict the Young's modulus, Poisson's ratio, film and film surface and substrate temperature coefficient or heat transfer coefficient, and material parameters, such as it does not need a special test environment and instruments. Therefore, it can meet the requirements of online detection.

The test structure consists of two curved beams of identical size and structure; the two curved beams are 2δ apart at the top, as shown in Figure 7.42. When two bending beams contact (Figure 7.42(b)), ohmmeter readings tend to infinity. A gradually increasing voltage V from zero is applied to the pressure solder blocks at both ends of the curved beam. The current flows through the bending beam produced by joule heat thermal expansion make it happen. Under the action of thermal stress, the two mutually vertical bending beams move. When the increase in the applied voltage to a value, they contact (Figure 7.42(a)), ohmmeter readings will mutate, recording the time value in the applied voltage V; because the two curved beams are completely symmetrical, the top of the temperature and voltage are the same, so the contact reliability is assured. Change the beam span 2δ, or beam span length L, or beam horizontal angle θ_A, record the corresponding applied voltage V, and further calculate the beam current density J, then by corresponding mathematical processing, we can obtain the thin film thermal expansion coefficient.

7.6 Exercises and Questions

(1) When measuring resistance with the four-probe method, the current should be limited to 0.050 A to avoid overheating and damage of the probe head. Traditional digital voltmeters measure potential drops in the range of 10 mV to 100 V. Which of the following 5000-Å thick thin film materials can be measured by the above method?

(a) Cu; $\rho = 1.73 \times 10^{-16}\Omega \cdot \text{cm}$; (b) Si; $\rho = 2\Omega \cdot \text{cm}$; (c) ZrO_2; $\rho = 10^{14}\Omega \cdot \text{cm}$; (d) $CoSi_2$; $\rho = 15 \times 10^{-6}\Omega \cdot \text{cm}$; (e) TiN; $\rho = 100 \times 10^{-6}\Omega \cdot \text{cm}$.

(2) A film window deicer resistor has dimensions exceeding 5 m by 1 mm. It is designed with a power output of 5 W and a power supply of 12 V. For a film of 5000 Å thickness, what is the required square resistance?

(3) There is a $2l$ long metallic thin film fringe produced by evaporation from a surface evaporation source h away from it. The metallic fringe is uniform in thickness and has the same number of atoms. Write the expression for its resistivity. The thin film conductor is assumed to be piecewise linear.

(4) At low temperatures, the resistivity of the evaporated metal film is higher than that of the bulk material (extrapolated to the same temperature). The possible reasons for this phenomenon are: (a) the size effect caused by surface scattering; (b) high concentration of defects; (c) fault defects caused by cooling. Design an experimental scheme to solve this problem.

(5) An insulating film $\Phi_B = 1\,\text{eV}$. Under the following conditions, determine how much the Schotton-based emission current will change:

(a) The temperature changes by 1%;
(b) The electric field changes by 1%;
(c) The contact area of the film changes by 1%.

(6) A conductive fringe is deposited on an insulator to form a resistance–capacitance structure whose RC time constant is used to limit the corresponding speed of the circuit.

   (a) Write the general formula for RC according to the length, width, thickness, and resistivity of the conductance, as well as the thickness and dielectric coefficient of the insulator in the corresponding region.
   (b) If a $0.5\,\mu$m thick SiO_2 thin film is deposited on a 1-μm wide, $1\,\mu$m thick, 1 cm long Al belt, with $\rho = 2.7 \times 10^{-6}\,\Omega \cdot \text{cm}$, then what is the RC (sec/cm)? [Note: the vacuum dielectric coefficient is 8.85×10^{-14} F/cm.]
   (c) If we use $TaSi_2$ ($\rho = 55 \times 10^{-6}\Omega \cdot \text{cm}$) and polysilicon ($\rho = 1000 \times 10^{-6}\,\Omega \cdot \text{cm}$) instead of Al, then what RC (sec/cm) will be?

(7) Two connected metal pads are apart by 1 cm, which can be used to monitor the resistance of the metal film deposited between the two metal pads. Make a resistance–time diagram from the beginning of evaporation to the completion of film deposition. What is the conductive mechanism at each stage of film forming? Contrast the resistance of the substrate at high temperature with that at low temperature. Contrast the resistance corresponding to the isolated and layered mechanisms.
(8) The Thomson approximation is used to calculate the mean free path of electrons in metal films. Mirror scattering occurs at the air interface, and scattering at the substrate interface.
(9) Consider an M/SiO_2/N–Si capacitor structure (assuming no oxide and interfacial charges)

   (a) Plot the band structure diagram to represent aggregation, consumption, and inversion mechanisms;
   (b) Plot the C–V figure;
   (c) Plot a series of C–V diagrams with different SiO_2 film thicknesses;
   (d) Qualitative description of C–V properties and Si doping degree.

(10) A contact electrode is made of aluminum on both sides of a silicon dioxide sheet and a 3000 Å thick silicon dioxide film. Assume that the silica body resistance coefficient is 10^{15} Ω-cm, and it is exerting strength under 300 K for an electric field of 10^5 V/cm.

(a) What is the current density?

(b) What is the current density in the membrane?

(11) A 1000 Å SiO_2 film is broken down when the applied electric field is 1.2×10^7 V/cm. A 1 μm damage channel and an Au electrode short circuit were observed on the film. This defect occurs for 10^{-4} s. If the current is mainly controlled by Schottky emission, what is the maximum temperature rise estimated for the film? (Its specific heat capacity is 0.24 cal/g°C).

(12) Coat a $Gd_3Ga_5O_{12}$ substrate (diameter 2 cm, thickness 0.75 mm) with a 1 μm thick $Y_3Fe_5O_{12}$ film. Why we fail to lift it with a strip magnet? But when suspended with a thin wire, it can be attracted to a magnet. Why?

(13) (a) In the zero magnetic field, the total energy (E_T) of the magnet film increases with the number of magnetic domains. Why is that? At the same time, EM decreases. Why? Make various contributions to the E_T diagram and indicate what is the most suitable domain number?

(b) The thin film in (1) has been partially magnetized, and the cylindrical magnetic chip interacts with the external magnetic field. Assume that only E_W and E_M contribute to E_T. Make reasonable assumptions about the relationship between energy contribution and domain radius, and plot the most suitable domain radius.

Chapter 8

Some Kinds of Important Functional Film Materials

Functional film materials, especially the new ones, are widely applied in areas like modern information technology, microelectronic technique, computer science, laser technology, and aerospace engineering etc. There are many kinds of functional film materials which are widely used. In this chapter, we will introduce several kinds of them that are already widely applied in microelectronics, photo-electronic technique, integrated optics, and MEME (microelectromechanical engineering), with the aim to provide the readers a basic knowledge of films. Other functional films, like separating membrane and biomedical film, are not included here, limited by space constraints.

8.1 Microelectronic and Photoelectronic Films

8.1.1 *Overview*

With the rapid development of modern technology and science, modern electronics technology is transferring from solid state electronics to microelectronics and photoelectronics. The processing precision or thickness of LSIC, MCM, and membrane passive network (apparent or embedded) has already reached the sub-micrometer scale. To meet the requirements of miniaturization, intelligence, and integration by modern microelectronic techniques, various kinds of electronic materials are developing toward nano, film, and composite materials. The most common used microelectronic and photoelectronic films at present are multifunctional semiconductor films. They are listed in Table 8.1.

Table 8.1: Commonly used microelectronic and photoelectronic film materials.

Categories	Film materials
Elementary semiconductors	Ge, Si, Se, Te
Group semiconductors	GaAs, GaP, GaN
Group semiconductors	ZnSe, ZnTe, ZnCdS, CdSe, CdS, PbS, HgCdTe, As_2S_3, As_2Se_3, As_2Te_3, GeTe
Others	ZnO, PbO_2, SiC, α-Si:H, SiC

8.1.2 *Fabrication method of semiconductor films*

There are many ways to fabricate semiconductor films. They can be mainly classified into CVD, VPE, MOCVD, PECVD, LPE, VBE, MBE, CBE, IBE, L-MBE, CSD, CSS, etc.

8.1.3 *Element semiconductor film*

8.1.3.1 *Silicon films*

(1) **Single crystal silicon films**

The crystal structure of single crystal silicon is diamond, and it is an indirect gap semiconductor. The band gap when $T = 0\,\text{K}$ is $E_g = 1.16\,\text{eV}$. It will decrease linearly with the increase of temperature when $T > 250\,\text{K}$. When it is in low temperature, E_g changes slowly with the change of temperature. The electrical properties greatly depend on the structure defects and doping condition of single crystal silicon. For high purity single crystal silicon which possesses a perfect crystal structure, the drift mobilities of electrons and holes are $1350\,\text{cm}^2/(\text{V}\cdot\text{s})$ and $500\,\text{cm}^2/(\text{V}\cdot\text{s})$, respectively, at $T = 300\,\text{K}$. The Hall mobilities of electrons and holes are $1900\,\text{cm}^2/(\text{V}\cdot\text{s})$ and $425\,\text{cm}^2/(\text{V}\cdot\text{s})$, respectively, in the same temperature. Owing to the existence of more defects in epitaxial silicon films, the mobility of carriers of it is slightly lower than that of single crystal silicon. In order to acquire a single crystal film with a perfect structure, the quality of the substrate must be high enough, and the epitaxy process must proceed under a highly purified condition, and the epitaxial temperature can be increased appropriately. For example, in the epitaxial method of using hydrogen to reduce silicon tetrachloride, the single crystal silicon film having the best perfect mirror surface can be acquired when the epitaxial temperature is 1270°C.

(2) **Polycrystalline silicon films**

Polycrystalline silicon films were first used as isolation films in 1964. The first MOS filed effect transistor made using polycrystalline silicon appeared in 1966. At present, polycrystalline silicon films are widely applied in semiconductor devices and integrated circuits.

Heavily doped low resistance polycrystalline films (resistivity can reach $10^{-3}\,\Omega \cdot \text{cm}$) can be used as the grid of a MOS field effect transistor. The best advantage of substituting the original Al films with polycrystalline silicon films is the realization of a self-aligned gate, namely the automatic alignment of source, drain, gate and gate SiO_2. Heavily doped polycrystalline silicon films can also be used as the interconnecting lead inside the integrated circuits, polar plate of the capacitors, polar plate of the charge storage component of the MOS random access memory, floating gate of floating gate devices, the electrode of charge coupling devices. Lightly doped polycrystalline silicon films are usually used to fabricate the load electrical resistor in MOS random access memory·and other resistors. Polycrystalline films are suitable for fabricating large-scale p–n junctions. That is why they are suitable for fabricating thin-film solar cells, with a much lower price. However, the grain boundary between polycrystalline silicon can affect the efficiency of energy conversion of the solar cells. The conversion efficiency at present is less than 10%. CVD is generally adopted to fabricate polycrystalline silicon films.

(3) **Amorphous silicon**

Amorphous silicon (a-i) is the research focus and core of amorphous semiconductor materials and devices. The structure of amorphous silicon differs a lot from single crystal silicon. There is a periodical spatial arrangement of atoms in single crystal silicon, which is described as long-range order. The alignment of atoms in amorphous silicon can be regarded as a continuous network, where there is no long-range order. However, in terms of individual silicon atoms, it shares basically the same environment with its nearest neighbor and next nearest neighbor. So, the bond length is roughly consistent, and the bond angle bias is not large. Therefore, amorphous silicon has a long-range disorder structure and a short-range order crystal structure, which has a conclusive effect on the energy state, band structure, and properties of amorphous silicon.

According to their characteristics, amorphous semiconductors can be divided into two categories: silicon compounds (C, Si, Ge, and their alloys) and sulfur compounds (S, Se, TE, and their alloys). From the current research and application, these materials can be presented in the form of thin films. The most studied and widely used are hydrogenated amorphous silicon films (a-Si:H) and silicon-based alloy films (such as a-SiC:H, a-SiN:H, a-SiGe:H, etc.).

Because there is a large amount of hydrogen in a-Si, which saturates the suspension bond in silicon and greatly improves the photoelectric properties of a-Si, so the general term *amorphous silicon* refers to the amorphous silicon containing hydrogen, or hydrogenated amorphous silicon (a-Si:H).

With the further study of amorphous silicon films, a series of new films have been obtained, including amorphous silicon based alloy films (such as a-SiC:H, a-SiN:H, a-SiGe:H, a-SiO:H and a-SiSn:H). Superlattice materials (such as a-Si:H/a-Ge:H, a-Si:H/a-SiC:H, a-Si:H/a-SiN:H, a-Si:H/a-C:H), microcrystalline silicon films (such as c-Si:H), polycrystalline silicon films (poly Si:H) and recently developed nano silicon films (nano-Si) all have very important application prospects.

Although there are similarities between amorphous silicon and amorphous silicon-based film materials and crystal materials, they are quite different. Their structural characteristics are as follows:

(1) In structure, the composition atoms of amorphous semiconductors have no long-range order. However, because the bonding force between atoms is very similar to that of a crystal, it usually keeps a short program in the range of several lattice

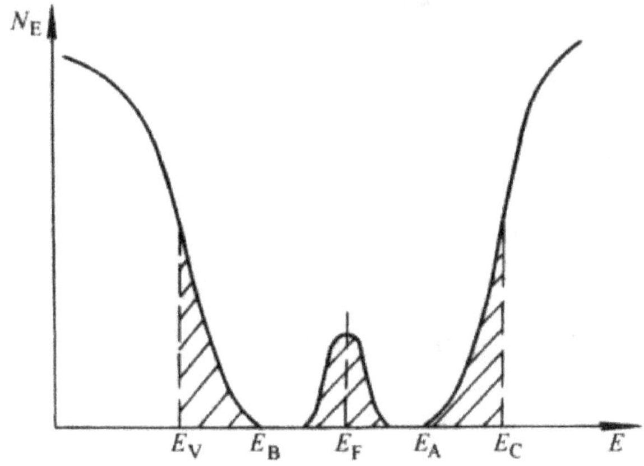

Figure 8.1: a-Si:H band model.

constants. The structure of an amorphous semiconductor is long-range disorder and short-range order. Reflected in its energy band structure, it has not only a conduction band, a forbidden band and a valence band, but also a conduction band tail state, a valence band tail state and defects in the band, which are localized, as shown in Figure 8.1. The jump conductivity mechanism is added to the electron transport, so its mobility has become very small and its resistivity at room temperature is very high.

(2) For most amorphous semiconductors, the atoms consist of covalent bonds, forming a kind of continuous random covalent bond network. All the valence electrons are bound in the bond and meet the ($8\sim N$) rule of the maximum number of bonds. N is the number of valence electrons of the atom. a-Si with four covalent bonds has a tetrahedral structure.

(3) Amorphous semiconductors can partially achieve continuous physical property control.

When the chemical composition of the amorphous semiconductor is changed continuously, its density, phase change temperature, conductivity, band gap width, and so on will change continuously, which provides a broad field for exploring new materials. The amorphous silicon-based alloy material is an important example.

(4) Amorphous semiconductors are in the metastable state in thermodynamics, and can be transformed into crystalline state under certain conditions (for example, thermal annealing and laser annealing can change amorphous silicon into polycrystalline or monocrystalline silicon). This is because amorphous semiconductors have higher lattice potential energy than their corresponding crystalline materials, so they are in a metastable state.

(5) The structure and electrical and optical properties of amorphous silicon and its alloy films are very sensitive to their preparation conditions and methods, so their performance repeatability is worse than that of their crystalline materials.

(6) The physical properties of an amorphous semiconductor are isotropic, because its structure is a kind of covalent bond random network structure, which is not restricted by periodic structure.

Amorphous silicon materials are formed by vapor deposition. According to the different methods of dissociation and deposition, the vapor deposition method can be divided into glow discharge decomposition (GD), sputtering (SP), vacuum evaporation, photochemical vapor deposition (photo-CVD), and hot wire (HW) deposition. Gas glow discharge decomposition technology plays an important role in the preparation of amorphous silicon-based semiconductor materials and devices. The following is a brief introduction to the principle of amorphous silicon-based thin film materials by glow discharge method.

The system of preparing amorphous silicon-based films by glow discharge is shown in Figure 8.2. According to the different frequency of the glow discharge power source, glow discharge can be divided into RF ($\sim$13.56 MHz) glow discharge, DC glow discharge, UHF (VHF-70–150 MHz) glow discharge, etc. Silane (SiH_4) and other raw gases are introduced into the vacuum reaction chamber, and are then decomposed by plasma glow discharge to produce plasma containing charged ions, neutral particles, active groups and electrons. They react chemically on the surface of the glass substrate with TCO film

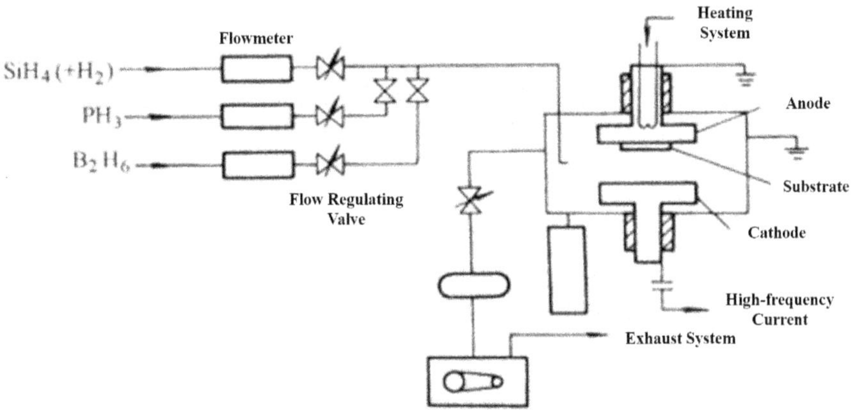

Figure 8.2: Schematic diagram of a glow discharge device for preparing a-Si:H film.

to form a-Si:H film. Therefore, this technique is also called plasma enhanced chemical vapor deposition (PECVD). If borane (B_2H_6) is mixed in SiH_4, p-type amorphous silicon (p-a-Si:H) can be produced; or if phosphane (PH_3) is mixed, p-type amorphous silicon (n-a-Si:H) can be grown. It can be seen from the above that only changing the raw gas can form pin junctions in turn.

8.1.3.2 *Germanium film*

Ge thin film materials can be divided into three categories including single crystal, polycrystal, and amorphous. Single crystal and polycrystal Ge thin films have for a long time attracted people's attention, and the processing and performance of them have already been researched. However, the importance of Ge can't be comparable to silicon films, and there is only a little application in some semiconductor devices. As for the amorphous Ge thin film, even though it can be fabricated by decomposing Ge using glow discharge, the performance of it is much worse than amorphous silicon film, narrowing its application area.

At present, single crystal Ge films fabricated using epitaxial method are mainly utilized to manufacture discrete devices like high frequency transistors or tunnel diodes, and are also used to fabricate application devices working in low temperature and radiation detectors. However, the application of Ge films in the microelectronics industry is still narrow. The main fabrication method of single crystal Ge is epitaxy, and only in rare cases is liquid phase epitaxy adopted.

8.1.4 *III–V group compound semiconductor films*

III–V group compound semiconductor films formed by III and V main group elements in the periodic table are widely applied for the fabrication of microwave devices like Gunn diode, Schottky diode, variable capacitance diode, tunnel diode, and photoelectron devices like luminous diode, laser, solar cells, and avalanche photodiode. At present, semiconductor single crystal films of group III–V compounds are paid attention to and widely studied and applied.

8.1.4.1 *Gallium arsenide film*

The epitaxial method for the deposition of the single crystal gallium arsenide (GaAs) film is mainly gas phase epitaxy, liquid phase epitaxy, and molecular beam epitaxy. The gallium arsenide crystal has a zinc-and-zinc structure. The two sets of surface centered cubic lattices that make up the crystal cells are not made of one atom but two atoms. The lattice constants of gallium arsenide are almost equal to that of germanium, slightly larger than silicon. Gallium arsenide is also a covalent crystal, but differs from silicon or germanium. Because of the negative energy of gallium and arsenic, the electron cloud distribution is asymmetric, and is polar. This polarity has a great effect on the performance of GaAs.

Owing to the existence of polarity and the strong bond between the two atoms, the band gap of GaAs is larger (with $E_g = 1.43\,\text{eV}$ at room temperature). So, a device made of GaAs can work even at 475°C, which is one of the advantages of gallium arsenide. Gallium arsenide is a direct band gap semiconductor, and the carrier can achieve a direct transition. Because of the high probability of a direct transition, gallium arsenide is high in the absorption coefficient of light and the radiative recombination rate of radiation, which makes gallium arsenide suitable for materials used as photodetectors, photovoltaic devices, and light-emitting devices. Gallium arsenide is a kind of multipotent semiconductor with negative resistance.

Gallium arsenide has a much higher electron mobility than silicon, and can be used as a high-speed transistor. Gallium arsenide has both the properties of semiconductor and the properties of insulators, and the resistivity can be more than $3 \times 10^8\,\Omega \cdot \text{cm}$. In this way, the device in the circuit can be fabricated with the semiconductor GaAs, while the isolation part of the circuit components can be made from the semi-insulated GaAs. This helps to reduce process steps, reduce chip area, reduce power consumption, and improve the speed of integrated circuits.

GaAs is high in absorption coefficient and is suitable for making optoelectronic devices. The theoretic photoelectric conversion efficiency of solar cells made from GaAs is large and the high temperature characteristic is fantastic along with good radiation resistance. The GaAs–GaAlAs heterojunction solar cell has the highest photovoltaic conversion efficiency (>24%) at present. GaAs has a high radiation recombination rate, which can be utilized to fabricate light-emitting diodes and lasers. Gallium arsenide is currently the semiconductor phototcathode material with the best quantum efficiency, and the epitaxial film in the transparent substrate can be used to fabricate a transparent photocathode. A single integrated circuit can be combined with GaAs or GaAs-based materials and optoelectronic devices.

8.1.4.2 *Gallium phosphide film*

There are two kinds of methods for the preparation of a single crystal gallium phosphide (GaP) thin film, including gas phase epitaxy and liquid phase epitaxy.

Gallium phosphide has a zinc blende structure and is a kind of indirect band gap semiconductor material with $E_g = 2.26\,\text{eV}$ at room temperature, and also a kind of multi-potent silicon semiconductor. GaP is a semiconductor material used to make light-emitting diodes, where the radiation transition associated with the center of the impurity is used to emit radiation, rather than an intrinsic radiation transition mechanism. The high luminous efficiency is due to the combined effect of some impurity in GaP as the radiation complex center, and the conversion from indirect transition to direct transition. For example, GaP with a higher concentration of N can be used to make a yellow light emitting diode. And a red light-emitting diode can be made from GaP doped with ZnO. In addition to the light-emitting diode, gallium phosphide can also be used to make devices such as avalanche diodes.

8.1.4.3 *Multi-component solid solution film*

The multi-component semiconductor (mainly ternary and a few quaternary) solid solution film materials of III–V group compound semiconductors have the advantages of adjusting the composition and controlling the performance artificially, thus receiving great attention and important application in the manufacture of semiconductor devices, especially light-emitting devices and lasers. There are many kinds of multi-component solid solution. What are mainly introduced in this paper include GaAsP, GaAlAs, InGaAsP, etc.

(1) **Gallium arsenide phosphorus thin film**

The performance of gallium arsenide phosphorus ($GaAs_{1-x}P_x$) varies with x. The typical x of $GaAs_{1-x}P_x$ material is 0.4. The main method for the growth of single crystal $GaAs_{0.6}P_{0.4}$ is gas phase epitaxy.

The ternary solid solution shares a common characteristic. Some of its properties such as the lattice constant and the band gap are between the two kinds of binary compounds (containing a common element), and change linearly as the composition varies. $GaAs_{1-x}P_x$ can be regarded as a ternary solid solution made up of GaAs and GaP, with a performance that is between GaAs and GaP, and varies with x. The band gap of $GaAs_{1-x}P_x$ varies with x, as shown in Figure 8.3. Because GaAs is a direct band gap semiconductor while GaP is an indirect band gap semiconductor, what kind of semiconductor $GaAs_{1-x}P_x$ is should be determined using x. As seen in Figure 8.4, the band gap of $GaAs_{1-x}P_x$ at Γ is less than at X when $x < 0.46$, and the band gap width is determined by the band gap at Γ, so it is a direct bandgap semiconductor, and will be an indirect band gap semiconductor when $x > 0.46$. For the

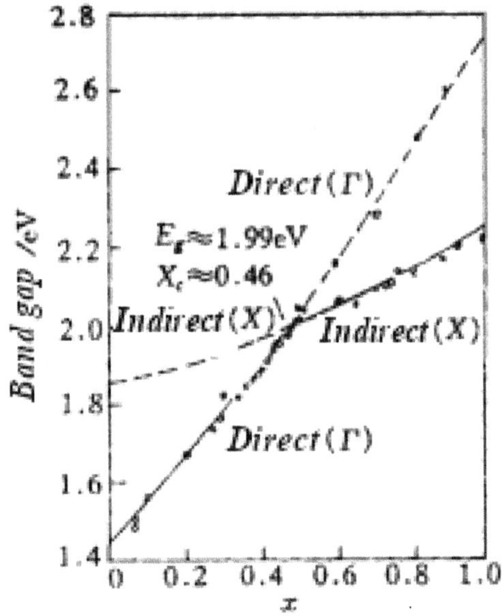

Figure 8.3: Change of band gap of $GaAs_{1-x}P_x$ with x.

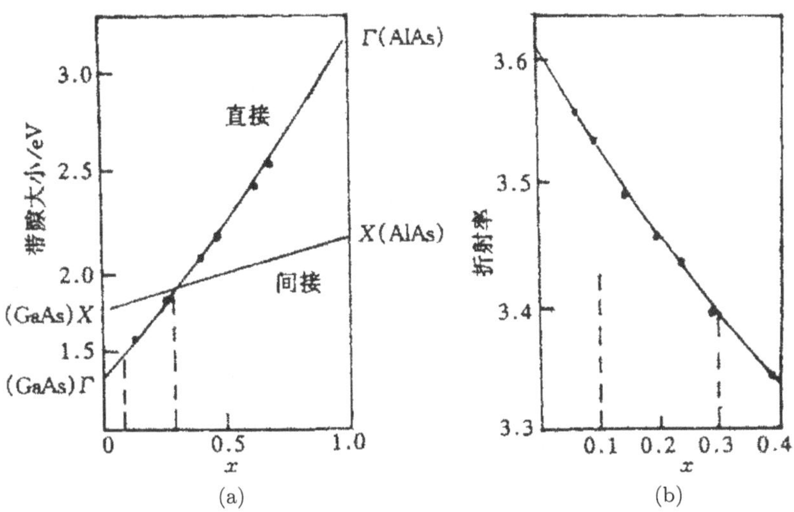

Figure 8.4: Change of the gap (a) and refractive index (b) of $Ga_{1-x}Al_xAs$ with x ($T = 300$ K, measuring the refractive index $hv = 2.4$ eV).

appropriate large E_g and luminescence efficiency not reduced to very low value, the commonly used value of x is 0.4, with $E_g \approx 1.91\,\text{eV}$. The corresponding eigen radiation spectral peak has a wavelength of about $0.65\,\mu\text{m}$. $\text{GaAs}_{1-x}\text{P}_x$ is a fantastic material for making red light-emitting diodes. N-doped $\text{GaAs}_{1-x}\text{P}_x$ is used to produce salmon, orange, yellow, and olivine light-emitting diodes.

(2) Gallium aluminum arsenide thin film

The gallium aluminum arsenide ($\text{Ga}_{1-x}\text{Al}_x\text{As}$) single crystal films of the optoelectronic devices are all grown by the liquid phase epitaxial method, and the sliding boat method is mostly used in industry. $\text{Ga}_{1-x}\text{Al}_x\text{As}$ can be thought of as a three-component solid solution composed of GaAs and AlAs, with the performance between GaAs and GaAs. The band gap and refractive index of $\text{Ga}_{1-x}\text{Al}_x\text{As}$ with respect to x are shown in Figure 8.5. Because AlAs is an indirect band gap semiconductor, $\text{Ga}_{1-x}\text{Al}_x\text{As}$ can be a direct or an indirect band gap semiconductor as x varies. When $x < 0.31$ $\text{Ga}_{1-x}\text{Al}_x\text{As}$ is a direct band gap semiconductor. On such a condition, the E_g is less than 1.90 eV, so it is similar to $\text{Ga}_{1-x}\text{Al}_x\text{As}$. $\text{Ga}_{1-x}\text{Al}_x\text{As}$ can be used to make red light-emitting diodes. A single crystal $\text{Ga}_{1-x}\text{Al}_x\text{As}$ film can be used to prepare semiconductor lasers that work continuously at room temperature.

$\text{Ga}_{1-x}\text{Al}_x\text{As}$ is also a very important material for the fabrication of high-efficiency P-Ga_{1-x}AlAs-n-GaAs hetero-junction solar cells, because the lattice of $\text{Ga}_{1-x}\text{Al}_x\text{As}$ matches well with GaAs. In addition, $\text{Ga}_{1-x}\text{Al}_x\text{As}$ can also be used to make integrated circuits.

(3) Indium gallium arsenide phosphorus ($\text{In}_{1-x}\text{Ga}_x\text{As}_y\text{P}_{1-y}$) thin film

Ternary solid solution can obtain the expected gap by changing the composition. But except for GaAlAs, the lattice match between ternary solid solution and the substrate is not very good. A defect in the epitaxial film has a negative effect on performance. In order to get the appropriate lattice constants on the adjustment of E_g, people are dedicated to the study and use of quaternary solid solution, of

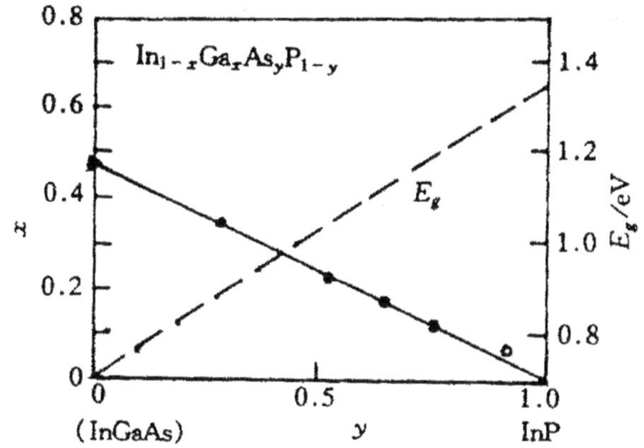

Figure 8.5: The composition and gap of the $In_{1-x}Ga_xAs_yp_{1-y}$ film that matches the InP lattice.

which indium gallium arsenide phosphorus ($In_{1-x}Ga_xAs_yP_{1-y}$) is a typical example (InP as the substrate).

The E_g of the $In_{1-x}Ga_xAs_yP_{1-y}$ epitaxy film varies between 2.25 and 0.365 eV with respect to different x and y, and the corresponding wavelength of light is 0.55–3.4 μm. If you consider lattice constants of $In_{1-x}Ga_xAs_yP_{1-y}$ films and the InP substrate to match well, you can achieve a E_g range of 1.34–0.76 eV, with the corresponding wavelength in the range of 0.93–1.6 μm. Figure 8.5 shows the composition of the $In_{1-x}Ga_xAs_yP_{1-y}$ epitaxial thin film, which has a good lattice match with the InP substrate lattice and the width of the corresponding gap. As a result, the single crystal indium gallium arsenide film can be used to make near-infrared LEDs. The film is also used in the manufacture of InGaAsP/InP dual hetero-junction (with source zone) laser from InGaAsP.

8.1.5 *II–VI group compound semiconductor film*

II–VI group semiconductors are direct band gap semiconductors, with a wide range of gap values (at room temperature, the E_g of ZnS is 3.6 eV and that of HgTe is −0.14 eV), and can be used to adjust and control the gap in ternary solid solution, so it can be used to make photo-electronic devices of various working wavelengths.

The epitaxial method is used to grow single crystal II–VI group ccompound semiconductor films and the vacuum evaporation deposition method is used to produce the polycrystalline counterparts. II–VI group compounds can be classified into two categories: the first kind is mercury compounds such as HgS, HgTe, and $Cd_xHg_{1-x}Te$. The characteristics of this kind of material is narrow band gap, high carrier mobility, high carrier concentration, low resistivity, and good parameter stability. The other kind is compounds that are composed of zinc and cadmium and are free of mercury, such as ZnS, ZnSe, ZnTe, CdS, CdSe, and CdTe. This kind of material is characterized by wide band gap, low mobility, low carrier concentration, high resistivity, and poor electrical parameter stability.

8.1.5.1 *Mercury-containing compound film*

Almost all vacuum technologies for deposition of films can be used to prepare II–VI group compound films, including various vacuum evaporation and sputtering methods. The most used is the simple single source evaporation method, the fast evaporation method, and the dual source evaporation method.

The polycrystalline compound semiconductor films prepared by vacuum evaporation can have two basic crystal structures, namely cubic zinc blende and hexagonal wurtzite. The mercury-containing HgS thin film can have both these structures, but the HgSe, HgTe, and $Cd_xHg_{1-x}Te$ films only have zinc blende structure. The optical properties and electrical properties of the film depend on the crystal structure, chemical composition, growth condition, and subsequent heat treatment of the film.

8.1.5.2 *Mercury-free compound film*

There are many methods for preparing mercury-free films, such as common single source evaporation, dual source evaporation, molecular beam epitaxy, etc. But the most common use of the single-source method is the ordinary one. The film is generally treated with heat treatment after deposition.

The mercury-free film has a relatively wide band gap, which has a significant impact on the optical, electro-optical, and electrical properties of the film. The band gaps of single crystal ZnS, ZnSe, ZnTe, CdS, CdSe, and CdTe are 3.6, 2.7, 2.26, 2.4, 1.67, and 1.44 eV, respectively. For thin films, the optical properties associated with E_g, such as the absorption limit of the absorption spectrum, are roughly similar to those of homogeneous single crystal materials.

The characteristics of the mercury-free compound film are high resistivity and low mobility. The highest transfer rate of the high-quality mercury-free films prepared with single source evaporation is 10–60% of the highest value of the same kind of single crystal, and varies with different compounds.

Because of the wide band gap of mercury-free compounds, they can be used to make visible, ultraviolet or even X-ray detectors. The detectors mainly utilize the light guide effect. Usually, mercury-free compounds are used to make photosensitive resistors. The most used are CdS, CdSe, and CdSSe films, which are sensitive to visible light. In addition, CdTe, ZnSe, ZnTe, and ZnS thin films can also be used to produce near-infrared, visible, and ultraviolet light sensitive resistors.

The main advantage of a CdTe material is that its spectral response is very consistent with the solar spectrum, which makes the theoretical conversion efficiency of a CdTe solar cell very high, 27% at room temperature (open circuit voltage V_{OC} is 1050 mV; short circuit current density J_{SC} is 30.8 ma/cm^2; filling factor FF is 83.7%). Moreover, CdTe has a very high direct transition (energy gap is about 1.44 eV) optical absorption coefficient (105 cm^{-1}). As far as the energy in the solar radiation spectrum is larger than the range of CdTe energy gap, the material with a thickness of 1 μM can absorb 99% light. Therefore, it can reduce the material consumption and reduce the cost, and is regarded as an ideal solar cell absorption layer material.

CdTe is a II–VI compound, which is a direct band gap material with a band gap of 1.44 eV. Its spectral response is in good agreement with the solar spectrum, and its electron affinity is very high, which is 4.28 eV. CdTe has a zinc blend structure, with lattice constant $a = 0.16477$ nm. Due to the direct band gap structure of CdTe thin films, the absorption coefficient of the light whose wavelength is less than the absorption edge is very large. The thickness of the film is 1 μm, which is enough to absorb 99% of the radiation energy larger than the CdTe band gap energy, so the requirement of the diffusion length of the material is reduced.

The structure of CdTe is similar to that of Si and Ge; that is to say, its crystal mainly depends on the covalent bond, but it has a certain ionic property. Compared with the group IV semiconductors of the same period, the binding strength of CdTe is very strong, and the energy required for electrons to get rid of the covalent bond is higher. Therefore, the conductivity of CdTe at room temperature is mainly determined by doping. The composition, structure, deposition condition, and heat treatment process of the film have a great influence on the resistivity and conductivity type of the film.

There are many methods to prepare CdTe polycrystalline films, such as near space sublimation, electrodeposition, screen printing, physical vapor deposition, spray pyrolysis, and so on. However, the preparation method to obtain the highest efficiency of the battery is to deposit CdTe thin film by near space sublimation system (see Figure 8.6). This method is simple in equipment, low in deposition rate, easy to control, low in pollution, and suitable for large-scale production, with uniform film formation, appropriate grain size, and excellent optical and electrical properties.

CdTe sublimates and decomposes into Cd and Te_2 when the source temperature is higher than 4500°C. When these are deposited

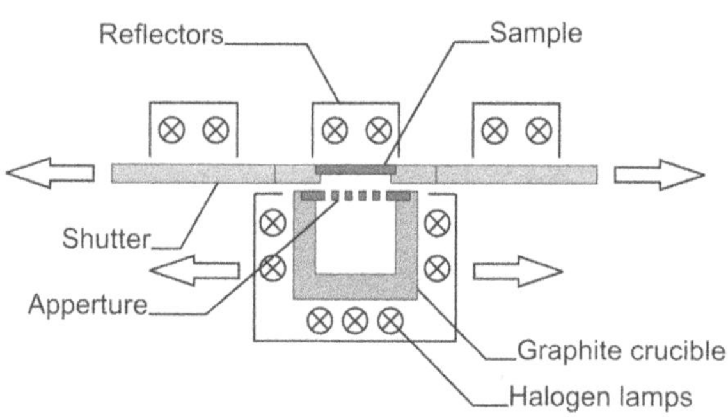

Figure 8.6: Near space sublimation system.

on the substrate with relatively low temperature, they are recombined to form CdTe. In order to prepare films with uniform thickness, chemical composition and grain size, it is necessary to maintain a certain pressure in the reaction chamber and make Cd and Te_2 not directly evaporate to the substrate. In this way, the distance between the source and the substrate must be small. This is the basic idea of near space sublimation. The key to the preparation is the type and pressure of the shielding gas, the temperature of the source, and the temperature of the substrate. The typical parameters of CdTe preparation are: gas: argon; spacing: 5 mm; source temperature: 620–680° C; substrate temperature: 520–580° C; deposition time: 8–15 min.

CdS is a very important class II–VI compound semiconductor material. The CdS thin film has a wurtzite structure, which is a direct band gap material with a wide band gap of 2.42 eV. It can pass through most visible light. Moreover, when the film thickness is less than 100 nm, CdS thin film can make wavelengths less than 500 nm pass through.

Figure 8.7 shows the spectral response curves of CdTe solar cells with different thicknesses (curve a, efficiency 13.38%; curve b, efficiency 15.8%). It can be seen that the shortwave response is extended after the CdS is thinned.

There are many methods to prepare CdS films, such as screen printing, electrodeposition, sputtering, near space sublimation, vacuum evaporation, spraying, and chemical water bath. The preparation of CdS thin films by a chemical water bath is simple and cheap and it is easy to achieve large-scale production, so people pay attention to it.

CdS thin films were deposited by the chemical water bath method. Thiourea was used as a sulfur source and the Cd $(NH_3)_4^{2+}]$ complex was used as a cadmium precursor. The reaction was carried out in ammonia aqueous solution at a temperature of about 82°C. In addition, NH_4Cl buffer was added to keep the pH value of the film-forming solution at 8–10. All the drugs used were analytical pure reagents, which were prepared with secondary deionized water. The concentration of each component was $[CdCl_2]$ = 0.0012M,

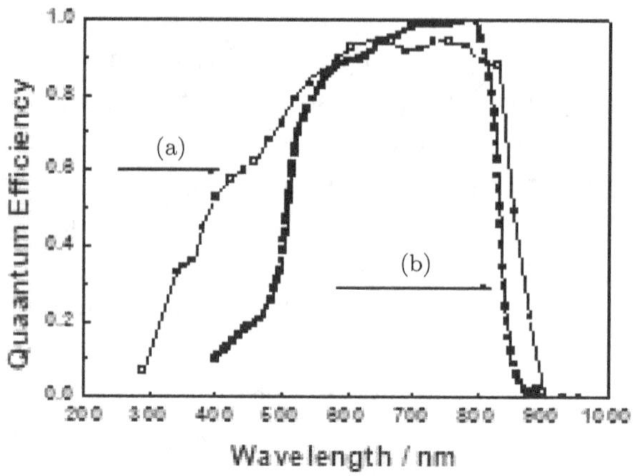

Figure 8.7: Spectral response curves of CdTe solar cells with different thicknesses: (a) 180 nm and (b) 60–80 nm.

[NH3H2O] = 0.1 M, [(NH_2)2Cs] = 0.004 M, [NH_4Cl] = 0.02 M. The reaction equation of the whole process is as follows: $Cd(NH_3)_4^{2+}$ + $(NH_2)_2$ CS + 2OHt → CdS + $4NH_3$ + $2H_2O$ + CN_2H_2.

8.1.5.3 *I–II–VI group semiconductor films*

I–II–VI group semiconductors, such as $AgGaS_2$, $AgGaSe_2$, $CuInSe_2$, etc., have unique photoelectric properties, and have a wide range of application prospects in nonlinear optics, infrared detection, solar cells, etc.

$CuInSe_2$ has two allotrope crystal structures of chalcopyrite and sphalerite. Its high temperature phase is sphalerite structure (phase transition temperature is 980°C), which belongs to the cubic crystal system. The lattice constant is $a = 0.58$ nm, and the density is 5.55 g/cm. The low temperature phase is chalcopyrite structure (the phase transition temperature is 810°C), which belongs to the orthorhombic crystal system. Bravais lattice is tetragonal with a space group of $I_4^{2d} = D_{2d}^{12}$. Each crystal cell contains four molecular groups. The lattice constant is $a = 0.5782$ nm, $c = 1.162l$ nm, and the lattice mismatch ratio with CdS ($a = 0.46$ nm, $c = 6.17$ nm) of wurtzite structure is 1.2%. This makes it superior to other Cu ternary compounds such as $CuInS_2$.

$CuInSe_2$ is a direct band gap semiconductor material, with a band gap of 1.04 eV at 77 K and 1.02 eV at 300 K. The band gap is not sensitive to temperature changes. Its band gap (1.04 eV) is close to the optimal band gap (1.5 eV) required for ground photovoltaic utilization. The electron affinity of $CuInSe_2$ is 4.58 eV, which is very little different from that of CdS (4.50 eV), which makes the heterojunctions formed by $CuInSe_2$ have no conduction peak and reduces the potential barrier of photogenerated carriers.

$CuInSe_2$ has an allowed direct intrinsic absorption limit of 0.95–1.04 eV, a forbidden direct absorption limit of 1.27 eV, and additional absorption in the low absorption region (long band) caused by the DOW Redfield effect. $CuInSe_2$ has an absorption coefficient of up to $6 \times 105\,\text{cm}^{-1}$, and is the material with a large absorption coefficient in semiconductor materials. With such a high absorption coefficient (i.e., a small absorption length), it is very favorable for the absorption of photons in the solar cell base area and the collection of a few carriers (i.e., the collection of photocurrents).

At room temperature (300 K), the direct band gap of single crystal $CuInSe_2$ is 0.95–0.97 eV. The polycrystalline film has 1.02 eV, and the optical absorption coefficient of a single crystal is larger than that of a polycrystalline film. The reason is that there are more single crystal materials, a more perfect stoichiometric ratio, and better component uniformity and crystallization. After heat treatment in inert gas, the absorption characteristic of polycrystalline films are close to that of the single crystal, which shows that the composition and crystallinity of polycrystalline films are improved after heat treatment.

The absorption property decreases with the decrease of working temperature, and the band gap slightly increases with the decrease of temperature. When the temperature drops from 300 K to 100 K, the band gap of single crystal $CuInSe_2$ is 0.98 eV and that of polycrystalline $CuInSe_2$ is 1.04 eV when the E_g increases by 0.02 eV, i.e., 100 K.

The electrical properties (resistivity, conductivity type, carrier concentration, mobility) of $CuInSe_2$ materials mainly depend on the component ratio of each element of the materials, as well as the inherent defects (such as vacancy, interstitial atom, and substitutional atom) caused by the deviation from the stoichiometric ratio, in addition to the non-intrinsic doping and grain boundary.

According to the defect conduction theory, when Se is insufficient, Se vacancy presents donor; when Se is excessive, it presents acceptor; when Cu is insufficient, Cu vacancy presents acceptor; when Cu is excessive, it presents donor; when it is insufficient, it presents acceptor; when it is excessive, it presents acceptor; when it is insufficient, it presents acceptor; when it is abundant, it presents donor.

When the composition ratio of the film deviates greatly from the stoichiometric ratio, the situation becomes very complicated. At this time, the composition of the film no longer has a single chalcopyrite structure, but contains other phases (Cu_2Se, $Cu_{2-x}Se$, In_2Se_3, InSe, etc.). In this case, the conductivity of the film is mainly determined by the ratio of Cu to In. Generally, with the increase of the Cu/In ratio, the resistivity decreases and the p-type conductivity increases. The conductivity type has little relationship with Se concentration, but p-type conductivity increases with Se concentration.

The key technology to prepare CIS films is to control the proportion of elements. The main growth methods are vacuum evaporation, selenium treatment of Cu in alloy films (including electrodeposition and chemical thermal reduction), near space vapor transport (CSCVT), spray pyrolysis, RF sputtering, etc.

Doping other elements on the basis of $CuInSe_2$ (CIS), such as making GA or Al partially replace in the atom, and replacing Se with S part, namely, preparing Cu $(In_{1-x}Ga_x)$ Se_2, Cu $(In_{1-x}Ga_x)$ $(Se_{2-y}S_y)$ [10], Cu $(In_{1-x}Al_x)$ $(Se_{2-x}S_x)$. The crystal structure of CIGS, CIGSS and CIASS is still chalcopyrite. By changing the atomic ratio of Ga/(Ga + In), the band gap can be changed from 1.04 eV to 1.72 eV, including the band gap range of 1.4–1.6 eV. The solar cells prepared with CIGS film have the following characteristics: (1) High photoelectric conversion efficiency. In 2008, the photoelectric conversion efficiency of small area CIGS thin film solar cells developed by NREL has reached 19.9%, which is the highest record of all kinds of thin film solar cells. (2) The stability of the battery is good, and the performance has basically no degradation in the use process. (3) It has a strong ability for anti-irradiation and has a strong competitiveness as a space power source. (4) CIGS is a direct band gap material, its visible light absorption coefficient is up to 10^5 cm^{-1} order of magnitude, which is very suitable for the thin film of solar cells.

8.1.6 *Zinc oxide films*

8.1.6.1 *Overview of ZnO films*

Zinc oxide (ZnO) is an important wide-band gap semiconductor material. The band gap is 3.37 eV at room temperature. ZnO is a hexagonal wurtzite structure. Each Zn atom and four O atoms are arranged in a tetrahedron. The lattice constants are a = 0.325 nm, c = 0.52 nm. The O atoms are arranged in the hexagonal packing position, while the Zn atoms occupy half of the tetrahedron position. Therefore, ZnO has a relatively open structure, and all the octahedral and half the tetrahedral positions are empty. The structural characteristics of ZnO make it easy to introduce foreign atoms into the ZnO lattice. ZnO doped with In, Ga, and Al is a kind of transparent conductive oxide with excellent photoelectric properties; ZnO doped with magnetic atoms such as Fe, Co, and Mn is a widely studied diluted magnetic semiconductor; ZnO doped with elements such as Li and Mg has piezoelectric and ferroelectric properties.

When the orientation of ZnO thin films perpendicular to the substrate surface is the same, they can have better anisotropic piezoelectric properties like single crystals. As a kind of piezoelectric material, they are widely used in ultrasonic transducers, spectrum analyzers, high-frequency filters, high-speed optical switches, and micromechanics because of their strong electromechanical coupling coefficient. They are the first choice material for the preparation of high-frequency surface acoustic devices.

ZnO has the characteristic of the resistivity changing with the type and concentration of the adsorbed gas. Generally, the conductivity increases when adsorbing the reducing gas and it decreases when adsorbing the oxidizing gas. When contacting the reducing gas, the conductivity will increase with the increase of gas concentration, while when contacting the oxidizing gas, the conductivity will decrease with the increase of gas concentration. A film doped with some elements has good sensitivity to a harmful gas, a combustible gas, an organic swallow gas, and so on. It can be made into various gas sensors.

The varistor properties of ZnO thin films mainly show that when the varistor material is affected by an applied voltage, there is an alarm value voltage, namely the varistor voltage ($V_{1\text{ma}}$). When the applied voltage is higher than this value, it will enter the breakdown region. At this time, a small change in the voltage will cause a rapid increase of the current, and the change range is characterized by the nonlinear coefficient α. The ZnO thin film has the outstanding characteristics of a low voltage sensitivity, a high nonlinear coefficient, a strong surge absorption capacity, and a stable performance, which opens up a wide prospect for the application of pressure-sensitive materials in the protection of microcircuits. ZnO and GaN are both hexagonal wurtzite structures with similar lattice constants and band gap widths. At the same time, their exciton activation energy is as high as 60 meV, much higher than the room temperature thermal ionization energy of 26 meV, and the exciton is not prone to thermal ionization. Because excitons with large binding energy are more likely to achieve high-efficiency laser emission at room temperature, ZnO is theoretically more likely to achieve UV-stimulated radiation at room temperature than ZnSe (22 meV), ZnS (40 meV), and GaN (25 meV). ZnO thin films have self-assembly properties in the deposition process. The cross-section of the columnar crystal is hexagonal, and the columnar crystal is perpendicular to the substrate surface. When photoinduced laser emission of ZnO thin films is done at room temperature, the corresponding crystal surfaces of the three parallel sides of the columnar crystal act as the reflecting surface during light injection, and the photons move back and forth between them to form standing waves and achieve light amplification. This property makes ZnO thin films suitable to realize a ultraviolet laser emission device.

8.1.6.2 *Preparation and doping of ZnO films*

The preparation methods of ZnO thin films include sputtering, PLD, sol–gel, MBE, and so on.

Due to the existence of O vacancy and Zn interstitial atoms, undoped ZnO films show n-type conductivity. n-Type doping of ZnO thin films is easy to achieve. The third group elements Al, Ga, In and the seventh group elements Cl, I can be used to dope ZnO thin films, replacing Zn and O atoms, respectively.

However, p-type doping of ZnO thin films is very difficult. One of the main reasons is the self-compensation effect caused by the more intrinsic donor defects in ZnO. There are six kinds of intrinsic point defects in ZnO: oxygen vacancy V_{O}, zinc vacancy V_{Zn}, inversion oxygen O_{Zn}, inversion zinc Zn_{O}, interstitial oxygen O_i, and interstitial zinc Zn_i. The oxygen vacancy V_{O} is a positive center, which has a negative Coulomb attraction potential. Its conduction band energy level moves to the low energy direction, and enters the band gap to form the donor energy level. The valence band energy level of zinc vacancy V_{Zn} moves to the high energy direction and enters the band gap to form the acceptor energy level. The defect of O_{Zn} is that the O occupies the position of Zn atom and produces the O anteposition of Zn. It attracts the valence electrons of the neighboring atoms to form the negative center, and the valence band energy level enters the band gap to form the acceptor defect. The Zn_{O} defect is the reverse defect of Zn and O, which becomes the positive center, and the conduction band energy level enters the band gap to form the donor defect. The gap Zn Zn_i is a positive center, and its conduction band energy level moves to low energy, and enters the band gap to form the donor energy level, while the O_i defect state is the acceptor energy level at the valence band top.

Figure 8.8 shows the energy levels of the above six defects. It is also clear from Figure 8.8 that O_i and V_{Zn} are shallow acceptors, while V_{O}, Zn_i and Zn_{O} are donor defects. The difficulty of defect formation can be indicated by the formation energy. The formation energy of V_{O} and Zn_i is very low under the condition of rich Zn and O (see Figure 8.9), which is easier to exist in ZnO than that of V_{Zn}

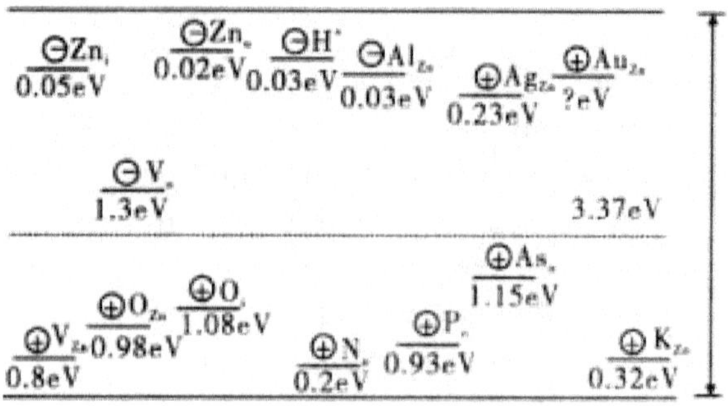

Figure 8.8: Intrinsic and doping energy levels of ZnO.

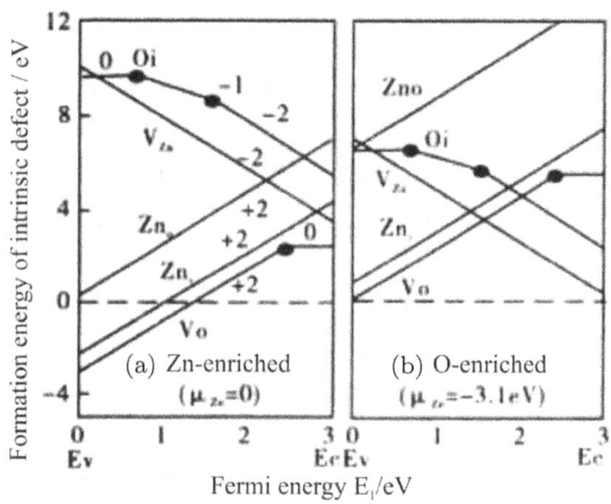

Figure 8.9: Formation energy of various intrinsic defects in ZnO under Zn- and O-rich conditions.

and O_i. The existence of these donors can compensate for the p-type shallow acceptor, which is the so-called self-compensation effect. The process of dopant forming inversion defect is the process of system energy reduction, so it is an inevitable result of the system tending to the equilibrium state that the larger the band gap width is, the more significant the energy reduction caused by self-compensation is, and it is easier to produce self-compensation when the dopant is the wide-band gap material, so it is difficult to realize the inversion of the material through the general dopant.

One of the difficulties in p-type doping of ZnO thin films is the monopolarity of doping elements on the surface of ZnO thin films, which can be described by Madelung energy. The higher the absolute value of Madelung energy is, the stronger the monopolarity is or the worse the activity of doped elements is. The more unstable the ZnO crystal structure is, the more difficult it is to obtain p-type ZnO films. The p-type ZnO film cannot be made without the method, which greatly limits the development and application of ZnO-based optoelectronic devices.

According to the current theoretical research and practical experience, in order to achieve effective p-type doping of ZnO, the following conditions must be met: first, increase the doping concentration of acceptor elements in ZnO; second, make the acceptor level shallower in ZnO; third, suppress the intrinsic donor defect concentration in ZnO and reduce the self-compensation effect. At present, three methods have been proposed to prepare p-type ZnO: (1) doping V-group elements into the oxygen vacancy; (2) Co-doping V-group elements and III-group elements into ZnO, or co-doping I-group elements and VII elements into ZnO; (3) using excessive oxygen to eliminate the self-compensation effect of oxygen vacancy, which is often carried out at the same time as the V-group element doping method.

Among the I-group elements, Li, K, Ag, Cu, and Au have been studied to some extent. Because AU has two valence states of +1 and +3, it can be used as both acceptor and donor in ZnO. The situation is complex, and its energy level position has not been measured experimentally. Ag and Cu exist as acceptors, and the acceptor energy levels are very deep, at 0.23 eV and 0.17 eV, respectively, at the bottom of the conduction band (as shown in Figure 8.8). Ag is preferred to replace Zn in ZnO, and Ag_{Zn} shows a very dominant behavior, so it is difficult to obtain p-ZnO by doping Ag in ZnO films. In Li-doped ZnO films, Li atoms replace Zn atoms as acceptors, but because of the small size of Li atoms, some of them will become gap atoms. At this time, Li is no longer acceptor, but will cause deep level hole traps, which exist as donors. Therefore, although Li is an acceptor in general, it is difficult to measure its energy level due to its high self-compensation. In ZnO films, the energy level of group I doped elements is very deep, and the doping degree is not high, so as acceptor doping is not very ideal. In the V-group element doping, there are mainly N, P, As doping, among which N doping is the best. The nitrogen sources used mainly include N_2, NO, N_2O, NH_3, Zn_3N_2, etc. The main preparation methods used are MBE, CVD, PLD, MOCVD, and DC reactive magnetron sputtering. The research on p-type ZnO films by co-doping technology mainly includes Al-N co-doping, Ga-N co-doping, In-N co-doping, N-Be co-doping, and so on. The theory shows that ZnO films with relatively high carrier concentration can be prepared by the method of N and AlGaln co-doping.

The aluminum-doped zinc oxide (AZO) film has the advantages of good conductivity, high light transmittance, strong absorption of ultraviolet, high infrared reflectance, and high attenuation of microwaves. It is a transparent conductive film with excellent performance. It can be used in flat displays, solar cells, transparent electric boards, and places where ultraviolet rays, thermal radiation and electromagnetic waves need to be blocked.

8.1.7 *Transparent conductive film*

The transparent conductive film is basically a kind of film with both high conductivity (low resistivity) and high transmittance in the visible light range. Transparent conductive films are widely used in microelectronics and optoelectronics, and the main applications are planar liquid crystal display (LCD), transparent electrode for camera target, transparent electrode of solar cell, electroluminescent display (ELD), electrochromic display (ECD), thermal mirror, anti-charge and electromagnetic shielding, transparent heat body, etc.

Transparent conductive films can be classified into four types: (1) metal film; (2) transparent conductive oxide film; this kind of film is characterized by high light transmission, high intensity of film, and excellent chemical stability; (3) non-oxide compound film; (4) conductive particle dispersion dielectrics. The categories are shown in Table 8.2.

Table 8.2: Categories and composition of transparent conducting films.

Category		Composition
Metal film	Single layer film	Ni, Pt, Au, Cu
	Multi-layer film	Au/Bi_2O_3/substrate, Au/Cu/substrate, ZnS/Ag/ZnS, SnO_2/Ag/SnO_2
Oxide	Single phase	SnO_2, In_2O_3,CdO, ZnO, Cd_2SnO_4
	Composite (doping) phase	In_2O_3:SnO_2(ITO), SnO_2:Sb, SnO_2:F, In_2O_3:W, In_2O_3:Mo, ZnO:Al
Non-oxide	Single phase	CuS, CdS, ZnS, LaB_6, TiN, TiC, ZrN, ZrB_2, HiN
	Dual layer film	TiO_2/TiN, ZrO_2/TiN
Electrically conductive particle dispersed dielectric		Ag, Au, Ru, SnO_2, ZnO, particle dispersed SiO_2

Among transparent conductive films, oxide transparent conductive films have the most extensive and important applications, including In_2O_3:SnO_2 (ITO) film. The ITO thin film has already been applied in many fields such as planar liquid crystal display, transparent electrode of solar cell, and transparent electrode for camera target, and the product has entered the market. The aluminum-doped ZnO thin film (AZO) is widely recognized as one of the most promising materials, and the research is continuing.

The methods of preparing oxide transparent conductive films include sputtering, CVD, vacuum evaporation, sol–gel reaction evaporation, coating, etc. In addition, the transparent conductive films are prepared by metal organic chemical vapor deposition (MOCVD), atomic layer epitaxy, and pulsed laser deposition. Different preparation methods have different advantages and disadvantages. The properties of the film are mainly determined by the preparation process And the improvement of the preparation technology is toward lowering the film's resistivity, escalating the transmission rate, and flattening the surface topography, making the film growth temperature close to room temperature, acquiring strong adhesion to the substrate, and lowering the cost.

8.2 Superconducting Thin Films

8.2.1 *Overview*

In 1911, Dutch physicist Onnes (H. K. Onnes) discovered that the low temperature material has zero resistance, called superconductivity. By the 1980s, 28 elements, and nearly 5000 alloys and compounds had been found to have superconductivity. At atmospheric pressure, the critical transition temperature of Nb (normal state to superconducting state) is the highest among the elements. Among the alloys and compounds, the highest superconducting critical transition temperature is held by Nb_3Ge, $T_c = 23.2\,K$. In addition, oxide superconducting materials and organic superconducting materials have been found. In 1986, Muller and others found evidence that the La–Ba–Cu–O compound shows signs of superconductivity at 35 K. In February 1987, Prof. Jingwu Zhu announced the discovery of oxide superconducting materials with $T_c = 93\,K$, and the presence of superconducting materials in the liquid nitrogen temperature field led to a worldwide upsurge in research on high-critical temperature superconducting materials. Muller and Bednorz won the 1987 Nobel Prize in physics for pioneering the new field of oxide superconductors.

8.2.2 *Preparation and performance of superconducting thin films*

According to the superconductor's critical temperature T_c, superconductors can be roughly divided into low-T_c superconductors and high-T_c superconductors. The low-T_c superconductor is mainly in the liquid helium and liquid hydrogen temperature zones with T_c generally less than 40 K. The high-T_c superconductor is mainly in the liquid nitrogen temperature zone, and the T_c is generally higher than 77 K. The preparation methods of superconducting films can be divided into several categories: sputtering, vacuum evaporation, chemical vapor deposition (CVD), surface diffusion, sol–gel, and pulsed laser deposition (PLD). At present, the electrical properties of the superconductive films prepared by sputtering and PLD are the best.

8.2.2.1 *Low-T_c superconducting thin films*

(1) **Single element superconductor**

There are 28 elements which have superconductivity at atmospheric pressure, and the preparation of the element superconductors is almost entirely by vacuum evaporation. However, for high-boiling point materials such as Nb and Ta, the sputter method should be given priority.

(2) **A15 compound superconducting thin films**

The β–W structure is known as the A15 structure in crystallography. There are 74 types of compound superconductors with A15 structures, many of which are mixed crystals, and are thermally balanced and can be prepared from solid or diffusion reactions. When preparing superconducting film, a superconductivity close to bulk materials of the superconductive material can be obtained by properly controlling the composition and the temperature of the substrate (700–900°C). However, the T_c at equilibrium is mostly a bit lower than the value for bulk. However, the non-equilibrium Nb_3Ge film is an exception, with the T_c as high as 23 K, while the T_c at equilibrium is 6 K.

(3) **Bl compound superconducting thin films**

NaCl structures are called B1 structures in crystallography. The compound superconductors with the Bl structure are materials with NaCl structure composed of C, N, O, and other elements that are combined with IIIb, IVb, Vb, and VIb transition elements. And the important ones are NbN and NbC_xN_{1-x}, MoN, and so on. NbN has excellent thermal stability and high mechanical strength, and can be produced at low temperature to form a thin film, which has the advantage of anti-neutron irradiation, and can be made into a high-reliability Josephson device.

(4) **Ternary compound superconducting thin films**

Ternary compound superconducting thin films mainly include sulfide superconducting thin films and oxide superconducting thin films. The former include $PbMo_6S_8$, $CuMo_6S_8$, and the latter include $BaPb_{1-x}Bi_xO_3$, etc.

$PbMo_6S_8$ is a superconducting material with the highest upper critical magnetic field H_{c2} (60 T, 0 K). The $PbMo_6S_8$ superconductive film can be prepared by the sputtering method at low temperature. The $CuMo_6S_8$ superconductive film is prepared by evaporation or sputtering. $BaPb_{1-x}Bi_xO_3$ is a solid solution composed of $BaPb_6O_3$ and $BaBiO_3$, with the crystalline structure of a perovskite. Superconductivity appears when x is between 0.05 and 0.3, and when $x = 0.3$, the maximum value of $T_c = 13\,K$ is obtained. Because the material is oxide, it is extremely stable in oxygen and can be used to make highly stable Josephson devices.

8.2.2.2 *High-T_c superconducting thin films*

So far, three generations of high-temperature superconductors have been discovered; the first generation is lanthanum high-temperature superconducting materials. The second generation is Y-system high-temperature superconducting materials. The third generation includes bismuth-base, thallium and other new high-temperature superconductivity materials. The superconducting phases in the material system of high-temperature superconductors are shown in Table 8.3.

(1) **Lanthanide series high-temperature superconducting thin films**

The superconducting transition temperature of the lanthanide superconductor, composed of lanthanum, barium, and copper oxide, or lanthanum, strontium, and copper oxide, is between 30 K and 40 K, and the structure is K_2TiF_4. Superconducting materials containing only lanthanum and copper oxide are superconducting at 91 K. The copper-free lanthanum, strontium, and nickel oxide superconducting materials are superconducting at 70 K.

Table 8.3: Superconducting phase of a high-T_c superconductor.

System	Composition	T_c (K)
La series	$(La_{1-x}M_x)_2CuO_4$ ($x \approx 0.080$)	
	M = Ba	30
	M = Sr, Ca	20,40
	M = Na	40
Y series	$Ba_2LnCu_3O_7$	90
	Ln = Y, La, Nd, Sm, Eu, Cd, Dy, Ho, Er, Tm, Yb, Lu	80
	$Ba_4Y_2Cu_8O_{15}$	
Bi series	$Bi_2O_2 \cdot SrO(n-1)Ca \cdot nCuO_2$	
	$Bi_2Sr_2CuO_6$(2201)	7–22
	$Bi_2Sr_2CaCu_2O_8$(2212)	80
	$Bi_2Sr_2Ca_2Cu_3O_{10}$(2223)	110
	$Bi_2Sr_2Ca_3Cu_4O_{12}$(2234)	90
Tl series	$Tl_2O_2 \cdot 2BaO \cdot (n-1)Ca \cdot nCuO_2$	
	$Tl_2Ba_2CuO_6$(2201)	20–90
	$Tl_2Ba_2CaCu_2O_8$(2212)	105
	$Tl_2Ba_2Ca_2Cu_3O_{10}$(2223)	125
	$TlO_2 \cdot 2BaO \cdot (n-1)Ca \cdot nCuO_2$	
	$TlBa_2CaCu_2O_7$(1212)	70
	$TlBa_2Ca_2Cu_3O_9$(1223)	110–116
	$TlBa_2Ca_3Cu_4O_{11}$(1234)	120
	$TlBa_2Ca_4Cu_5O_{13}$(1245)	< 120
Others	$(Nd_{0.8}Sr_{0.2}Ge_{0.2})_2CuO_4$	27
	$(Nd_{1-x}Ge_x)_2CuO_4$($x \approx 0.07$)	25
	$(B_{1-x}M_x)BiO_3$($x \approx 0.4$)	30
	$M \equiv K, Rb$	

(2) **Yttrium-system high-temperature superconducting thin films**

The structure of yttrium-system superconducting materials, composed of yttrium, barium, and copper oxide, is layered perovskite, and the superconductivity transition temperature is 98 K. All advanced film methods have been used in order to make the high-quality high-T_c YBaCuO superconducting thin film. It is proved that multi-source co-evaporation, the sequence evaporation, multi-target co-sputterring, and sequential sputtering are better approaches. Laser evaporation (also known as pulse laser deposition, PLD) and magnetic control sputtering are the most rapidly developed and applied.

(3) **Bismuth-base superconductive films**

The bismuth-base high-temperature superconductor is composed of bismuth, strontium, calcium, and copper oxide, and the structure is bismuth titanate. Its superconducting critical temperature can reach 114–120 K, which is higher than the yttrium system. It can be used to make electronic components and wires, and its disadvantages include the difficulty to form a crystalline structure, bad reproducibility, and unstable superconducting transition temperature. Because the BiSrCaCuO high-T_c superconductor has multiple superconducting phases, it is difficult to synthesize bismuth superconductive films with a single phase. Especially the (2212) phase and the (2223) phase will usually coexist. For the bismuth-base high-T_c superconducting thin film, the multi-target (Bi, SrCu, CaCu) magnetic control sputtering method is used for stack deposition.

(4) **Thallium-based high-temperature superconducting thin films**

Thallium-based superconducting materials are composed of thallium, calcium, barium, and copper oxide, and the structure is also bismuth titanate. The critical temperature of superconductivity transition is 106–125 K. Its advantage is fantastic reproducibility and its disadvantage is strong toxicity. The TI is volatile, so Thallium-based superconducting thin films can only be deposited at a relatively low substrate temperature (from room temperature to 200°C). The attendant phenomenon of thallium-based superconductors is serious, and it is difficult to produce a highly pure single-phase (2223) high-T_c superconducting thin film. There are usually other superconducting phases in the films, such as (2212) and (2234).

Figure 8.10 shows the relationship between the T_c, J_c, and H_c of the superconductor. As shown in Figure 8.5, the performance of the YBaCuO high-T_c superconductor is much higher than that of the

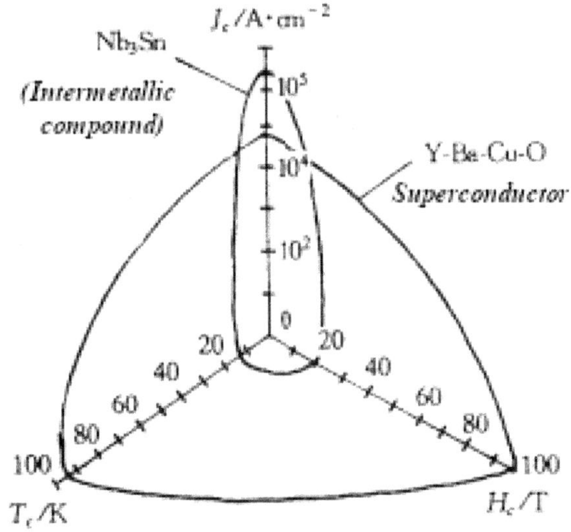

Figure 8.10: The relationship between T_c, J_c and H_c.

metal compound Nb_3Sn, except the critical current density J_c. The high-T_c superconducting thin film is much more powerful than the YBaCuO superconducting block material.

Based on the method of forming high-T_c films, the process technology of wet carving, dry carving, peeling, and multi-layer engraving was developed, and the multilayer structure and three-terminal devices of high-T_c superconducting thin films were preliminarily tested in the device structure design. In addition, the high-T_c superconducting block sample and the particle structure in the film are also utilized to study the performances of superconducting quantum interference devices (SQUID) and high-frequency mixer and detector.

8.2.3 *Research progress of superconducting thin films*

8.2.3.1 *Magnesium diboride films*

(1) **Overview**

MgB_2 has been a familiar material since the 1950s. However, it was not discovered by Japanese scientists until March 2001 that MgB_2 is a superconductor with a superconducting transition temperature of 39 K. Subsequently, scientists from all over the world have carried out in-depth research on MgB_2, including the preparation of bulk, film, wire, and strip samples, the influence of various alternative elements on the transition temperature, the isotope effect, the measurement of the Hall effect, the study of thermokinetics, the relationship between critical current and magnetic field, the study of microwave and tunnel characteristics, etc.

Because MgB_2 is a conventional superconductor, its superconducting mechanism can be explained by BCS theory. At present, the critical temperature of MgB_2 is the highest. The chemical elements that make up the oxide high-temperature superconductor are expensive, and the synthetic superconductor material is brittle, which is difficult to process into a wire. Boron and magnesium are cheap and easy to turn into a wire. The oxide high-temperature superconductor is a complex compound composed of oxygen and more than two kinds of metal elements. The discovery of the MgB_2 superconductor made the study of simple compound superconductors heated up. Scientists believe that simple compound superconductors with higher critical temperatures will eventually be discovered. Moreover, the anisotropy of MgB_2 is small and it has a high critical current density, so it has a wide application prospect.

MgB_2 has a simple hexagonal AlB_2 structure, P6/mmm space group, and its structure is shown in Figure 8.11. This structure consists of a graphite like boron layer separated by a tightly packed hexagonal magnesium layer. The magnesium atom is in the center of the hexagon formed by the boron atom and provides electrons for the boron atom surface. MgB_2 has strong anisotropy in the long direction of B–B chain, and the distance between the boron atom planes is obviously larger than that between the boron atom planes. The superconducting transition temperature of MgB_2 is almost twice that of the Nb_3Ge superconductor, which is the lowest superconductor with the highest superconducting transition temperature. In the framework of BCS theory, the higher frequency phonon modes produced by low-mass elements can lead to higher transition temperature. The discovery of MgB_2 once again confirmed

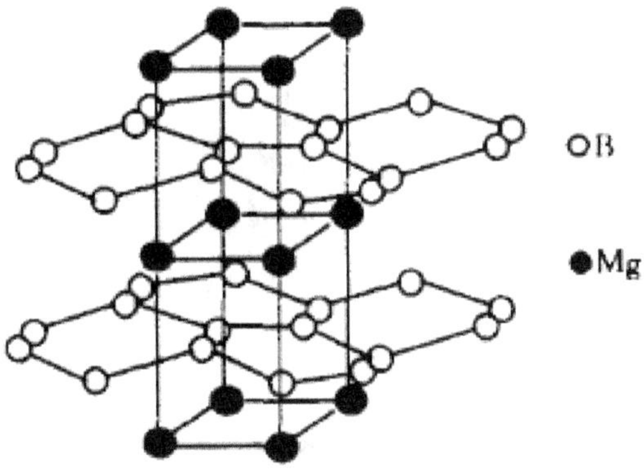

Figure 8.11: MgB_2 structure diagram.

the prediction that compounds containing light elements have higher T_c. The superconducting transition temperature of MgB_2 up to 39 K is close to or higher than the theoretical upper limit predicted by BCS theory.

(2) **Preparation of magnesium diboride** (MgB_2) **films**

So far, to prepare MgB_2 thin films with uniform density, smooth surface, and excellent superconductivity, researchers have tried many methods; they mainly include hybrid physical–chemical vapor deposition (HPCVD), pulse laser deposition (PLD), magnetron sputtering (MS), molecular beam epitaxy (MBE), electron beam evaporation (EBV), etc. Here we mainly introduce the HPCVD method. For other methods, please refer to Chapters 2 and 3.

The principle of MgB_2 superconducting thin films prepared by hybrid physicochemical vapor deposition is shown in Figure 8.12. The most important thing is to produce a high-Mg vapor pressure (about 44 mTorr at 700°) in a clean environment. By heating the base (550–760°C), a high vapor pressure of Mg is formed. The deposition rate of MgB_2 (0.1–5 nm/s) is controlled by controlling the inflow of B_2H_6 (diborane) and H_2 (5–250 sccm). Finally, highly oriented along the c-axis MgB_2 films were grown on sapphire and SiC substrates. The orientation relationship was (0001) [1120] MgB_2 ||| (0001) [1120] SiC. Superconductivity transition temperature $T_c = 41$ K, resistivity

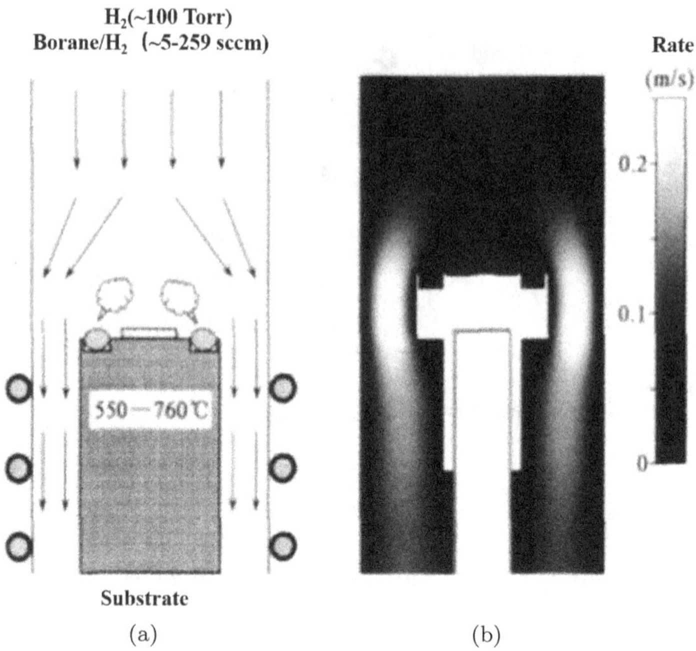

Figure 8.12: (a) Schematic diagram of HPCVD; (b) distribution diagram of the reaction gas rate near the base.

$\rho_{40K} = 0.1\,\mu\Omega \cdot \text{cm}$, residual resistivity RRR $(R_{300K}/R_{40K}) = 80$, critical current density J_c $(4.2\,\text{K}, 0\text{T}) = 4 \times 10^7\,\text{A/cm}^2$, surface resistance R_s $(18\,\text{GHz}) = 230\,\mu\Omega$, penetration depth $\lambda(0) = 50\,\text{nm}$, upper boundary magnetic field H_{c2} (carbon doping) $= 60\,\text{T}$, and irreversible magnetic field intensity$H_{irr} = 45$ T. The magnetoresistance of MgB_2 superconducting thin film is anisotropic. In the direction of the parallel plane, $\Delta\rho/\rho_0 = 1.36$ (18 T). At present, the HPCVD method of the MgB_2 thin film is expected to use the widely used CVD thin film growth equipment to explore the best process parameters in the field of application, and to prepare high-quality MgB_2 thin film and thick film, which will lay a solid foundation for the *in situ* preparation of MgB_2 superconducting electronic devices and the second generation of MgB_2 superconductors.

Among the many deposition processes of MgB_2 thin films, the biggest advantage of pulse laser deposition is that the composition of the MgB_2 superconducting thin films prepared is prone to be consistent with that of the target materials, and it is easy to make multilayer and heterogeneous films. In particular, the heterojunction of multiple oxides can be made by simple target exchange.

The MgB_2 superconducting thin film prepared by magnetron sputtering has high quality, less bombardment times of electron and ion film formation during deposition, avoiding high substrate temperature, and good adhesion between the sputtering film and the substrate. The equipment of MBE is the most complex, but it can realize the growth of artificial atomic layer by layer. It can accurately control the Mg/B atom proportion in the experimental process, and can strictly control the growth process and growth rate. In the growth process, the growth situation can be observed. The quality of the epitaxial MgB_2 film is the best and there is no pollution problem.

CVD method can control the composition of the film arbitrarily, and prepare new structure and composition materials. The film is uniform and has small directivity. This method has been improved and further improved continuously. The HPCVD system can produce a high vapor pressure of Mg and provide a clean deposition environment, and the deposition speed is fast, the equipment is simple, the superconductivity of MgB_2 film obtained by HPCVD method is the best, and the T_c is the highest. Carbon doping is essential for the preparation of high-quality and high-performance MgB_2 films, but the best raw materials and process parameters for carbon doping (C) have not been found. Table 8.4 compares the properties of MgB_2 films prepared by several main preparation methods. It can be seen from Table 8.4 that at the current technological level, the comprehensive properties of MgB_2 superconducting thin films prepared by the mixed physicochemical vapor deposition method are the best.

8.2.3.2 *New superconducting materials*

(1) Organic superconductive material

In 1979, the first organic superconductor, based on tetramethyltetraselenafulvalene (TMTSF), was discovered by Jerome of Paris

Table 8.4: Comparison of the properties of MgB2 films prepared by several methods.

Characteristic	HPCVD	PLD	MS	MBE	E-Beam
T_c (max)	~41 K	~38 K	~35 K	~35 K	~38 K
ρ_{300K} ($\mu\Omega \cdot$ cm)	~8.5	150	200	8.9	350
J_c (A/cm^2)	3.5×10^7	1.6×10^7	$<10^6$	1.6×10^5	2.4×10^6
RRR = R_{300K}/R_{40K}	~80	~1.4	~1.6	~1.7	~1.35
Grain size	400–500 nm	Several to dozens nm	~100 nm	~180 nm	100–200 nm
Working gas	B_2H_6, H_2	Argon	Argon	—	Argon
$H_{c2}(T)$	60 (Carbon doping)	27.5	19	39	37.78
$H_{irr}(T)$	45	9 (21 K)	3	—	19.2
Boron source	B_2H_6	Boron powder	Boron powder	Boron powder	Boron powder

University and Bichigarde of Copenhagen University. Its molecular formula is (TMTSF)-PF6, and its transition temperature is 0.9 K. Since 1979, efforts have been made to find organic superconductors with higher transition temperatures. In a practical sense, an important difference between organic superconductors and other superconductors is that the density of organic materials is low, about 2 g/cm^3; that is, their density is only 20–30% of that of general metals (such as niobium), because of the large spacing between atoms and molecules and the small mass of carbon atoms. More than 40 kinds of charge transfer salts with superconductivity have been found, but their transition temperature is generally low, and many of them can only appear superconductive under high pressure. Before 1991, most of the organic superconductors with higher transition temperature were related to the salt bis(ethylenedisulfide)tetrathiafulvalene (ET) of organic molecules. In 1983, scientists at the IBM laboratory in California, USA, discovered the rhenium compound (ET)-ReO_4, whose transition temperature was 2 K under high pressure. The following year, Soviet scientists discovered the first type of ET superconductor, iodized salt $\beta_2(ET)_2I_3$, at atmospheric pressure, with a transition temperature of 1.5 K. By 1988, the salt of thiocyanin κ-(ET)–Cu[(CN)]Cl transition temperature reached 13 K. The later modified superconductors, such as $(EDT–TTF)_4$ $Hg_{3-\delta}I_8$,

$\delta = 0.1$–0.2 ($T_c = 8.1\,\text{K}$), did not exceed this record. K_3C_{60}, a potassium salt of C_{60}, was discovered in 1991, with a transition temperature of 19 K. The T_c values of rubidium, cesium, and C_{60} compounds (Rb_2C_{60}) were all 33 K. At present, the highest record for this kind of superconductor is the C_{60} single crystal with a porous surface discovered by Lucent Technology Company of the United States, whose critical temperature has reached 117 K.

(2) **Cobalt oxide**

In 2003, Takada of the National Institute for Materials Science, Japan, found that cobalt oxide is a new superconducting material. The structural formula of this compound is $Na_xCoO_2 \cdot yH_2O$ ($x = 0.35$, $y = 1.3$). It is composed of a two-dimensional CoO_2 surface separated by a thick Na^+ and H_2O molecular insulating layer. Superconductivity is realized at about 5 K. This structure is similar to that of copper oxide, indicating that it has the same potential physical properties as copper oxide.

This was another breakthrough since the discovery of magnesium diboride. As a material with a high efficiency of thermoelectric conversion, cobalt sodium oxide does not possess the characteristics of superconducting material in the dry state. It is a layered mineral structure, through the oxidation of bromine to remove sodium. When the temperature is −268°C, water molecules are injected into the interlayer of cobalt oxide to fill the interlayer with water molecules, which act as an insulator, and the magnetic susceptibility and resistance will drop sharply, and it will become a superconducting material. The new superconductor has the same superconductivity as the high temperature superconductor, but the atomic arrangement of copper oxide is square, while the atomic arrangement of cobalt oxide is triangle. Therefore, a study of the arrangement of cobalt oxide atoms can further enrich the theory of superconductivity.

(2) **Iron-based superconductor**

In the last decade of the 20th century, rare earth transition metal phosphides with ZrCuSiAs structure were discovered, but no superconductivity. In 2006 and 2007, Professor Hideo Hosono of the Center for Advanced Cooperative Science and Technology of Tokyo University of Technology and other professors successively found that LaOFeP and LaNiPO exhibited superconductivity at low temperature, but because the critical temperature was below 10 K, they did not attract special attention and interest.

At the beginning of January 2008, Prof. Hideo Hosono *et al.* found that in the Fe–O–P group element compound LaOFeA, part of the oxygen can be substituted by fluorine in the way of doping, which can make the critical temperature of $LaO_{1-x}F_xFeAs$ reach 26 K. This breakthrough opened a new round of high-temperature superconductivity research upsurge in the scientific community. In March 2008, the first hole-doped Fe based superconductor, $La_{1-x}Sr_xOFeAs$, was successfully synthesized by Prof.Wen Haihu of the Institute of Physics of Chinese Academy of Sciences. Prof. Chen Xianhui of the University of Science and Technology of China and Prof. Wang Nanlin of the Institute of Physics of Chinese Academy of Sciences independently found iron-based superconductors with critical temperature over 40 K; Prof. Zhao Zhongxian of the Institute of Physics of Chinese Academy of Sciences found that the superconducting transition temperature of $PrO_{1-x}F_xFeAs$ can reach 52 K. In April 2008, Prof. Zhao Zhongxian and other researchers found that the transition temperatures of $SmO_{1-x}F_xFeAs$ and $ReFeAsO_{1-\delta}$ superconductors synthesized under pressure were further increased to 55 K. A new type of iron-based superconducting thin film was prepared by Prof. Hideo Hosono. At present, according to the composition ratio and crystal structure of the parent compounds, the new iron-based superconductors can be roughly divided into the following four systems: (1) "1111" system, including LnOFePn (Ln = La, Ce, Pr, Nd, Sm, Gd, Tb, Dy, Ho, Y; Pn = P, As) and DvFeAsF (Dv = Ca, Sr); (2) "122" system, including aFe_2As_2 (a = Ba, Sr, K, Cs, Ca, Eu); (3) "111" system, members include aFe_2As_2 (a = Li, Na, etc.; (4) "11" system, members include FeSe(Te), etc. The research and discovery of iron-based superconductors was rated as one of the top ten scientific and technological advances in the world in 2008 by *Science* magazine of the United States. At present, scientists generally believe that the pairing mechanism and superconducting mechanism of iron-based superconductors are still unclear and need further study. On the other hand, the exploration of new iron-based superconductors is in the ascent. Along with the idea of designing new structures and multilayer, iron-based superconductors with higher critical transition temperature are likely to be discovered in the future.

8.3 Ferroelectric Films

8.3.1 *Overview*

When the applied electric field does not exist, ferroelectric materials have spontaneous polarization, and the direction of spontaneous polarization can be changed by the applied electric field; at the same time, there is a P–E hysteresis relationship between the polarization strength P and the electric field E similar to the B–H hysteresis loop of ferromagnet. Now, hundreds of ferroelectric materials have been found and applied in ultrasonic transducer, pressure sensor, filter, resonator, and so on. Table 8.5 shows representative ferroelectric materials.

Table 8.5: Typical ferroelectric materials.

Ferroelectrics	Abbreviation	T_c(°C)	P_s($\mathrm{C \cdot m^{-2}}$)	Structure
$BaTiO_3$	BT	120	0.26	Perovskite
$PbTiO_3$	PT	492	0.57	Perovskite
$PbZr_xTi_{1-x}O_3$ ($x = 0.52$)	PZT	386	0.39	Perovskite
$KNbO_3$	KN	435	0.30	Perovskite
$LiNbO_3$	LN	1210	0.71	Lithium niobate
$LiTaO_3$	LT	620	0.50	Lithium niobate
$Sr_{1-x}Ba_xNb_2O_6$ ($x = 0.25$)	SBN	75	0.32	Tungsten
$Ba_{0.8}Na_{0.4}Nb_2O_6$	BNN	560	0.40	Tungsten
$Pb_{1-x}Ba_xNb_2O_6$ ($x = 0.57$)	PBN	316	0.30	Tungsten
KH_2PO_4	KDP	−150	0.05	Hydrogen bond
KD_2PO_4	DKDP	−60	0.062	Hydrogen bond
$PbHPO_4$	LHP	37	0.018	Hydrogen bond
$(NH_2CH_2COOH)_3H_2SO_4$	TGS	49	0.03	Hydrogen bond
$NaNO_2$	NN	163.6	0.115	Hydrogen bond
$NaKC_4H_4O_6 \cdot 4H_2O$	RS	24	0.0025	Hydrogen bond

Note: T_c — Curie temperature; P_s — maximum spontaneous polarization intensity.

8.3.2 *Preparation of ferroelectric films*

As early as the 1950s, research on the preparation of ferroelectric thin films was carried out. In the late 1970s and early 1980s, modern film preparation technology made great breakthroughs. Using various membrane preparation techniques, such as radiofrequency magnetic control sputtering (RF magnetron sputtering), sol–gel, metal-organic chemical vapor deposition (MOCVD), pulsed laser deposition (PLD), and molecular beam epitaxy (MBE), it is possible to prepare ferroelectric films with a complete structure and excellent performance. They were also used for research on the preparation of the devices.

Since ferroelectric thin films are mostly multi-element metal oxide film material with extremely complex chemical composition, it is more difficult to prepare ferroelectric films than to prepare a single component or dual component films. At present, the most widely used ferroelectric thin film preparation technology mainly involves sputtering, pulse laser deposition, sol–gel, and chemical vapor deposition.

At present, there are two kinds of ferroelectric films which have been studied in depth and have been applied in practical application, namely titanate series and niobate and borates series. The titanates series of ferroelectric films include lead titanate ($PbTiO_3$), lead zirconium titanate [$Pb(Zr, Ti)O_3$, or PZT], and La doped lead zirconium titanate[$(Pb, La)(Zr, Ti)O_3$, PLZT], barium titanate ($BaTiO_3$), barium strontium titanate [$(Ba, Sr)TiO_3$, BST], and bismuth titanate ($Bi_4Ti_3O_{12}$) and so on. The ferroelectric films of niobate and borate series include lithium niobate ($LiNbO_3$), potassium niobate ($KNbO_3$), strontium barium niobate [$(Sr, Ba)Nb_2O_6$, SBN], potassium tantalate niobate, [$(K(Ta,Nb)O_3$, KTN], and lithium triborate (LiB_3O_7). The titanate series ferroelectric thin films have important applications in microelectronics and photonics, and niobate and borates series of ferroelectric films are mainly used in optoelectronics.

In the application of ferroelectric random access memory based on the switch effect, $Pb(Zr, Ti)O_3$ (PZT) is a more commonly used material. Due to the poor fatigue resistance of PZT, the new material system has been developed and studied in recent years, and the $SrBi_2Ta_2O_9$ (SBT) ferroelectric thin film has been found, which is also called the Y1 thin film. The Y1 thin film has good anti-fatigue property. And FRAM made from it will not show significant fatigue phenomenon after 1012 repeated switch polarizations, and has good lifetime and low leakage current. Dynamic random access memory (DRAM) based on high capacitance usually uses ferroelectric films with dielectric constant up to 10^3–10^4 as capacitance media, which can greatly reduce the area of the planar storage capacitor, and facilitate the preparation of ultra-large-scale integration (ULSI) DRAM. The ferroelectric films studied currently include PZT, $SrTiO_3$ (ST), $BaTiO_3$ (BT), and (Ba, Sr)TiO_3 (BST). Because of the disadvantages of lead series ferroelectric films working in ferroelectric phase, such as fatigue, aging, large leakage current, and instability, the research on dielectric films is mainly focused on the BST film in paraelectric phase with high dielectric constant. In the application of optoelectronics, the (Pb, La)(Zr, Ti)O_3 (PLZT) ferroelectric thin film is the most studied material. Because of its good optical and electrical properties, the adjustment of the chemical composition of the system can meet the demands of electro-optic, elastoptoptic, and nonlinear optics. In addition, PLZT can also be used for integrated optics, and is a promising optical waveguide material. But the chemical composition of PLZT is complex, and its performance is sensitive to changes in chemical composition, which is not good for the preparation of thin films. $KTa_xNb_{1-x}O_3$ (KTN) is also a type of thin film material that has a promising prospect for optical electronics. The KTN is better than PLZT in terms of optical nonlinearity and photo-refraction effects, and it is not as demanding as PLZT in terms of film preparation. PLZT and KTN are both materials with perovskite structure. The ferroelectric crystal like (Sr_{1-x}Bax)Nb_2O_6(SBN) with tungsten bronze structure is an important electro-optical material, and the preparation of the films of such materials has been reported. With the development of optoelectronics, this kind of ferroelectric film will become more and more important.

8.3.3 *Research progress of ferroelectric thin films*

8.3.3.1 *Bismuth layered perovskite ferroelectric thin films*

Although Pb (Zr, Ti) O_3 (PZT) material has always been one of the hot spots in ferroelectric thin film research, due to the existence of lead and fatigue problems in PZT material, people have been looking for new lead-free ferroelectric materials in the application of ferroelectric memory. Bismuth layered perovskite ferroelectric oxide (BLSF) is a kind of special Perovskite ferroelectric oxide, also known as aurivllius structure, which has attracted great attention in recent years. BLSF has a general structural formula of $A_{n-1}Bi_2B_nO_{3n+3}$, which is composed of a perovskite-like structure layer $(A_{n-1}\text{-}B_nO_{3n+1})^2$ and bismuth oxide layer $(Bi_2O_2)^{2+}$ stacked alternately along the *c* axis, where A represents Bi, Ba, Sr, CA, Pb, K, or Na, etc.; B can be Ti, Nb, Ta, Mo, W or Fe, etc.; *n* represents the number of perovskite structures between $(Bi_2O_2)^{2+}$ layers. This kind of layered structure can be regarded as a kind of natural ferroelectric superlattice. The SBT ferroelectric thin film with BLSF structure has attracted the attention of many scholars since 1994 because of its characteristics of being lead-free, fatigue-free, and high Curie temperature.

In 1995, J. F. Scott *et al.* published a "fatigue free" ferroelectric material with the $SrBi_2Ta_2O_9$ (SBT) structure of layered perovskite in *Nature*. SBT not only has an outstanding fatigue performance and a low turnover voltage, but also has a high synthesis temperature (above 750°C), which is incompatible with the current IC process, limiting its application. Moreover, the residual polarization value of the SBT film is low ($2\,P_r = 4\text{–}16\,\mu C/cm^2$), which is not conducive to high-density storage and can't fully meet the requirements of human beings for the preparation of FeRAM materials.

Park *et al.* from South Korea reported in *Nature* in October 1999 that La^{3+} partially replaced Bi^{3+} near the titanium oxide octahedral layer in $Bi_4Ti_3O_{12}$(BIT) with a bismuth layered perovskite structure, and the $Bi_{4-x}La_xTi_3O_{12}$(BLT) film obtained has excellent properties such as high residual polarization, high Curie temperature, and superb fatigue resistance, and the preparation temperature is about 650°C, which is lower than the synthesis temperature of SBT 100–200°C BIT is also a typical BLSF structure ferroelectric material. When the Curie temperature is $Tc = 675°C$, the symmetry of the BiT crystal is monoclinic point group m when it is lower than T_c, and it is tetragonal paraelectric phase 4/mmm when it is higher than T_c. The ferroelectric paraelectric phase transition belongs to the first order. The crystal structure of BiT is composed of a calcium like titanium ore layer and a $(Bi_2O_2)^{2+}$ layer. The calcium-like titanium ore layer contains three TiO_6 octahedrons and the A-site Bi^{3+}, TiO_6 octahedron forms a O–Ti–O linear chain through the vertex angle, the Bi (A-site) ion is located in the center of the TiO_6 octahedron network, and the Ti (B-site) ion is located in the octahedron composed of oxygen ions in each surface center. The ferroelectric polarization of this kind of material mainly comes from the displacement of Bi^{3+} at a position relative to the octahedral chain of TiO_6 along the *a*-axis and *b*-axis. So, the spontaneous polarization vector is located in the *ac* plane, which is about 4.5° from the *a*-axis, but the spontaneous polarization components along the *a*-axis and *c*-axis are very different, which are $50\,\mu C/cm^2$ and $4\,\mu C/cm^2$, respectively.

It is found that other lanthanide elements, such as Nd, Pr, Sm, and so on, can also achieve similar effects. Therefore, lanthanide rare earth ion-doped BIT ferroelectric thin films have become one of the hotspots in recent years. In 2002, Kojima *et al.* prepared (104)-oriented epitaxial $Bi_{3.54}Nd_{0.462}Ti_3O_{12}$ films by MOCVD. After 2×10^{10} turns, no obvious fatigue phenomenon was found, and the residual polarization reached $25\,\mu C/cm^2$. Chon *et al.* prepared a $Bi_{3.15}Nd_{0.85}Ti_3O_{12}$ thin film with *c*-axis orientation by the sol–gel method. Its 2 P_r reached $103\,\mu C/cm^2$. In 2009, Zhou Yichun *et al.* prepared $Bi_{3.15}Nd_{0.85}Ti_3O_{12}$ thin films by the chemical solution precipitation method (CSD) with the 2 P_r and the E_c of about $65.4\,\mu C/cm^2$ and $1146\,kV/cm$, respectively. A large number of research results show that among all lanthanide-doped BiT films, the effect of Nd doping is the most significant. Therefore, the

Bi_{4-x}–$Nd_xTi_3O_{12}$ (BNT) film is considered to be one of the most likely materials to replace the PZT film because of its large residual polarization, good fatigue resistance, and high Curie temperature.

8.3.3.2 *Ferroelectric multilayer films*

Ferroelectric multilayers have great potential of being used in various electronic devices due to their controllable electrical properties. Compared with superlattice, the preparation of multilayered films does not need complex equipment, and the process is relatively simple, which is suitable for large-scale production. The purpose of research on ferroelectric multilayers is to explore new physical phenomena of materials and get high performance or new performance that a single material does not have through the combination of multilayers and the physical effects such as the stress or strain effect and the interlayer coupling effect when the thickness of the film is small. The synthesized ferroelectric multilayers are generally composed of two or more kinds of film materials with different properties, or the same material, but with different component proportion, and sandwich structure. Therefore, ferroelectric multilayer is an effective way to obtain ferroelectric thin film materials with good properties and to explore new physical phenomena.

In 2005, Ho Nyung published an article in *Nature* which reported that the asymmetry of the $BaTiO_3/SrTiO_3/CaTiO_3$ three-component ferroelectric superlattice improved the polarization intensity of ferroelectric thin films. The preparation and characterization of ferroelectric superlattices and nano ferroelectric multilayers, such as $PbTiO_3$/PZT, $BaTiO_3/SrTiO_3$, $Ba_{0.2}Ti_{0.8}O_3$, $Ba_{0.8}Ti_{0.2}O_3$, $Pb(Zr_{0.8}Ti_{0.2})O_3/Pb(Zr_{0.2}Ti_{0.8})O_3$, have been studied by MBE, sol–gel, and sputtering techniques. Compared with general oxide superlattices and multilayers, both ferroelectric superlattices and multilayers exhibit certain dielectric and ferroelectric anomalies; that is, the dielectric and ferroelectric properties of ferroelectric superlattices or multilayers show extreme values under a certain period structure.

In 2006, Longhai Wang *et al.* compared PZT, PZT/Pt, Pt/PZT/Pt films and nano ferroelectric multilayers. It was found that Pt/PZT/Pt films have the largest residual polarization, the smallest coercive field, the best fatigue resistance, and the lowest dielectric constant, dielectric loss and leakage current density. In 2007, Liu Hong *et al.* studied $Pb(Zr_{0.8}Ti_{0.2})O_3/Pb(Zr_{0.2}Ti_{0.8})O_3$ ferroelectric multilayers. It was found that the dielectric constant of PZT multilayers reached the maximum value (328) and the dielectric loss reached the minimum value (0.0098) when the thickness of the tripartite phase was $d_R = 33$ nm and the ratio of $d_R/d_T = 1:3$ (here d_T is the thickness of the tetragonal phase). Under the same conditions, the dielectric constant of a PZT80 single-layer film is only about 98, and the dielectric loss of a PZT80 single-layer film is about 0.036.

In 2008, Lucian Pintilie *et al.* used the PLD method to prepare $PbZrO_3/Pb(Zr_{0.8}Ti_{0.2})O_3$ epitaxial multilayer films. When the total thickness is the same, the dielectric properties are enhanced when the interface is increased. In 2009, Ren Tianling *et al.* prepared $BiFeO_3/Bi_{3.15}Nd_{0.85}Ti_3O_{12}$ (BFO/BNdT) multilayer ferroelectric thin films by the chemical solution deposition method, with a residual polarization of $11.05\,\mu C/cm^2$ and a coercive field of $50\,kV/cm$. The BNdT layer improves the (111) orientation of BFO, and the spontaneous polarization of BFO is along the [111] direction, so it improves the residual polarization of multilayer. In 2009, Hyeong-Ho Park and others prepared PZT/BLT multilayered films by light-sensitive precursor (photochemical metal organic deposition). The upper and lower electrodes were Pt, and the films in contact with the electrodes were BLT. PZT/BLT multilayered films have enhanced dielectric and ferroelectric properties, especially fatigue resistance and less leakage current. In 2009, L. Feig *et al.* prepared $PbZr_{0.2}Ti_{0.8}O_3/PbZr_{0.4}Ti_{0.6}O_3$ nanoferroelectric multilayers on SRO/STO (100). It was found that with the increase of interface density, the *a* domain extended to the whole multilayers, making the domain wall movement easier and $P_r \cdot \varepsilon_r$ of multilayers increased. However, when the interface density continues to increase to an extreme value, defects and dislocations increase, while $P_r \cdot \varepsilon_r$ decreases.

At present, four theoretical models have been developed to explain the dielectric enhancement effect of ferroelectric multilayers;

The interface polarization model, the interface charge model, the space charge model, and the Maxwell–Wagner model.

8.3.3.3 *Dielectric (ferroelectric)/semiconductor composite films*

In order to realize the microminiaturization and monolithic of electronic information systems, continuously promote the thin film of electronic materials and the chip of electronic devices, functional oxide materials with ferroelectric, piezoelectric, pyroelectric, high-k dielectric, soft magnetic, magnetoelectric, electro-optical, acousto-optical and nonlinear optical properties and semiconductor materials are grown together in the form of a solid film to form dielectric electric (ferroelectric)/semiconductor artificial composite structure (single-layer, multilayer, or even superlattice). With the integrated characteristics of this integrated film, the dielectric passive device can be integrated with the semiconductor active device to realize the multi-functional integration and modularization of the active–passive and enhance the intensive system function.

At present, there are many research studies on the integration of high-k gate dielectric on Si substrate, and there are also reports on the research on dielectric materials, semiconductor ZnO, GaAs, and other composite films. However, due to the limitation of the thermal stability of Si and GaAs, the amorphous layer SiO_x or GaO_x is prone to form at the interface, which has a negative effect on the epitaxial growth and transport properties of the interface induced dielectric films. Therefore, there are two main problems in the composite growth of dielectric and semiconductor: one is how to grow together, the other is the performance change after composite growth. Dielectric films are generally grown in a high-temperature and oxygen atmosphere, while semiconductors are grown in low temperature and high vacuum without oxygen. The difference in growth temperature and vacuum degree is several orders of magnitude. In addition, their lattice mismatch is large (>10%), and the growth mechanism is inconsistent. The growth method and interface behavior of dielectric/semiconductor integrated films are greatly different from that of single materials. In 2004, Ahn *et al.* from Yale University predicted that the composite of oxide film and semiconductor film would produce new effects and devices through the first principle calculation; in 2006, Edge *et al.* prepared amorphous $LaAlO_3$ and $CaZrO_3$ dielectric layers with nanometer thickness on Si by MBE, and in 2006, Goncharova *et al.* epitaxially grew $SrTiO_3$ or $BaTiO_3$ films on Si. In 2004, Motorola's researchers explored the growth behavior of perovskite oxide films on second-generation GaAs. In 2007, Penn State University prepared epitaxial BiFeO3 films on GaN. In 2002, Shen *et al.* prepared $Pb(Zr_{0.53}Ti_{0.47})O_3$ films on Al_xGa_{1-x}/GaN substrate, hoping to use ferroelectric polarization to control the concentration of a two-dimensional electron gas in the semiconductor channel. In 2005, Cao *et al.* prepared $Pb(Zr_{0.3}Ti_{0.7})O_3$ on GaN/Al_2O_3 substrate. It was found that the oxide $Pb(ZrTi)O_3$ directly deposited on GaN was polycrystalline, and had no obvious positive effect on the semiconductor carrier. In 2009, Li Yanrong *et al.* used the laser molecular beam epitaxy (L-MBE) method, self-buffering, and several nano-TiO_2 transition layer induction methods to epitaxially grow $SrTiO_3$ thin films on GaN.

8.4 Magnetic Films

8.4.1 *Overview*

The research and application of magnetic films started early. In the middle of the 20th century, magnetic films have been widely applied in electronic, microelectronic, communication, aerospace, medical treatment, laser and other high-tech fields. In the 1950s, the Ni–Fe film was successfully used as computer memory. In the 1960s, epitaxial oxide film was created and then bubble memory was developed. In the 1970s, yttrium garnet doped epitaxial film technology was further improved and amorphous alloy films were widely used in magnetic recording and magneto-optic memory devices. There are many series of magnetic film materials, and they can be roughly divided into three series: magnetic recording film, magneto optical film, and magnetoresistance film.

Magnetic recording films have been developing for 40 years and the storage density continue to double almost every year. Magnetic recording thin films like Fe_2O_3, γ-Fe_2O_3, Co–γ–Fe_2O_3, CrO_2, Ni–Co–P, Ni–P, Co–Cr, and barrium ferrite were researched and developed successively. Nowadays, to improve the recording density, the research direction is perpendicular magnetic recording thin films. Meanwhile, the development and application of a thin film magnetic head has promoted the progress of the disk and video fields.

Magneto-optic recording is a combination of optical recording and magnetic recording; it has high recording density and rewritable function ($>10^{16}$). A magneto-optic disk with TbFeCo amorphous thin film magneto-optical recording medium is characterized by portability, large storage capacity (>600 Mb), long life, and contact-free rewritable feature, so it has been used to backup data of computers, store and inspect data online, compute at work station, process text and signal, and so on. Many performance parameters of the second-generation magneto-optic thin films such as Bi-substitution garnet magneto-optic film, MnBiAl magneto-optic thin film, and Pt/Co modulated magneto-optic film are superior those of to TbFeCo amorphous thin films.

Magnetic reluctance thin films are widely used for preparing magnetic sensors. Typical magnetic reluctance thin films are Ni–Co, Ni–Fe, Ni–Fi–Co, and so on. The strength of the magneto-resistance effect of thin films is influenced by the size of the thin film, shape, and deposition process. In recent years, giant magnetoresistance was found in magnetic multilayer thin films the resistivity of which is 1 order of magnitude more than monolayer thin film. The multilayer thin films have periodicity with the nonmagnetic layer between magnetic layers with nanoscale thickness, such as Fe/Cr, Co/Cr, Fe/Cu, Fe/Ag magnetosphere, and so on. From the magnetoresistance effect, magnetic multilayer thin films are the development direction of magnetoresistance thin films and it provides the favorable conditions for developing new magnetoresistance sensors and new magnetoresistance heads.

8.4.2 *Magnetic recording thin films*

With the rapid development of information science and technology, high storage density, high capacity and miniaturization have become the direction of research and development. The magnetic recording medium is progressing from a thick thin film with discrete particles to a continuous thin film. At present, some international companies have made a hard disk with a continuous metal film; it has a high capability of 500 MB; besides, a perpendicular magnetic recording disk such as Co–Cr has been put into market. The methods to prepare a continuous metal magnetic film can be divided into physical methods and chemical methods. Vacuum evaporation, sputtering and ion plating are physical methods and electroplating and chemical plating are chemical methods. Several main magnetic recording films are briefly described below:

8.4.2.1 γ-Fe_2O_3 *thin films*

Because Fe can't be oxidized to γ-Fe_2O_3 directly, so we must compound it by a complicated process. Normally, γ-Fe_2O_3 is prepared by the sputtering method. Using Fe or Fe_3O_4 as the sputtering target, Co, Ti, Cu can be added to improve the properties of the film.

8.4.2.2 $Co_xFe_{3-x}O_4$ *thin films*

Reactive sputtering and electron beam vapor deposition can be used to prepare $Co_xFe_{3-x}O_4$ thin films. First, deposit Fe film onto substrate in a high vacuum and oxidize the Fe film into an α-Fe_2O_3 thin film in air at about 400°C. Then, deposit Co film on the α-Fe_2O_3 film, and the thickness is decided by the amount of Co doped in it. Next, put it into an environment of 2×10^{-3} Pa and 250–400°C to anneal; at this moment, Co ions spread to the α-Fe_2O_3 thin film and a $Co_xFe_{3-x}O_4$ ferrite film is generated subsequently. Finally, clean the surface of film to remove excess Co in 0.1–0.15% NHO_3 solution. If we use a Fe–Al–Co alloy as the target in an Ar+O_2 atmosphere, we can prepare an oxygen-containing cobalt ferrite film. There is spray pyrolysis besides evaporation coating and sputter coating. Mix Fe and Co organic slat into solution and spray them on the substrate, and then heat them until decomposition.

8.4.2.3 $BaFe_{12}O_{19}$ *thin films*

As the perpendicular recording medium, barium ferrite is a kind of widely used oxide material and is prepared mainly by secondary sputtering and DC magnetron sputtering. The Magnetism of a typical single phase $BaFe_{12}O_{19}$ thin film: $M_s = 1.28 \times 10^7$ A/m, $H_{c\perp} = 54112$ A/m, $M_{r\perp}/M_{sa} = 0.209$, $M_{r\perp}/M_{r//} = 3.39$, $K_u = 1.67 \times 10^{-2}$ J/cm^3.

8.4.3 *Magneto-optic thin films*

The reason that magneto-optic recording is superior to ordinary magnetic recording is that magneto-optical films have magnetic uni-axial anisotropy perpendicular to the film surface. The magnetization intensity orients parallel spontaneously in the direction perpendicular to the surface of the film and a very small magnetic domain can form a high surface density, the process of information writting is realized by using the coercivity of the thin film with the change of temperature, and the readout process is realized by the magneto-optic effect of the magnetic medium. There are the Kerr effect and the Faraday effect in the magneto-optic effects of the magnetic medium. The Kerr effect and the Faraday effect are the effects of polarization plane deflection of linearly polarized light after it is reflected or transmitted through the magneto-optic medium, the strength of which is decided by the dielectric tensor and the complex index of refraction. The structure, ingredient, thickness of magneto-optic film and interface effects of matching film layers will change the property of the magneto-optic disk.

To achieve the requirements of magneto-optic recording, the magnetic film should have magnetic anisotropy in the vertical direction of the surface of film, and $K_u > 2\pi M_s^2$. It should have rectangular hysteresis loop ($M_r/M_s = 1$) and high room temperature coercivity; besides, it needs high magneto-optic recording sensitivity, large magneto-optic effect (greater Kerr angle θ_K or greater Faraday angle θ_F), low noise when writing information into the disk (no grain or small grain), high enough cycle times (10^6 times), good oxidation resistance, corrosion resistance and long-term stability; the Curie temperature is 400–600 K, and compensation temperature is around room temperature. At present, the films satisfying the above conditions are RE-TM amorphous magneto-optic film, Bi-substitution garnet magneto-optic film, and multilayer films adjusted by Pt/Co series.

8.4.3.1 *RE-TM magneto-optic films*

In the RE-TM amorphous film, the magnetic moment of RE atoms and TM atoms are antiparallel, its saturated magnetic moment M_s, Curie temperature T_c and compensation temperature T_{comp} are related to the composition of the film, and the uniaxial magnetocrystalline anisotropy constantK_u is closely related to the preparation process of the film. The magneto-optic effect comes from exchange of electrons with d–f energy levels of rare earth metal atoms and d–d energy levels of transition metal atoms. The magnetic properties of several typical rare earth transition metal amorphous films are shown in Table 8.6.

As a new generation of magneto-optic disc material, the RE-TM magneto-optical film has met some market requirements, but in terms of cost, stability, and magneto-optic effect, it still has much insufficiency. So we need to enhance the magneto-optic effect by adopting multiple elements (like some elements among Nd, Dy, Tb,

Table 8.6: Magnetic properties of several rare earth transition metal amorphous alloy films.

Composition of film	$M_s(T)$	$K_u(10^4\ \text{J}\cdot\text{m}^{-3})$	Curie temperature (K)	Kerr rotation angle θ_K (°)	Faraday rotation angle $\theta_F/105$ (°)
$Cd_{24}Fe_{76}$	0.006	2.5	480	0.38°	1.8°
$Tb_{18}Co_{82}$	0.025	1.6	>600	0.45°	2.9°
$(Cd_{95}Tb_4)_{24}(Fe_{95}Co_5)_{76}$	0.003	1.2	580	0.36°	1.9°
$(Gd_{95}Tb_5)_{24}Fe_{76}$	0.008	3.5	460	0.30°	1.0°

Gd, Ho, Pr), changing the properties and corrosion resistance by doping Ti, Ta, Ga, Cr, Pt, Na, etc., and strengthening the carrier-to-noise ratio of the magneto-optic disk by adopting multilayer chemistry.

8.4.3.2 *Oxide and manganese bismuth series magneto-optic films*

Alloy films based on Mn of MnBi, such as PtMnSb alloy film ($\theta_K \approx 1.9°$), have a strong magneto-optic effect. The study of the Mn–Bi alloy film began in the 1960s; it will have strong anisotropy and magneto-optic effect perpendicular to the surface of film when it has a hexagonal structure, and Curie temperature is 360°C, $H_c \approx 57,750$–79,500 A/m. But magnetism may be poor as it will turn into tetragonal structure (called the high temperature phase) because of rapid cooling above Curie temperature. Besides, because of its large size of grain, the noise of the grain boundary is difficult to reduce, so it is difficult to put into practice. The MnBi series alloy film can be obtained by adding various metal impurities into MnBi film to inhibit the formation of high-temperature phase. For example, the highest θ_K of MnBiAlSi ($\theta_K \approx 2.04°$) and MnBi RE (RE = Ce, Pr, Nd, Sm) is 2.8°, reflectivity $R \geq 0.4$, so the magneto-optical figure of merit ($\theta_K\sqrt{R}$) can reach 1.5–1.8 and 3 times more than the normal magneto-optic film; therefore, it can be a second-generation film medium of the magneto-optic disk.

The magneto-optic film of garnet type with chemical formula $R_3Fe_5O_{12}$ has a great magneto-optical effect, such as $Y_3Fe_5O_{12}$(YIG), $Dy_3Fe_5O_{12}$(DyIG), and so on. The Bi-substitution garnet is a major magneto-optic material among the garnet series, for example, Bi, Ca:DyIG and Bi, Al:DyIG are expected to become a new generation of magneto-optical materials. The Bi-substitution garnet film has excellent stability, and is mostly prepared by radio-frequency sputtering and pyrolysis (for example, by coating an Al, Cu or Ti reflective heat-absorbing layer). Currently, the structure of the garnet-type thin film magneto-optical disk is GGG/BiGa:DyIG/Al (or Cr), and the recent development is multilayered structure.

8.4.3.3 *Multilayer modulation magneto-optic films*

Multilayer modulation films such as Pt/Co are a new magneto-optic material which is being researched. In multilayer modulation, Co provides magnetism and the magneto-optic effect, and the interface effect between layers of Pt and Co provides uniaxial magnetic anisotropy. Because the anisotropic stretching distance between the interfaces is short, the Co layer must be very thin. The thickness of the Co layer is about 0.36–0.5 nm, and the thickness of the Pt layer is about 1–2 nm. The Pt/Co multilayer modulation films are prepared by electron beam evaporation or DC magnetron sputtering and the film thickness is required to be very accurate. The Pt/Co thin film magneto-optic disk has a signal-to-noise ratio of over 40 dB. It has a good magneto-optical effect in the range of near ultraviolet, and its signal-to-noise ratio is close to that of the RE-TM film magneto-optic disc, but it is more resistant to corrosion compared to the latter one.

Moreover, CeTe and CdSb have a strong magneto-optic effect at very low temperature. Alloys composed of lanthanide (such as VSb_xTe_{1-x}) have large θ_K; however, these materials are difficult to apply at the moment.

8.4.4 *Magnetic resistance films*

The magnetoresistance effect of magnetic thin films is the effect that the resistance of them changes when magnetization changes relative to current direction. If the angle between magnetization M and current density$J$ in single-domain films is θ, the $\rho_{\|}$ is the resistivity component parallel to the direction of magnetization, and $\rho_{\perp}$ is the resistivity component perpendicular to the direction of magnetization, there are formulas

$$\Delta\rho = \rho_{\|} - \rho_{\perp}; \quad \rho(\theta) = \rho_{\|}\,\mathrm{con}^2\theta + \rho_{\perp}\sin^2\theta = \rho_{\perp} + \Delta\rho\,\mathrm{con}^2\theta.$$

For different kinds of magnetoresistance materials at room temperature (300 K), the relative change rate of the anisotropic magnetoresistance Δ is shown in Table 8.7. At present, NiFe, NiCo, and NiFeCo alloys are the most practical magnetoresistance films.

Table 8.7: $\Delta\rho/\rho_0$ (300 K) of some magnetoresistance materials.

Composition of alloy	$\frac{\Delta\rho}{\rho_0}/(\%)$	Composition of alloy	$\frac{\Delta\rho}{\rho_0}/(\%)$
Ni	2.66	99Ni–Fe	2.7
99.4Ni–0.6Co	2.1	99.8Ni–8.3Fe	3
97.5Ni–2.5Co	3	85.0Ni–15Fe	5.4
94.6Ni–5.4Co	3.6	83.0Ni–17Fe	4.6
90.0Ni–10.0Co	5.02	76.0Ni–24Fe	4.3
80.0Ni–20.0Co	6.48	70.0Ni–30Fe	3.8
70.0Ni–30.0Co	5.53	90.0Ni–10Cu	2.5
60.0Ni–40.0Co	5.83	83.2Ni–16.8Pd	2.6
50.0Ni–50.0Co	5.05	97Ni–3.0Sn	2.32
40.0Ni–60.0Co	4.3	99.0Ni–1.0Al	2.28
30.0Ni–70.0Co	3.4	98Ni–2.0Al	2.4
97.8Ni–2.2Mn	2.93	95Ni–2.0Al	2.18
94.0Ni–6.0Mn	2.48	95Ni–5.0Zn	2.6
92.2Ni–2.4Fe–4.5Cu	3.65	80Ni–16.3Fe–3.8Mn	2.2
69.0Ni–16.0Fe–14Cu	3.3	35.5Ni–49.2Fe–15.3Cu	3.3

The properties of magnetoresistance films are closely related to the preparation method and technological parameters. The common preparation method is magnetron sputtering; pulsed laser deposition has also been widely used in the preparation of magnetoresistance films in recent years. The properties of magnetoresistance films mainly depend on the thickness, grain size, surface state, doping, and substrate type of films, and they are also closely related to the preparation process.

In recent years, the giant magneto-resistance (GMR) effect has been found in magnetic multilayers and nanoparticle films; its resistance variation was enhanced by one order of magnitude than that of conventional magnetoresistance reluctance films. For example, the $\Delta\rho/\rho_0$ of the Fe/Cr superlattice can reach above 50%. The structure of multilayer films is to clip a thin non-magnetic film between two layers of magnetic films (such as Fe, Co, Ni), and the typical materials are Fe/Cr, Co/Cr, Co/Cu, Co/Ru, etc. The arrangement of the magnetization vector in two layers of magnetic films can be ferromagnetic, that is, parallel orientation, or antiferromagnetic, that is, antiparallel orientation, and it is decided by the thickness of film. The relative change rate of resistance of multilayer films with antimagnetic orientation is more than 100%, and the change of anti-ferromagnetic structure is related to the thickness and interface state of the non-magnetic film, and is also affected by the external magnetic field. Nanogranular films are Ni–Co and Ni–Fe–Co films doped with nonmagnetic atoms like Cu, Ag, and Al; both the surface and impurity scattering affect the resistance change of the film with magnetic field.

In 1992, Helmolt *et al.* found that the magnetoresistance effect of $La_{2/3}Ba_{1/3}MnO_3$ perovskite material was as high as 60%. In 1995, Raveau *et al.* found an unusually large magnetoresistance, up to 2.5×10^7%, in the $Pr_{0.7}Sr_{0.05}Ca_{0.25}MnO_{3.8}$ sample. This important discovery opens up a new direction for the study of the magnetoresistance effect.

In 1995, Professor J. S. Moodera of Massachusetts Institute of Technology first observed a large magnetoresistance signal of another GMR structure at room temperature. He sandwiched a very thin layer (about 0.7 nm) of insulating material between the magnetic films, which has a greater relative change in magnetoresistance and a higher sensitivity to the magnetic field. The electron can tunnel through the thin insulating layer and keep its spin direction unchanged, so it is called the tunneling magnetoresistance (TMR) effect. For example, the TMR signal of $Fe/Al_2O_3/Fe$ structure can reach 70%. When the magnetization direction of the two ferromagnetic layers in the tunnel structure is parallel, most of the self-spin bands in one layer tunnel into the empty state of most of the self-spin bands in the other layer. When the magnetization direction of the two ferromagnetic layers is antiparallel, a few of the self-spin band electrons in one electrode will tunnel into the empty state of a few of the self-spin band in the other electrode. In the two cases, the tunneling resistance is different, which leads to the tunneling magnetoresistance effect.

Later, IBM's Parkin and Japan's Yuasa used another new insulating material, MgO, respectively, and observed up to 180% and 200% TMR signals. In 2006, it was reported that the TMR signal at room temperature was up to 410%. These important findings open up a new direction for the study of the magnetoresistance effect.

GMR, CMR and TMR effects are not only related to temperature and external magnetic field, but also related to the preparation process. At present, ultra-high vacuum electron beam evaporation, magnetron sputtering, multi-target co-sputtering, pulse laser deposition, and other methods are mostly used to prepare Fe/Cr, Co/Cu multilayer films and manganese oxide films. The GMR film has a good application prospect in magnetic recording and magnetic sensors, which can increase the surface recording density of the hard disk over 1000 MB/in^2, greatly exceeding that of writable disks. A lot of magnetoresistive devices have been fabricated by using giant magnetoresistive films, for example, magnetoresistive head, rotary encoder, displacement magnetoresistive sensor, non-contact magnetoresistive switch, etc. The 2007 Nobel Prize in physics was awarded to two physicists: Albert Fert of Paris Sud University in France and Peter Grünberg of Ulrich Research Center in Germany for their contribution to the discovery of the GMR effect in 1988.

8.4.5 *Oxide magnetic film*

There are three kinds of oxide magnetic films: spinel, lead magnetite, and garnet ferrite. Although research on oxide magnetic films has made great breakthrough, there are still many problems to be solved. Therefore, in the future, the main aims of research on oxide magnetic films are to improve the consistency and repeatability of the film materials and explore new technology; to study new phenomena and new effects in the film and their theoretical mechanism; on the basis of theoretical research and technological exploration, to improve the performance of the oxide magnetic film and promote its application in electronic systems.

8.4.5.1 *Spinel ferrite films*

(1) **Mn series ferrite films**

In 2000, the Engineering Department of Nagasaki University and TDK Company of Japan prepared about 1 μm thick MnZn ferrite films on glass, crystallized Mo, Al and $ZnAl_2O_4$ substrates by pulse laser deposition (PLD). The substrate temperature was room temperature. It is found that the crystallized MnZn ferrite film can be obtained at room temperature by crystallizing $ZnAl_2O_4$ substrate, and the coercive force of the film deposited on the crystallized $ZnAl_2O_4$ substrate can be reduced by about 25% compared with that on the glass substrate.

In 2003, MnZn spinel ferrite thin films were prepared by high-speed target sputtering in the Department of Physical Electronics, Tokyo Institute of Technology, Japan. Firstly, a 12-nm thick Pt substrate was deposited on the glass ceramic substrate, so that the ferrite film could be oriented in the (111) direction. When the total pressure of Ar_2+O_2 is 2 mT and the partial pressure of O_2 is 25%, the saturation magnetization $\mu_0 M_s$ of the films is 480 mT. The deposition rate of this method is 16 times higher than that of the traditional method, which greatly improves the film forming efficiency.

In 2005, MnZn ferrite thin films with high magnetic conductivity in the GHz band were prepared by the spin spray plating method shown in Figure 8.13 on low temperature (<100°C) and glass substrate. The composition was $Mn_xZn_yFe_zO_{4-\delta}$ $(0.23 < x < 0.30, 0.0 < y < 0.38, 2.39 < z < 2.64)$. The results show that the cut-off

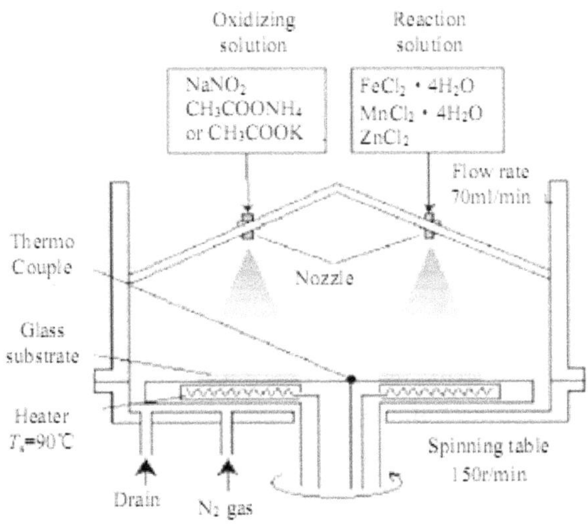

Figure 8.13: Schematic diagram of a rotary plating device.

frequency of MnZn ferrite film can reach about 300 MHz. When the film is placed on the microstrip line and tested at GHz, it is found that the transmission loss of the film is very strong, which is 10 times that of the commercial film with the same component, and the film also has a very low reflection coefficient ($S_{11} < -10\,\text{dB}$). Therefore, the film can be deposited in the interlayer of multilayer PCB as a noise suppressor.

In 2005, Xu Zuo *et al.* prepared $MnFe_2O_4$ ferrite thin films by the PLD method. It was found that only about 20% of $MnFe_2O_4$ thin films grown by the traditional process occupy the octahedral (b) position of spinel lattice, and the distribution of Mn^{2+} in spinel lattice can be controlled by stacking growth. When the oxygen partial pressure is lower than 5 mT, the magnetocrystalline anisotropy field μH of $MnFe_2O_4$ is more than 500 mT, and when the oxygen partial pressure is lower than 8 mT, the total magnetic anisotropy field H_a is in the membrane surface; when the oxygen partial pressure is higher than 8 mT, H_a is vertical to the membrane surface. Based on the high vertical anisotropy of the film, $MnFe_2O_4$ can be used in X-band or higher, such as phase shifter, filter, isolator, and circulator.

(2) **Ni series ferrite films**

In 2002, Matsushita *et al.* of Tokyo Institute of Technology, Japan, prepared a series of NiZn ferrite films on glass substrates at 90°C. Among them, the real part of the film with the component of $Ni_{0.28}Zn_{0.18}Fe_{2.54}O_4$ and the thickness of 0.4 μm is about 42, and the resonance frequency can reach 1.2 GHz. The resonance frequency of the ferrite film is 9 times that of the bulk under the same composition (the resonance frequency of the bulk is 13 MHz). Therefore, the film can be used not only in RF inductors, but also as electromagnetic interference absorbing materials.

In 2005, Munetoshi Seki *et al.* of the Institute of Science and Industry, Osaka University, Japan prepared Zn and Ti substituted Ni ferrite films on α-$Al_2O_3(0001)$ substrates by the PLD method, and measured their photoinduced magnetization (PIM) (illuminated by a Xe lamp). The results show that the photoinduced magnetization is essentially caused by the electron migration caused by photons, and not by the radiant heat of the Xe lamp. The analysis of the relationship between activation energy and photoinduced magnetization shows that the enhancement of PIM in Ti substituted Ni ferrite films is due to the covalent charge transfer from Ti^{4+}+Fe^{2+} $\rightarrow$ Ti^{3+}+Fe^{3+}.

(3) **Zn series ferrite film**

In 2002, $Zn_xFe_{3-x}O_4$ thin films with a Zn/Fe ratio of 0.36–0.76 were prepared on glass substrates by rotary spray plating at Carnegie Mellon University and Washington Naval Research Center. The films are uniform and dense with an average grain size of about 0.3 μM. In 2006, single-phase $ZnFe_2O_4$ thin films with a grain size of 8–80 nm were deposited on SiO_2 substrates by PLD and RF sputtering. The saturation magnetization M_s of the nanocrystalline film is strongly dependent on the grain size. When the grain size is large, the M_s of the material decreases sharply. The saturation magnetization μ_0 M_s of $ZnFe_2O_4$ is 556 mT at 500°C and 7.5×10^{-3} mT. In the same year, Katsuhisa Tanaka of the Institute of Engineering, Kyoto University, Japan, and others prepared 85 nm thick $ZnFe_2O_4$ thin films on SiO_2 substrates by RF sputtering. The results show that there is a large Faraday rotation angle at the wavelength of 390 nm for the films annealed at 300°C and not annealed at 300°C especially for the films annealed at 300°C with the largest Faraday effect, and the largest Faraday rotation angle of 1.65°/μm is obtained at the wavelength of 386 nm.

(4) **Li series ferrite films**

In 1997, PLD was used to prepare Li ferrite films on sapphire substrates by University of Florida, New York University, and University of Colorado. The film thickness was 4–8 μm, oxygen partial pressure was 200–400 mT, substrate temperature was about 1000°C, and measured at 9.5 GHz, the linewidth of ferromagnetic resonance was 33.5 mT, which was significantly larger than that of single crystal and large grain materials, and its $\mu_0 M_s$ in-plane reaches 360–400 mT, which is close to the bulk material, but the in-plane coercive force $\mu_0 H_C$ is too large. In addition, Li ferrite films were prepared by RF sputtering on Al_2O_3 substrates. The effects of Mn,

Ti, Zn elements and the process conditions (temperature, oxygen partial pressure, etc.) were studied.

8.4.5.2 *Magnetoplumbite ferrite films*

In 2002, BaM thin films were deposited on Si (100) substrates with various substrates by RF/DC magnetron sputtering at Kangwon National University of Korea. The substrates include Fe, Cr, Al_2O_3, Fe_2O_3, $ZnFe_2O_4$, and TiO_2. The results show that the BaM films are randomly oriented without substrate layer; except BaM/Fe/Si films, other BaM films have almost the same coercive force in the in-plane and vertical in-plane directions, among which BaM/TiO_2/Si has the highest coercive force, 0 μH_c can reach 440 mT; when $ZnFe_2O_4$ is used as the substrate layer, the diffusion and permeation of Si can be inhibited to some extent. The microstructure of BaM films deposited on TiO_2 substrate is related to the microstructure of TiO_2 substrate and the total pressure of sputtering.

In the same year, a BaM film was prepared by the PLD method in Northeastern University of the United States, on which $CoFe_2O_4$ (CoF) was deposited on MgO (111) substrate. It is found that when the substrate temperature of CoF is 400°C, the deposited BaM/CoF/MgO has the maximum coercive force, which is higher than the coercive force of BaM/MgO deposited directly on MgO substrate ($0\,\mu H_c = 40\,\text{mT}$). This conclusion provides a new idea for the deposition of BaM with high coercive force in the future.

In 2003, BaM and SrM thin films were prepared by the opposite target sputtering method at Tokyo Institute of Technology. A Pt (111)-oriented substrate was deposited on the SiO_2 substrate. The substrate temperature was 420–550°C, and the working gas was Ar and Kr. The results show that the Pt substrate layer is very effective for the preparation of SrM films with *c*-axis orientation. When Ar is used as working gas only, the coercivity of BaM and SRM ferrite films with thickness less than 40 nm is relatively low in the vertical in-plane direction. When Ar and Kr are used as working gas, SrM has 300 mT of $\mu_0 H_c$ even if the film thickness is within 20–40 nm. In addition, when Ar and Kr are used as working gas, the coercivity of

SrM films with thickness less than 40 nm can be effectively reduced. The SrM thin films with *c*-axis preferred orientation can be prepared at 450°C.

$Ba_{0.5}Sr_{0.5}TiO_3/BaFe_{12}O_{19}$ (BSTO/BaM) thin films were prepared on Al_2O_3 substrates by magnetron sputtering at South Florida University. The magnetization curves (M–H) of BSTO/BaM films show two transitions, but the above phenomena do not occur in BaM of the same composition. This abnormal phenomenon of M–H may be caused by the magnetoelectric effect between grains in the medium of the BSTO polarization layer.

In 2005, Pd, Pt, and Pd–Pt alloys were deposited on the SiO_2/Si substrates by the opposite target sputtering method in the School of Engineering, Shinshu University, Japan. The substrate temperature was 400–600°C, and the BaM ferrite films with a thickness of 15–35 nm were prepared. It is found that the coercive force in the vertical in-plane direction of the films deposited on the Pd–Pt substrate is greater than that of the BaM films deposited on the Pt and Pd substrates. When the substrate temperature is 450°C, the 30-nm thick BAM film deposited on the Pd–Pt liner has a good *c*-axis orientation, and the coercive force perpendicular to the film surface is 0 $\mu H_c = 210\,\text{mT}$.

8.4.5.3 *Garnet ferrite films*

In 2002, a garnet film with the composition of $(Gd, Bi)_3(Fe, Ga)_5O_{12}$ was deposited on (111)-oriented gadolinium gallium garnet substrates by liquid phase epitaxy at Osnabruck University in Germany. The film thickness is about $2\,\mu\text{m}$, saturation magnetization M_s is about $10\,\text{kA/m}$, compensation temperature is less than 250 K, and Faraday rotation angle ($\lambda = 1.3\,\mu\text{m}$) is $-8.21°/\text{m}$.

In 2005, Dumont *et al.* deposited YIG ferrite film on SiO_2 substrate by the PLD method, and measured the stoichiometric ratio of the film by the Rutherford backscatter method. When the partial pressure of oxygen is lower than the pressure of the metering ratio, it is in the state of anoxia, and the Curie temperature and saturation magnetization $\mu_0 M_s$ (about 12.6 mT) decrease; when the partial pressure of oxygen is higher than the pressure of the metering ratio, Fe and Y are insufficient, Curie temperature increases by 10%, saturation magnetization increases by 20%, and lattice constant decreases. Only when the stoichiometric specific pressure is 30 mT, YIG films show the magnetic and structural characteristics of the bulk.

8.5 Magnetoelectric Thin Films

8.5.1 *Overview*

The magnetoelectric effect is the property that can produce dielectric polarization (i.e., positive magnetoelectric effect: $P = \alpha H$) or magnetic polarization (i.e., reverse magnetoelectric effect: $M = \alpha E$) under the action of an external magnetic field. It not only has two kinds of ordered structures with coexistence of magnetic order and ferroelectric order, but also has some form of coupling between the two kinds of ordered structures. Here, α is the magnetoelectric coefficient that characterizes the performance of magnetoelectric materials. It can be expressed as $E = \alpha EH$). The larger the magnetoelectric coefficient is, the higher the magnetoelectric conversion efficiency is, that is, the stronger the coupling between magnetoelectric order and ferroelectric order.

As early as 1894, after studying crystal symmetry, Mr. Curie thought that an asymmetric molecular crystal would be polarized directionally under the action of a magnetic field. Half a century later, Landau and Lifshitz believed that there was linear magnetoelectric induction in the magnetic ordered crystal. Based on the theoretical analysis, Dzyaloshinskii predicted that the antiferromagnetic material Cr_2O_3 could produce magnetoelectric induction. Astrov measured the induced electric field of Cr_2O_3 under the action of a magnetic field. Rado and Folen detected the magnetic field induced by the polarization of Cr_2O_3. In 1958, Smolensky and Joffe synthesized the ferroelectric perovskite ceramic material $Pb(Fe_{1/2}Nb_{1/2})O_3$ (PFN). Subsequently, many magnetoelectric compounds have been found, such as $BiFeO_3$, $BaMnF_4$, $HoMnO_3$, and so on. In addition, some other compounds with perovskite structure, hydromanganese structure, pseudo ilmenites structure and spinel structure also have magnetoelectric effect. In addition, some other compounds with perovskite structure, hydromanganese structure, pseudo ilmenites structure and spinel structure also have magnetoelectric effect. Usually, magnetoelectric induction occurs in the material where the magnetic dipole and the electric dipole coexist. In the process of magnetoelectric induction, the ordered magnetic sublattice and the ordered ferroelectric sublattice interact to complete two transitions: one is the transition from ferroelectric state to paramagnetic state; the other is the transition from ferromagnetic state (ferromagnetic state or antiferromagnetic state) to paramagnetic state. Ferroelectric order and ferromagnetic order are often accompanied by the occurrence of ferroelasticity, resulting in structural channels that can satisfy the movement of ferroelectric ions, magnetic interaction paths, and symmetry environments that are usually superexchange type. Because of the coexistence of ferroelectricity, ferromagnetism and ferroelasticity in magnetoelectric materials, it is beneficial to the coupling of spontaneous polarization, spontaneous magnetization, and spontaneous deformation. Therefore, magnetoelectric materials have a very broad application prospect in information recording memory, magnetic field detection, sensors and so on.

In terms of composition, magnetoelectric materials can be divided into single-phase magnetoelectric materials and composite magnetoelectric materials. A single-phase magnetoelectric material refers to a kind of multiphase multiferroic material which has a magnetoelectric effect, while a composite magnetoelectric material refers to a kind of multiphase multiferroic material which has magnetoelectric effect through some coupling between different components.

Because the temperature T_N or T_C of single-phase magnetoelectric materials (such as Cr_2O_3) is much lower than room temperature, the magnetoelectric effect is obvious only at a lower temperature, and the magnetoelectric effect is relatively small (the magnetoelectric coefficient a_E is about $20\,\mathrm{mV/cm \cdot Oe}$), so it is difficult for these single-phase magnetoelectric materials to find practical application. From the 1970s, onwards Philips laboratory began to study ferroelectric ferromagnetic solid-phase sintered ceramics, ferroelectric/ferromagnetic/polymer and other magnetoelectric composite materials. In the nineties of the last century, GE Company first developed the piezoelectric/piezomagnetic composite magnetic sensor and achieved great success. In 2001, the Terenol-D/PZT sandwich composite magnetoelectric material was reported by the University of Pennsylvania of the United States. Its magnetoelectric coefficient a_E can reach ~6.0 V/cm · Oe. In recent years, piezoelectric materials and magnetic materials such as piezomagnetic materials, and magnetostrictive materials and their preparation methods have made new breakthroughs. Magnetoelectric composite materials have attracted more and more attention. This section briefly introduces the materials, preparation methods, and basic performance characterization of magnetoelectric thin films.

8.5.2 *Single phase magnetoelectric films*

Many single-phase magnetoelectric materials were studied recently:

(1) Orthogonal structure $RMnO_3$ (R = Gd, Tb, Dy) ($T_N \sim 40\,K$, $T_C \sim 25\,K$);
(2) Orthogonal structure RMn_2O_5 (R = Gd_2Lu) ($T_C \sim 40\,K$, $T_N \sim 35\,K$);
(3) Tripartite structure (R3c) $BiFeO_3$ ($T_C \sim 1110\,K$, $T_N \sim 640\,K$);
(4) Hexagonal structure (P63 cm) $RMnO_3$ (R = Y, Ho_2Lu) ($T_C = 570$–$990\,K$, $T_N = 70$–$130\,K$);
(5) ACr_2X_4 (A = Co, Zn, Cd, Hg; X = O, S, Se) with cubic (Fd3m) spinel structure;
(6) $Ba_{2-x}Sr_xZn_2Fe_{12}O_{22}$ with galena structure;
(7) $Ni_3V_2O_8$ with Kagome staircase structure;
(8) $MnWO_4$ with wolframite structure.

However, as a single-phase magnetoelectric thin film, the $BiFeO_3$ thin film is the only material that exhibits ferroelectricity and magnetism at room temperature at the same time. At room temperature, $BiFeO_3$ has rhombic distorted perovskite structure, the space group is $R3_c$, the lattice constant is $a = b = c = 5.633\,Å$ and

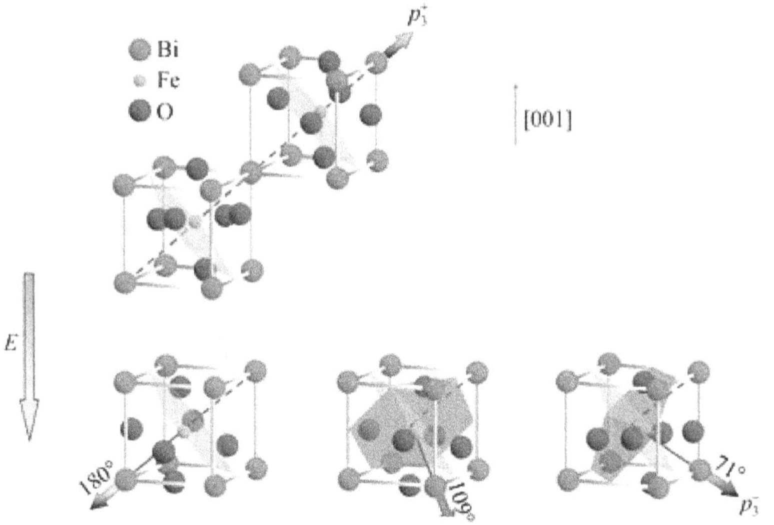

Figure 8.14: Structure diagram of $BiFeO_3$.

$\alpha = \beta = \gamma = 59.4°$, and the lattice structure is shown in Figure 8.14. Compared with the cubic perovskite structure, with the surrounding oxygen ions as the coordinates, Bi ions will move along the [111] direction, while the oxygen octahedron will twist around the [111] axis, resulting in polarization along the [111] direction, and its ferroelectric phase transition occurs at $T_c \sim 1103\,\text{K}$. It is predicted theoretically that the saturated electrode can reach $100\,\mu\text{C/cm}^2$ at room temperature.

The magnetic structure of $BiFeO_3$ is quite special, its ground state is antiferromagnetic state, and the magnetic phase transition occurs at $T_N \sim 643\,\text{K}$. However, the neutron scattering experiment reveals that its antiferromagnetic spin order is uneven, showing a spatial modulation structure, and its spin shows an incommensurate sine curve arrangement, with a wavelength (or period) of $\sim 62\,\text{nm}$, as shown in Figure 8.15. This modulation structure causes the magnetic moment of each ion to cancel each other, so the macroscopical size of $BiFeO_3$ only shows weak magnetism. It can be imagined that if the characteristic size of the microstructure is smaller than the wavelength of this sinusoidal structure, the cancellation of

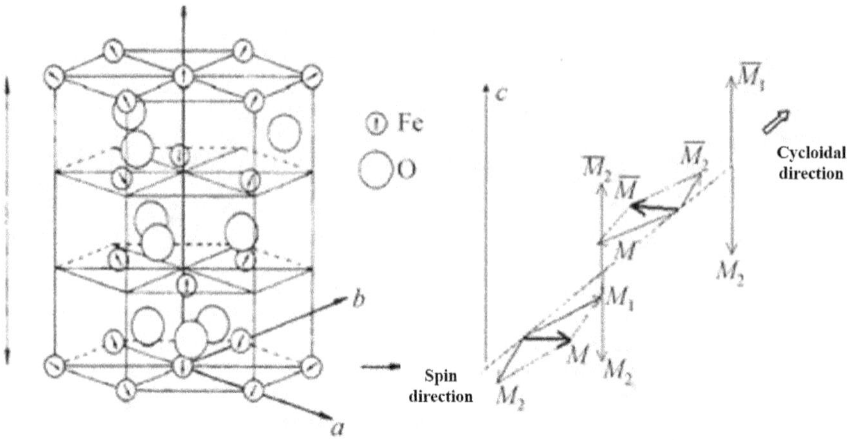

Figure 8.15: Spin structure of $BiFeO_3$.

the ionic magnetic moment will not be complete, thus showing enhanced magnetism. Not only the thin film samples show enhanced polarization, but also enhanced magnetism. Similar to $BiMnO_3$, the coupling of magnetism and ferroelectricity in $BiFeO_3$ is weak. However, there is coupling between the antiferroelectric domain and the ferroelectric domain and the external electric field to some extent.

$BiFeO_3$ thin films can be prepared by PLD, sol–gel, sputtering and hydrothermal methods. The structure of the $BiFeO_3$ thin film is closely related to the orientation of the substrate and the thin film. For example, the $BiFeO_3$ thin film is prepared on the $SrTiO_3$ single crystal substrate with different orientations. The $BiFeO_3$ thin film grown on the (111) substrate has a tripartite structure, which is consistent with the structure of bulk single crystals, and is in a single domain state without stress. However, the structure of $BiFeO_3$ thin films grown on (101) and (001) substrates is greatly affected by the substrate stress, which changes from three-way structure to monoclinic structure. In addition, due to the influence of substrate stress, the structure of the film is closely related to the thickness. Under the condition of 200 nm thickness, the lattice constant is $a = 0.3935$ nm, $c/a = 1.016$. In the range of 10–400 nm, the c/a value decreases with the increase of film thickness.

Claude *et al.* studied the strong magnetism of the $BiFeO_3$ thin film, and thought that the spiral spin structure would produce strong magnetism because of the inhibition of epitaxial stress or enhanced anisotropy. From the microstructure point of view, the magnetism of the thin film originates from the tilt of the magnetic sublattice caused by the antisymmetric spin coupling, which makes the spin alignment of (111) in-plane collinear tilt, resulting in the non-zero net magnetic moment.

Therefore, although the pure $BiFeO_3$ shows weak antiferromagnetism at room temperature, most of the magnetic measurements at room temperature are linear. The structure of the film is closely related to the orientation of the substrate and the film. The results show that the $BiFeO_3$ thin film grown on the (111) bottom has a tripartite structure, which is consistent with the structure of bulk single crystals and is in a single domain state without stress. However, the structure of the films grown on (101) and (001) substrates is greatly affected by the substrate stress, which changes from three-way structure to monoclinic structure. In addition, due to the influence of substrate stress, the structure of the film is closely related to the thickness. Under the condition of 200 nm thickness, the lattice constant is $a = 0.3935\,\text{nm}$ and $c/a = 1.016$. In the range of 10–400 nm, the c/a value decreases with the increase of film thickness.

8.5.3 *Multiphase composite magnetoelectric thin films*

The coupling between ferroelectrics and ferromagnetics can be realized through the stress–strain coupling transfer between ferroelectrics/piezoelectric materials and magnetostrictive materials. This kind of magnetoelectric material composed of ferroelectrics/piezoelectric materials and magnetic materials is a multiferroics magnetoelectric composite material. Since Van Run and others reported that the magnetoelectric coefficient of $BaTiO_3$–$CoFe_2O_4$ composite ceramics is nearly two orders of magnitude larger than that of Cr_2O_3 in 1974, people have been paying close attention to composite magnetoelectric materials. Compared with block magnetoelectric composite materials, the magnetic composite film has unique properties. For example:

(1) The ferroelectric/piezoelectric phase and the magnetostrictive phase in the composite can be controlled and adjusted on the nanometer scale, and the magnetoelectric coupling mechanism can be studied on the nanometer scale.
(2) The interface loss is a problem that cannot be ignored. In the film, the combination of different phases can be realized at the atomic scale, which can effectively reduce the interface coupling loss.
(3) By controlling the experimental conditions, different phases with similar lattice parameters are combined together; highly preferred orientation or even superlattice composite films can be obtained, which is conducive to the study of the physical mechanism of magnetoelectric coupling.
(4) Multiferroic magnetoelectric films can be used to manufacture integrated magnetic/electrical devices, such as micro sensors, MEMS devices, high-density information storage devices, etc.

Because magnetoelectric composite films are composed of various components, the laser pulse deposition (PLD) and sol–gel spin coating methods are more commonly used. These two methods can easily control the composition of the composite film. By adjusting the technological parameters in the preparation process, different structures can be obtained. According to the classification of the composite structure, the magnetoelectric composite film can be divided into 1-3, 0-3, 2-2, and other structural types.

8.5.3.1 *1-3 Type cylindrical composite magnetoelectric thin films*

In 2004, Zheng *et al.* selected $BaTiO_3$ with high-voltage electrical properties and $CoFe_2O_4$ with high magnetostriction as the target material, and deposited it at high temperature above 900°C by PLD method on (001) $SrTiO_3$ single crystal substrate. The heteroepitaxial self-assembly of $BaTiO_3$–$CoFe_2O_4$ results in phase separation.

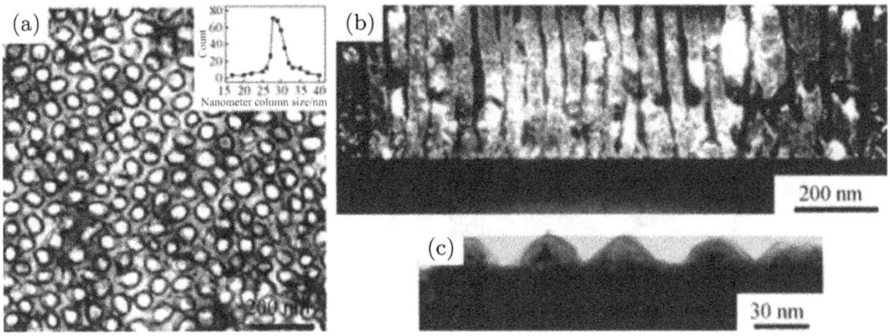

Figure 8.16: Morphology of $BaTiO_3$–$CoFe_2O_4$ nano-1-3 composite films grown on $SrTiO_3$ single crystal substrate. (a) TEM profile; (b) TEM cross-section dark field image; (c) TEM cross-section bright field image, showing $CoFe_2O_4$ nanocolumn outcrop.

The $CoFe_2O_4$ in the composite film is embedded in the $BaTiO_3$ matrix as an array, forming a 1-3 structure composite film, as shown in Figure 8.16.

Then, Zheng *et al.* prepared 1-3 composite $BiFeO_3$–$CoFe_2O_4$ structure on $SrTiO_3$ single crystal substrate by the same method. In this structure, the phenomenon of magnetization reversal induced by the applied electric field was observed by a scanning probe microscope, which indicated that there was magnetoelectric coupling effect in this 1-3 cylindrical nanostructure, thus providing direct evidence of magnetoelectric coupling effect on the film. They observed the magnetic domain inversion of the composite film after the electrode by a magnetic force microscope (MFM), which proved that there was a coupling between the two composite phases, as shown in Figures 8.17(a) and 8.17(b). At the same time, the change in the magnetization of the composite film after the electrode was measured by the SQUID (Figure 8.17(c)), and the magnetoelectric sensitivity coefficient of the composite film was estimated to be $\alpha_{33} = 1.0 \times 10^{-2}\,\mathrm{G\,cm/V}$.

Levin and others have grown $x\mathrm{CoFe_2O_4}$–$(1-x)\mathrm{PbTiO_3}$ composite films with similar structures and morphologies. They think that different substrates make the films in different stress states, so that the final nanocomposite films have different structures and morphologies. The growth of the composite film was simulated by

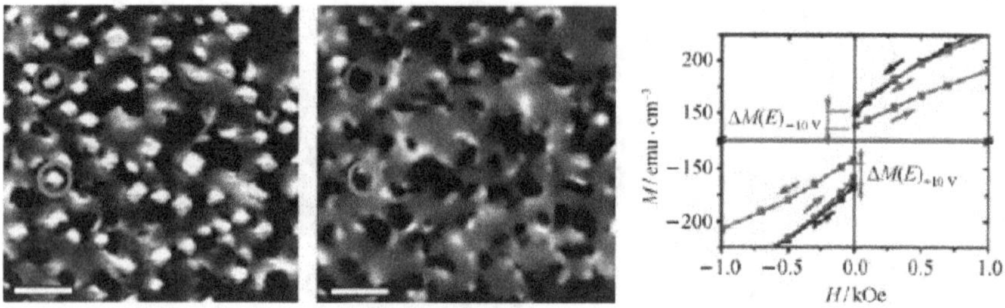

(a) (b) (c)

Figure 8.17: Electric field-induced domain inversion (a,b) and magnetization change (c) in composite films.

the phase field method considering the different stress states, and the results were similar to the experimental observation. Although the 1-3 columnar nanostructure composite film shows strong magnetoelectric coupling, the growth of this film needs more stringent conditions, such as a rather high growth temperature (greater than 900°C), and the growth of the nanocolumnar array is also difficult to control. On the other hand, because the magnetic phase with poor resistance runs through the whole film, the leakage conductivity of the film is too large, and it is not easy to directly observe the positive magnetoelectric effect, so its application will be limited.

8.5.3.2 *0-3 type particle composite magnetoelectric thin films*

The 0-3 type composite magnetoelectric thin film is a kind of particle composite polycrystalline thin film. Wan *et al.* used sol–gel method to spin the $CoFe_2O_4$ and $Pb(Zr, Ti)O_3$ precursor sol on $Pt/Ti/SiO_2/Si$ substrate and then annealed it at 650°C for 6 min. During the annealing process, two phases were separated and recombined, and a suspected granular composite film was obtained, as shown in Figure 8.18. The measurement shows that the film has good ferromagnetic and ferroelectric response, and the magnetoelectric coefficient of the composite film was measured. Recently, Zhong *et al.* obtained particle composite films of the $x\mathrm{Bi_{3.15}Nd_{0.85}Ti_3O_{12}}$–$(1-x)\mathrm{CoFe_2O_4}$ system by similar chemical methods, and also observed good ferromagnetic and ferroelectric properties.

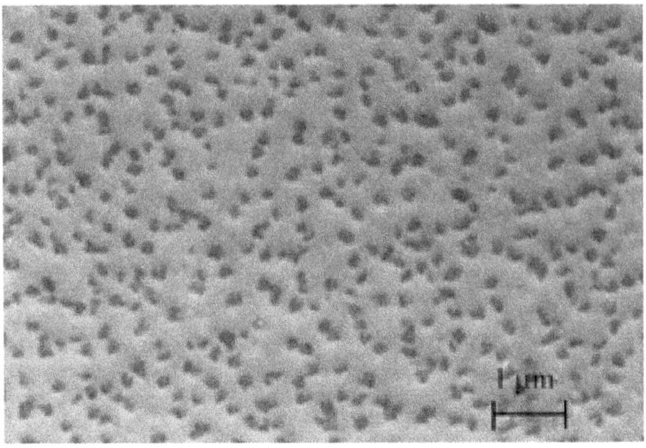

Figure 8.18: SEM morphology of magnetoelectric PZT/CFO composite films prepared by sol–gel method.

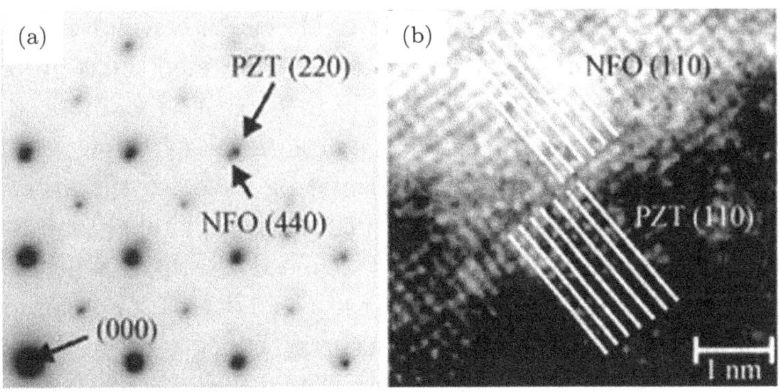

Figure 8.19: Surface morphology and lattice diffraction pattern of the PZT–NFO composite film. (a) Lattice images of PZT–NFO films; (b) selected area electron diffraction patterns of the films.

Ryu *et al.* prepared PZT–$NiFe_2O_4$ composite films on (110) $SrTiO_3$ substrates doped with 0.5% Nb by PLD method. $NiFe_2O_4$ is randomly distributed in the PZT matrix as nanoparticles, as shown in Figure 8.19. The experiment shows that the magnetoelectric coefficient reaches saturation under the electric field of about 0.8 MV/cm. It is observed that the magnetoelectric coefficient first increases with the bias static magnetic field, then decreases to a maximum value, and the magnetoelectric coefficient α_{E31} reaches the maximum value at a smaller static magnetic field compared with α_{E33}. These changes are similar to those of bulk magnetoelectric composite ceramics. However, in PZT–$NiFe_2O_4$ composite films, the maximum value of α_{E31} is about $4\,\text{mV}/(\text{cm}\cdot\text{Oe})$, and the maximum value of α_{E33} is about $16\,\text{mV}/(\text{cm}\cdot\text{Oe})$, which is obviously smaller than the magnetoelectric coefficient of bulk magnetoelectric composite ceramics. This may be due to the confinement effect of the substrate on the film in the composite film.

8.5.3.3 *2-2 type laminated composite magnetoelectric thin films*

By depositing the ferroelectric and ferromagnetic materials on the substrate layer by layer, laminated magnetoelectric composite films can be obtained. Takeuchi *et al.* used the improved PLD method to modulate the thickness by adding a baffle which can move at a constant speed in the deposition chamber, and obtained the composition gradient composite films of $BaTiO_3$ and $CoFe_2O_4$. Compared with 1-3 and 0-3 composite films, the magnetic layer with low resistance in 2-2 structure is isolated by the insulating ferroelectric layer in the out of plane direction, so 2-2 structure can completely eliminate the leakage conduction problem. He *et al.* used chemical solution spin coating method to spin PZT and $Co_{1-x}Zn_xFeO_3$ (CZFO) precursor sol on Pt/Ti/SiO_2/Si substrate in turn, which can easily obtain composite films with multilayer structure. Also, according to the different deposition sequences of two phases, two kinds of multilayer structures can be obtained (as shown in Figures 8.20(a) and 8.20(b)): that is, substrate/PZT/CZFO/PZT/CZFO (referred to as PCPC) and substrate/CZFO/PZT/CZFO/PZT (referred to as CPCP).

The electrical properties of the films are shown in Figures 8.20(c) and 8.20(d). The results show that, compared with pure PZT films, the composite films have higher leakage current due to the lower resistivity of ferrite phase in the composite films; however, the leakage current characteristics between the composite films of the two structures do not show obvious differences and are not sensitive to the deposition sequence. However, the ferroelectricity of the composite

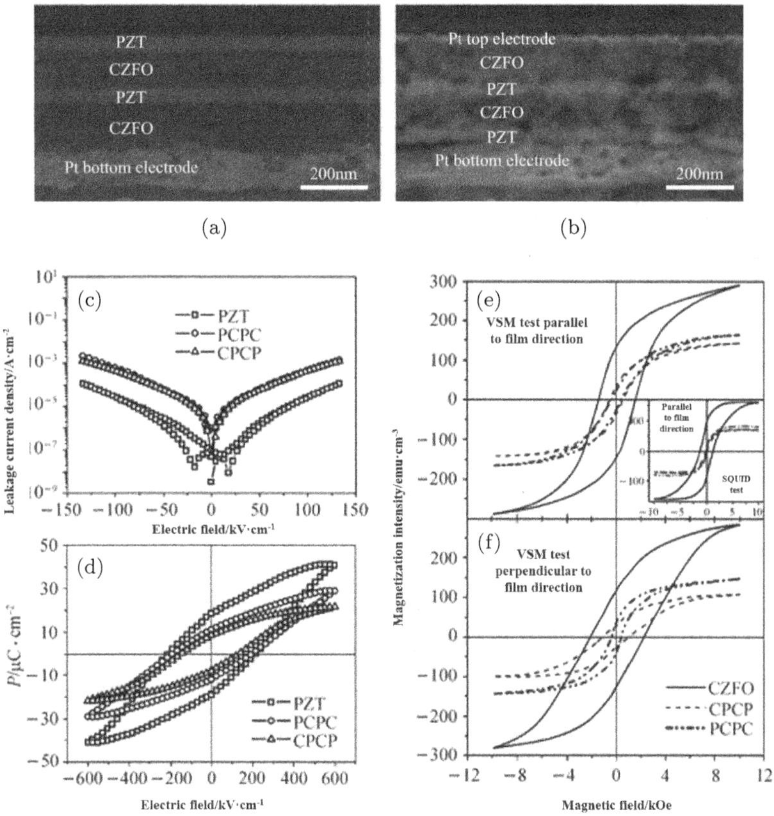

Figure 8.20: Microstructure (a) and (b), electrical properties (c) and (d) and magnetic properties (e) and (f) of CPCP and PCPC composite films.

film is weaker than that of the single-phase PZT due to the Co ferrite layer in the paraelectric phase. The magnetic properties of the composite films are shown in Figures 8.20(e) and 8.20(f), and the results show that they have good ferromagnetic properties. Due to the existence of a non-ferromagnetic PZT layer, the magnetization of the composite film is significantly lower than that of single-phase ferrite. Because there are more stresses in the composite film, the coercive force of the composite film is lower than that of the single-phase ferrite according to the piezomagnetic effect.

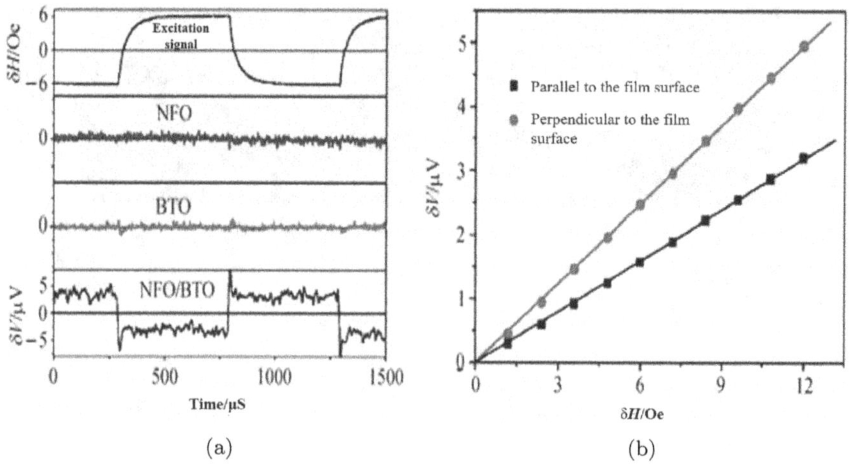

Figure 8.21: Magnetoelectric properties of $BaTiO_3/NiFe_2O_4$ composite films prepared by PLD.

In 2007, Deng *et al.* epitaxially grew $BaTiO_3/NiFe_2O_4$ two-layer structure thin films on $SrTiO_3$ single crystal substrate by PLD method. This kind of thin film has good epitaxial characteristics, and the interface between layers is clear. In particular, they also observed a significant magnetoelectric response directly in the composite film, as shown in Figure 8.21(a). By applying alternating square wave magnetic field excitation to the film, the corresponding square wave response voltage signal appears in the composite film, which is not available in two single-phase films, which intuitively provides clear evidence for the magnetoelectric response signal. At the same time, the response signal changes linearly with the amplitude of the alternating magnetic field, and its slope is the magnetoelectric voltage coefficient. Using the definition of the magnetoelectric voltage coefficient in bulk magnetoelectric materials: $\alpha_E = (\delta V/t\delta H)$ (where t is the thickness of the composite film), from the slope of the linear relationship in Figure 8.21(b), we can get that the magnetoelectric voltage coefficient of the composite film in the direction perpendicular to the film is $12\,\text{mV}/(\text{cm}\cdot\text{Oe})$. $PZT/Ni_{0.8}Zn_{0.2}Fe_2O_4$, $Fe_3O_4/BaTiO_3$ composite films were also prepared by PLD, with the magnetoelectric coefficient of $15–30\,\text{mV}/(\text{cm}\cdot\text{Oe})$. $Pr_{0.85}Ca_{0.15}MnO_3/Ba_{0.6}Sr_{0.4}TiO_3$,

$La_{0.7}Ca_{0.3}MnO_3/BaTiO_3$ and other superlattice composite magnetoelectric thin films were also prepared by PLD method, showing good permittivity, ferromagnetism, magnetoresistance effect, and magnetocapacitance effect.

8.6 Integrated Optical Thin Films

The optical communication system with optical fiber as the transmission medium has a great influence on information transmission technology; this is because optical fiber not only has low loss, but also has the characteristics of light weight, small size, anti-electromagnetic interference, and wide bandwidth. Integrated optics is a new field which has been developed gradually to meet the requirements of optical communication. Using technology similar to integrated circuit technology to let optical components (such as light emitting devices, optical amplifier, optical switch, optical logic components, optical path components, various kinds of optical modulation components, optical coupling and receiving components, etc.) integrate on the same substrate, so the micro-optical system with independent functions is formed. This integrated optical path has the characteristics of small volume, high efficiency, low power consumption, stable and reliable performance, easy to use and so on. The integrated optical film is different from the optical film. For optical films, the beam passes through the film; for integrated optical thin films, the beam propagates along the thin film.

8.6.1 *Optical waveguide films*

The medium that guides the light is called the optical waveguide; its basic structure is that high refractive index film 2 (refractive index n_2) is sandwiched between low refractive index media 1 and 3 (refractive indices n_1 and n_3, respectively), and $n_2 > n_1 \geq n_3$. The light guide in the dielectric optical waveguide is realized through repeated total reflection of light and propagation in high refractive index film 2. High refractive index film 2 is called optical waveguide region; medium 1, 3 is called the cladding region.

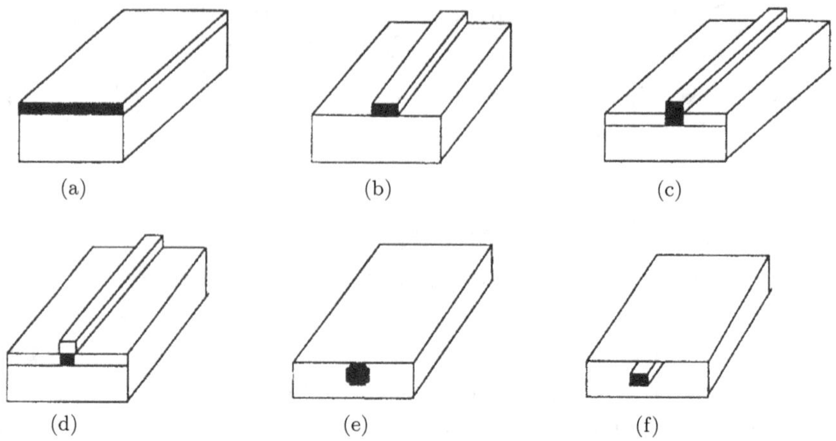

Figure 8.22: Different optical waveguides: (a) flat-plate waveguide; (b) convex-stripe waveguide; (c) loaded waveguide; (d) ridged waveguide; (e) diffused waveguide; (f) buried waveguide.

The types of optical waveguides in thin-film systems are shown in Figure 8.22. The shaded part, which is the waveguide, has higher refractive index than the surrounding part. Figure 8.22(a) is the flat-plat waveguide which is made up of the film formed on the substrate by coating, evaporating, sputtering and thermal diffusion. The flat-plat waveguide of Figure 8.22(b) is made by photoetching and re-sputtering. Figure 8.22(c) shows the flat-plat waveguide in which the light path is made of different materials (photoresist, etc.) than film, and the light is enclosed in it because the equivalent refractive index of the lower part of the waveguide is increased. Figure 8.22(d) shows the ridge waveguide. Figure 8.22(e), manufactured by ion exchange and ion implantation, obtains a cylindrical refractive index distribution, thereby reducing scattering losses. Figure 8.22(f) is the cladding of the same material of the substrate and it is grown on Figure 8.22(b).

The materials used in the optical waveguide should be transparent, stable, and easy to fabricate and process. The materials that satisfy these requirements are crystal, amorphous, liquid crystal, organic material, inorganic material, etc. Table 8.8 lists the commonly used optical waveguide materials and their basic properties.

Table 8.8: Typical materials for optical waveguides.

Materials		Preparation method	Refractive index (Δn)	Absorption coefficient ($dB \cdot cm^{-1}$)
Waveguide region	Substrate			
Polyurethane	SiO_2 glass	Spin coating	$\approx 10^{-2}$	0.1–0.5
Corning 7059 glass	SiO_2 glass	Sputtering	$\approx 10^{-2}$	0.1–0.5
SiO_2		Ion implantation	$\approx 10^{-2}$	≈ 0.4
$LiNbO_3$		Diffusion	5×10^{-4}–10^{-2}	0.1–0.2
		N^+, O^+, Ne^+ ion implantation	5×10^{-4}–10^{-2}	<1
$LiNbO_3$	$LiTaO_3$	Diffusion	10^{-2}–10^{-1}	≈ 1
n-GaAs	N^+-GaAs	Liquid-phase epitaxy (LPE) Gas-phase epitaxy (VPE) Molecular beam epitaxy (MBE)	$\approx 10^{-3}$	≈ 24
$Ga_{1-x}Al_xAs$	$Ga_{1-y}Al_yAs$	LPE, MBE	≈ 0.4 (y–x)	≈ 10
CdS_xSe_{1-x}	CdS	Diffusion	$\approx 10^{-2}$	10–15

8.6.2 *Optical switch film*

Switching devices for switching optical circuits are important devices for the integrated optical circuit, and it is a device that allows light to switch in spacetime. Optical switching devices are mainly realized through electro-optic effect, acousto-optic effect and magneto-optic effect. The materials to make optical switch by using electro-optic effect are dielectric materials, such as $LiNbO_3$, $LiTaO_3$, $Bi_{12}SiO_{20}$(BSO), GaAs, GeSi/Si, Si single crystals, and so on. In $LiNbO_3$, the optical switch is made by thermal diffusion of Ti, while in BSO, the optical switch is made by doping Ga through liquid phase epitaxy (LPE).

Figure 8.23 is a directional coupling modulation switch with a single electrode structure; it is called the cross working state and is

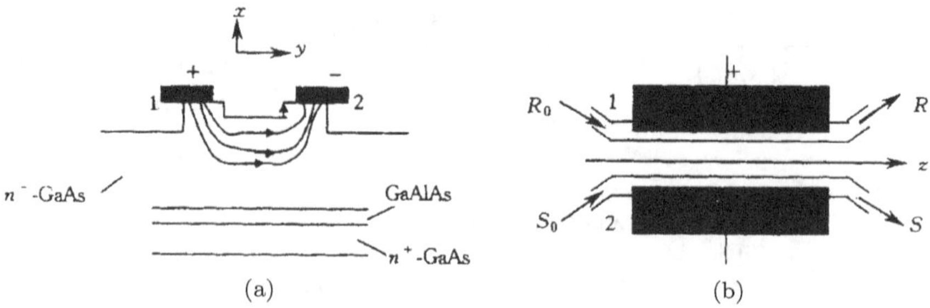

Figure 8.23: Sketch map of a directional coupling modulation switch with a single-electrode structure: (a) cross-sectional view; (b) top view.

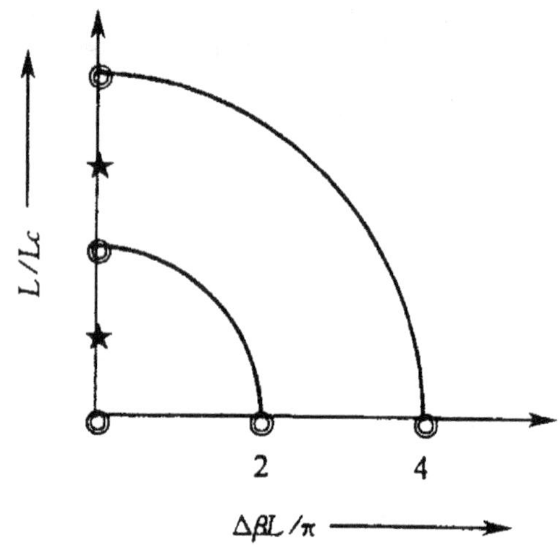

Figure 8.24: Working curve of a directional coupling switch with a single-electrode structure.

expressed by ⋆ if light is input from waveguide 1 and output from waveguide 2; if it is still output from waveguide 1, it is called the straight-through working state and is expressed by ◎. Figure 8.24 shows the working curve of a directional coupling modulation switch with a single-electrode structure. L is the length of the active area, L_C is the shortest coupling length, and $\Delta\beta$ is the difference of propagation constant and it changes with the modulation voltage. As can be seen from Figure 8.24, the directional coupling switch with a single electrode can always reach the straight working state by adjusting the voltage for any value of L/L_C. In order to obtain the cross-working state for use as a switch, the length of the active area of the coupling switch must be odd times of the coupling length. This kind of dual-channel directional coupling switch was first realized in GaAs material with a three-electrode structure. Subsequently, a directional coupling switch with a dual-electrode structure was realized on $LiNbO_3$ and GaAs substrates.

The materials used to make optical switch devices with acousto-optical effect are TeO_2, $LiNbO_3$, etc. The optical switch device is obtained because of the situation that periodic change of refractive index is caused by the surface elastic wave due to its structure, and this results in Bragg diffraction or Naman–Nath diffraction.

8.6.3 *Film lens*

Lens plays an important role in integrated optics: such as emitting the light produced by the collimating light source; focusing the light on the detector and the output device; completing Fourier transform by converging diffracted light, and so on. Thin-film waveguide lenses include short-range lens, Luneburg lens, mode refractive index lens, grating lens, Fresnel lens, Prague lens, distributed refractive index Fresnel lens, etc.

The mode refractive index lens shown in Figure 8.25 is made by making a part with the shape of a classical lens on a flat-plate waveguide and realizing the convergence and divergence functions by using the change in equivalent refractive index caused by the change in thickness.

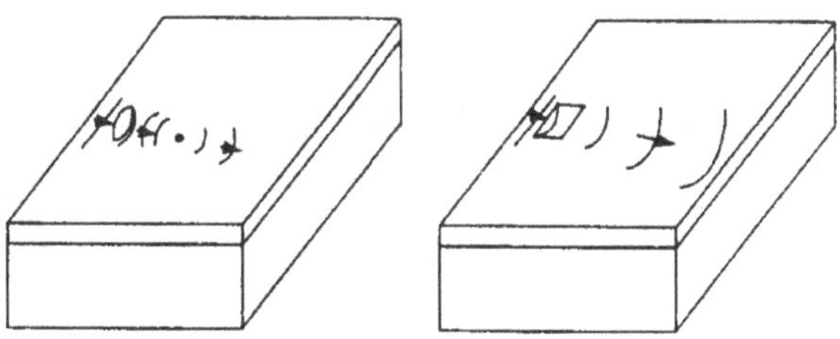

(a) (b)

Figure 8.25: Schematic diagram of mode refractive index lens: (a) aggregation function; (b) divergence functions.

A Luneburg lens has a graded refractive index which leads to special focusing properties, and it has centrosymmetric thickness change (equivalent to refractive index change) and centrosymmetric object plane and image plane. The manufacturing method is as follows: set the mask controlling the shape above the substrate, leave the substrate slightly, and deposit the film by sputtering and other methods.

Short-range lens utilize the principle that light travels along the shortest distance to bend the waveguide to achieve convergence effect. However, the junction between flat-plat waveguide and curved part has strong scattering; to overcome this, the edge can be made into a circle; although the aberration can be corrected, it is difficult to fabricate.

Since the 1980s, the planar microlens array has been one of the most important devices. Figure 8.26 shows a process diagram for making a spherical self-focusing plane microlens array. The basic idea is that A^+ ions with high electric polarizability (such as Ti^+ and Ag^+) replace B^+ ions with low electric polarizability (such as K^+ and Na^+) in the substrate to form an approximately hemispherical

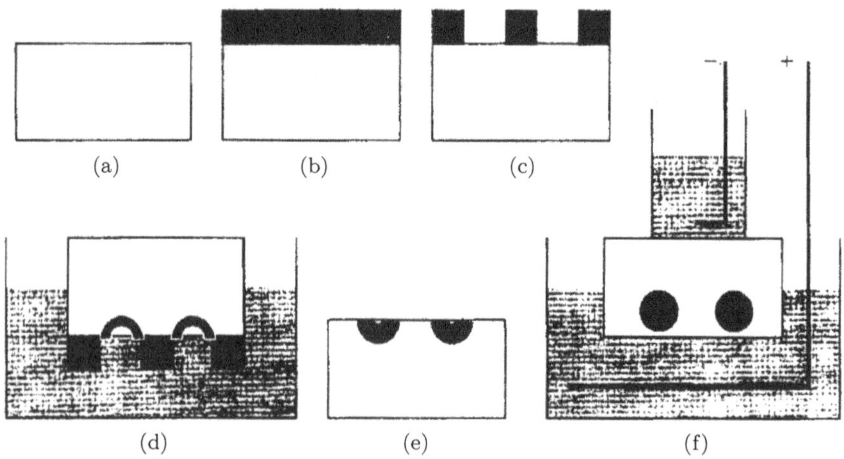

Figure 8.26: Preparation process of a spherical self-focusing plane microlens array: (a) glass substrate; (b) preparation of mask; (c) photolithographic window; (d) ion exchange of first step; (e) remove the mask; (f) ion exchange of second step.

refractive index distribution with the maximum refractive index located at the window and gradually decreasing inward. Then, the existing distribution is diffused inward with the field-assisted ion-exchange method; meanwhile, C^+ ions with low polarizability in the molten salt are diffused into the substrate, so the region with high refractive index moves below the substrate, forming an approximately spherical refractive index distribution.

8.6.4 *Film laser*

Thin film lasers in optical integration and optical communication systems are usually realized by growing GaAlAs, InGaAsP, GaInAsSb, and other films on GaAs, InP, GaSb, and other substrates.

These films can be grown by liquid-phase epitaxy (LPE), vapor-phase epitaxy (VPE), metal organic chemical vapor deposition (MOCVD), molecular beam epitaxy (MBE), etc. Figure 8.27 shows several typical structures of thin film lasers:

Figure 8.27(a) shows a distributed feedback (DFB) type laser, and Figure 8.27(b) shows a distributed Bragg reflection (DRB) laser; both of them use diffraction gratings as reflectors. Figure 8.27(c) shows a double-waveguide (ITG) laser, whose resonator is composed of cleaving planes, emitting light into its adjacent waveguide and taking it out from the outside through it. Whether long wavelength or short wavelength, the lifetime of lasers can reach more than 1×10^6 h.

So far, semiconductor lasers with an emission wavelength of 1.3 and 1.5 microns, a threshold current of about 10 mA, and an output power of about 200 mW have been developed, and their

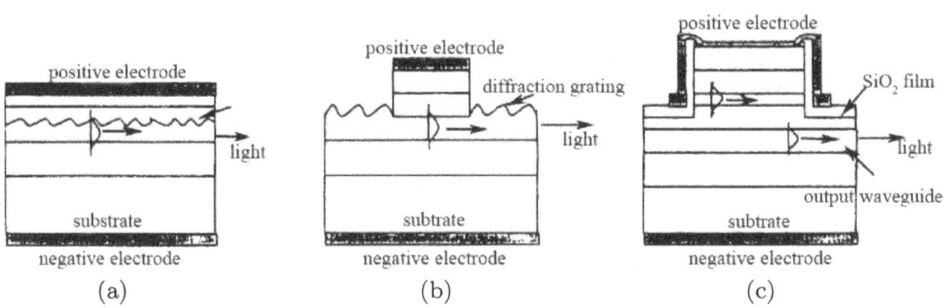

Figure 8.27: Several typical theoretical structures of thin-film lasers.

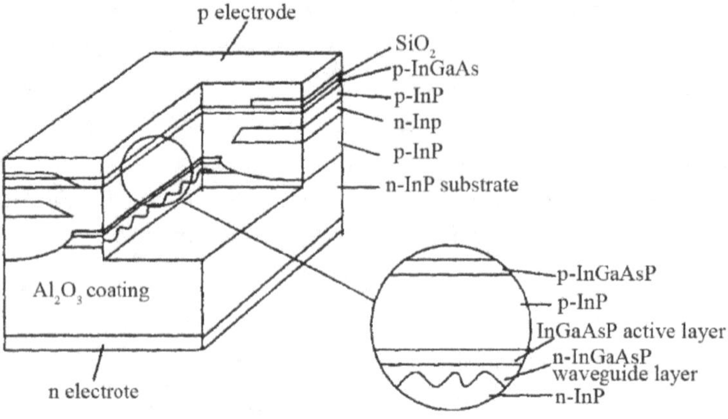

(a) DFB laser; (b) DRB laser; (c) ITG laser;

Figure 8.28: Structure diagram of an InGaAs/InP distributed feedback (DFB) laser: (a) DFB laser; (b) DRB laser; (c) ITG laser.

response bandwidth is 17 GHz. Figure 8.28 shows the structure of an InGaAsP/InP DFB laser.

8.7 Diamond Films

8.7.1 *Overview*

Diamond is a kind of coordination type, natural nonmetallic mineral, composed of carbon. It often contains Si, Al, Ca, Mg, Mn, Ti, Cr, N, and other impurities, and the coordination number of C atoms is 4. The C–C spacing is 0.154 nm; the whole structure can be regarded as a tetrahedron connected with common angular vertices and covalent bonds. Diamond has the following characteristics: (1) the highest hardness in nature; (2) a high modulus elasticity; (3) the highest thermal conductivity in the known materials; (4) the width of the forbidden band is 5.45 eV, so intrinsic diamond is an excellent electrical insulator; (5) good transparence, the transmission spectrum can be extended from the ultraviolet band to the infrared band; (6) good corrosion resistance, it can resist acid, alkali, and a variety of gas erosion; (7) an important semiconductor material. The main physical properties of diamond are shown in Table 8.9.

Table 8.9: Main physical properties of diamond and CVD diamond films.

Physical property	Natural diamond	CVD diamond film
Lattice constant (Å)	3.567	
Density ($g \cdot cm^{-3}$)	3.515	2.8–3.5
Melting point (°C)	4000	Close to 4000
Elasticity modulus (GPa)	$1220^{(1)}$	
Hardness (10^{-1} MPa)	$7000\text{–}10000^{(1)}$	9000–10000
Longitudinal wave velocity ($m \cdot s^{-1}$)	$18200^{(1)}$	
Thermal expansion coefficient ($10^{-6}/°C^{-1}$)	$1.1^{(3)}$	
Thermal conductivity ($W \cdot cm^{-1} \cdot K^{-1}$)	$20^{(1)}$	10–20
Bandgap (eV)	5.45	5.45
Saturated electron velocity ($cm \cdot s^{-1}$)	$2.7^{(1)}$	
Carrier mobility ($cm^2 \cdot Vs^{-1}$)		
Electron	$2200^{(2)}$	
Gap	$1600^{(1)}$	
Breakdown field strength (10^5 $V \cdot cm^{-1}$)	100	
Permittivity	5.5	5.5
Resistivity ($\Omega \cdot cm$)	10^{16}	$>10^{12}$
Optical absorption edge (μm)	0.22	
Refractive index ($5900 \cdot nm$)	0.241	0.241
Optical transmission range	225 nm ≈ far infrared	Close to natural diamond

Note: (1) Rank first in all known substances; (2) rank second in all substances; (3) comparable to Invar alloy.

8.7.2 *Preparation method of diamond films*

After Bundy and his collaborators successfully synthesized diamond under high temperature and pressure with graphite and a catalyst in the 1950s, people began to study the synthesis of diamond films by chemical vapor deposition. From the perspective of thermodynamics, graphite is a thermodynamically stable phase and diamond is a thermodynamically unstable phase in the range of temperature and pressure of CVD diamond. However, due to dynamic factors, the activated groups formed by carbon compounds under the action of plasma or high temperature heat source will form diamond and graphite simultaneously when they come into contact with the substrate. Since atomic hydrogen etches graphite much faster than does diamond, thermally unstable diamond, rather than thermally stable graphite, will eventually be deposited on the substrate in the presence of enough atomic hydrogens. In the early 1980s, high-quality polycrystalline diamond films were prepared by hot filament CVD (HFCVD) method, which caused a great response all over the world. Besides HFCVD, microwave plasma CVD (MWCVD), DC arc plasma jet, flame deposition, and other methods have been developed very quickly. The common features of all these methods are the following: plasma or high temperature heat source is required that can break down carbon-based compounds to form activated carbon-based groups and dissociate hydrogen into atomic hydrogen; the substrate must also be maintained in a temperature range (800–1000°C) suitable for the vapor phase growth of diamond; the higher the temperature of the activation source (plasma or high temperature heat source) or plasma density, the higher the deposition rate of diamond, whereas too high or too low substrate temperature is not conducive to the deposition of diamond films. Atomic oxygen can also etch graphite carbon selectively, so diamond film deposition can be realized in the C–H–O ternary system. The phase diagram of vapor phase growth shows that diamond can only be deposited in a specific component range of C, H, and O components, as shown in Figure 8.29. The purity of chemical vapor deposition diamond films has reached a level that can't be detected by spectroscopy, it exceeds that of high-quality type-II a natural diamond, and the optical transmission is equivalent to that of type-II a natural diamond.

8.7.3 *Properties of diamond films*

8.7.3.1 *Characterization of diamond films*

In general, diamond film deposited by chemical vapor deposition is a kind of dense polycrystalline film composed of diamond grains.

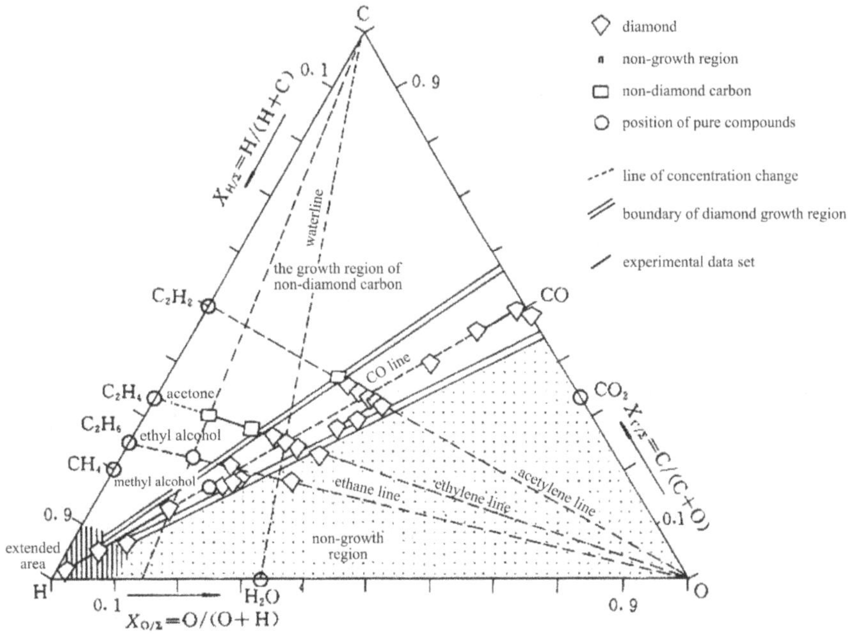

Figure 8.29: Phase diagram of low pressure deposition of diamond films.

In addition to homogeneous epitaxy on the surface of diamond single crystals, there is a large area of heteroepitaxy, so single-crystal diamond films cannot be prepared at present. The diamond film is usually characterized by SEM, Raman scattering spectrum analysis, and X-ray diffraction.

The CVD diamond film can be used in the thin film state (attached to the substrate) or as a self-supporting film (no thick film as substrate). The orientation characteristic is determined by X-ray diffraction, usually with (111) and (100) orientation growth, and in special situations with (100) orientation growth. The intrinsic quality of the diamond film is mainly determined by Raman scattering spectrum analysis. The Raman characteristic peak of diamond is located at roughly $1332\,\text{cm}^{-1}$, while the non-diamond peak usually appears between $1350\,\text{cm}^{-1}$ and $1550\,\text{cm}^{-1}$ and presents a wide diffuse peak, and the specific position is related to the composition of non-diamond carbon in the diamond film. The relative area (or peak height) between the diamond characteristic peak and non-diamond carbon peak indicates the relative content of non-diamond impurities in the diamond film. Since the sensitivity of non-diamond carbon to Raman scattering is 50 times higher than that of diamond, this factor should be considered when evaluating the intrinsic quality of diamond film based on the Raman spectrum.

The exact position of the characteristic peak of diamond is related to the internal stress in the diamond film. Under stress, the characteristic peak of diamond will deviate from the position of 1332 cm^{-1}; the peak will move to the position with higher wave number when under compressive stress and to the position with lower wave number when under in tensile stress. The degree of deviation between the Raman characteristic peak and the standard peak can be used to estimate the magnitude and character of stress in the diamond film. The defects in diamond grains, including grain boundary, microtwins, and dislocation, will increase the full width at half maximum (FWHM) of the diamond characteristic peak.

Therefore, there should be no non-diamond carbon peak in high-quality diamond films, and the FWHM of the diamond Raman characteristic peak should be very small (Figure 8.30). The FWHM of the Raman characteristic peak of the highest quality diamond films (optical diamond films) is the same as that of natural type-II a diamond single crystals.

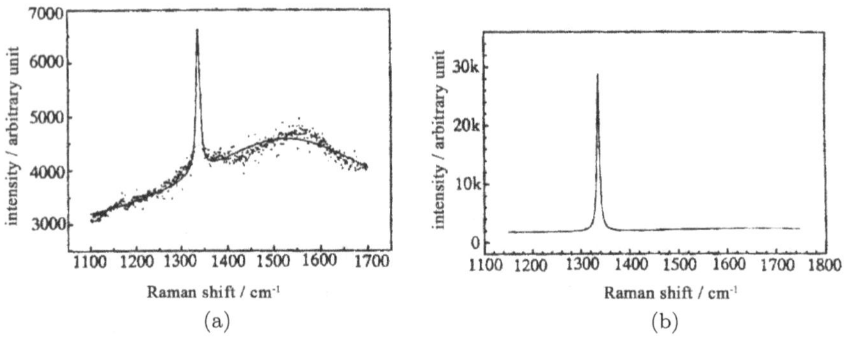

Figure 8.30: Typical Raman Spectra of CVD diamond films: (a) lower-quality diamond film; (b) high-quality diamond film.

8.7.3.2 *Polishing of diamond films*

The surface of CVD diamond thick film is very rough and it must be polished for many applications. The diamond is extremely hard and chemically stable (it doesn't even react in aqua regia), so polishing is extremely difficult. At present, we can only use the traditional mechanical polishing method of polishing diamond by diamond. The high processing cost is one of the limiting factors in industrial application of diamond films.

Many methods of polishing diamond films have been studied in recent years, among which polishing the surface with laser and thermal chemical polishing based on the reaction–diffusion principle are most likely to be practical. The principle of thermochemical polishing is to make use of the high solubility and diffusion coefficient of carbon in transition metals or rare earth metals at high temperature, so that the carbon on the diamond surface in touch with these metals at high temperature can dissolve quickly and diffuse into the metal material, so as to achieve the purpose of polishing the surface of diamond film. To achieve the same effect, it may take several days with mechanical polishing method. Besides, using the laser beam to polish the surface of a diamond film is also an effective method. YAG lasers with a wavelength of 1.06 μm are generally used, and an excimer laser with a shorter wavelength can also be used. Whether using thermochemical polishing or laser polishing, it is a kind of rough polishing, so to achieve the required polishing accuracy the mechanical polishing method is used. In addition to polishing, laser is almost the only means for processing a diamond film (cutting, punching, patterning, etc.).

8.7.3.3 *Metallization of diamond films*

Metallization of diamond films is the process of forming a layer of metal on the surface of diamond films, and this is an important step for many applications. Metallization plays an important role in the applications of diamond films in electronics, tools, and radiators. Although it is not difficult to form ohmic contact or rectifying contact on the surface of natural diamond, it is very difficult to form it on the surface of a CVD diamond film; at present, there is no commercial process or material.

8.8 Exercises and Questions

(1) Try to cite the main types and applications of semiconductor films.
(2) What are the main preparation methods of semiconductor films? Try to give 2–3 examples.
(3) What is superconductivity? What are the main types of superconducting materials and thin films?
(4) What is ferroelectricity? What are the main types of ferroelectric materials?
(5) What are the main types of ferroelectric thin film materials? What are the main applications?
(6) What are the types of magnetic films? Try to give an example.
(7) What is an optical waveguide? Try to give the structure of optical waveguide.
(8) Try to explain the meaning of thin film optical switch and thin film laser.
(9) What are the important characteristics of diamond and diamond films?
(10) How to determine the properties of diamond films by Raman peak?

Chapter 9

The Application of Thin Films

With the rapid development of high-tech industries such as integrated circuits, solid-state luminescence, magnetic recording materials and devices, etc., thin film materials are widely used in various important fields of modern science and technology and the national economy, such as aerospace, medicine, energy, transportation, communication, and information. Thin film materials and their devices are developing towards the goal of becoming comprehensive, intelligent, compound, environmental friendly, energy-saving, and long-lifetime, and nano-sized. This chapter mainly introduces the application of several typical thin film materials.

9.1 The Application of Semiconductor Films

The new semiconductor thin film materials have become an important part of the material science. With the rapid development of the amorphous semiconductor in science and technology, it is applied widely to high-tech fields and is emerging as a new industry. For example, the power made by using the highly efficient and large area amorphous silicon (a-Si:H) thin film solar cells at a power station has been connected to the state grid (it is a green power source without pollution); large-screen liquid crystal displays and flat-panel televisions made of a-Si thin film transistors have been sold as commodities; amorphous silicon copier drums have been used earlier; a-Si sensors and camera tubes, amorphous silicon electroluminescent devices, and high-recording speed and large-capacity optical discs are also developing to become practical commercialized productions.

The polycrystalline silicon film, which combines the advantages of crystalline silicon materials and amorphous silicon alloy films, is a new type of functional film material widely used in the microelectronics technology of energy science and information science. Compound semiconductor thin films have been widely used in new energy science and information science.

9.1.1 *Applications of amorphous silicon semiconductor thin films*

Due to its unique physical properties and processing advantages in manufacturing technology, the a-Si thin film can be used as large-area, high-efficiency solar cell materials, large-screen liquid crystal displays, and planar imaging television, as well as for making a-Si sensors and camera tubes, amorphous electroluminescent devices, etc. For the application of amorphous silicon semiconductor thin film materials, the following characteristics are required:

(1) Forming the film on any substrate;
(2) Easy to achieve films with large area and not limited by shape;
(3) Simple synthesized process and low cost;
(4) Excellent optical and electrical properties, especially high light absorption coefficient.

The research and application of amorphous silicon films is mainly focused on the hydrogenated amorphous silicon (a-Si:H) film, which has better performance than unhydrogenated amorphous silicon. The doping of amorphous semiconductors and the fabrication of p–n junctions were also first achieved in the hydrogenated amorphous silicon film, which has a wide application prospect. The a-Si:H film can be used to manufacture photoresistors, photodiodes, camera targets, image sensors, color discriminators, xerographic drums, etc. Amorphous silicon and silicon-based alloy films also have some potential in the manufacture of light-emitting devices; although they have a very low luminous efficiency currently, they have achieved a lot of achievements in a-Si:H thin film field effect transistors. In addition, the amorphous silicon films can also be used as the material for the surface passivation of a semiconductor device to reduce the surface leakage current of p–n junctions.

Amorphous silicon films have been successfully used in the manufacture of thin film solar cells. The a-Si:H film has excellent photoconductivity, and its absorption coefficient in the visible range is larger than that of single crystal silicon; the film thickness is only about 1 μm, and the formation process of amorphous silicon films is simple; amorphous silicon films can be deposited on the substrate material of glass, stainless steel, or polyacyl to form a large film area. Therefore, amorphous silicon solar cells are more widely used than single crystal silicon solar cells. At present, the energy conversion efficiency of single crystal silicon solar cells has reached 20%, but its cost is very high. Amorphous silicon solar cells have attracted great attention since the advent of the mid-1970s. The first amorphous silicon thin-film cells were reported by Carlson and Wronsky in 1976 and were commercialized in 1981. In 1991, they accounted for 50% of the photovoltaic cell market. However, their development was hampered by the fact that the conversion efficiency of amorphous silicon cells declined with the history of illumination, i.e., the photoinduced attenuation effect. Around 1995, they accounted for only about 10% of the photovoltaic cell market. A variety of new processes have been developed to improve their stability and efficiency, such as the use of a-SiC window layers, the use of graded band gap designs, and the development of multi-junction layering techniques. When combined, the conversion efficiency of the three-junction, small-area amorphous silicon cells reaches 14.6%, and the maximum stable efficiency of a-Si/a-SiGe modules with an area of 1.2 m^2 is 9.5%. Sanyo, a Japanese company, has prepared a-Si/c-Si solar cells with an area of about 101 cm^2 based on CZ–Si, with a conversion efficiency of 20.7%. In China, in 1990, the efficiency of the amorphous silicon single-junction solar cell developed by the Institute of semiconductors of the Chinese Academy of Sciences was 11.2%, and the efficiency of the 20 cm × 20 cm amorphous silicon single-junction module developed by Nankai University reached 9.1%. The annual outdoor test efficiency attenuation was less than 15%. Due to the low-temperature technology (about 200°C), fewer consumables

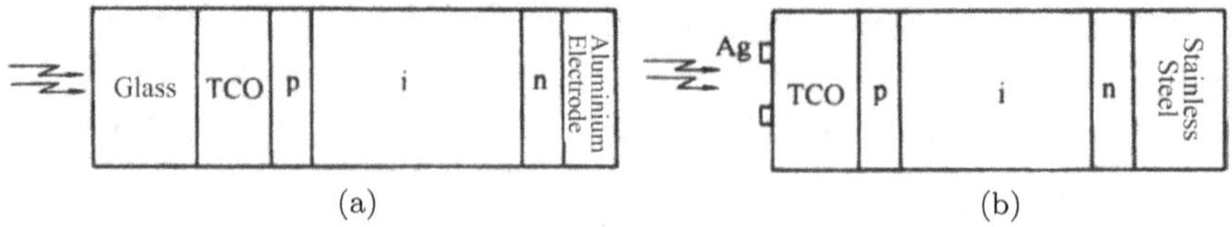

Figure 9.1: Schematic diagram of the a-Si solar cell structure.

(battery thickness is less than 1 nm), materials, and devices are completed at the same time, which is convenient for large-area continuous production.

Amorphous silicon solar cells are thin-film solar cells with glass, stainless steel and special plastics as the substrates. The structure is shown in Figure 9.1.

For a-Si solar cells with glass substrates, light is incident from the glass surface, and the battery current is drawn from the transparent conductive film (TCO) and aluminum electrode. The electrodes of a stainless steel substrate solar cell are similar to those of a c-Si cell. A silver (Ag) gate electrode is prepared on a transparent conductive film. The battery current is drawn from the stainless steel and the gate electrode. According to how solar cells work, light has to pass through the p layer to the i (active) layer to contribute to the photogenic current. Therefore, the p layer should absorb as little light as possible, called the window layer.

Amorphous silicon solar cells work similarly to monocrystalline silicon solar cells, using the photovoltaic effect of semiconductors. Unlike monocrystalline silicon solar cells, the photogenic carriers in amorphous silicon solar cells have drift motion but have no diffusion motion. Because of the long-range disorder in the structure of amorphous silicon, the carrier diffusion length is very short due to the strong scattering caused by the random network. If no electric field exists at or near the source of the photogenic carrier, the photogenic carrier will soon recombine and cannot be collected due to the limited diffusion length. In order that the photogenic carriers are collected efficiently, it is necessary to cover the electric field as far as possible within the range of the light injection in the a-Si solar cell. Therefore, the battery is designed to be pin type (p layer is the incident light surface). Layer is the intrinsic absorption layer, located in the built-in electric field generated by p and n.

The a-Si battery can also be designed to be nip structure, i.e., the n-layer is the incident light surface. Experiments have shown that the characteristics of cells with a structure of pin are better than cells with a structure of nip, so the actual batteries are made of pin structure. Table 9.1 shows the main manufacturers of amorphous silicon thin film solar cells and their production capacity.

Table 9.1: Production capacity of amorphous silicon solar thin film cells.

No.	Company	Capacity (MWp/year)	Remark
1	Japan Kaneka Company	25	910 × 910 mm battery component, glass substrate
2	Germany RWE Schott Solar Company	30	The largest component 100 × 605 mm, glass substrate
3	Japan Sanyo Solar Company	5	Polyimide substrate, efficiency was more than 8.5%
4	Japan TDK Company	5	Stainless steel substrate, efficiency was over 8%
5	Japan Mitsubishi Electric Company	10	Glass substrate amorphous solar cell
6	Japan Fujitsu	10	Polymer substrate flexible amorphous solar cells
7	USA United Solar	25	Stainless steel, polyimide substrate, efficiency was over 10%
8	England Intersolar Company	5	305 × 915 mm glass substrate, efficiency was over 7%
9	Germany Esole Company	40	
10	USA ECD Company	15	
11	China Tianjin Jinneng Company	5	
12	China Harbin Corolla Solar Company	1	

9.1.2 *The application of polycrystalline silicon semiconductor films*

The polycrystalline silicon film is composed of a lot of small crystal grains having different sizes and different crystal face orientations (the grain sizes are generally several tens to several hundreds of nanometers; even large particle sizes have micron dimensions). Many of the performance parameters of high-quality semiconductor polycrystalline silicon films can be replaced by parameters of single crystal silicon (c-Si) and hydrogenated amorphous silicon alloy (a-Si:H) films. The polycrystalline silicon film has a high photosensitivity in the long wavelength band, can effectively absorb visible light, and has good light stability just like crystalline silicon. Therefore, it is considered the most ideal material for photovoltaic devices with high efficiency and low consumption. On the other hand, large-grain polycrystalline silicon films also have a high mobility compared with crystalline silicon, and can be made into a large-area, field-effect thin film transistor, sensor, and other optoelectronic devices with fast response; thereby they have a broad application prospect in the new-generation large-array liquid crystal display technology and microelectronics technology, especially in solar cells.

In 1964, the polycrystalline silicon film was used as the isolation film of integrated circuits. In 1966, the first polycrystalline silicon MOS field effect transistor was created. Currently, the polycrystalline silicon film is widely used in semiconductor devices and integrated circuits.

Polycrystalline silicon films that have low resistance and are heavily doped (resistance about $10^{-3}\,\Omega\cdot\text{cm}$) are used as the gate of the MOS transistor. The silicon gate N-channel technology developed on this basis promotes the rapid development of integrated circuits. After the polycrystalline silicon film replaces the original aluminum film as the gate of the MOS transistor, its biggest advantage is forming the self-aligned gate, namely the automatic arrangement of the source, drain and gate, and alignment of the gate and gate SiO_2 automatically. Heavily doped polycrystalline silicon films can also be used as an internal interconnected lead for integrated circuits, which greatly increases the design flexibility of the integrated circuit and simplifies the process.

The heavily doped polycrystalline silicon films are often used as a plate of a capacitor, a plate of the storage element of a MOS random storage charge, a floating gate of a floating gate device, an electrode of a charge coupled device, etc. The lightly doped polycrystalline silicon films are commonly used to prepare load resistors and other resistors of MOS random access memories in integrated circuits. Polycrystalline silicon films are also suitable for the fabrication of large-area P-N junctions and are therefore suitable for the fabrication of thin film solar cells at a much lower cost than monocrystalline silicon. However, the grain boundaries existing in polycrystalline silicon would affect the energy conversion efficiency of solar cells, which is currently less than 10%.

The resistivity of polycrystalline silicon films is largely different from that of single crystal silicon films. The undoped polycrystalline silicon films have a high resistivity of 10^6–$10^8\,\Omega \cdot \text{cm}$, which is several orders of magnitude higher than that of the undoped single crystal silicon films. The resistivity of the doped polycrystalline silicon films decreases with increasing doping concentration, and the minimum resistivity of phosphorus heavily doped polycrystalline silicon films is $4 \times 10^{-4}\,\Omega \cdot \text{cm}$, which is close to the resistivity value of heavily doped single crystal silicon films. The square resistance (film resistance) of a film of thickness $0.4\,\mu\text{m}$ is $10\,\Omega/\text{cm}$, which can fulfill the interconnected and MOS transistor gate for material requirements. But the minimum resistivity of boron-doped and arsenic-doped polycrystalline silicon films is $2 \times 10^{-3}\,\Omega \cdot \text{cm}$.

Integrated circuit technology is used to manufacture p–n junctions and insulating layers on a single crystal silicon substrate using an extremely high-precision approach, and a large number of circuit components on this basis are composed, such as diodes, transistors, resistors, capacitors, etc. which are connected by wire to same electronic devices that can perform complex functions. Figure 9.2(a) shows a structural cross-sectional view of a representative MOS field effect transistor, and Figure 9.2(b) shows the most basic

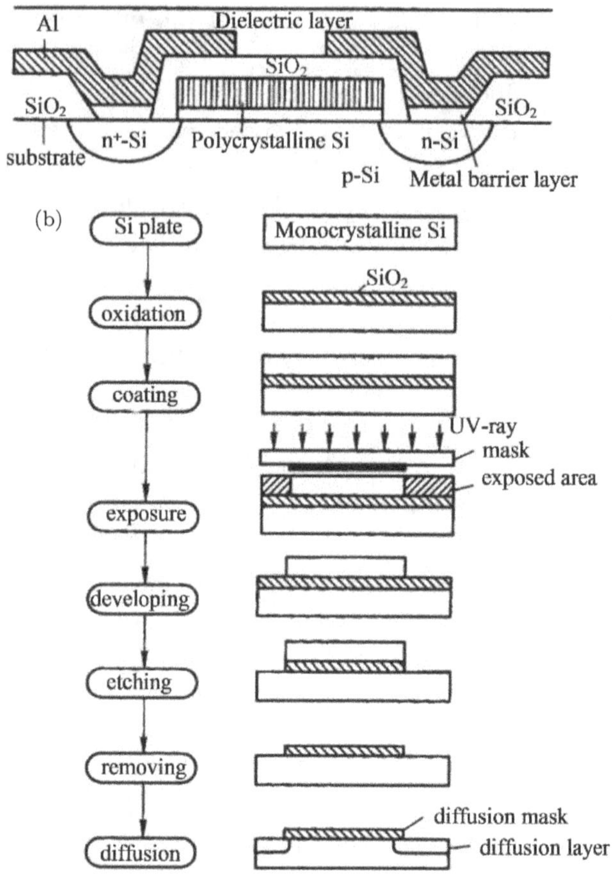

Figure 9.2: Structural section of a MOS field effect transistor (a) and the manufacturing process of an integrated circuit (b).

manufacturing process flow diagram of an integrated circuit, where polycrystalline silicon films are used as the gate of the MOS field effect transistor.

Polysilicon thin films can also be used in a variety of sensors, such as pressure sensors, acceleration sensors, strain gauges, thermoelectric sensors, and actuators. For the first several kinds of mechanical sensors, the theoretical basis is the piezoresistive effect of polysilicon. In order to improve the strain coefficient of a polysilicon film, the carrier mobility, the thickness of the film, and the specular reflection coefficient of the film should be increased. In order to make the semiconductor pressure sensor meet the requirements of high-temperature pressure measurement, the polysilicon pressure sensor uses SiO_2 layer instead of the p–n junction in the traditional diffused silicon pressure sensor to achieve electrical isolation. This kind of pressure sensor has the characteristics of a simple and mature process, strong controllability of the key process, good sensor performance, and large piezoresistive coefficient of polysilicon material, so the pressure sensor made is of high sensitivity.

9.1.3 *The application of compound semiconductor films*

Compound semiconductor thin films are mainly used to prepare thin film solar cells, such as CdTe thin film solar cells, CIS thin film solar cells, and CIGS thin film solar cells.

9.1.3.1 *CdTe thin film solar cells*

In 1963, the first heterojunction CdTe thin film battery was developed, with the structure of n-CdTe/p-$Cu_{2-x}Te$ and an efficiency of 7%. In 1972, Bonnet and Rabenhorst used a CdSTe film with a gradual energy gap as an absorption layer to produce a solar cell with an efficiency of 5–6%. In 1982, Chu *et al.* reported a solar cell with the ITO/CdTe structure with an efficiency of 7.2%. In 1993, an n-CdS/p-CdTe polycrystalline thin film battery prepared by the University of South Florida of the United States achieved a conversion efficiency of 15.8%. In 1997, Matsushita of Japan developed CdTe batteries with a conversion efficiency of 16%. The Chinese scholar Xuanzhi Wu developed a CdTe battery with a conversion efficiency of 16.5%, the highest in the world, using $ZnSnO_4/CdSnO_4$ materials instead of SnO_2 in the US Renewable Energy Laboratory (NREL). The efficiency of its commercial components has reached 10%.

Because using CdTe it is difficult to produce solar cells with high conductivity and shallow homojunction, heterojunction structure is generally adopted. The basic structure of the CdTe solar cell is glass/SnO_2:F/CdS/CdTe (Figure 9.3). Light is emitted from the

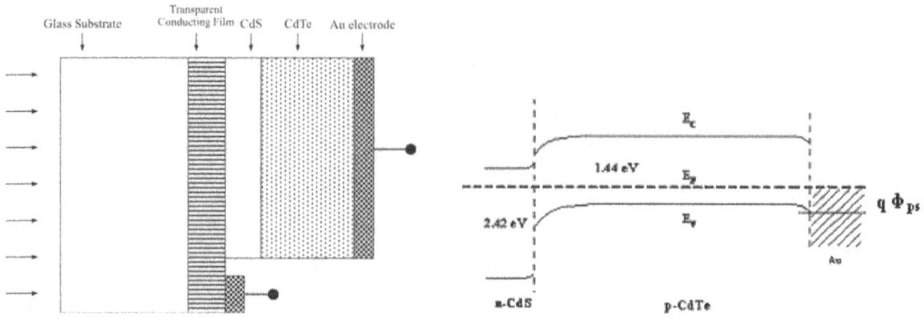

(a) (b)

Figure 9.3: Basic structure of a CdTe solar cell (a) and the energy band diagram (b).

glass surface. Because CdTe has a high work function ($\sim$5.5 eV), it is difficult to form ohmic contact with most metals (Figure 9.3(b)), and it is difficult to achieve heavy doping, and it cannot solve the ohmic contact problem in tunnel transportation. Now, a more successful method is to deposit a layer of heavily doped materials on p-CdTe, such as ZnTe, and HgTe, to realize ohmic contact.

9.1.3.2 *CIS thin film solar cell*

Homojunction solar cells and heterojunction solar cells based on $CuInSe_2$ thin film materials mainly include n-$CuInSe_2$/p-$CuInSe_2$, $(InCd)S_2/CuInSe_2$, CdS/$CuInSe_2$, ITO/$CuInSe_2$, GaAs/CuInSe, ZnO/$CuInSe_2$, etc. Among these photovoltaic devices, the CdS/ $CuInSe_2$ battery is the most valued.

(1) **n-CdS/P-$CuInSe_2$ solar cells**

Batteries composed of low-resistance n-type CdS and high-resistance p-type $CuInSe_2$ generally have a high short-circuit current, a medium open-circuit voltage, and a low fill factor.

In order to further improve the performance of the battery, the series resistance of the battery must be reduced. Therefore, the resistance of the CdS layer should be reduced, or the resistance of the $CuInSe_2$ layer should be greatly reduced to maintain the CdS layer resistance. To dramatically decrease the resistance of $CuInSe_2$ and at the same time ensure the growth of a single phase material is very difficult, and the growth of CdS of low resistance on $CuInSe_2$ is also difficult, especially the growth of a CdS layer of resistivity less than $1\Omega \cdot cm$ is almost impossible. Therefore, the performance of this kind of structure batteries has not seen a big breakthrough so far.

(2) **PIN-type CdS/CuInSe$_2$ solar cell**

In order to obtain a CdS/$CuInSe_2$ battery with good performance, a low resistance $CuInSe_2$ layer needs to be formed. The experiments found that when a low-resistance $CuInSe_2$ material comes into contact with CdS, a large number of copper nodules would be generated at the interface. These copper nodules make the battery much less efficient. The PIN-type CdS/$CuInSe_2$ battery solves this problem. Layer i is composed of high-resistance n-type CdS and high-resistance p-type $CulnSe_2$, avoiding the formation of Cu nodules. The n-layer is formed by low-resistance n-type CdS, which has low volume resistance and low contact resistance with the upper electrode. The p-layer is composed of low-resistance p-type $CuInSe_2$, which has low volume resistance and back contact resistance. Moreover, the p-layer and the high-resistance p-layer form a back field, which is conducive to the improvement of V_{OC}.

9.1.3.3 *CIGS thin film solar cell*

Cu(In, Ga)Se_2(CIGS) are I-II-III compound semiconductor materials; the family has a chalcopyrite crystal structure layer for it to absorb solar cells known as CIGS thin film solar cells. A CIGS thin film battery has the following characteristics:

(1) The photoelectric conversion efficiency is high. In 2008, America's National Renewable Energy Laboratory (NREL) developed the small area CIGS thin film solar battery and the photoelectric conversion efficiency has reached 19.9%, which is the current record for all kinds of thin film solar cells.
(2) The battery stability and performance is good. In the process of using, no failure occurs.
(3) Irradiation resistance is strong; used as a space power that uses the strong competitive properties of weak light.

(4) Low-light performance is good.
(5) The band gap width of the ternary $CuInSe_2$ (CIS) thin film is 1.04 eV. By replacing In with an appropriate amount of Ga, a CIGS quaterary polycrystalline solid solution is formed. The bandgap width can be continuously adjusted within the range of 1.04–1.67 eV.
(6) CIGS is a direct bandgap material with a visible light absorption coefficient of up to the order of $105\,cm^{-1}$, which is very suitable for thin-film solar cells.
(7) When the technology becomes mature, the battery manufacturing cost and energy repayment time will be lower than those of crystalline silicon solar cells. Therefore, upon mass production of CIGS thin film photovoltaic modules, they will certainly occupy an important position in the photovoltaic market.

In order to make full use of the solar spectrum, since the late 1980s, the elements Ga and S have been added to $CuInSe_2$ to increase the bandgap width, make it more compatible with the solar spectrum, and achieve higher photoelectric conversion efficiency. In 1994, the United States NREL invented the three-step co-evaporation method; the prepared CIGS thin film's grain size increased significantly. But improving the quality of the CIGS thin film not only improves the open circuit voltage of the battery, but also improves the collection of photogenerated carriers because the elements of Ga on the longitudinal gradient concentration form a band gradient, and the short circuit current also increased, and the photoelectric conversion efficiency reached 16.4%. Since then, the efficiency record for small-area CIGS solar cells has been held by NREL. In 1999, the conversion efficiency of the CIGS thin film battery was increased to 18.8% ($0.449\,cm^2$). In 2008, the photoelectric conversion efficiency of the battery reached 19.9% ($0.419\,cm^2$), which is the highest to date.

Figure 9.4 shows the typical structure of the CIGS thin film solar cell and its photovoltaic modules. In addition to glass or other flexible substrates, there are seven thin films including Mo back electrode layer, CIGS absorption layer, CdS buffer layer (or other cadmium-free materials), i-ZnO and ZnO:Al window layer, MgF_2

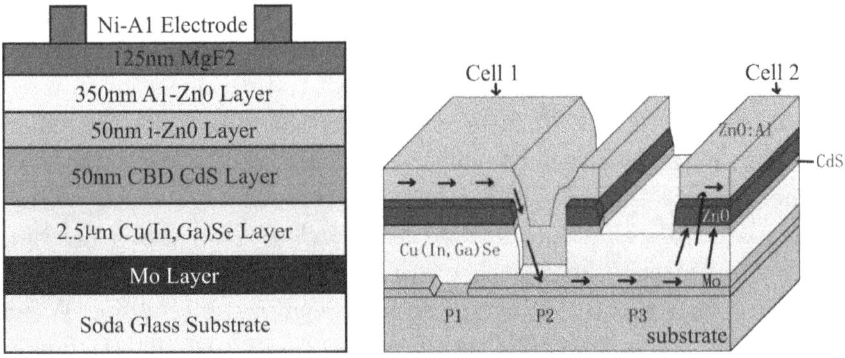

(a) (b)

Figure 9.4: Schematic diagram of CIGS thin film solar cells (a) and the components (b).

minus reflection layer, and Ni–Al top electrode layer. Among them, p-type CIGS and n-type CdS and high-resistance n-type ZnO form a p–n heterojunction, which is the core layer of the CIGS thin film solar cell. The Mo back electrode layer should not only ensure good adhesion with the substrate, but also ensure good ohmic contact with the CIGS layer, and must have high conductivity and appropriate crystal orientation. As the window layer, ZnO:Al must have high light transmittance and conductivity. The MgF_2 minus reflection layer is not required for the CIGS photovoltaic module, and the upper and lower electrodes are only used on the two sides of the module.

The CIGS photovoltaic modules are made up of a series of 5–7-mm wide strip cells. Three cutting lines complete the definition, division, and interconnection of sub-batteries. The three cutting lines and their spacing do not contribute to the photoelectric conversion efficiency, and are called the dead zone. Therefore, the width and spacing of these lines should be as small as possible. The Mo back electrode and high and low resistance zinc oxide were prepared by sputtering, the Ni–Al layer of the upper electrode was prepared by evaporation, and the CdS layer was prepared by chemical water bath (CBD).

The photoelectric conversion efficiency of the CIGS polycrystalline thin film solar cell is higher than that of the CIS single crystal solar cell; the main reason is: the grain size of the CIGS polycrystalline film is large and columnar, and the cross-section of the film is almost monolayer. Just like the copper-poor ordered defect compound (ODC) layer formed on the surface of the CIGS thin film, there is also an ODC layer on the surface of each grain of the polycrystalline CIGS. Between this ODC and the CIGS inside the grain, the valence zone is shifted to form a hole barrier, which prevents the holes in the grain from spreading to the grain boundary, thus greatly reducing the interfacial recombination. When the content of Ga in CIGS accounts for 25–30% of the III family composition, only 1.2% of the lattice mismatch between it and the CdS interface compound is low; grain boundaries have the effect of impurity absorption on grains, making the internal purity of grains higher than that of single crystal materials.

Years of outdoor experiments have shown that CIGS thin-film solar cells and their photovoltaic modules have excellent stability. A study of the outdoor stability of the CIS module by the Solar Energy Research Center (SERI) in Colorado, USA, showed that the performance of the CIS module did not degrade when it was operated in outdoor conditions for 5 months. Siemens Solar conducted eight years of outdoor testing of the CIS modules and found that the average efficiency of the battery remained the same or even increased. Showa Shell of Japan conducted outdoor testing of the 11-kW CIGS battery array and regularly took out the same components from the array and tested them under standard conditions (25 C, AM1.5) for 3 years, which also showed that the efficiency of the CIGS component had almost no attenuation.

Table 9.2 lists the performance parameters and test conditions of current CIGS solar cells and components. Among them, AM 1.5D (direct beam AM 1.5 spectrum) only contains direct radiation sunlight, and AM 1.5G (global AM 1.5 spectrum) contains not only direct radiation sunlight, but also scattered radiation sunlight. Table 9.3 shows the major manufacturers of CIGS photovoltaic modules in the world and their current status.

Although CIGS thin film solar cell research has made great progress, but because of the complexity of its materials and device

Table 9.2: $Cu(In,Ga)Se_2$ thin film solar cell performance.

Constituent	Efficiency (%)	Area (cm^2)	Performance parameter	Test condition	Institution (date)
Cu(In,Ga) Se_2 glass substrate (cell)	19.9 ± 0.5	0.419	$V_{sc} = 0.690$ $J_{sc} = 35.5\,mA/cm^2$ $FF = 0.812$	AM1.5G 1000 W/m^2 25°C	NREL (2007.11)
Cu(In,Ga) $(S,Se)_2$ (cadmium-free component)	15.2	900	$V_{sc} = 34.25$ $I_{sc} = 0.543\,A$ $FF = 0.700$	AM1.5G 1000 W/m^2 25°C	Univ. Uppsala (2000.03)
$Cu(In,Ga)Se_2$ glass substrate (cell)	19.9 ± 0.5	0.419	$V_{sc} = 0.690$ $J_{sc} = 35.5\,mA/cm^2$ $FF = 0.812$	AM1.5G 1000 W/m^2 25°C	Showa Shell (2008.06)
$Cu(In,Ga)Se_2$ glass substrate (photospot)	21.5 ± 1.5	0.102	$V_{sc} = 0.7359$ $J_{sc} = 510\,mA/cm^2$ $FF = 0.8047$	AM1.5D 25°C 14.05 times condensation	NREL (2001.02)

Table 9.3: Major manufacturers of CIGS photovoltaic modules.

Company	Annual production capacity (MW)	Substrate area/m × m	Maximum efficiency/ average efficiency	Marketization
Johanna Solar, Germany	30 (2008)	0.5 × 1.2	—/9.4%	No
Würth Solar, Germany	14.8 (2007)	0.6 × 1.2	<13%/11.7%	Yes
Global Solar, USA	4.2 (2006)	1-foot-wide foil	10%/8%	Yes
Showa Shell, Japan	20 (2007)	0.6 × 1.2	14.2%/11.8%	Yes
Honda Soltec Co. Ltd., Japan	27 (2008)	0.8 × 1.3	13%/10%	No
Sulfur Cell, Germany	5 (2008)	0.65 × 1.25	8.2%/∼7%	Yes
AVANCIS, Germany	20 (2008)	0.6 × 1.2	13.1%/12.2%	No
Solibro GmbH (O. Cells), Germany	25–30 (2009)	0.6 × 1.2	—/—	No

structure, there are still many unsatisfactory reports. Scientists are making unremitted efforts to explore, it such as dry preparation of the CdS transition layer, Cd-free process, less In and less Ga process, non-vacuum low-cost process, etc.

(1) **CdS transition layer prepared by the dry process**

At present, the CBD process is widely used in making the CdS buffer layer, which has low cost and stable and reliable performance. However, it belongs to the wet process, which is inconsistent with the dry process of the whole CIGS solar cell. At present, the main idea of dry film deposition is to use MOCVD, ALCVD (atomic layer chemical vapor deposition), and evaporation to deposit a CdS buffer layer. The efficiency of the CIGS battery in the *in situ* evaporation process buffer layer can exceed 15%. The CIGS thin film battery prepared with the CdZnS buffer layer achieved an efficiency of 19.5%. ZnSe and $ZnIn_2Se_4$ buffer layers can be prepared by the "dry process". The efficiency of preparing CIGS thin film batteries with ZnSe and $ZnIn_2Se_4$ is around 15%.

(2) **Cd-free technology**

Since Cd can bring problems related to environmental pollution, Cd-free process should be actively developed in the preparation of CIGS. Considering that the buffer layer material should be high resistance n type or intrinsic to prevent short circuit of the pn junction, there should be good lattice matching between buffer layer and absorption layer to reduce interface defects and reduce interface recombination. The buffer layer needs a higher band gap to absorb the least light. The process between the buffer layer and the absorption layer should match. Currently, the cadium-free buffer layer can be divided into two categories: Zn-containing sulfide, selenide, or oxide, and In sulfide or selenide. Cadmium-free buffer layers that have been used to produce large-area CIGS thin film batteries are mainly ZnS prepared by chemical water bath and In_2S_3 prepared by chemical vapor deposition in the atomic layer.

(3) **Less In, less Ga process**

Considering the scarcity of raw materials, limited resources of In and Ga will limit the long-term development of CIGS thin film solar cells. In and Ga elements should be used as little as possible in the absorption layer of batteries, so relevant research on alternative absorption layer materials should be carried out. Some compound materials, such as the $Cu_2ZnSn(S,Se)_4$ crystal which belongs to kesterite structure, where zinc and Sn replace half of In in the chalcopyrite structure, respectively, can be used in the absorption of thin-film batteries layers, and other technologies of batteries, such as metal Mo electrode heterojunction buffer layer and the window on the materials and electrode materials and preparation process are the same as the current CIGS thin film solar cells. By using $Cu_2ZnSn(S,Se)_4$ as the absorption layer, the thin-film solar cell has achieved a photoelectric conversion efficiency of 6%. The $Cu(In_{1-x}Al_x)Se_2$ compound semiconductor material can be formed by replacing the element Ga in CIGS with the element Al. By changing the ratio of Al/(Al+In), the bandgap width is adjustable within the range of 1.0–2.6 eV. Compared with $Cu(In_{1-x}Gax)Se_2$, Al replaces the rare metal Ga, both reducing the material cost and widening the band gap.

(4) **Non-vacuum low-cost process**

For CIGS thin-film solar cells, if they are prepared by the non-vacuum method, the input of expensive vacuum equipment can be reduced, which can further reduce the battery cost. The non-vacuum method can be divided into two main methods: coating method and electrodeposition method.

In the coating method, Cu-In-Ga-Se is mixed in a certain proportion to prepare a nanoparticle coating, which is then coated on the substrate and dried to form the absorption layer of the battery. The current coating method can obtain high-quality CIGS thin film solar cells with an efficiency of about 14.5%.

Electrodeposition is a method of depositing CIGS films in an acidic solution with the substrate as the cathode. Bhattacharya of the National Renewable Energy Laboratory (NREL) deposited a CIGS film in a one-step process in a chloride solution system to produce a copper-rich CIGS film, which was heat-treated in a vacuum selenium atmosphere and evaporated a small amount of In and Ga to adjust the CIGS film composition to approximate the chemical measurement ratio, resulting in a cell efficiency of 15.4%.

9.1.4 *The application of silicon carbide films*

With the development of semiconductor materials Si and Ge are generally referred to as first-generation electronic materials; GaAs, InP, IuAs and their alloys are called second-generation electronic materials; meanwhile, high-temperature and wide-band gap semiconductors such as SiC, GaN, AlN, diamond, etc. are also regarded as second-generation semiconductor materials. With the development of science and technology, studying the electronic devices operating under extreme conditions (such as high temperature, high frequency, high power, and strong radiation) is becoming a key research point and difficulty. Conventional semiconductors such as Si, GaAs, etc. have been facing some severe challenges; thus the development of wide-band gap semiconductor materials ($E_g > 2.3\,\text{eV}$) has become an urgent matter. SiC is one of the core components of third-generation semiconductor materials. Compared with Si and Ga, SiC, which has a lot of advantages (such as wide band gap, high thermal conductivity, high electron saturation drift rate, good chemical stability, etc.), is very suitable to manufacture high temperature, high frequency, high radiation resistance, high power, and high density-integrated electronics. Blue, green, and ultraviolet light-emitting devices and photodetectors can be produced using its unique forbidden band width (2.3–3.3 eV). In addition, compared with other compound semiconductor materials such as GaN, AlN, and the like, SiC with its unique property can form a natural oxide layer of SiO_2 on its surface. It is very advantageous for making various MOS-based devices (from integrated circuits to gate-insulated power transistors).

In 1955, Lely synthesized SiC crystals by the sublimation method, which is the foundation for the development of SiC. In 1960, Conner and Smillens revealed the potential properties of SiC as a luminescent material. Although it has long been recognized that SiC has excellent electrical properties, its development has been constrained by the growth process of single crystals. It was not until the early 1980s that Tairov *et al.* used an improved sublimation process to grow SiC crystals. Then SiC as a practical semiconductor has attracted a lot of attention. Some advanced countries and research institutions in the world have invested heavy funds to research SiC. Subsequently, Cree Research Inc. developed a commercial 6H-SiC wafer using the improved Lely method in 1991 and obtained a 4H-SiC wafer in 1994. This breakthrough enabled immediately SiC devices and related technologies to become a research hotspot. The United States and Japan have made some plans, which have enabled SiC to become an important research topic. At present, research on SiC is mainly focused on the bulk single crystal growth of 4H-, 6H- and 3c-SiC, film growth, and device application.

Due to the breakthroughs of single crystal preparation and epitaxial growth technology in the mid-1990s, research on SiC devices has received extensive attention in recent years, which is based on three reasons: first, high-temperature (>300°C), high-power, and low-loss electronic devices can be developed; second, they can make high-brightness and light-emitting tubes to enable humans to obtain high-reproducibility and long-life full-color (including white light) light sources; and third, they can make short-wavelength lasers (using small-size beam spots to realize high-density data light storage) and UV detectors.

Compared with the traditional semiconductors Si and GaAs, SiC with good properties is especially suitable for fabricating electronic devices with high temperature, high frequency, high power, radiation resistant and corrosion resistant characteristics. For example, the MOSFET has many attractive advantages such as fast conversion speed, large peak current, easy to drive, wide safe working area (SWA), excellent avalanche and dv/dt performance, etc. which are partially affected by its conduction characteristics, which are strongly dependent on the rated voltage and temperature. This problem can be solved using SiC MOSFET. In fact, SiC monopole devices have greatly exceeded the theoretical limits of Si in the field of power conversion.

Exhibiting high output power and power density (maximum 4.6 W/mm), SiC MESFETs are the greatest used wide bandgap semiconductors currently in the X-band (8–12 GHz). In addition, Cree Company has reported that a P-I-N diode with breakdown voltage greater than 5.9 kV (close to the theoretical value) has been made using a 50-μm thick 4H-SiC epitaxial layer. This device can be widely used in many important fields such as satellite, rocket, radar, and communication, fighter, aircraft marine exploration, oil drilling, automotive electronics, etc. Using SiC, high-frequency and high-power devices operating in the X-band has been made in recent years and applied to military radar and satellite communications. High-frequency, high-power devices operating in the S-band (1.55–5.2 GHz) have been successfully applied to the transmission field of high-definition television images. With the support of the US Naval Research Section, some research institutes with Purdue University as its core have already researched and developed high-power microelectronic SiC devices and bipolar transistors, monolithic digital integrated circuits, CCD devices, dual-injection MOS devices, CMOS, DMOS integration circuits, etc. Meanwhile, their work is also supported by the Ballistic Missile Defense Office and semiconductor research companies. Their findings are also praised highly by the US Defense Advanced Research Projects Agency.

At present, the main institutions conducting SiC research in the international society include Cree Research Inc., Epitronics, SiCrystal AG, TDI, Acreo, Sterling Semiconductor, etc. Interestingly, research on device design and processing technology has been carried out simultaneously in the case where some of the most important electrical parameters are not yet clear. The research on SiC devices needs to overcome the problems of severe red shift of the output light wavelength and large leakage current when devices are working at high temperature. These problems are caused by the bandgap narrowing of the material due to the various defects existing in the materials. Therefore, how to reduce harmful defects in the growth of SiC crystals is an extremely important research topic for the development of such semiconductor materials, and is also critical to the research on SiC devices.

9.2 The Application of Optical Films

Optical films are widely used in the national economy and national defense. A wide variety of optical thin film devices have become indispensable components of optical systems and optical instruments. With the development of optical film technology, optical films have been developed and applied to many fields from traditional optical instruments to astrophysics, infrared physics, aerospace, laser, electronics, communications, materials, construction, biomedicine, and agriculture.

9.2.1 *Antireflection film*

The antireflection film is used to reduce the reflection loss of light energy on the surface of optical components. When the light through air ($n_0 = 1$) is in vertical incidence to the surface of the optical component with the refractive index n, the reflectance of one surface is

$$R = \left(\frac{1-n}{1+n}\right)^2. \tag{9.1}$$

The reflection losses at the interface of air/glass ($n = 1.52$) and air/Ge ($n = 4.0$) were 4% and 36%, respectively.

The reflected light impairs seriously the characteristics of the optical system. The losses of light energy not only reduce the imaging intensity, but also reduce the contrast of the image and the resolution for the stray light to reach the image plane. Therefore, it is necessary to deposit an appropriate film layer on the surface of the optical component.

The antireflection film can be composed of a single-layer film or a multilayer film system with 20 or more layers. The former can enable the reflectance of a certain wavelength to decline to zero, and the latter has no reflectance in a certain wave band. In a particular application, using a type of antireflective film is related to a variety of factors, such as substrate material, wavelength region, cost, etc. According to the nature of the substrate, antireflection films are classified into a low refractive index substrate and a high refractive index substrate, which roughly correspond to the visible light region and the infrared region, respectively.

9.2.1.1 *Antireflection film of low refractive index substrate*

(1) **Single-layer antireflection film**

The single-layer antireflection film is the simplest antireflection film, as shown in Figure 9.5, where n_0, n_1, and n_s are the refractive indices of the incident medium, the film material, and the substrate, respectively; r_1 and r_2 are the amplitude reflection coefficients of interfaces 1 and 2, respectively; and δ_1 is the phase thicknesses of the film. The single-layer antireflection film refers to depositing a film layer on the substrate only when the refractive index of the film is smaller than that of the substrate, and its reflection can be reduced.

To reduce the reflection effect, the two beams of light reflected should be equal in size and opposite in direction, namely $r_1 = r_2$, $2\delta_1 = \pi$. To obtain $r_1 = r_2$, the following relationship should

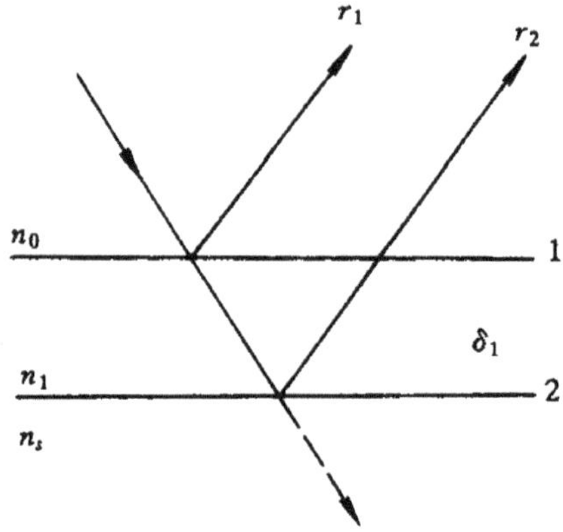

Figure 9.5: Single-layer antireflection film.

be satisfied:

$$n_1 = \sqrt{n_0 n_2}. \tag{9.2}$$

To obtain $2\delta_1 = \pi$, n_1 and d_1 should satisfy the following relationship:

$$n_1 d_1 = \frac{\lambda_0}{4}. \tag{9.3}$$

Obviously, the condition for forming the ideal single-layer antireflection film is that the thickness of the optical film layer is $\lambda_0/4$ and the refractive index is the square root of the product of the refractive index of the incident medium and the refractive index of the substrate.

MIAN-brand glass (K9 glass) with a refractive index of 1.52 is commonly used as the substrate material in the visible light region, where the refractive index would be 1.23 for completely eliminating reflection. At present, the widely used depositing film material with a low refractive index is magnesium fluoride ($n = 1.38$). At this time, it can be calculated that the residual reflectance at the center wavelength is $R = 1.3\%$.

Figure 9.6 shows the reflection curve of a single-layer antireflection film deposited on MIAN-brand glass. In Figure 9.6, n_1, n_2, and n_3 are the refractive indices of three different film materials, respectively. The average reflectance of the magnesium fluoride antireflection film in the entire visible light region is about 1.5 %, and its refractive index increases as the wavelength increases or decreases at the center wavelength. Therefore, the surface of the film presents purplish red due to having a relatively large refractive index in the near-blue and red light regions.

The center wavelength is also called the designed wavelength. Considering the human eye for spectral sensitivity, the center wavelength for the visual instrument is usually 550 nm, and the center wavelength for the photographic instrument is usually 500 nm.

The calculation shows that the reflectivity with increasing incident angle can be neglected in general applications when the angle of the incident light is less than 50°. Although the film producing the effect is not ideal, it has a simple process and is used widely.

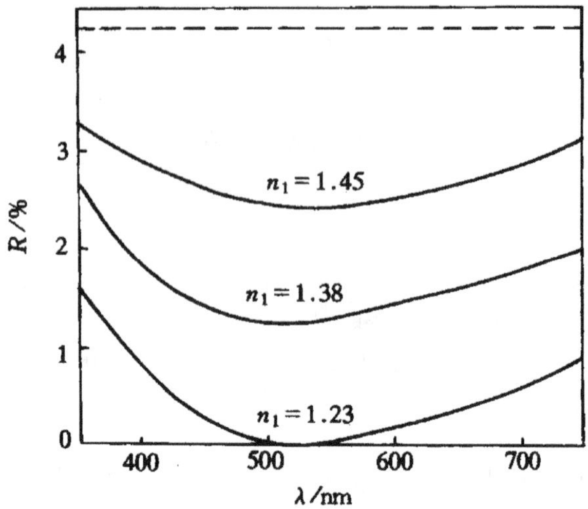

Figure 9.6: Reflection curve of a single-layer magnesium fluoride reflective film.

(2) **Double-layer antireflection films**

Since the residual reflectance of a single-layer antireflection film is generally high, it cannot meet the requirements of a complicated optical system. Therefore, a film having a refractive index r_2 higher than that of the substrate is first deposited on the substrate, and the thickness of the forming film is $\lambda_0/4$, and then the low-refractive index material is deposited on it to form a thick film with thickness $\lambda_0/4$.

To enable the reflected light at the center wavelength of the "$\lambda/4 - \lambda/4$" film system to decline to zero, the refractive index should satisfy the following relationship: $n_2 = n_1\sqrt{n_s/n_o}$. For a substrate with a refractive index of 1.52, silicon oxide (SiO) with a refractive index of 1.70 and a thickness of $\lambda_0/4$ was deposited firstly on the substrate to form a layer film, which is equivalent to the refractive index of the substrate increasing from 1.52 to 1.90 when $Y = n_2^2/n_s = 1.9$. Therefore, the ideal antireflection condition can be just satisfied using the magnesium fluoride film, so that the reflectance of the wavelength λ_0 is close to zero. As the wavelengths deviating from the center wavelength don't meet the interference

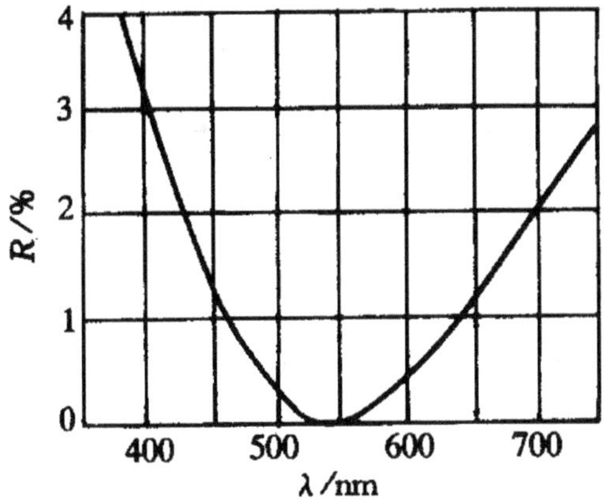

Figure 9.7: The reflection curve of $\lambda/4 - \lambda/2$ antireflection film.

cancellation condition, the reflectance of the structure is large, and its reflection characteristic curve is V-shaped, which is also called the V-shaped antireflection film, as shown in Figure 9.7.

The double-layer antireflection films with the "$\lambda/4 - \lambda/4$" film system are generally applicable to optical systems such as visual optical instruments, lasers, or other sources using monochromatic light as a light source.

Since the materials that can be used for depositing films are limited, it is very difficult to obtain the desired refractive index. Therefore, according to the existing materials, two kinds of high-and low-refractive index materials can be selected, and then the thickness of each layer optical film can be calculated based on the required antireflection film. Thus, it is not limited by the refractive index of the material, but the film thickness is difficultly controlled at the time of depositing the film.

V-shaped films have a better effect of antireflection in a narrow spectral range and vice versa, so they cause a worse effect in a broad spectral range; for application in color photography, laser television, etc., therefore, it is necessary to deposit a broadband antireflection film. The simplest way is to use double-layer antireflection films with the "$\lambda/4 - \lambda/2$" film structure. When a film of thickness with $\lambda_0/2$ is deposited on the substrate, it has no effect on the reflectance

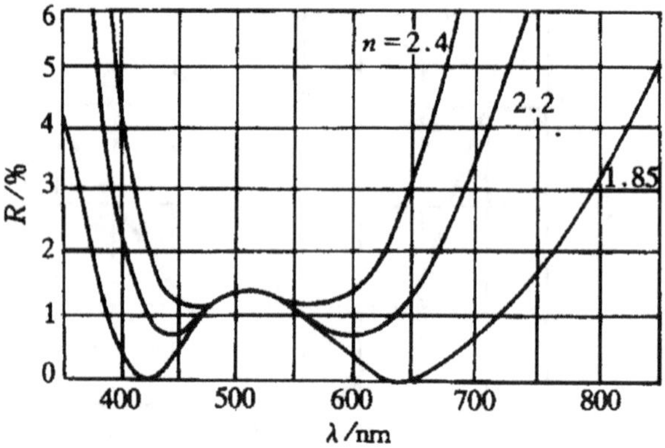

Figure 9.8: The reflection characteristic curve of "$\lambda/4 - \lambda/2$" antireflection films.

at the center wavelength, but the reflectance at other wavelengths would be changed, enabling the antireflection band at both ends of the center wavelength to become widened. The reflection spectrum curve of this film structure looks like W-shape, which is called W type film, as shown in Figure 9.8. The average reflectance of the $\lambda/4 - \lambda/2$ film structure in the visible light region is sometimes less than 1%, which is a simple process, but its bandwidth is still not ideal.

(3) **Multi-layer antireflection films**

Three or more layers structure of antireflection films is required in some applications, because two-layer films still don't meet the low reflection requirements in a wide spectral range. The commonly used three-layer antireflection films are the "$\lambda/4 - \lambda/2 - \lambda/4$" film structure. For the center wavelength, the optical film of thickness $\lambda_0/2$ is regarded as a dummy layer that has no effect on the reflectivity, and is the same as the antireflection action of the "$\lambda/4 - \lambda/4$" V-shaped film. Therefore, it should meet the relationship $n_3 = n_1\sqrt{n_s/n_o}$ that enables the reflectance at the center wavelength to decline to zero. However, the film layer with $\lambda_0/2$ would affect other wavelengths, and selecting the appropriate refractive index value can enable the reflection characteristic curve to become flat.

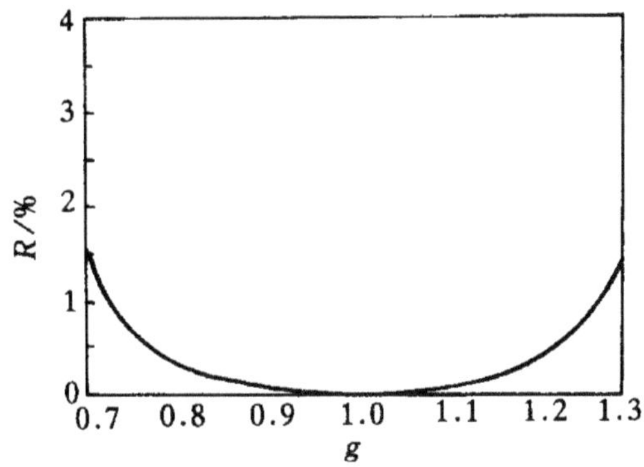

Figure 9.9: The reflection characteristic curve of "$\lambda/4 - \lambda/2 - \lambda/4$" antireflection films.

The antireflection film structure of a typical three layer film with "$\lambda/4 - \lambda/2 - \lambda/4$" is Sub/M2HL/Air, where Sub is the substrate, Air refers to the incident medium being air, and the refractive index of the high-refractive index material is $n_H = 2.0$ (hafnium oxide), the refractive index of the intermediate-refractive index material is $n_M = 1.62$ (cerium fluoride), the refractive index of the low-refractive index material is $n_L = 1.38$ (magnesium fluoride), and the refractive index of the substrate is $n_s = 1.52$ (K9 glass), and the curve of its remainder reflectance with the relative wave number g is shown in Figure 9.9.

After using two $\lambda/4$ film layers to replace the inner film with the intermediate refractive index, the performances of the three layer antireflection films with the "$\lambda/4 - \lambda/2 - \lambda/4$" structure and the width at the low reflection region can be further improved.

The structure of the typical multilayer antireflection films is substrate/NM2HL/air, wherein $n_N = 1.48$, $n_H = 2.25$, $n_M = 1.62$, $n_L = 1.38$, $n_S = 1.52$, its spectral reflection characteristic curve is shown in Figure 9.10, and the abscissa refers to the relative wave number g.

9.2.1.2 *Antireflection film of high refractive index substrate*

In general, most of the practical glass varieties in the visible region are no longer transparent after incident wavelength is greater

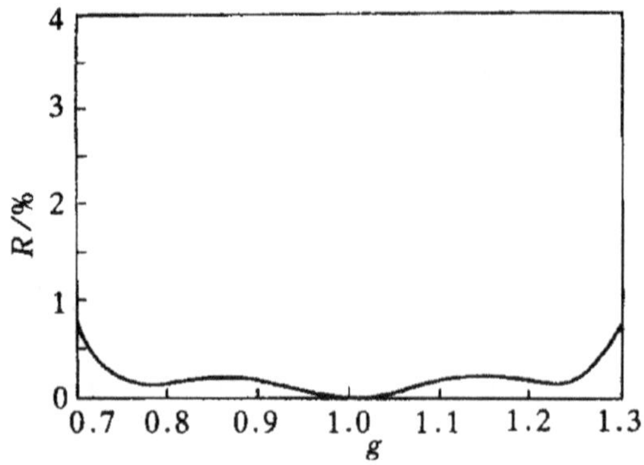

Figure 9.10: The reflection characteristic curve of "$\lambda/4-\lambda/4-\lambda/2-\lambda/4$" antireflection films.

than 3 μm. Therefore, other materials should be used in the infrared region. Certain special glasses and many crystalline materials, particularly semiconductor materials, have the infrared transparent property. However, these infrared substrate materials have a relatively high refractive index, such as arsenic sulfide ($n = 2.2$), silicon ($n = 3.5$) and germanium ($n = 4.0$), which can be widely used only when the antireflection films are deposited on its surface.

The foregoing considerations regarding a single-layer antireflection film are also applicable to the high refractive index substrates. For substrates such as silicon, germanium, gallium arsenide, indium arsenide, etc., their transparent properties can be effectively improved by using a single layer of zinc sulfide, cerium dioxide, or silicon monoxide. The silicon monoxide has an absorption peak larger than 9 μm, which can be used as an antireflection film for the first and second atmospheric infrared windows only when the wavelength is smaller than 8 μm. Zinc sulfide can be used as an antireflection film of three atmospheric windows in the 2–6 μm bands. Figure 9.11 shows the transmission curves of a single-layer antireflection film on the germanium substrate. To widen the antireflection band, it is also possible to deposit two, three, or more layers of anti-reflection film on the substrate.

The infrared broadband antireflection film is generally designed using the decreasing method. The selected thickness of all the film

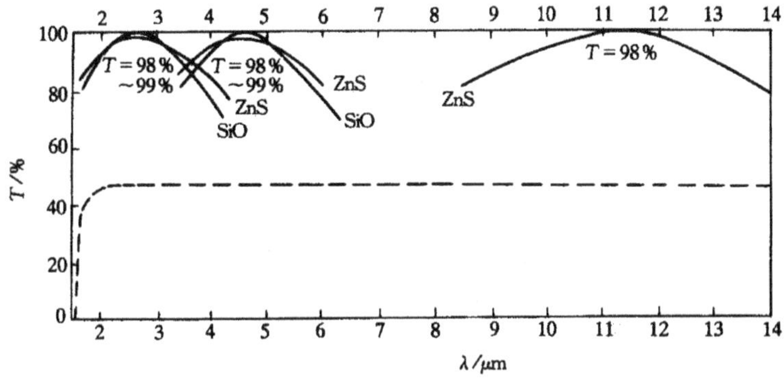

Figure 9.11: Transmission curve of a single-layer antireflection film on the germanium substrate.

layers is $\lambda_0/4$, and the refractive indices of these film layers are arranged in descending order, namely $n_s > n_m > n_{m-1} > \cdots > n_1 > n_0$, where n_s is the refractive index of the substrate, and m refers to the number of film layers.

The refractive indices of the film layers should satisfy the condition of zero reflection, namely

$$\frac{n_o}{n_1} = \frac{n_1}{n_2} = \cdots = \frac{n_{m-1}}{n_m} = \frac{n_m}{n_s}$$

and there are m wavelength positions of these zero reflection points, which are

$$\frac{m+1}{2m}\lambda_0, \quad \frac{m+1}{2(m-1)}\lambda_0, \quad \frac{m+1}{2(m-1)}\lambda_o, \ldots, \quad \frac{m+1}{4}\lambda_0, \quad \frac{m+1}{2}\lambda_o,$$

respectively.

Figure 9.12 shows the reflection characteristics curve of the two-layer antireflection films and the three-layer antireflection films deposited on the germanium substrate ($n = 4.0$) as described above, where the abscissa is the relative wave number g (λ/λ_0). The solid line in Figure 9.12 refers to films of the structure substrate/HL/air, wherein $n_s = 4.0$, $n_H = 2.50$, $n_L = 1.59$; its reflectance at the center wavelength is 5.6%. The dashed line refers to films of the structure substrate/HML/air, where $n_s = 4.0$, $n_H = 2.83$, $n_M = 2.0$, $n_L = 1.41$.

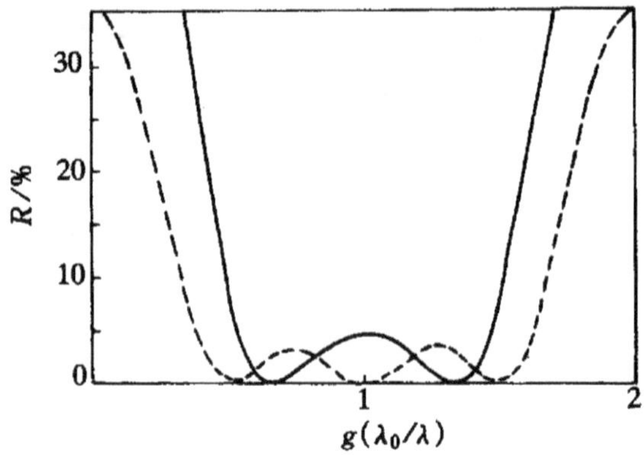

Figure 9.12: The reflection characteristic curve of antireflection films on the germanium substrate

9.2.2 *The reflective film*

Contrary to the effect of the antireflection film, the reflective film is used to reflect most or almost all of the incident light energy. For example, to make reflective mirrors in optical instruments and lasers, we need to deposit reflective films. Reflective films can be classified into metal films, metal-dielectric films, and full dielectric films.

9.2.2.1 *The metal reflective film*

The metal reflective film has a high reflectance and a certain absorbance. Therefore, the metal high-reflection film is used only in the case where there is no special requirement for the absorption loss of the film.

Figure 9.13 shows the reflectance spectra characteristic curves of several metals. In general, the reflectance of the metal film in the infrared regions is relatively high, and its reflectance in the ultraviolet regions is low, and even becomes transparent, which shows its non-metallic properties. For example, the reflectance of silver in both the red light and the infrared light regions is larger than 0.9, and its the reflectivity in the ultraviolet region is very low, and even the reflectance of silver at nearly 316 nm drops to 0.04, which is equivalent to the reflection of glass.

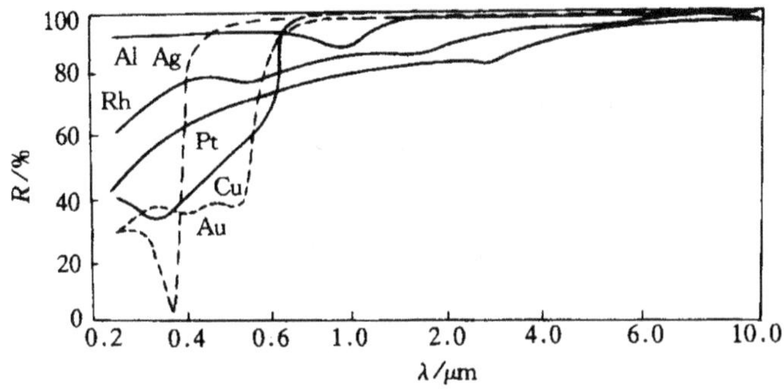

Figure 9.13: The reflectance spectral characteristic curves of several metals.

The most commonly used metal film materials with high reflection are aluminum, silver, gold, etc. It can be seen from Figure 9.13 that the aluminum film has a high reflectivity in the ultraviolet regions, the visible regions, and the infrared regions, and is easy to evaporate and has good adhesion on the glass substrate. The aluminum film can form a layer of aluminum oxide film in the air to protect the film layer. However, the thickness of the natural aluminum oxide film is relatively thin, being only about 5 nm, so a layer of protective film must be deposited on its surface. It is worth noting that the thickness of the protective films would be one half of the operating wavelengths. The commonly used protective materials are silicon oxide, silicon dioxide, and aluminum oxide. SiO is not used as a protective film of aluminum mirrors in the ultraviolet region because of it having a large absorption in the regions. Protective film materials used in the ultraviolet regions are silicon dioxide, magnesium fluoride, and lithium fluoride. The best conditions for preparing the aluminum film are high-purity aluminum (99.99%), rapid evaporation rate in high vacuum (50–100 nm/s), and substrate temperature below 50°C; the aluminum reflective film used in the UV region would be deposited at a degree of vacuum higher than 1.33×10^{-3}Pa.

The silver film has a high reflectance in both the visible and the infrared regions, which would make a small polarization when the light is obliquely incident on its surface. However, the silver film on a glass substrate has a poor adhesion, mechanical strength, and chemical stability, and is easy to react with sulfides in the air. Silicon monoxide and magnesium fluoride cannot be used as a protective film owing to their having a poor adhesion with the silver film. Therefore, the silver film can only be deposited on the back surface of glass as an internal reflection film, and then it is protected using a protective film. The silver film can also be deposited on the front surface of the glass, where the film structure would consist of glass substrate/$A1_2O_3$(30 nm)–Ag–$A1_2O_3$(30 nm)–SiO_x (100–200 nm)/air. Aluminum oxide has a good adhesion with the aluminum film, which can be used as a transition layer to improve adhesion between silver and the substrate, and silver and the protective film. The silicon oxide film layer has a good function of resistance to moisture and erosion. It has been proved that even if this film structure is exposed to an environment of the most severe sulfide and moisture, it still has a high reflectivity; the reflectivity of this film structure can be kept above 95% when the wavelength of the vertical incident light is 450 nm. As Figure 9.14 shows, the absorption band of 3 μm is caused by the absorption of water in SiO_x film, and the absorption band at 9.6 μm is produced by SiO_x intrinsic absorption.

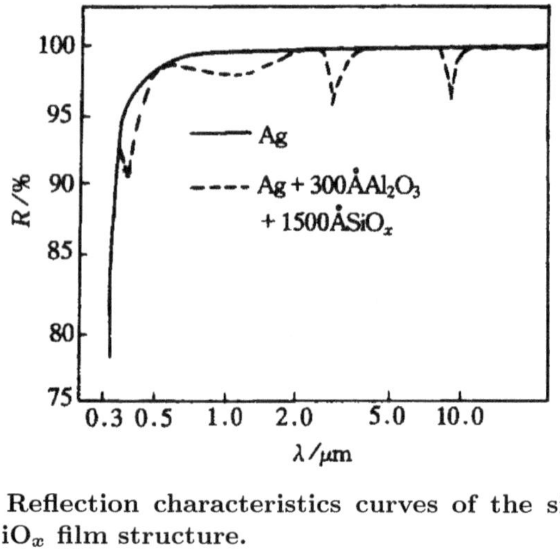

Figure 9.14: Reflection characteristics curves of the silver film and $Ag+Al_2O_3+SiO_x$ film structure.

In addition, the following various membrane structures can be used:

- Glass substrate/Al_2O_3(900 nm)–Ag–Al_2O_3($\lambda_0/4$)–TiO_2($\lambda_0/4$);
- Glass substrate/Al_2O_3(900 nm)–Ag–[Al_2O_3(30 nm)–MgF_2]($\lambda_0/4$) –Al_2O_3 ($\lambda_0/4$)/air;
- Glass substrate/Al_2O_3(900 nm)–Ag–[Al_2O_3(30 nm)–SiO_2]($\lambda_0/4$)–Al_2O_3 ($\lambda_0/4$)/air.

The optimum conditions for preparing the silver film resemble those for preparing the aluminum film, which mainly include high vacuum degree, rapid evaporation rate, and lower substrate temperature. The gold film has a high reflectance in the infrared regions, so it is often used as an infrared mirror. Between the gold film and the glass substrate there is poor adhesion, but there is good adhesion with the chromium film or the nickel-chromium film. Therefore, a chromium film or a nickel-chromium film is commonly used between the gold film and glass as a liner. The gold film is quite stable in air, but it is very soft and sometimes needs depositing of a protective film, such as Bi_2O_3 protective film or a pair of TiO_2/SiO_2 protective films.

Hypothesize that the plural refractive index of the metal film is $n - ik, i = \sqrt{-1}$ being an imaginary unit. When light is incident vertically in air, the reflectance of the metal film is

$$R = \frac{(1-n)^2 + k^2}{(1+n)^2 + k^2}.$$

The reflectivity of the metal film is limited by the optical constant of the metal material and cannot make a high value. In order to increase the reflectance of the metal film, a pair of or a plurality of dielectric films with alternating high and low refractive indices may be deposited on the metal film to form the metal and dielectric film structures, which are also referred to as metal enhanced film structures. It not only improves the reflectivity of the metal film, but also prevents the metal film from the affected atmosphere.

If dielectric films of two layers having optical thickness $\lambda_0/4$ are deposited on the metal and their the refractive indices are n_1 and n_2, respectively, where a dielectric film with n_2 is in close contact with the metal film, its reflectance at vertical incidence light wavelength λ_0 is

$$R = \frac{[1 - (n_1/n_2)^2 n]^2 + (n_1/n_2)^4 k^2}{[1 + (n_1/n_2)^2 n]^2 + (n_1/n_2)^4 k^2}.$$

When $(n_1/n_2)^2 > 1$, the reflectivity of the metal-deposited dielectric film is greater than that of the pure metal film, and the reflectance would increase with increasing the ratio n_1/n_2. For example, the reflectance of metallic aluminum is about 91.6% at a wavelength of 550 nm, wherein $n - ik = 0.85 - i5.99$. When pairs of MgF_2/CeO_2 film layers were deposited on the aluminum film, the reflectance of the structure was increased to about 97%. Figure 9.15 shows the reflection characteristics of the aluminum film deposited with two pairs of MgF_2/ZnS films and that without deposition. It can be seen from Figure 9.15 that the reflectance of the structure can be increased to nearly 99% by depositing the dielectric film, and doing so also improves the reflective characteristics of the entire visible region. However, increasing the region of reflectance is still limited, and the reflectance is decreased outside these regions. Typically, these additional dielectric films don't usually exceed three pairs.

9.2.2.2 *All-media high-reflection films*

In some applications, such as multi-beam interferometers, reflective films of high-quality lasers, etc., due to the large absorption loss of

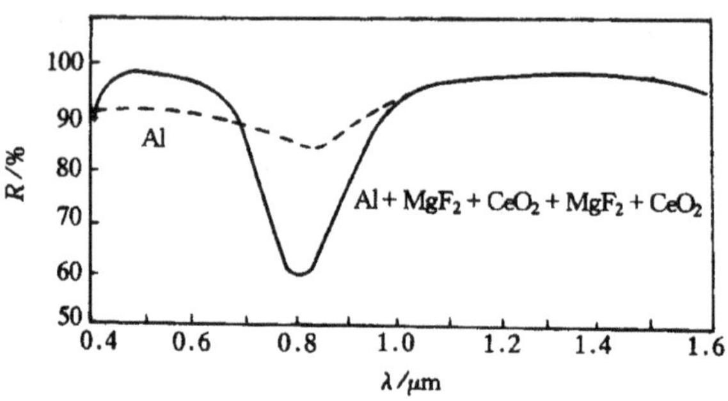

Figure 9.15: Reflection characteristics of aluminum films with and without dielectric films deposited.

the above metal high-reflection film, cannot meet the requirements, so low-absorption and high-reflectivity all-media high-reflection films should be used. The structure is to alternately plate a film having an optical thickness of $\lambda/3$ and a low refractive index on the substrate. Multilayer dielectric films can be expressed as substrate/ HLHL...HLH/air or simplified to substrate/(HL)^{s}H/air, where s is the number of basic cycles, and (HL) is the number of repetitions, or membrane logarithm.

The reflectance at the center wavelength when light is incident vertically in air is

$$R = \left[\frac{1-(n_H/n_L)^{2s}(n_H^2/n_s)}{1+(n_H/n_L)^{2s}(n_H^2/n_s)}\right]^2,$$

since

$$\left(\frac{nH}{nL}\right)^{2s}\frac{n_N^2}{n_S} \gg 1,$$

then

$$R \approx 1-4\left(\frac{nL}{nH}\right)^{2s}\frac{n_s}{n_H^2}, \tag{9.4}$$

$$T \approx 1-R \approx 4\left(\frac{nL}{nH}\right)^{2s}\frac{n_s}{n_H^2}. \tag{9.5}$$

Equations (9.4) and (9.5) show that when the reflectance of the film system is high, the additional plating of two layers can reduce the transmittance of the film system by $(n_{\mathrm{L}}/n_{\mathrm{H}})^2$ times. The reflectivity of the dielectric film can be close to 100% when the loss is neglected. In fact, due to the absorption and scattering loss in the film layer, when the film system reaches a certain number of layers, continuing to add two layers can't increase the reflectivity. On the contrary, due to the increase of absorption and scattering loss, the reflectivity will decrease. Therefore, the absorption and scattering losses in the film system limit the maximum number of layers in the dielectric film system.

Figure 9.16 shows the reflection characteristics of a highly reflective film with different numbers of layers, and the abscissa is the

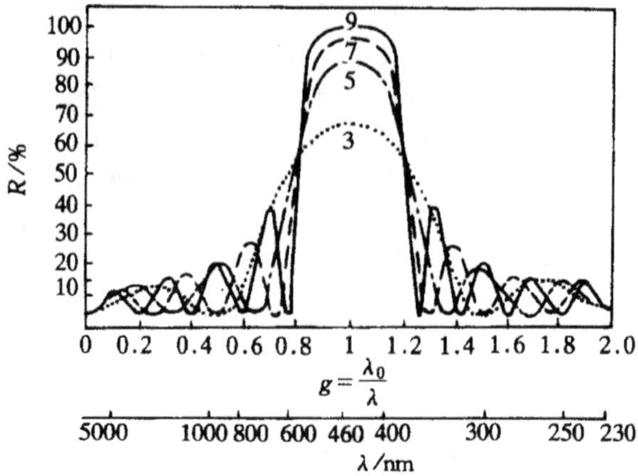

Figure 9.16: Reflection characteristic curve of the high reflection film.

relative wave number $g = \lambda_0/\lambda$. As can be seen from Figure 9.16, at $g = 1$, that is, $\lambda = \lambda_0$, the reflectance is the maximum. In the main peak of reflection, the reflectance slowly decreases as the wavelength deviates from the center wavelength. Once the wavelength exceeds the main peak, the reflectance drops abruptly. This shows that the width of the high reflection band is limited, and as the number of layers increases, only the reflectance in the reflection band and the number of oscillations outside the high reflection band can be increased.

The relative wavenumber width of the high reflection band is calculated by Equation (9.6). The corresponding wavelength bandwidth is $2\Delta\lambda = 2\Delta g\lambda_0/m^2$.

$$2\Delta g = \frac{4}{\pi}\sin^{-1}\left(\frac{n_H - n_L}{n_H + n_L}\right). \tag{9.6}$$

Equation (9.6) shows that the width of the reflection band depends on the ratio of the high and low refractive indices. The larger the ratio, the wider the high reflection band.

High-reflection films with an optical thickness of 20/4, with high-order sub-high reflection bands, appear at wavelengths of $\lambda_0/3$, $\lambda_0/5, \lambda_0/7$, the relative wavenumber width of each reflection band is the same, the corresponding wavelength width is approximately

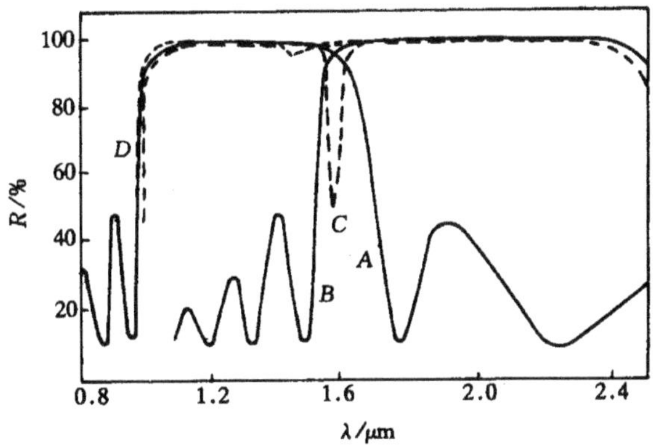

Figure 9.17: Reflection characteristic curve of the $\lambda/4$ multilayer film with the two super-reflecting bands slightly overlapping.

reduced by a ratio of 1/9, 1/25, and 1/49. Therefore, if a narrow-band high-reflection film of a certain wavelength has to be prepared, a higher-order reflection band can be utilized.

The high-reflection band of the all-media high-reflection film is limited and sometimes cannot meet the requirements of some practical applications. In order to broaden the high-reflection band, the easiest way is to superimpose two (curves A and B shown in Figure 9.17) or more highly reflective film stacks of different central wavelengths, and the film system is substrate/$[(H_1L_1)^{sl}H_1]\lambda_1[(H_2L_2)^{s2}H_2]$/air. However, after the superposition, interference with each other occurs, and thus a transmission peak appears at the center of the widened high-reflection band, as shown by curve C in Figure 9.17. If a low-refractive index $\lambda'/4$ layer is interposed between two high-reflection film stacks, the film system is substrate/$[(H_1L_1)^{sl}H_1]\lambda_1L\lambda'$ $[(H_2L_2)^{s2}H_2]\lambda_2$/air, and the transmission peak can be eliminated. See curve D in Figure 9.17.

9.2.3 *Beam splitting film*

The function of the beam splitting film is to divide the energy of the incident light into transmitted light and reflected light. The ratio of the transmittance of the film to the reflectance is called the split ratio of a split film. According to the requirements, the splitting ratio

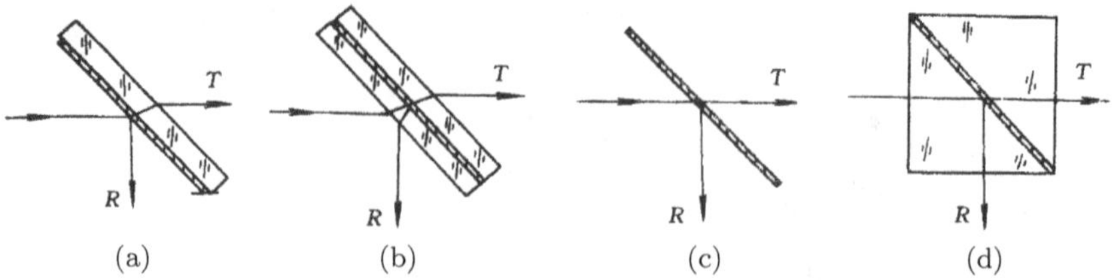

Figure 9.18: Beam splitter types: (a) single board type; (b) glued flat type; (c) film type; (d) glued prism type.

has different values, and the most commonly used spectroscope has a split ratio of 1:1. The product of the transmittance and reflectance of a split film is called the efficiency of beam splitting of the split film. A non-absorbent medium beam splitting film having a split ratio of 50:50 has the highest efficiency of beam splitting.

There are four most common structures of beam splitters, as shown in Figure 9.18.

9.2.3.1 *Metal beam splitter*

A thin metal film is deposited on a glass substrate to form a metal beam splitter. Most metal films have a substantially constant transmittance and reflectance in the visible range, and the plating process is relatively simple.

The material of the metal film is nickel–chromium alloy, chromium, silver, aluminum, or the like. Nickel–chromium–iron alloy is a kind of nickel–chromium alloy which has good neutrality in a wide wavelength range and has good mechanical strength and chemical stability (the neutrality of the nickel–chromium–iron film is better than that of the nickel–chromium film). The chrome film also has properties similar to the nickel–chromium alloy film. Although its neutrality is not as good as that of the nickel–chromium alloy film, it is also ideal. The splitting curve of the chrome film is relatively flat; in the visible region, the reflectance of the long wave end is generally about 10% higher than that of the short wave end. The absorption of the chromium film is large and that of the silver film is minimal, but its mechanical strength and chemical stability are poor, so it is generally used in cemented cubic beam splitters.

When using a metal spectroscopic film, pay attention to the position of the film surface facing the incident beam. The correct usage is shown in Figure 9.19.

The metal beam splitting film has a high absorptivity and a large loss of light energy. To reduce absorption, a layer of high refractive index film of $\lambda/4$ can be inserted between the metal film and the glass plate. For example, insert a layer of $\lambda/4$ zinc sulfide film between the chromium film and the glass substrate so that the absorption of the chromium film with a split ratio of 1:1 is reduced to 0.13 from the original 0.38.

Table 9.4 lists the performance comparison of the films of Cr and Cr $+\lambda/4$(ZnS).

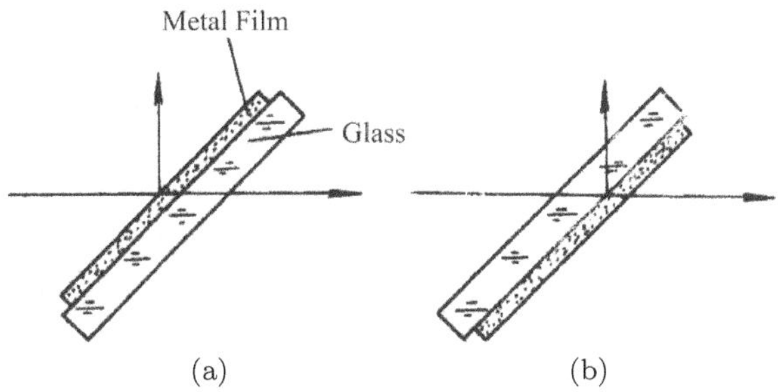

Figure 9.19: Instructions for use of metal beam splitters: (a) correct; (b) Incorrect.

Table 9.4: The performance comparison of the films of Cr and Cr + $\lambda/4$(ZnS).

R/T	Film system	D/nm	R	T	A	A/T
1:2	Cr	7.2	0.220	0.440	0.340	0.77
	Cr+$\lambda/4$(ZnS)	2.5	0.330	0.660	0.020	0.03
1:1	Cr	12.4	0.310	0.310	0.380	1.22
	Cr+$\lambda/4$(ZnS)	7.4	0.435	0.435	0.130	0.25
2:1	Cr	18.3	0.405	0.202	0.392	1.94
	Cr+$\lambda/4$(ZnS)	16.2	0.530	0.265	0.205	0.77

Note: D is the film thickness; R is the reflectance; T is the transmittance; A is the absorption rate.

9.2.3.2 *Medium beam splitting film*

Since the absorption loss of the metal spectroscopy film is large, a medium beam splitting film is used in the case where the light energy is required to be high. The absorption of the medium beam splitting film is small and negligible. Therefore, its efficiency of beam splitting is high, but its neutral spectral wavelength range is not as wide as that of the metal film, and its polarization effect on light is larger than that of the metal film.

(1) **Single layer medium beam splitting film**

The refractive index of a single-layer $\lambda/4$ high-refractive index dielectric film varies slowly with wavelength in a relatively wide wavelength range around the center wavelength. The refractive index at the center wavelength is shown in Equation (9.7).

$$R = \left(\frac{\eta o - \eta_1^2/\eta_s}{\eta o + \eta_1^2/\eta_s}\right)^2. \tag{9.7}$$

In the above equation:

$$\eta_0 = n_0/\cos\theta_0,\ \eta_1 = n_1/\cos\theta_1,\ \eta_s = n_s/\cos\theta_s\ (p\text{ polarization}),$$
$$\eta_0 = n_0/\cos\theta_0,\ \eta_1 = n_1/\cos\theta_1,\ \eta_s = n_s/\cos\theta_s\ (s\text{ polarization}).$$

where θ_0 is the angle of incidence, and θ_1, θ_s are the angles of refraction in the film layer and the substrate, respectively.

For example, zinc sulfide with a refractive index of 2.35 is a commonly used material for a beam splitting film. If $n_o = 1$, $n_s = 1.52$, when incident at 45°, $R_s = 0.460$, $R_p = 0.185$; for extreme reflection of natural light

$$R = \frac{R_S + R_P}{2} = 0.323.$$

(2) **Multilayer medium beam splitting film**

The maximum efficiency of beam splitting of the beam splitter is 50:50. If the natural light reaches 50:50, the single layer film is not enough, and a multilayer film is required.

The design of the medium multilayer film can be divided into two steps: the first step is to find a $\lambda_0/4$ film system so that its refractive index at the center wavelength is the required value in the case of normal incidence. For a broad spectrum neutral beam splitter, the second step is achromatic, that is, inserting a half-wave layer in the film system. Inserting a high-refractive index half-wave layer increases the reflectance on both sides of the center wavelength more than inserting a half-wave layer with low refractive index.

The 50:50 flat-plate spectroscopic film designed by the above method is glass substrate/HLHL/air. $n_H = 2.35$ (zinc sulfide), $n_L = 1.38$ (Magnesium fluoride). The reflectance at the center wavelength is 50.5%. The reflectivity at both ends will decrease. This beam splitting film system is suitable for monochromatic light systems. In order to improve the neutrality of the beam splitting film in the visible region, a glass substrate/2LHLHL/air film system is used. Figure 9.20 shows the splitting reflection curve of the flat beam splitting film.

The film system of the prism beam splitter are: glass substrate/HLHL2H/glass, glass substrate/LHLHL2H/glass, glass substrate/2LHLHL2H/air. Its spectral reflection curve is shown in Figure 9.21.

If further improved, the modified beam splitter film system is: glass substrate//LH$\frac{2}{3}$L$\frac{4}{3}$HL2H/air; its splitting reflection curve is shown in Figure 9.22.

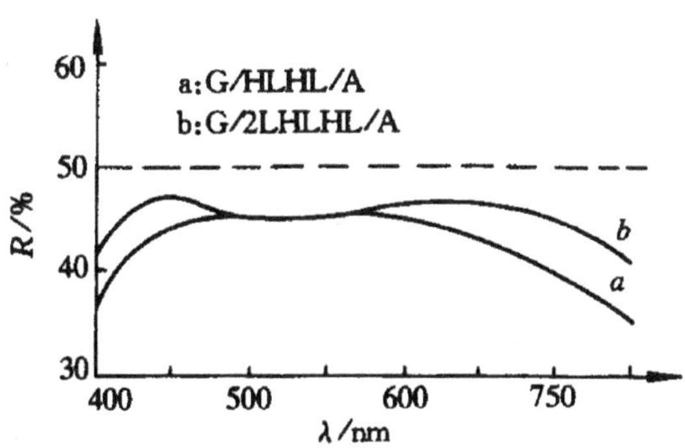

Figure 9.20: Reflection characteristic curve of the plate beam splitting film, in which G is glass, and A is air.

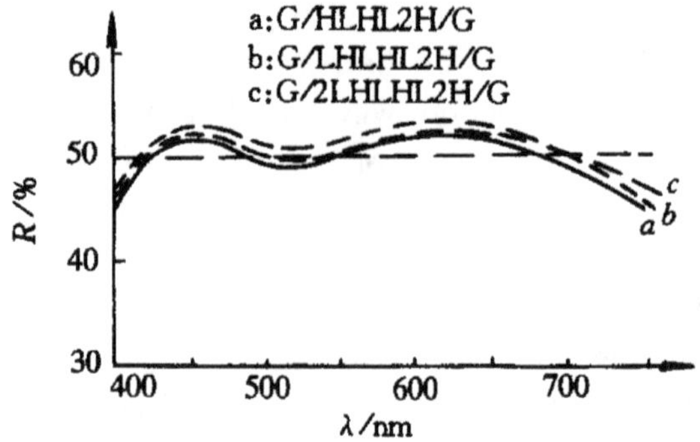

Figure 9.21: Reflective characteristic curve of the prism beam splitter, where *G* is glass.

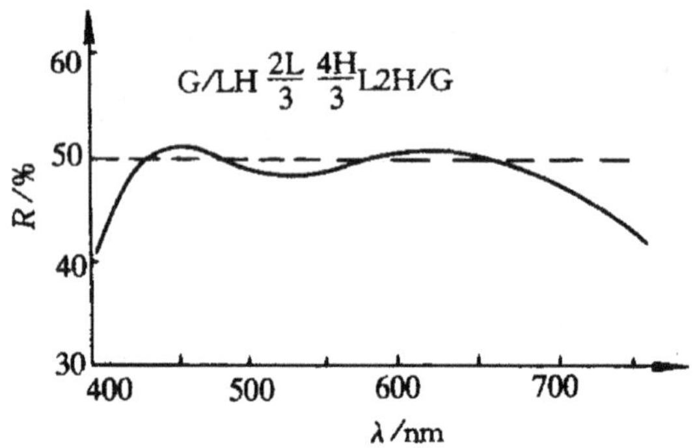

Figure 9.22: Reflecting characteristic curve of the prism beam splitter after modification.

When obliquely incident, the spectral characteristics of the film system are subject to polarization effects, and the s and p components are separated. For the medium film beam splitter, they modify the designed prism beam splitter reflective characteristic curve (Figure 9.19), the reflectance of the p-polarized component is always lower than the reflectance of the s-polarized component. In the cubic beam splitting prism, this polarization effect is more significant, so it cannot be used in the case where the polarization effect is restricted, and a metal beam splitter with a small polarization effect must be used instead.

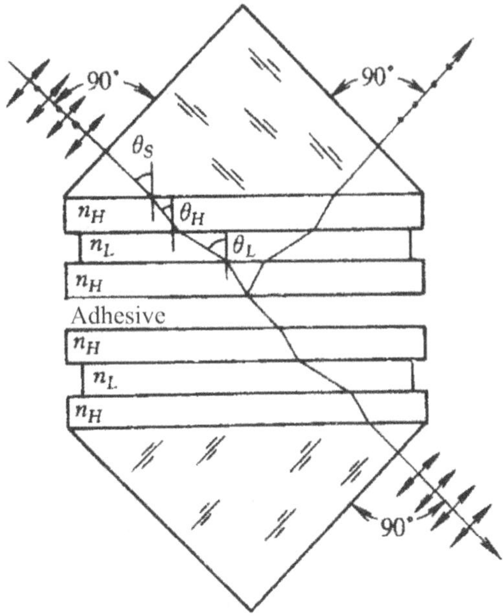

Figure 9.23: The structure of a prism polarizing beam splitter.

9.2.3.3 *Polarizing beam splitter*

(1) **Prism polarizing beam splitter**

The prism polarizing beam splitter totally reflects s-polarized light and transmits p-polarized light to achieve a neutral splitting with an inverse transmission ratio of 50:50. Figure 9.23 shows the structure of a prism polarizing beam splitter.

In order to make the p-polarized light transmit completely, the incident angle must satisfy the Brewster angle condition at each interface of the polarizing beam splitting film. This prism polarizing beam splitter must meet

$$n_s \sin\theta_s = \frac{n_H n_L}{\sqrt{n_H^2 + n_L^2}}. \tag{9.8}$$

For a given high- and low-refractive index material, there are two ways to satisfy Equation (9.8)

(1) Select the refractive index of the prism to determine the angle of the prism.
(2) Select the angle of the prism (usually 45°), and then calculate the refractive index of the prism.

In order to totally reflect s-polarized light, a film system uses a medium high-reflection film, that is, a film layer having an effective optical thickness $n_H d_H \cos\theta_H$, $n_L d_L \cos\theta_L$ of $\lambda_0/4$ with high and low refractive indices alternately. As long as the number of layers in the film layer is sufficient, the reflectance of s-polarized light can approach 100%. Figure 9.24 shows the characteristics of a cubic prism polarizing beam splitter made of zinc sulfide and crystal stone. The film system was substrate/$(HL)^5H$/substrate, where $n_s = 1.65$, $n_H = 2.30$, $n_L = 1.35$, $\theta_H = 45°$, $\lambda = 500\,\text{nm}$.

As can be seen from Figure 9.24, the reflection band of s-polarized light covers almost the entire visible region. Therefore, if the incident light is natural light, the polarizing beam splitter can also be used as a neutral beam splitter with an inverse transmission ratio of 50:50. Wideband polarization splitting characteristics can be achieved by superimposing two $\lambda/4$ film systems with different center wavelengths.

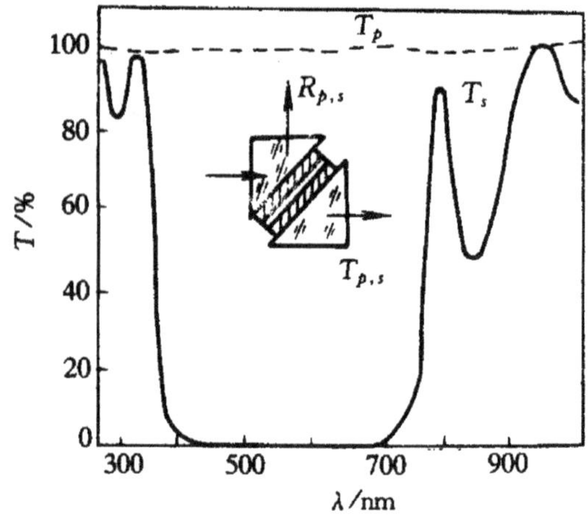

Figure 9.24: Spectral characteristics of a cubic prism polarizing beam splitter.

(2) **Flat polarizing beam splitter**

When a reflective film composed of two different refractive index materials plated on a flat plate is obliquely placed in a parallel optical path, the effective refractive index of the film for p polarization and s polarization is different: $n_p = n/\cos\theta$, $n_s = n/\cos\theta$. The film system is like two film systems for p-acoustic polarization and s-polarized light, and their reflection bandwidths for the two polarization components are different. Generally speaking, the reflection bandwidth of s-polarized light is larger than the reflection bandwidth of p-polarized light. At this time, if the operating wavelength or the wave segment is between the edge of the p-polarized and the s-polarized reflection band by selecting the center wavelength, the p-polarized light is highly transmitted, and the s-polarized light is highly reflected. In order to obtain a large operating wavelength bandwidth, a large n_H/n_L film material should be selected and the incident angle should be as large as possible. When the direction of incident light is not required, the Brewster angle of the substrate can be selected as the working angle (for a substrate with a refractive index of 1.52, the incident angle is 57°).

The basic structure of the plate polarizing beam splitter film system is substrate/$(0.5\text{HL}0.5\text{H})^{\text{s}}$/air or substrate/$(0.5\text{LH}0.5\text{L})^{\text{s}}$/air.

Figure 9.25 shows the spectral characteristics of a plate polarizing beam splitter with an incident angle of 57° and an operating wavelength of 1.06 μm.

9.3 Application of Magnetic Films

9.3.1 *Magnetic recording film*

Magnetic storage technology plays a pivotal role in modern technology. Since the recording density of the magnetic signal largely depends on the width and flying height of the magnetic head, and the thickness of the recording medium, in order to further increase the density and capacity of magnetic storage, it is necessary to continuously reduce the volume of the magnetic head while reducing the thickness of the magnetic recording medium. Therefore, thin film

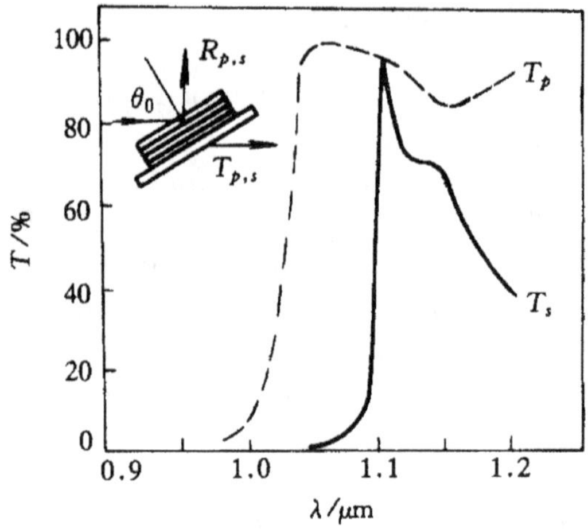

Figure 9.25: Spectral characteristics of a plate polarizing beam splitter.

magnetic head materials and thin film magnetic storage media are one of the main development directions of magnetic materials.

First, compared with the $\gamma-Fe_2O_3$ magnetic powder coating type magnetic recording medium, a thin film recording material is composed of an alloy of a transition element, and thus has a high saturation magnetization. Moreover, since it does not contain any binder, the magnetic medium allowed to be used is smaller in thickness and more uniform in nature. These are all beneficial to increase the recording density of data and reduce the cost. At the same time, the exploration of new thin-film magnetic head materials with high magnetic induction, good soft magnetic properties, and excellent wear resistance is also important for further improving the recording density and capacity.

There are two ways to read and write magnetic records, namely parallel recording and vertical recording as shown in Figures 9.26(a) and 9.26(b) respectively. In the case of parallel recording, the magnetic moment vectors in the magnetic recording medium are arranged in a direction parallel to the plane of the recording medium, and the change in the magnetization direction corresponds to the change in the recorded data. While the read/write magnetic

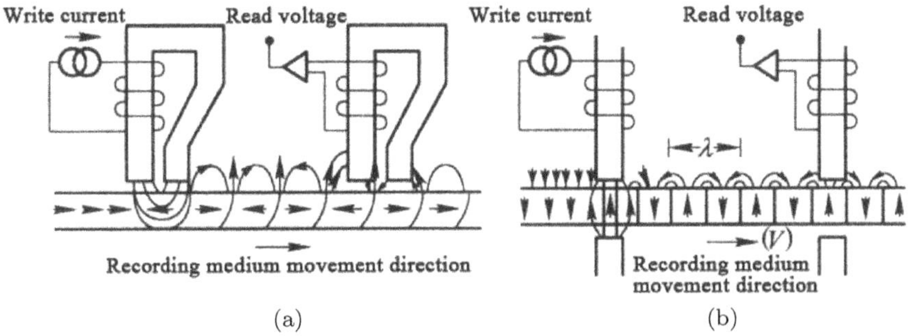

Figure 9.26: Schematic diagram of magnetic recording mode: (a) parallel recording; (b) perpendicular recording.

head moves relative to the magnetic recording medium, the head constantly changes the magnetization state to either change the magnetization direction of the magnetic recording medium (that is, write data), or generate an induced potential in the head coil due to relative motion (that is, read data). Obviously, the material performance requirements for both the magnetic recording medium and the magnetic head are not the same. For the magnetic head material, it must have typical soft magnetic characteristics, that is, high saturation magnetization, low coercive force, high magnetic permeability, low magnetostriction coefficient, and high allowable frequency of use. For magnetic recording medium, it must have typical hard magnetic properties, that is, high saturation magnetization, high residual magnetic induction, and proper coercive force level.

Another important storage technology of equal importance to magnetic recording technology is optical storage technology; in the past 20 years its development has changed the pattern of information storage technology. It uses the focused laser beam to change the physical characteristics of the storage medium to realize information storage, and uses the low-power laser to scan the storage medium to detect the intensity or polarization state of the reflected light to realize information reading. Below, we first introduce the magnetic head and magnetic recording medium film materials separately.

For a read/write magnetic head operating on the principle of electromagnetic induction, data reading and writing can be

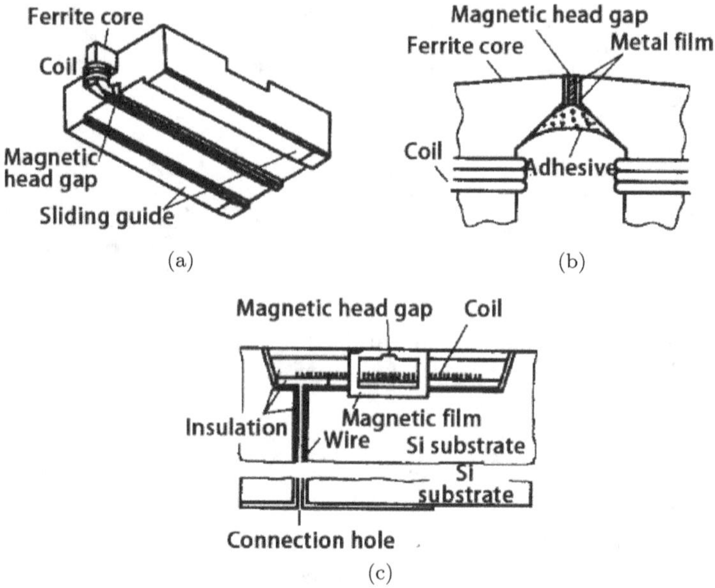

Figure 9.27: Schematic diagram of (a) a sintered ferrite magnetic head, (b) a ferrite, thin film composite magnetic head, and (c) a metal thin film magnetic head.

performed by the same magnetic head: a composite magnetic head or thin film magnetic head, and its structure is shown in Figure 9.27. This magnetic head uses a sintered ferrite having a high magnetic permeability, and has the advantages of excellent soft magnetic properties and wear resistance, and high electrical resistivity, so that its high frequency characteristics are good. However, its magnetization is much lower than that of alloy soft magnetic materials, such as permalloy (79% Ni–Fe alloy) or 85Fe–9Si-5–4Al alloy (Sendust).

Table 9.5 compares the performance characteristics of the above three magnetic head materials. In order to further improve the performance of the magnetic head, on the one hand, plating or sputtering, evaporation, etc. may be used to deposit a layer of alloy film having a soft magnetic property with a thickness of several micrometers at the gap of the magnetic head to form a ferrite-alloy thin film composite head, as shown in Figure 9.27(b). On the other hand, it is also possible to completely use the thin film technology to deposit both the magnetic material and the magnetic field coil on

Table 9.5: Comparison of the performance of three kinds of magnetic head materials.

Performance	Material		
	Mo-permalloy	Hot pressing Mn–Zn ferrite	FeSiAl alloy
Saturation flux density (T)	0.8	0.4–006	1.0
Coercivity ($A \cdot m^{-1}$)	2	12–16	2
Magnetic permeability/$H \cdot m^{-1}$	11000	3000–10000	8000
Resistivity ($\Omega \cdot cm$)	100×10^{-6}	5	85×10^{-6}
Vickers hardness	120	700	480

Note: These data are obtained in the case of 1 kHz and a material thickness of 0.2 mm.

a specific substrate to form a thin film magnetic head, as shown in Figure 9.27(c). Obviously, compared with the ferrite magnetic head in Figure 9.27(a), the magnetic head prepared by thin film technology will have higher sensitivity, and can effectively reduce the size of the magnetic head and increase the magnetic recording density.

In order to further improve the sensitivity of the magnetic head, an important measure is to continue to increase the saturation magnetization of the magnetic film material. Pure Fe has a saturation magnetization of 2.1 T, which is more than twice that of the above-mentioned film materials, but it has a large magnetostriction coefficient and is easily corroded in the atmosphere, so it is not suitable for use as a magnetic head material. For a long time, people have been looking for new magnetic head materials with high saturation magnetization. In recent years, $Fe_{16}N_2$, a metastable nitride of Fe, has been epitaxially grown on a GaAs substrate by molecular beam epitaxy. It is said to have a saturation magnetization of 2.6 A/m, and is a promising magnetic head material. Figure 9.28 is a high-resolution electron micrograph of the $Fe_{l6}N_2$ epitaxial layer on a Fe/GaAs substrate. Obviously, the thin film epitaxy method plays a considerable role in stabilizing the metastable $Fe_{l6}N_2$ structure.

Another important type of magnetic head is a magnetic head made by the phenomenon that the magnetic field is changed according to the magnetic resistance effect, and the resistivity of the

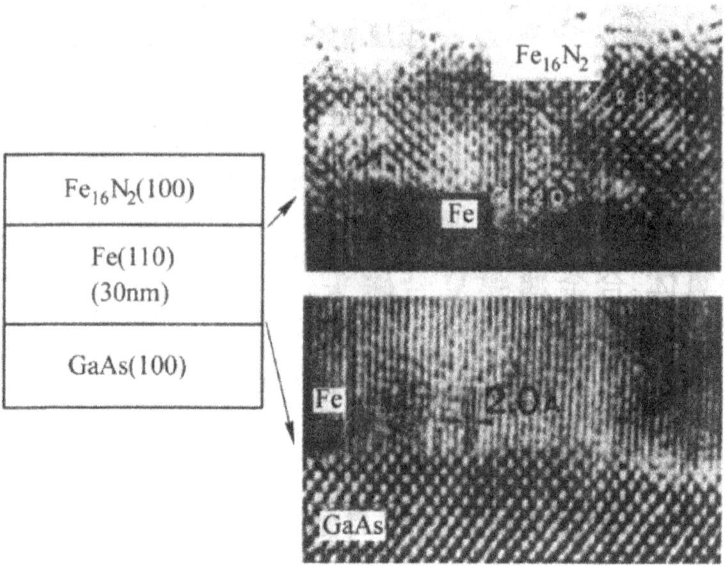

Figure 9.28: Epitaxial structure of the $Fe_{l6}N_2$ film on a GaAs substrate.

material is changed correspondingly. These types of heads do not have a write function and are therefore read-only magnetic heads. The magnetic head is made by the magnetoresistance effect, which has the advantages of high signal readout sensitivity, signal strength not affected by the moving speed of the magnetic head, and no need to prepare an induction coil. Therefore, by combining it with the above-mentioned thin film magnetic head technology, the volume of the magnetic head can be effectively reduced, and the magnetic recording density can be improved. Permalloy not only has good soft magnetic properties, but also has a certain level (about 2.5%) of magnetoresistance effect, so it has been used to make such a magnetic head.

In recent years, one of the important advances in magnetic thin film research is the discovery of ultra-high magnetoresistance (giant magnetoresistive) materials. As shown in Figure 9.29(a), the Fe–Cr multilayer film material prepared by the sputtering method exhibits a very high magnetoresistance effect.

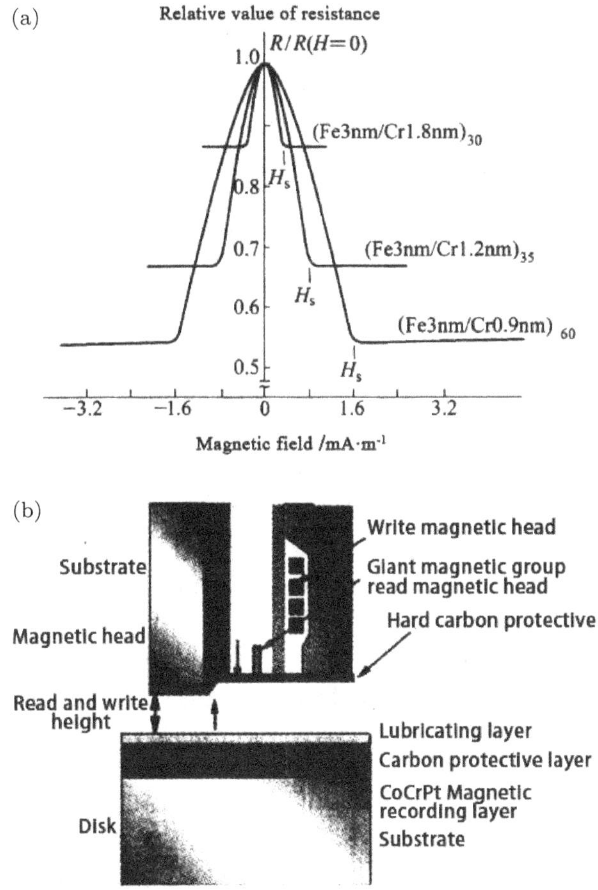

Figure 9.29: Schematic diagram of (a) giant magnetoresistance effect of a Fe–Cr multilayer film material; (b) the structure of a giant magnetoresistive head.

At a low temperature of 4.2 K, the resistivity will decrease as the magnetic field strength increases, and the maximum reduction can reach nearly 50%. According to analysis, the generation of this giant magnetoresistance effect is related to the antiferromagnetic coupling between ferromagnetic Fe constituent layers. When the magnetic field causes the magnetization vector of the adjacent Fe composition layer to reverse, its scattering ability to the conduction electrons also changes accordingly. It is found that the magnitude of the giant magnetoresistance effect of the above film material changes periodically as the thickness of the non-magnetic layer increases. At present, such giant magnetoresistive film materials are widely used in hard disk read/write magnetic head technology. Figure 9.29(b) is a schematic view showing the structure of a hard disk read/write magnetic head prepared using a giant magnetoresistance material. In the magnetic head, the read head of the giant magnetoresistive film is combined with the write head of the thin film induction coil. This higher sensitivity, smaller size thin film magnetic head increases the data recording density of hard disks by 100% per year.

9.3.2 *Magneto-optical thin film*

There are two properties of magnetic materials that magneto-optical memory technique relies on:

(1) Thermomagnetic effect, that the magnetization state of materials changes along with temperature.
(2) Kerr magneto-optical effect, that the polarization direction of polarized light that reflects from the surface of materials changes by the magnetization states.

In the magneto-optical disc, the direction of magnetization vectors is perpendicular to the thin film surface, that is, up or down along the normal. As in the write-in process shown in Figure 9.30(a), part of the magnetic medium was heated by a laser beam to the temperature at which ferromagnetism disappears. This area is exposed to a magnetic field; meanwhile, the temperature drops down and ferromagnetism reappears, so that the magnetization vectors of this area can be arranged by the information to record, to complete the write-in process.

As the condition of the polarized light normal incidence, the polarization direction of the reflected light will get a slightly but completely different change, caused by the areas that have different magnetization directions, which is the Kerr magneto-optical effect. Thus, in the read-out process shown in Figure 9.30(b), the polarization of the laser beam that reflects back will be corresponding to the

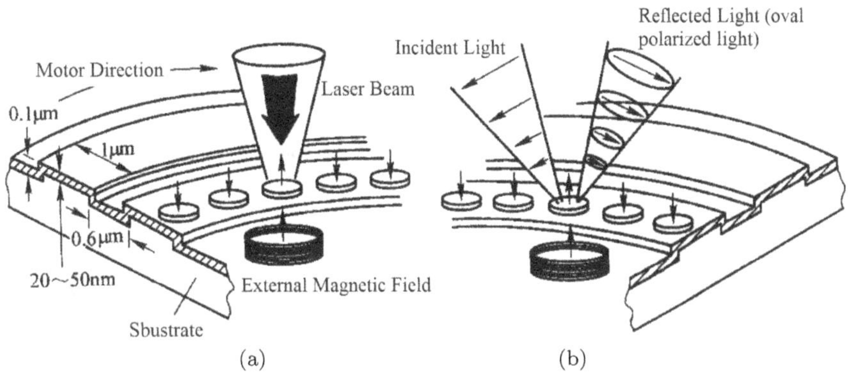

Figure 9.30: Writing (a) and reading (b) processes of a magneto-optical storage medium.

information, so that we can read-out the information by detecting the change of the laser polarization plane.

Thus, besides the normal requirements mentioned above, there are two more requirements of the magneto-optical thin film; they are:

(1) Appropriate magnetic transition temperature, to make sure that the laser power is not too high when rewriting the information, and the magnetization state of the thin film has enough stability.
(2) Strong Kerr magneto-optical effect, which means the polarization direction can be changed greatly when the magnetization directions of materials are different.

The materials used for making magneto-optical storage thin films at present are mostly amorphous alloys of rare earth elements and transition elements. The materials are shown in Table 9.6. The compensation point temperature, which is related to the heating temperature when the laser is written-in, is the characteristic temperature that the magnetization intensity of this kind of alloy is approaching to 0, and the magnetic anisotropy constant is proportional to the coercivity that materials can get. This kind of thin film is often prepared by various kinds of sputtering methods, because there are elements that have different properties in them.

Table 9.6: Properties of commonly used magneto-optical storage thin film materials.

Component	Curie temperature (°C)	Compensation point temperature (°C)	Magnetic wizard performance constant ($J \cdot m^{-3}$)	Kerr angle (°)
$Tb_{23}Fe_{77}$	400	300	3×10^5	0.23
$Tb_{21}Co_{79}$	—	300	1.4×10^5	0.33
$Gd_{26}Fe_{74}$	480	300	2.5×10^4	0.29
$Gd_{21}Co_{79}$	—	300	1×10^4	0.33
$Tb_{23}Fe_{66}Co_{12}$	About 500	<300	1×10^5	0.38
$Gd_{22}Tb_4Fe_{74}$	450	300	4×10^4	0.30
$Gd_{16}Tb_6Co_{78}$	—	280	4×10^4	0.32

9.4 Application of Super-hard Thin Films

9.4.1 *Application of diamond films*

The diamond film is considered to be an ideal material because it has the highest hardness and heat conductivity, with a low friction coefficient (0.05–0.1, similar to Teflon) and good chemical stability. A CVD diamond film tool is one of the main application goals of research studies. It has been applied widely in cutting machines, drawing dies and in various situations with friction, and has held a sweeping market. Compared with the traditional polycrystalline diamond tool (PCD), the CVD diamond film contains totally diamond without any binder, so that it has longer life with better machining precision and roughness. Compared with the diamond single crystal tool, it has comparable properties but lower prices. Thus, it has a very broad market prospect.

The CVD diamond film tool can be divided into the diamond thick film tool and the diamond thin film coated tool. The former is made from a free-standing diamond thick film (0.3–1 mm) while the latter is deposited on the tool substrates (usually cemented carbide), with thickness less than 30 μm.

9.4.1.1 *Tool grade diamond films*

This is the kind of free-standing film that is used for making a diamond thick film tool. It is black with low mass and high mechanical strength. The CVD thick film was classified after 1995; different uses and mass of diamond free-standing film are divided into optical grade diamond film, heat sink grade diamond film, tool grade diamond film, etc. After that, tool grade diamond films were sold all around the world as a commodity that is mainly used for making diamond thick film tools (cutting machines, drawing dies, etc.).

9.4.1.2 *Diamond thick film tool*

This kind includes diamond thick film brazed cutting tool, diamond film wire drawing dies, etc. It is made by first using a CVD diamond free-standing film to make the embryo.

Other tool products include grinding wheel correction pens, high-pressure water processing nozzles, etc.

A diamond welding knife can be used longer than PVD, has excellent machining precision and roughness, but is expensive; it is usually used to do some ultra-precision cutting. It can be used for process materials like high silicon aluminum alloys, non-ferrous metals, ceramics, composites, plastics, and so on, taking the place of PCD and the diamond single crystal tool. The diamond film shows even better properties than the single crystal and has a very broad market prospect, because it is quasi-isotropic, and can be used better in making drawing dies.

With the support of the "863" project, China has activated the production line of CVD diamond films, with a certain scale of production. China has at least 4 departments selling the products at present.

Other companies besides those of China, like Sumitomo of Japan, has done research since the 1980s, but hasn't made mass sales yet.

The cost of manufacturing and processing is still a big factor that limits the application. The cost (especially the processing of diamond thick films) must be reduced to enlarge the market scale.

9.4.1.3 *Diamond thin film coated cutting tool*

Diamond thin film coated cutting tool was usually grown on cemented carbide. In the process of CVD, the diamond nucleation is delayed, which does harm to the adhesion property and deposition of the film, because the Co, as the binder in the alloy, can promote the growth of the graphite, and carbon have a great solubility as well as diffusion coefficient in Co. In addition, the thermal expansion coefficient of cemented carbide and diamond is very different, so the adhesion of diamond film and cemented carbide substrate is very poor. Researchers around the world have tried all means to this end, including:

(1) use the cemented carbide that has the lowest amount of Co.
(2) acid leaching treatment to remove the layer of Co on the surface of the cemented carbide.
(3) block the diffusion of Co and alleviate the thermal stress caused by the difference in thermal expansion coefficient by applying various transition layers.
(4) use heat treatment, plasma etching, laser treatment, and other surface treatment methods to remove surface layer Co, and make the surface rough, in order to promote the adhesion of diamond film and cemented carbide substrate.
(5) Apply the appropriate gradient compound transition layer.

At present, the adhesion between diamond film and cemented carbide substrate has been basically solved. Diamond thin film coated cutting tools have entered the market and have formed a certain scale. The thickness of diamond film is generally less than $30\,\mu$m, depending on the specific application.

The properties of the diamond-coated cemented carbide tool is roughly equal to or slightly lower than PCD, but the preparation cost is much lower than that of PCD, and can be coated on the surface of the tool with complex shapes (such as tools designed for various chip grooves), so the market prospect is very good. Wide-ranging industrial coating equipment has been made which can paint hundreds of carbide blades each time, but due to the mass production it has the problems of quality control and inspection technology; the size of the market is not very big. It can be expected, that once technical issues are resolved, diamond film tools will find wide industrial application.

9.4.2 *Applications of diamond-like films*

There are several applications of diamond-like films:

9.4.2.1 *Mechanical processing industry and wear-resistant parts*

The DLC film is very suitable for tool coating, due to its low friction coefficient, high hardness, and good anti-abrasive wear properties. IBM has been working hard, in recent years, at developing DLC film-plated microdrills to drill microholes in circuit boards. By using them, the drilling speed has been increased by 50%, the service life has been increased by 5 times, and the drilling processing cost has been reduced by 50%. In Japan, the patented technology of DLC/Ti and Si coating is applied on the cemented carbide matrix of a precision stamping and shearing die in the microelectronics industry, which can improve the die life and is widely applied. The film thickness of DLC is 1.0–1.2 μm, the ratio of the content of Ti and Si is 0.4, and the Vickers hardness values are 39.2–44.1 GPa. In China, is a DLC film with about 1 μm thickness is deposited on the cemented carbide; this increases the life by 1.5 times when cutting eutectic Al–Si alloy and by 8 times when cutting wear-resistant aluminum bronze.

9.4.2.2 *High fidelity speaker diaphragm*

The electroacoustic field is the earliest application of diamond and DLC films, especially the speaker diaphragm. In 1986, Sumitomo deposited a DLC film on a titanium film to produce high-frequency loudspeakers with a high frequency response of up to 30 kHz. Subsequently, AIWA launched small high-fidelity headsets with a DLC film, with a frequency response range of 10–30,000 Hz.

9.4.2.3 *Field emission plane display device*

As the DLC film has a low electronic affinity potential and is electronically easy to launch, it can be used to make field emission plane displays (FEDs), which are a new generation of excellent-performance display devices. A FED containing DLC films is called a diamond field emission plane display (DFED).

9.4.2.4 *Mask*

The DLC film can be used as the step substrate and finally to make step junctions. The DLC film has good corrosion resistance, resulting in a high frequency characteristic in the YBC step junctions made from it. Sun *et al.* used DLC films as masks to prepare step junctions on the $SrTiO_3$ substrate and obtained a good result. The DLC film can also be used to make Si planar device lithographic masks.

9.4.2.5 *Magnetic media protection film*

With the development of computer technology, hard disk storage density is becoming higher and higher; this requires that the magnetic head and disk clearance is very small, which can lead to magnetic head and disk wear on frequent contact and collision. To protect magnetic media, depositing a film on the disk that is both durable and thin enough not to affect its storage density is very much needed. A DLC film of 40 nm was deposited on a hard disk by RF-PCVD, and it was found that the film with Si transition layer showed good protective effect, high bonding strength with the substrate, and no adverse effect on the electromagnetic characteristics of the hard disk. Someone deposited a layer of DLC film on videotape and achieved good protection.

9.4.2.6 *Antireflection protective film*

Ge is the most common window and lens material in the range of 8–13 μm, but it is prone to become scratched by sand and seawater erosion. By using a DLC film as an anti-reflection film, good optical performance and corrosion resistance can be obtained at the same time. It has a good performance in the MgF_2 infrared detection window as well. Since MgF_2 refractive index is only 1.37, the monolayer DLC film will reduce the transmittance of MgF_2 to varying degrees. The adoption of an appropriate double-layer or gradient DLC film can improve its optical transmittance (the highest peak transmittance can reach 99%) and obtain excellent corrosion resistance. In addition, some domestic units in China deposited a DLC film on the ZnS matrix to improve its infrared transmittance.

9.4.2.7 *Optical protective film*

The transparent DLC film can guarantee the optical properties of optical components and improve wear resistance and corrosion resistance significantly at the same time. It has been applied in optics lens protective film, DC protective film, watch glass surface protective film, glass lens (glass, resin) protective film, and car windshield protective film, etc., and has a huge market potential.

9.4.2.8 *Heart valve*

Because of the excellent biological compatibility of a DLC film, depositing a layer of DLC film on the metal ring of an artificial heart valve can greatly improve its biological compatibility.

9.4.2.9 *High frequency scalpel*

At present, the high-frequency scalpel is made of stainless steel; it will stick to the muscle when used and give off a bad smell when heated by electricity. By taking advantage of the characteristics of small surface energy and non-wetting in DLC films, ART Company of the United States introduced a SiO_2 mesh and transition metal elements to regulate its electrical conductivity, and produced a non-stick meat high-frequency scalpel and put it on the market; this significantly improved the working conditions of medical personnel.

9.5 Application of Luminescent Films

9.5.1 *Thin-film luminescent displays*

The display device is the main part of the whole information system, and is really important and responsible for man–machine dialogue. With the deepening of the information revolution, it will be more and more closely linked with people's daily life and have broad market prospects. The traditional display device is the cathode ray tube (CRT). It has the disadvantages of a large volume and a high power consumption. The main development direction of display technology is to be flat. Among the many flat panel displays, the thin film electroluminescent display has attracted extensive attention and developed rapidly due to its active luminescence, total solid, impact resistance, large view angle, wide applicable temperature, simple working procedure, etc. Sharp first introduced commercial thin-film electroluminescent displays in 1983, and Grid produced portable computers with 6 inch, 320 × 240 pixel thin-film electroluminescent displays the same year. In the late 1980*s*, Sharp and Planar successively launched 9 inch orange thin-film electroluminescent displays and their series of products. More recently, the orange thin film electroluminescent display with ZnS: Mn as a luminescent layer has been developed and widely used in many fields, especially the military, aerospace, etc., because it has the unique advantages of a total solid, and can withstand up to 50 gravitational accelerations. The biggest problem of the thin film electroluminescent display is that it is difficult to obtain panchromatic display. In fact, only the blue brightness out of the three primary colors can't meet the minimum requirements for panchromatic display, which is the so-called "blue light problem" and greatly limits the application range of thin-film electroluminescence display devices. If the thin film electroluminescent display device can achieve panchromatic display, it will bring a breakthrough in display technology. Therefore, the issue of blue light has been a hot topic in this field. Over the past decade, scientists around the world have carried out many studies on blue light and have made some progress.

In order to solve the problem of blue light, it is necessary to explore the materials, device preparation, structure, operation mode, and electroluminescence mechanism. This section mainly describes the research progress on the structural, material, and physical processes of thin film electroluminescent display devices.

So far, many kinds of electroluminescent materials with good properties have been obtained, and the exploration of new materials is also deepening. Several main materials are introduced here.

9.5.1.1 *ZnS:Mn*

ZnS:Mn is the earliest electroluminescent material with the highest brightness and efficiency. Its luminescence band spectrum is about 540–680 nm, the peak is about 585 nm, and color is orange. The red and green glow required for panchromatic display can be obtained by adding a color filter. The conventional preparation method of ZnS:Mn films is by electron beam evaporation. In recent years, other methods for preparation have been tried in order to obtain better performance or lower cost.

9.5.1.2 *ZnS:Tb*

ZnS:Tb electroluminescence brightness and efficiency is second only to ZnS: Mn. It has a narrow luminescence spectrum, with a peak of around 540 nm in green. The performance of a device prepared by the sputtering method is much better than that of a device prepared by electron beam or heat evaporation.

9.5.1.3 *SrS:Ce*

Due to the short wavelength of blue light, the requirement that the substrate material band gap width should be large, which is difficult for ZnS, CaS, and SrS that are similar to ZnS but with larger forbidden band width, was discovered naturally. Thus, SrS:Ce is the earliest blue electroluminescent material with better performance.

A good electroluminescent material is a necessary condition for obtaining a good electroluminescent device.

The luminous layer is the key in TFELD, but the insulation layer also has a large impact on the overall performance of the device, so it is necessary to optimize the insulating layer. Ryu *et al.* adopted multilayer SiNO with gradient refractive index as the insulating layer, and optimized its structure to maximize the transmittance of the whole insulating layer, thus achieving higher efficiency than the traditional structure. Yoon *et al.* prepared $BaTiO_3$, as the insulating layer by MOCVD, and obtained high dielectric constant and breakdown field strength by optimizing the preparation conditions. Prased *et al.* prepared TiO_2:Ce as the insulating layer, and obtained ZnS:Mn TFELD, which has the advantages of low threshold voltage and high brightness.

At present, TFELD mostly adopts a double insulation layer structure. Although this structure has achieved good results, other types of device structures are being explored for better performance.

9.5.2 *Organic electroluminescent films*

Since invented, inorganic luminescent plates (ZnS and GaP) have been widely used in many fields and have made remarkable achievements. However, they still have many shortcomings, such as a small variety of luminous materials, complex device-manufacturing process, high cost, large energy consumption, and difficulty in providing panchromatic display. On the contrary, organic materials have a wide range of choices and can be used for molecular design, so organic electroluminescent devices have the following advantages over inorganic luminescent plates: (1) they can get all colors that the spectrum covered, especially the blue light that inorganic materials can hardly get; (2) the production process is simple, and they can achieve ultra-thin large-area plate display; (3) dc low voltage drive, so that it can match the integrated circuit. It is these potential advantages that have aroused great interest from many researchers all over the world.

The study of electroluminescence in organic compounds began in the mid-1930*s*. In 1936, Destriau dispersed organic fluorescent compounds in a polymer to get a thin film, that is, the first electroluminescent device. Pope first reported the electroluminescence of anthracene single crystals in 1963. Subsequently, the electroluminescence of anthracene, naphthalene, and clove oil was reported. In order to reduce the driving voltage, the film of the device is the key factor. Therefore, Vincett *et al.* made thin films by vacuum deposition and LB film respectively, and obtained electroluminescent devices with a driving voltage of less than 30 V, but the quantum efficiency of the devices was still very low. The electroluminescence of organic materials remained stagnant for more than 20 years after that.

In 1987, C. W. Tang *et al.* of Kodak Company made an organic thin film electroluminescent device driven by dc voltage (less than 10 V) with high brightness (more than 1000 cd/m^2) and high optical efficiency (1.5 lm/W) by using ultra-thin film technology and the I'PD organic hole transport layer with better hole transmission effect. That was a breakthrough and led to a new era. Subsequently, Kyushu University, in Japan, Saito's team reported the device they got, which was made of a multilayer structure with stability, low driving voltage and high brightness. In 1990, J. H. Burronghes *et al.* from Cambridge University proposed for the first time to make polymer organic electroluminescent devices with the conjugated polymer PPV, which made research on organic electroluminescent devices grow in depth and become a research hotspot in the world.

Organic electroluminescent materials must first meet the following two requirements: (1) the material is easy to form a dense amorphous film and does not change to crystals with time; (2) the material has solid-state fluorescence. Ionization energy and electron affinity should also be considered; the appropriate wavelength of luminescence should be considered when choosing the luminescent material; meanwhile, the charge injection and transmission should be considered when choosing the transmission material. The material should also have good thermal stability.

Luminescent materials with electron transport properties include metal chelates, polycyclic condensation or conjugated aromatic compounds, azole derivatives, and coumarin derivatives. Luminescent materials with hole transport properties include triphenylamine derivatives, vinyl carbazole derivatives, etc.

The organic electroluminescent device developed by C. W. Tang uses Alq chelate. It is one of the best organic electroluminescent devices in terms of comprehensive performance, and almost meets all the requirements of organic electroluminescent devices. This is because Alq has a high electron mobility of 10^{-5}cm^2/(V $\cdot$ s), good film-forming performance, high glass transition temperature, and high luminescence efficiency in the solid state.

9.6 Applications of Ferroelectric Thin Films

9.6.1 *Ferroelectric memory*

The applications of ferroelectric thin films in memory have many different forms, and the materials used and their properties are very different, respectively, as listed below:

9.6.1.1 *FeRAM*

It is a kind of non-volatile memory prepared by electric hysteresis loop, which is the bistable polarization characteristic of ferroelectric thin films. It has broad application prospects in computers, aerospace, military and other fields. FeRAM is the main research and development direction of ferroelectric thin film memory, and is valued by many companies. For example, Ramtron developed FeRAM products with 4–6 kbit parallel and serial structure in 1995. The company formed a consortium with Symetrix and launched the 16 Mbit FeRAM product in 1998.

9.6.1.2 *DRAM*

DRAM has the largest amount of semiconductor memory in a computer at present. In order to improve its storage density, a high-density plane layout must be adopted. Currently, the key factor of limited DRAM storage capacity is the charge storage density per unit area of the storage unit (condenser), that is, $Q'_c(Q'_c = \varepsilon_0\varepsilon_r\Delta V d$, where d is the thickness of the dielectric film, and ΔV is the voltage change from "0" state to "1" state). Traditional DRAM adopts SiO_3/Si_3N_4 or Ta_2O_5 as the capacitor medium, and the maximum capacity that can be achieved is 16 Mbit at present. In order to improve the integration of DRAM, it is necessary to reduce the area occupied by the planar capacitor and maintain the comprehensive electrical characteristics of the dielectric film. In recent years, the application of BST with high-ε_r dielectric films to the new generation DRAM (Gbit) has become a hot topic.

9.6.2 *Infrared pyroelectric detector*

The thin-film infrared pyroelectric unit detector and array detector, in addition to having the advantages of the pyroelectric detector prepared by bulk materials, also has high detection sensitivity, expected to be integrated with the semiconductor integrated circuit, and other outstanding advantages. If make it into an infrared focal plane array, the pixel is expected to be very high, thus improving the resolution of the pyroelectric image. At present, internationally the pixel with advanced level has reached 384 × 288 units, and can detect temperature changes below 0.1 K.[4]

We take $PbTiO_3$ as an example to introduce the structure of several unit detectors.

9.6.2.1 *$PbTiO_3$ thin film infrared pyroelectric detector*

Figure 9.31 is a schematic diagram of a simple pyroelectric infrared detector of a $PbTiO_3$ thin film. It is prepared by rF sputtering method, which means depositing a $PbTiO_3$ thin film of about 20 μm thickness on a mica sheet with a Pt film. After evaporation of the Al electrode on $PbTiO_3$ thin film, a layer of black gold is evaporated as the infrared absorption electrode, and then add a voltage of 20 V in 20 min for polarization. When an external infrared light is irradiated to the black gold electrode by the chopper and Ge filter, the temperature of the $PbTiO_3$ film changes and the pyroelectric

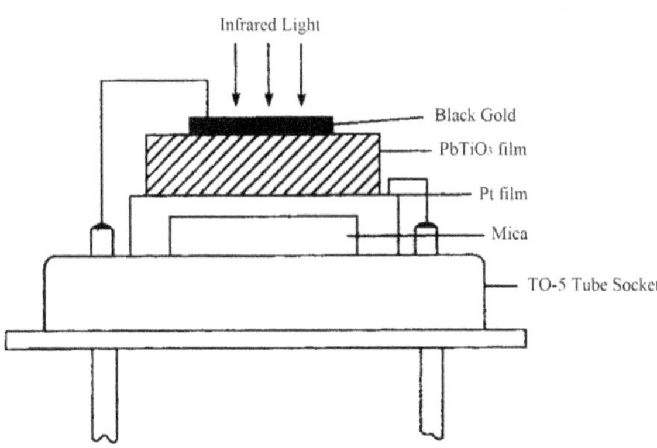

Figure 9.31: $PbTiO_3$ thin film infrared detector.

voltage is generated. The amplifier circuit consists of a field effect transistor with high input impedance, an input resistor, and a lock is amplifier. Because the heat capacity of the $PbTiO_3$ thin film is small, it has a high response sensitivity, which is proportional to the strength of the input infrared light. The voltage response rate (R_V) was 465 V/W, and the detection rate D^* was 1.7×10^8 cm $Hz^{1/2}$/W. If a MgO monocrystalline substrate is deposited with Pt films and then sputter deposition of c-axis-oriented $PbTiO_3$ films, is done the R_V and D^* values can be further improved.

9.6.2.2 *Highly sensitive $PbTiO_3$ thin film pyroelectric infrared detector*

Figure 9.32 shows a schematic diagram of this kind of highly sensitive $PbTiO_3$ thin film pyroelectric infrared detector with no actual substrate. $PbTiO_3$ thin films were deposited on a MgO single crystal substrate by rF planar magnetron sputtering. Since the presence of a substrate has a great influence on the voltage response rate of the device, the MgO single crystal substrate under the $PbTiO_3$ films is chemically corroded away, and then a NiCr thin film electrode

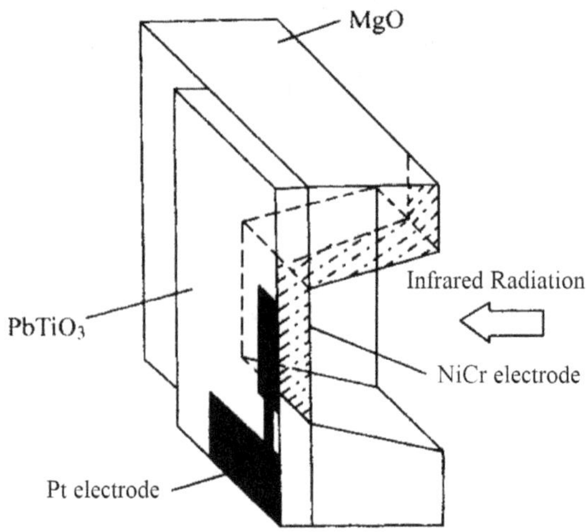

Figure 9.32: High-sensitivity $PbTiO_3$ thin film pyroelectric infrared sensor without a substrate.

with 5 μm thickness is deposited on the film by evaporation. The substrate-free infrared sensor can effectively prevent the absorbed infrared radiation heat transfer through the film to the substrate, thus improving the sensitivity. Reference [5] describes the manufacturing process of this kind of detector (sensor) in detail.

9.6.2.3 *Field-effect tube type pyroelectric infrared detector (sensor)*

A $PbTiO_3$ thin film is combined with a silicon integrated circuit; based on the strong pyroelectricity of $PbTiO_3$, this kind of silicon monolithic infrared detector is made. It is also known as the infrared field effect transistor (IR-OPFET), and the structure is shown in Figure 9.33. Under infrared irradiation, the polarization state of the $PbTiO_3$ film changes due to the temperature variation of the gate electrode, and so the channel above the Si semiconductor surface potential changes, and the carrier concentration in the current channel depletion layer changes, so that the leakage current is modulated, and the voltage at the ends of the load resistance is changed, and at last the intensity of the infrared light can be detected.

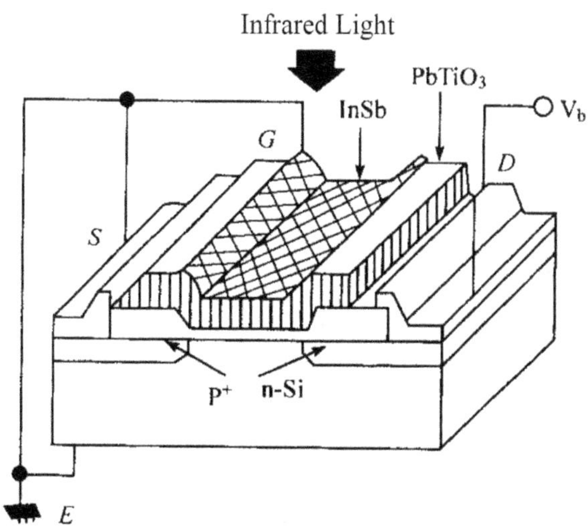

Figure 9.33: Structure of an infrared field effect transistor (IR-OPFET).

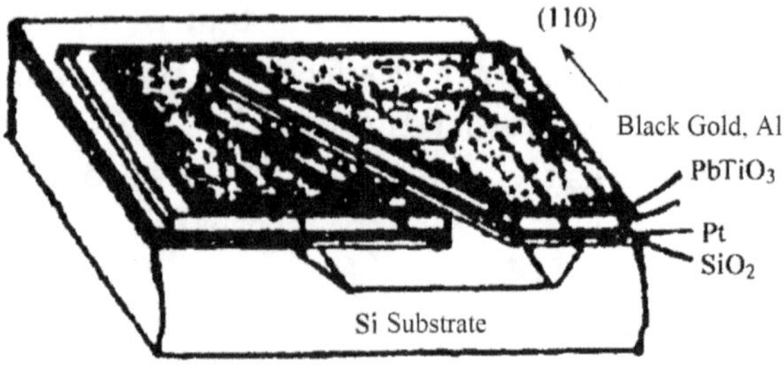

Figure 9.34: Bipolar transistor infrared sensor with the suspension SiO_2 membrane-sensitive region.

If other active devices such as transistors and charge-coupled devices (CCD) are combined on the same silicon chip, multifunctional pyroelectric infrared sensors can be made.

Since there will be a loss when heat is transferred from IR–OPFET to the silicon chip through the $PbTiO_3$ thin film, and the big time constant will cause an output drop in the high frequency region, it is necessary to reduce the heat capacity in the infrared-sensitive region to improve the output signal. For this purpose, a bipolar transistor infrared sensor with the suspension SiO_2 film-sensitive region was developed, as shown in Figure 9.34. The experimental results show that the pyroelectric current of the device in the suspension SiO_2 film-sensitive region is about one order of magnitude larger than that in the other region. This phenomenon greatly improves the sensitivity of the infrared detector.

9.6.3 *Ferroelectric thin film microelectromechanical system*

A microelectromechanical system (MEMS) is composed of tiny mechanical and electromechanical devices fabricated on the surface of a Sic, by which devices can be improved using ferroelectric and piezoelectric films. In a MEMS, micro smart sensors and micro actuators are two important components. A micro smart sensor requires some intelligent characteristics in the process of sensing, which requires the combination of sensing technology and microelectronics technology. The compatibility between ferroelectric thin film preparation technology and semiconductor technology has become a reality, leading to the possibility of developing the micro smart sensor. Integrated micromachining technology (such as photolithography, corrosion, etc.) can be used to process PZT ferroelectric thin films deposited on the silicon surface (for instance, a cantilever beam structure can be formed by the corrosion method), and micro acceleration sensors with high sensitivity can be made. In recent years, due to the development of integrated ferroelectrics, a new direction for preparing micro actuators with ferroelectric thin films has been developed. One of the most remarkable developments in this area is the appearance of piezoelectric thin film micro-ultrasonic motors. The piezoelectric thin film micro-motor is a result of the combination of piezoelectric ceramic thin film technology and silicon microelectronics technology. It has the advantages of a low drive voltage (3–5 V), a small size, a large torque, etc., and can be compatible with IC. Muralt *et al.* prepared a 20–30 μm thick PZT piezoelectric film on the Pt/Ta/Si_3N_4/SiO_2/Si bottom electrode by sol–gel method, and it was used as the vibrator of a micro-motor to generate rotating surface standing sound waves. Meanwhile, the micro-rotor was made by silicon micromachining technology and assembled into a PZT piezoelectric film micro-motor, the structure of which is shown in Figure 9.35. The micro-motor has a driving voltage of only 1–3 V, an operating frequency of 25–100 kHz, and a rotating speed of

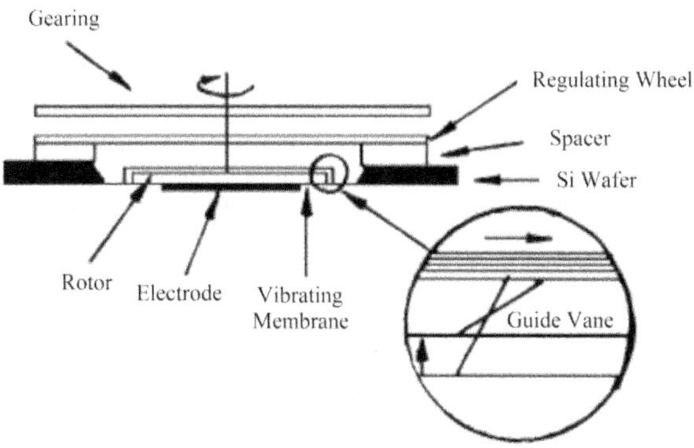

Figure 9.35: PZT piezoelectric film micro motor structure.

100–200 r/min, and can operate at the voltage of standard IC. This progress has made research on ferroelectric thin film micro actuators extremely active and it has become one of the hotspots of MEMS research.

9.6.4 *Ferroelectric waveguide and ferroelectric superlattice*

The development of modern information technology requires opto-electronic information processing systems with high speed and large capacity. Ferroelectric materials have excellent dielectric, piezoelectric, pyroelectric, and ferroelectric properties, as well as electro-optical, nonlinear optical, photoelastic, and photorefractive properties, so they can be widely used in different optoelectronic devices.

For bulk materials, making them into thin film devices has many advantages: (1) they can be integrated with microelectronic and optoelectronic devices that are used widely, so that monolithic integrated devices are made. Integrating thin film and semiconductor diode lasers to make small, compact, monolithic integrated lightning or frequency-doubling devices is an obvious example; (2) the large refractive index difference between the film and the substrate makes the film and the substrate themselves constitute an optical waveguide, and it is expected that the device can realize the limitation of a high-energy beam, thus enabling the device to have a high power density, which is very useful in device design; (3) miniaturization of integrated optical waveguide devices can be realized; (4) the anti-light damage ability is better than the same type of body materials, etc.

In order to render ferroelectric thin film integrated optical waveguide devices practical, a lot of research studies have been done, including material selection and thin film preparation, performance testing and characterization, optical loss mechanism research and measures to reduce optical loss, device physics research and device design, etc. In terms of material selection, PLZT and $LiNbO_3$ are most often reported. In order to improve the orientation of materials and integrate them with semiconductor integrated circuits, many barrier materials and their preparation have also been studied. For example, in order to prepare $LiNbO_3$ films with good properties on Si or GaAs substrates, MgO are usually selected as the transition layer. From the test and characterization of thin film optical performance, the basic parameters required for ferroelectric thin film integrated optical waveguide devices, such as second harmonic generation (SHG) coefficient, electro-optical coefficient, optical loss coefficient, and refractive index, have been basically satisfied.

The main factor affecting the practical application of ferroelectric thin films in integrated optoelectronic devices is that the optical loss of the ferroelectric thin films prepared is large at present (the optimal optical loss that can be achieved is about 2 dB/cm). Optical loss of thin films mainly comes from optical absorption loss, leakage loss, scattering loss, and surface roughness. Since optical absorption is an intrinsic characteristic, leakage loss is easy to solve. Therefore, the most important way to reduce optical loss of thin films is to reduce scattering loss and surface roughness. Ferroelectric thin films prepared by the improved MOCVD technique, the shadow mask PLD technique, and the liquid phase epitaxy technique have made important progress in reducing optical loss. If the optical loss of ferroelectric thin films can be reduced to less than 1 dB/cm, it is expected to make practical devices.

The basic characteristic of ferroelectric materials is spontaneous polarization. The regions in the material with uniform spontaneous polarization are called electric domains. Domains can be modulated in different ways. The artificially modulated domain structure, namely the periodic lamellar domain, developed by Chinese scholars has attracted great attention. The lamellar domain structure can be realized by diffusion method, pulsed electric field treatment, proton exchange method, and special crystal pull method, and can also be realized by controlling the fabrication of the film. Therefore, we discuss this special artificially modulated domain structure as a new type of iron electrical appliance.

Figure 9.36 shows a schematic diagram of a periodic lamellar domain structure. The period of the structure is $a + b$, where a and

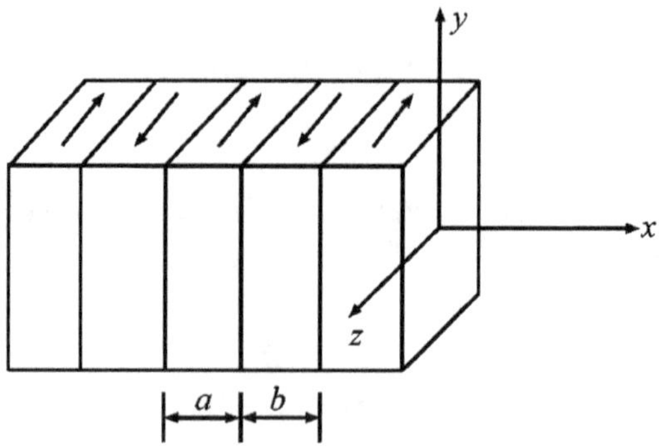

Figure 9.36: A kind of periodic lamellar domain structure.

b are the widths of the positive domain and the reverse domain, respectively (spontaneous polarization reverses when transiting from one domain to another). If the period of the domain structure is comparable to the wavelength of light wave or ultrasonic wave, the materials with periodic modulation structure are called optical superlattice or micron superlattice, respectively. They have important applications in laser doublers and UHF piezoelectric resonators, respectively. For example, using a 180° sheet domain of $LiNbO_3$ crystal, the quasi-phase matching frequency doubling laser output has been successfully realized.

Different from the domain structure shown in Figure 9.36, another kind of periodic lamellar domain structure is shown in Figure 9.37. In this structure, the spontaneous polarization is perpendicular to the domain wall. At the domain wall, the spontaneous polarization is reversed, thus reversing the piezoelectric coefficients as third-order tensors and leading to unique acoustic properties. Therefore, the periodic lamellar domain structure in Figure 9.37 is called acoustic superlattice. Theoretical analysis shows that the pitch resonance frequency of the acoustic superlattice is

$$f_r = \nu/(a + b), \tag{9.9}$$

where, $a + b$ is the period of the lamellar domain; ν is the speed of sound. Obviously, the resonant frequency is only determined by the period of the superlattice and is independent of the total length

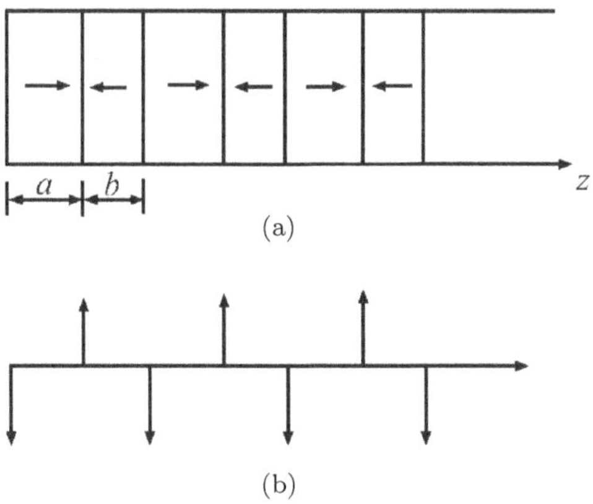

Figure 9.37: Another periodic lamellar domain structure.

of the crystal, so the ultrahigh frequency piezoelectric resonator can be fabricated by using bulk crystals. Acoustic supercrystals with a period of about $7\,\mu$m have a pitch resonance frequency of about 1 GHz. If ferroelectric thin film preparation technology is adopted to decrease the period of acoustic superlattice to about $1\,\mu$m, the pitch resonance frequency can reach 10 GHz.

9.7 Application of Superconducting Thin Films

Superconducting thin films are mainly used in weak current fields, especially in electronic components and integrated circuits. They can be used to make passive devices such as microstrip transmission lines, microstrip antennas, microstrip resonators, filters, delay lines, circulators, directional couplers, infrared detectors, and switching devices. Active devices such as Josephson devices are based on superconductivity. Due to the development of the artificial weak connection technology and the multilayer epitaxy technology, a variety of prototype superconducting tunnel junctions have been produced, such as point contact tunnel junction, neck-shrink tunnel junction, double crystal tunnel junction, double epitaxy grain boundary junction, edge tunnel junction, step edge junction, metal bridge junction, split tunnel junction, sandwich junction, shape junction, etc. The high-T_c superconducting thin film can also be used to measure the weak magnetic field of the human body. The application of microwave devices requires the use of double-sided superconducting thin films (in the filter, devices made of double-sided thin films have insertion losses of more than 30%, lower than those with metal grounding surfaces). In addition, a relatively large area, usually 2–3 inches, is required for the thin film, either because the frequency is not very high, or because the multipole structure is used to improve performance, or because the integrated production of multiple devices is required. Therefore, a variety of technologies such as multiple co-evaporation, inverted tube, target sputtering and PLD have been used to prepare large-area double-sided superconducting thin films for making high-performance devices. Superconducting quantum interference devices (SQUIDs), microwave devices, superconducting computers, and other aspects of superconducting thin films have been developed particularly rapidly.

9.7.1 *SQUID equipment*

SQUID is developed based on the superconducting quantum interference effect, and usually contains one or two weakly connected superconducting rings. The former is called the single junction SOUID, or RFSQUID. The latter is called the double junction SQUID, which is biased by the dc current when it works, so it is also called DCSQUID. SQUID is the most sensitive sensor to detect magnetic signals, and can distinguish 10^{-5}T magnetic fields. The application range of SQUID and other magnetic measurement methods is shown in Figure 9.38. SQUID is not only the most sensitive detector of the magnetic field, but also it can be used to detect all physical quantities that can be converted into magnetic field, such as current, voltage, resistance, inductance, susceptibility, magnetic field gradient, temperature, displacement, etc. In addition, a SQUID can have a very wide frequency response, ranging from zero dc frequency to a few tens of megahertz or even higher, and its response range is mainly limited by its readout circuit. All of these properties can be put on a chip a few millimeters in size.

Currently, superconducting thin films have been used to develop all kinds of SQUID instruments used in magnetic measurement such

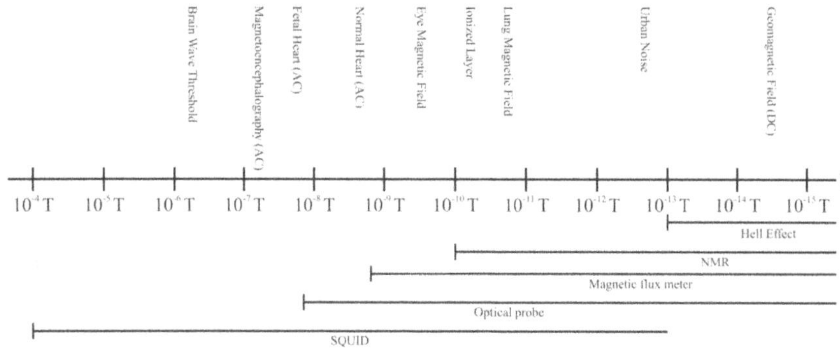

Figure 9.38: Application range of various magnetic measuring instruments.

as SQUID magnetometer, magnetic field gradient meter, SQUID gradiometer, magnetic susceptibility instrument (susceptometer), etc. Galvanometer, SQUID voltmeter, and so on are used for electrical measurement. In addition, there are SQUID microscope, SQUID amplifier, SQUID temperature telemeter, and other quantum interference instruments.

SQUID has a great deal of potential applications. People are expanding its application scope to use it for precision measurement of geomagnetic field, determination of underground water layer distribution, research on earth formation and transition, search for mineral deposits and underground resources, etc. In the field of biomedicine, the magnetometers used in biology and medicine can detect the changes in weak electromagnetic signals caused by the activities of the heart, lung, and nerves. Through non-contact detection, magnetoencephalogram, magnetocardiogram, pulmonary magnetogram, and even neuromagnetogram can be obtained for medical diagnosis without the influence and interference of human body conditions. In addition, earthquake prediction, space, and military fields will also widely use SQUID.

9.7.2 *Superconducting microwave devices and superconducting infrared detectors*

Based on Josephson effect and quantization of magnetic flux, superconducting microwave devices can be divided into microwave passive devices and microwave active devices. As the surface of superconducting thin films have low resistance, which is, under the same temperature and frequency, 1–3 orders of magnitude lower than conventional conductors, when used to constitute a microwave device, its performance will be improved compared with traditional components. Therefore, superconducting thin films are most used in microwave passive devices. So far, microwave passive devices have been developed, such as microwave resonator, filter, antenna, multiplexer, delay line, and so on. In the Josephson effect, on the basis of in-depth study, people developed a series of active microwave devices, including detector, mixer, oscillator and parametric amplifier and signal generator, etc.

Because the surface resistance of the superconducting thin film is very low in the microwave frequency band, the developed superconducting filter has the characteristics of very low band insertion loss, steep edge, very good suppression of out-of-band interference, small size, light weight, etc. It has important application value in the military, especially in the field of communication. Superconducting filters of various types and applications have been developed, which will have broad application prospects in the fields of cellular phone base stations, mobile communications, military affairs, and astronomical observation, and some have reached the level of commercial application.

As an important branch of microwave devices, superconducting antennas, such as electric dipole antenna, spiral antenna, microstrip antenna, and microstrip antenna array, have been studied theoretically. When using superconducting materials below the critical temperature, the resistance is close to zero, which can reduce the wastage of the antenna by several orders of magnitude, thus making the antenna's radiation efficiency increase one hundred times or more; at the same time, the antenna loss decrease lowers the thermal noise of the antenna itself; and the antenna is under low temperature and noise, lower than room temperature and low several times, also beneficial to the improvement of the antenna sensitivity.

The application of superconducting technology, especially superconducting microwave devices, to radar can solve the problems of frequency stability, low noise reception, fast signal processing, and thermal consumption of phased array radar. High short-term stability can be achieved by using a superconducting cavity stabilized quantum oscillator, and high sensitivity can be obtained by using a quasi-particle mixer.

Using superconducting thin films, superconducting thin film infrared detectors with high sensitivity, low power consumption, low noise, and wide frequency band, the far-infrared and submillimeter wave spectrum area, can be made. The superconducting thin film infrared detector is the best detector to detect 10–16 T magnetic signal 10 and 15 V electrical signal; the detection range can cover almost the entire electromagnetic spectrum, but the existing detector cannot detect submillimeter bandgap; in many fields such as the military, aerospace, and energy, it has a very large application prospect.

9.7.3 *Superconducting filter*

Superconducting thin films (such as YBCO) have extremely low surface resistance to microwaves compared with conventional metals such as copper. Compared with copper, in the superconducting state (77 K), the surface resistance of the YBCO thin film resistor (10 GHz, 77 K) is 0.125 mΩ, and surface resistance of copper (10 GHz) at 77 K is 817 mΩ; at 300 K is 2611 mΩ; so the YBCO thin film has obvious advantages. The surface resistance of YBCO is 10–1000 times lower than that of copper in the 1–20 GHz frequency range. Compared with conventional (such as copper) filters, the superconducting thin film filter has five advantages:

(1) small passband loss;
(2) the resistance band is highly inhibited;
(3) the sideband is steep, or the transition zone is narrow;
(4) extremely narrow band filters can be made. For example, the relative bandwidth of conventional filters can hardly reach 2%, while the relative bandwidth of superconducting filters can reach less than 0.2%.
(5) the superconducting thin film filter is small in size and light in weight.

The design of the superconducting filter is based on the theory and method of the traditional planar microstrip filter. By increasing the number of resonators, the performance and technical indicators of the superconducting filter are likely to be superior to all kinds of metal filters (including 3D filters). In 1995, Zhang *et al.* in the United States developed the 19th order Chebyshev high-temperature superconducting filter. In 2001, Ueno *et al.* in Japan developed a 32nd order Chebyshev filter, whose performance was far from that of conventional metal filters. In 1998, Hong Jiasheng and Lancaster in the UK first applied quasi-elliptic functions to the design of superconducting filters, which are characterized by setting one or more pairs of zeros near the passband so that a very steep sideband can be obtained with fewer resonant units. In 2003, Tsuzuki *et al.* made a superconducting filter with 5 pairs of zeroth points as an extremely narrow band filter using the theory of quasi-elliptic functions. So the quasi-elliptic function filter has become a kind of superconducting filter.

When the superconducting filter is developed successfully, if the superconducting filter is placed in a small mechanical refrigerator, this simple and fast method can ensure the low-temperature environment necessary for the normal operation of the superconducting device. If the preamplifier is also integrated into the refrigerator, the noise coefficient of the amplifier will be greatly reduced due to the low temperature, so that the system's anti-interference ability can be improved and its reception sensitivity can be improved at the same time. This gives birth to the superconducting microwave receiver front-end subsystem, referred to as the superconducting subsystem. The subsystem is generally placed in the front end of various microwave receivers and has been widely used in aerospace, mobile communication, radar, and guidance.

9.7.4 *Superconducting digital computer*

The superconducting digital computer is a new generation high-speed, low-consumption, large-capacity and high-performance computer based on the Josephson junction rapid single flux quantum (RSFQ) mechanism. Josephson devices have the characteristics of extremely high switching speed (10–1000 times that of silicon devices) and low power consumption (only about one-thousandth of that of silicon devices), so they can greatly improve the computing speed of digital computers, and reduce their size, and are highly integrated without heat dissipation, signal interference, and switching rate barriers.

The RSFQ digital circuit chip is the core of the superconducting supercomputer. The RSFQ digital circuit can generate, transmit, remember, and reproduce quantized low voltage pulses of very small width, which are the area of a single magnetic fluxon and form a binary logic circuit. The RSFQ logic storage circuit can realize the logic functions of all digital circuits, and can be made into superconducting memory and superconducting large-scale integrated circuits.

Various RSFQ digital circuits have been designed using superconducting thin film materials, including RSFQ A/D conversion circuit, RSFQ Multiplexer, RSFQ Shift Register, and RSFQ Readout. At present, the United States are planning to develop a superconducting computer with one hundred billion floating point operations per second; the superconducting computer has under liquid nitrogen temperature superconducting network processor and chip, built-in SRAM under liquid nitrogen temperature (SRAM PIMs), built-in dynamic random access memory (DRAM PIMs) and Holographic Data Storage, Holographic Data Storage, etc. Because the superconducting digital computer has a high computing speed and a large computing ability, in the national economy and national defense it will have a wide range of applications; especially it can meet the nuclear reactor control, space systems, fluid dynamics modeling, global economic modeling, long-term weather forecast, genetic analysis, strategic defense system, the urgent needs of large engineering calculations, and other fields; many developed countries are going to speed up the pace of the research.

9.8 Application of LB Films

Due to their natural orientation and controllability of the molecular structure, LB films have wide application prospects in such fields as surface catalysis, photoelectric conversion, gas sensitivity, photoconductivity, pyroelectricity, and superconductivity. The LB film has the characteristics of accurate control of film thickness, low requirement for membrane preparation, simple and easy operation, highly ordered molecular arrangement in the film, etc. Therefore, it can realize assembly at the molecular level and has a broad application prospect in the fields of materials science, optics, electrochemistry, and bionics.

9.8.1 *Application of LB films in biofilm or biomimetic simulation*

Due to its special physical structure and chemical properties, the LB film has great application value in the field of biomimetic simulation of biofilms. People have made in-depth studies on chemical simulation and biomineralization of biofilms by using LB technology.

9.8.1.1 *Chemical simulation of biofilms*

The biofilm is the most basic tissue unit in the life system. It separates cells and organelles from the surrounding media, forming many tiny compartments with specific functions, which play a role in maintaining the concentration difference and potential difference of the concentration on both sides of the membrane. The physical structure and chemical properties of LB films are similar to those of biofilms, with excellent biocompatibility and the ability to fix functional molecules in a given location. Therefore, monomolecular and LB films are often used as simplified models of biological cells. Scientists used the "protective plate" method to deposit biocatalysts with active layer of PGA (penicillin G acyltransferase). The easy selectivity and adsorption layer of LB technology created a suitable environment for PGA to maintain its function. On testing the activity value of the enzyme and the degree of PGA separation in solution, the results showed that it could meet the needs of biocatalysis. Various organic molecules such as phospholipids and proteins were assembled by LB membrane technology to mimic the structure of biofilms. The study of the physical and chemical characteristics of biofilms in the biological conversion of energy and the mass transport process in a variety of roles further reveals the essence of life.

9.8.1.2 *Biomineralization and membrane controlled crystal growth*

The process of biomineralization refers to the formation of minerals by the selective precipitation of inorganic elements from the environment on a specific organic matrix with the participation of cells in the organism. LB films can provide a biomimetic system to induce the formation of mineralization. For example, bilirubin and cholesterol mixed monomolecular membranes were prepared by using LB film technology to simulate the formation process of mixed stones *in vitro*. By changing the proportion of the two substances when forming the films, the compatibility of the two substances was found to change to some extent under different membrane pressures. It is of great significance to further reveal the formation mechanism of gallstones. The formation of cyclic and fractal structures in the uric acid system in the Agar gel medium was studied, and the factors affecting the ring growth were investigated. By investigating the different factors (including ion species and ion concentration, different kinds of additives, and glue-line position) that impact on the formation of urinary stone structure, it was found that the system with Fe^{2+} is prone to form multi-ring. Different kinds of ion concentrations that affect the speed of the ring formation and the ring width, the additives containing system that is easy to form a multi-ring, and the position of the glue-line affects the shape of the ring. Biomineralization suggests an important way for the synthesis of inorganic materials to form the organic self-assembly body first. Inorganic precursor produces chemical reaction at the interface between self-assembly aggregate and solution phase, and the organic/inorganic complex is formed under the action of template of the self-assembly body. After removing the organic template, organic and inorganic materials can be obtained.

$CaCO_3$, $CuSO_4$, NaCl, CdS, and PbS have been successfully induced by using the ordered molecular membrane as the template for inducing the growth of inorganic crystals. If the dipalmitoyl phosphatidylcholine (DPPC) LB film is used as a simple simulation of the biological cell membrane, heterogeneous nucleation of CaC_2O_4 crystals in metastable solution can be induced, and the crystal of CaC_2O_4 can be induced to produce crystal face oriented growth. C_4S transforms the crystal of CaC_2O_4 from hexagonal prismatic crystal to hexagonal thin crystal, and l-Glu transforms the crystal of CaC_2O_4 from hexagonal prismatic crystal to "indefinite" crystal.

Heywood *et al.* studied the crystal growth of $BaSO_4$ and $CaCO_3$ induced by long chain alkyl sulfonic acid membrane. The results showed that under the long chain alkyl sulfonate monomolecular membrane, $CaCO_3$ crystal nucleated along (111) crystal plane and $BaSO_4$ along (100) crystal plane. At the same time, they studied the crystallization process of gypsum ($CaSO_4 \cdot 2H_2O$) on monolayer of LB, revealing the importance of hydrogen bond and ionic bond in the nucleation process of inorganic crystal. The study on crystal growth shows that nucleating the crystal surface does not necessarily lead to oriented crystal growth. If the electrostatic field generated by a particular crystal surface forming a crystal nucleus matches the polarity and charge density at the nucleation site on the membrane, the crystal will grow in the preferred orientation when the crystal surface is in contact with the membrane.

The crystal nucleation and growth processes are related to the LB structure adopted. The matching of the structure and stereochemistry of the membrane promotes the nucleation growth of the crystal along the specific crystal surface, and the factors such as charge accumulation, concentration of nucleating species, and pH value of the solution also play an important role in controlling the crystal morphology, structure, and orientation.

9.8.2 *Ultra-thin films prepared by LB technology*

Using LB technology to prepare ordered nanomaterials superthin films has many advantages: monolayer nanofilms can be prepared, or layer by layer accumulation, forming a multilayer or superlattice structure, assembly method can be optional; different nanomaterials can be selected to accumulate different nanomaterials to form alternating or mixed membranes, so that it has a variety of functions; film formation can be carried out at room temperature and pressure, without time limitation, and basically without damaging the structure of membrane-forming nanomaterials. Film thickness and film uniformity can be controlled; the self-assembly ability of nanomaterials can be effectively utilized to form new materials. The LB film structure is easy to determine and easy to obtain.

Antimony-doped SnO_2 (ATO) thin film is a kind of thin film material with great application value and potential. Due to its good light transmittance and electrical conductivity, it has been widely used in building glass, liquid crystal display, transparent electrode, and solar energy utilization. $SnCl_4.5H_2O$ and $SbCl_3$ can be selected as the basic materials. Antimony-doped tin oxide (ATO) precipitation was obtained by co-precipitation method, and ATO nanohydrosol was obtained by gel solution, which was dissolved in pure water and used as a subphase. ATO composite film of $10\,mm \times 30\,mm$ was prepared by LB film technology, and ATO superthin film was prepared after sintering. The results show that the ATO ultrathin film has the advantages of both monolayer and ATO powder, and has great application value for the preparation of nanoscale electronic devices. So far, many nanoparticle super-thin films have been studied, such as Au, Ag, Pt, Si, CdS, Ag_2S, SiO_2, TiO_2 and so on.

9.8.3 *LB film application in optics*

The LB film is the supramolecular film with least defects that can be prepared by people at present. It has great application prospects in photodetector, photocell, photoelectric switch, photoelectric information storage, processing and simulation of photosynthesis, composition of nonlinear optical materials, and so on. The optical properties of Y-type cyanine dye LB film in the UV-visible range were characterized by elliptical polarization spectroscopy (SE), and the optical constants (complex dielectric constant, extinction coefficient, absorption coefficient, reflection coefficient, refractive index, etc.) of the film were obtained. On the ITO conductive glass, for the preparation of the 38 layer Z bacteria rhodopsin (bR) LB films, control the average transfer ratio to 0.93 or above, measuring the UV-Vis absorption spectrum of LB films, and use Z scanning technology in the output laser wavelength of 400 nm and 800 nm of bacterium violet LB films of third-order nonlinear optical properties are studied; at 800 nm, it's the third order nonlinear optical susceptibility for 10^{-9} esu. However, it is 10^{-8} esu at 400 nm, which indicates that purplish LB film has potential application prospect in nonlinear optical devices. The spectra and second-order nonlinear optical properties of two kinds of "push-pull" azobenzene molecular LB films were studied by UV-vis absorption and second harmonic generation. The 4-nitro-4'-amino-azobenzene (NAA) and 4-carboxyl-4'-amino-azobenzene (CAA) molecules can form stable monolayers on the subphase surface, and can be transferred to solid substrates to form LB multilayer films. Since $-NO_2$ has a stronger electron absorption capacity than that of –COOH, electrons can be transferred more easily in NAA molecules and form larger molecular dipole moments, so the molecules have higher first-order hyperpolarizability. The experimental results show that both azobenzene compounds have good nonlinear optical properties in LB films, while the cis-isomerism of azobenzene chromophore can cause optically induced birefringence, so they have the potential to develop optical memory devices.

9.8.4 *LB film application in semiconductor materials*

Induction of semiconductor nanomaterials by LB monolayer is a new method that was developed in the 1990s. The nucleation ions supported in the double layer of ionic insoluble monolayer react with appropriate reagents (such as H_2S, H_2Se, etc.) to produce a semiconductor material. Appropriate monolayers and reaction conditions are selected to control the product particles within the nanometer size range. In addition, the prepared nanoparticles can be reassembled by LB. When the synthesized film is exposed to a H_2S environment, the CdS mole fraction in the film will increase. If the film is immersed into $CdCl_2$ aqueous solution in a H_2S environment, the CdS will not lose. Therefore, through the repeated cycle of this vulcanization–intercalation reaction, the size quantized CdS particles can be further generated between the hydrophilic layers. If the LB film is covered on the gold electrode, and then inserted into a certain proportion of a mixed $CdSO_4$, and $Na_2S_2O_3$ solution, then another platinum electrode is inserted, and the two electrodes are connected to the potentiometer. After a period of time, a layer of uniform CdS film is deposited on the surface of the LB film. The type of CdS film can be controlled by changing the LB monolayer. In addition, PbS, CdSe, PbSe, ZnS, Cu_2S, SnO_2, and other crystals were prepared by the technology, and monolayer and multilayer films of TiO_2 clusters with quantized dimensions were synthesized.

The spectral properties, electrical properties, and electrochemical properties of crystals grown in LB films are obviously dependent on their geometric morphology, which may be a phenomenon worth utilizing for semiconductor materials. For example, equilateral triangle PbS(PbS I) and right triangle PbS(PbS II) were obtained by membrane epitaxy. The absorption intensity of PbS-I and PbS-II nanofilms in the ultraviolet region will be increased by applying a bias pressure ranging from 0.5 V to 1.1 V, respectively. However, when the external bias of PbS obtained by non-epitaxy changed from 0 V to −1.5 V, the absorption with a wavelength greater than 700 nm showed no change.

9.8.5 *LB film application in ferroelectric materials*

A. Bune and l. Blinov *et al.* were the first to study the ferroelectric characteristics of LB films. In 1995 and 1996, they used LB film technology to produce the superthin films of the polymer PVDF and its copolymer P(VDF-TrFE 70:30). The number of layers of LB film was between 2 and 120, and the thickness of each layer was about 0.5 nm. In 1998, A. Bune and l. Blinov *et al.* reported the ferroelectric phase transition of LB films by measuring pyroelectric and permittivity in *Nature* and *Physical Review*, respectively. In the thin polymer [P(VDF-TrFE 70:30)] ferroelectric LB film, the surface ferroelectric first-order phase transition occurred at low temperature (about 20°C), while the bulk ferroelectric first-order phase transition occurred at high temperature (about 80°C). Through analysis, it is concluded that there is no obvious minimum thickness for ferroelectric properties of polymer LB films, and the ferroelectric properties of such a two-dimensional system cannot be calculated by the average field theory of a three-dimensional system. The critical temperature of ferroelectric characteristics of 30 layers of polymer LB film was studied. When no electric field was applied, the temperature of ferroelectric phase transition was about $T_c0 = 80 \pm 10°C$, and a single hysteresis loop appeared below this temperature. When the electric field is added, the critical temperature of the ferroelectric phase transition is $T_{cr} = 145 \pm 5°C$. A double hysteresis loop occurs between the two temperatures. The ferroelectric characteristics disappear when the critical temperature T_{cr} is above. In 2001, Batirov *et al.* reported the ferroelectric characteristics of the photoconductor ferroelectric LB film, and studied the changes of its ferroelectric characteristics with and without light. In 2002, Yu G. Fokin and O. A. Aktsipetrov *et al.* reported the use of second harmonic generation.

L. Blinov *et al.* first reported the pyroelectric effect of LB multilayer films. They reported the pyroelectric effect of type X and type Z amphiphilic azo compounds. R. Capan *et al.* studied the alternating LB films of siloxane polymer and eicosaniline by using Fourier transform spectroscopy, and provided the theoretical model of pyroelectric response from the microscopic mechanism. It was found that, although the pyroelectric coefficient of LB films was not very large, their relative capacitance r and dielectric loss tangent were both very small, so for the quality factor $p/(r$ tangent) 1/2 of the pyroelectric detector, the energy of LB films reached the same order as that of existing common pyroelectric materials.

9.8.6 *Application of LB films in sensors*

LB film has important applications in the sensor field because of its highly ordered molecular arrangement. Phthalocyanine is a kind of organic semiconductor material sensitive to gas, and has good thermal stability and chemical stability. The LB film of phthalocyanine has a great application prospect in high-sensitivity gas detection. As with sandwich type rare earth element praseodymium metal phthalocyanine complexes ($Pr/Pc(OC_8H_{17})_8]_2$) for gas-sensitive material, the use of LB film technology will be ($Pr/Pc\ (OC_8\ H_{17})_8]_2$) with the ratio of 1:3 and stearyl alcohol (OA) of the mixed LB multilayer control on MOSFET, formed a new type of LB films instead of the usual effect of MOSFET gate metal in the field of pipe devices. When the device is put into NO_2 gas, the leakage current of the device will change from 0.05×10^{-6}A to 1.5×10^{-6} A with the change of gas concentration and the number of LB film layers, and the detection sensitivity can reach 5 PPM NO_2, so this gas sensor has high sensitivity.

With the use of LB film technology, different characteristics of lipid thin film were deposited to the light addressable potentiometric sensor (LAPS) system, which was prepared to approximate the human taste way detection sensors, and test the sensor to sweet, sour, bitter, salty, and fresh responses of five basic taste materials such as solution, show that the sensor can distinguish different taste substances.

In addition, LB film has important application value in the preparation of electrochemical sensor, biosensor, humidity sensor, etc.

9.9 Exercises and Questions

(1) What are the types of thin-film solar cell materials? Please briefly review the characteristics and advantages of amorphous silicon thin films, polysilicon thin films and compound semiconductor thin films.
(2) How to design a reflective film system and anti-reflection film system in the range of 2–3 microns?
(3) Please give a brief example to illustrate the application of YBCO superconducting thin films.
(4) Please review the preparation and application of perovskite ferroelectric thin films and layered perovskite ferroelectric thin films.
(5) Briefly introduce the application field and development trend of magnetic thin films.
(6) What are the characteristics and applications of diamond films and diamond-like films?
(7) Please briefly introduce the mechanism and application scope of luminescent films.

References

1. Maissel LI and Glang R (Eds), *Handbook of Thin Film Technology*, McGran-hill Book Company, 1970.
2. Tian MB *et al.*, *Thin Film Science and Technology Manual*, Beijing: Machinery Industry, 1991.
3. Araujo CP, Scott JF and Tailor GW, *Ferroelectric Thin Films: Synthesis and Basic Properties*, Amsterdam, Gordon and Breach Publishers, 1996.
4. Wu ZQ and Wang B, *Film Growth*, Beijing: Science Press, 2001.
5. Yang BC, *Electronic Thin Film Materials*, Beijing: Science Press, 1996.
6. Ciureanu P and Gavrila H, *Magnetic Heads for Digital Recording*, Amsterdam, Elsevier Science Publishers, 1990.
7. Li JG, Xiao DQ, *Modern Surface Engineering Design Manual*, Beijing: National Defense Industry, September 2000.
8. Wu JL, Wu QD, (Eds.) *Several New Thin Film Materials*, Beijing: Peking University, 1999.
9. Tang WZ, *Principle of Preparation of Thin Film Materials*, Beijing: Metallurgical Industry Press, 2003.
10. Tian MB, *Film Technology and Film Materials*, Beijing: Tsinghua University Press, 2006.
11. Zheng WT, *Film Materials and Film Technology*, Beijing: Chemical Industry Press, 2004.
12. Zhang R *et al.*, *Modern Methods of Material Analysis* (Engineering Applications Edition), Beijing: Chemical Industry Press, 2007.
13. Freund LB, Suresh S, *Thin Film Materials — Stress Defection Formation and Surface Evolution*, London: Cambrige University Press, 2003.
14. Zhu JG, Sun XS, Li W, *Electronic and Optoelectronic Materials*, Beijing: National Defense Industry, 2007.
15. Ning Z *et al.*, *Solid Film Materials and Preparation Technology*, Beijing: Science Press, 2008.
16. Cai X *et al.*, *Modern Thin Film Materials and Technology*, Shanghai: East China University of Science and Technology Press, 2007.

17. Fu CL, *Ferroelectric Thin Film Materials and their Applications*, Beijing: Science Press, 2009.
18. Mai ZH, *X-ray Characterization of Thin Film Structure*, Beijing: Science Press, 2009.
19. Kagan, *Thin Film Transistor (TFT) and its Application in Flat Panel Display*, Electronic Industry Press, 2008.
20. Shi DL *et al.*, *High Temperature Superconductor Application Research*, Shanghai: Shanghai Science and Technology Press, 2009.
21. Milton O, *Materials Science of Thin Films: Deposition and Structure* (2nd edn.), World Scientific Press, 2006.
22. Machlin ES, *Materials Science in Microelectronics*, Elsevier, 2005.
23. Ronald R. Willey, *Practical Design and Production of Optical Thin Films*, New York: Marcel Dekker Inc., 2002.

Index

www.ingramcontent.com/pod-product-compliance
Lightning Source LLC
LaVergne TN
LVHW010338220926
848484LV00004BB/921
* 9 7 8 9 8 1 1 2 2 3 9 8 3 *